VOLUME NINETY

VITAMINS AND HORMONES

Adiponectin

VOLUME NINETY

VITAMINS AND HORMONES

Adiponectin

Editor-in-Chief

GERALD LITWACK, PhD
Toluca Lake, North Hollywood, California

ELSEVIER

AMSTERDAM • BOSTON • HEIDELBERG • LONDON
NEW YORK • OXFORD • PARIS • SAN DIEGO
SAN FRANCISCO • SINGAPORE • SYDNEY • TOKYO
Academic Press is an imprint of Elsevier

Cover photo credit:
Juhl, C., and Beck-Sickinger, A. G.
Molecular Tools to Characterize Adiponectin Activity
Vitamins and Hormones (2012) **90**, pp. 31–56

Academic Press is an imprint of Elsevier
32 Jamestown Road, London, NW1 7BY, UK
Radarweg 29, PO Box 211, 1000 AE Amsterdam, The Netherlands
Linacre House, Jordan Hill, Oxford OX2 8DP, UK
225 Wyman Street, Waltham, MA 02451, USA
525 B Street, Suite 1900, San Diego, CA 92101-4495, USA

First edition 2012

ISBN: 978-0-12-398313-8
ISSN: 0083-6729

For information on all Academic Press publications visit our website at store.elsevier.com

Printed and bound by CPI Group (UK) Ltd, Croydon, CR0 4YY

Transferred to digital print 2012

CONTENTS

CONTRIBUTORS

Richard R. Almon
Department of Pharmaceutical Sciences; Department of Biological Sciences, and New York State Center of Excellence in Bioinformatics and Life Sciences, State University of New York at Buffalo, Buffalo, New York, USA

Olga Astapova
Department of Pathology, The Cardiovascular Research Institute, Wayne State University School of Medicine, Detroit, Michigan, USA

Annette G. Beck-Sickinger
Faculty of Biosciences, Pharmacy and Psychology, Institute of Biochemistry, Universität Leipzig, Leipzig, Germany

V. Vilceu Bordignon
Department of Animal Science, McGill University, Ste-Anne-de-Bellevue, Quebec, Canada

Florence Caldefie-Chézet
Clermont Université, Université d'Auvergne, UFR Pharmacie, Laboratoire SVFp, 28 place Henri Dunant, and INRA, UMR 1019, ECRIN, CRNH Auvergne, F-63000 Clermont-Ferrand

Štefan Čikoš
Institute of Animal Physiology, Slovak Academy of Sciences, Šoltésovej 4, Košice, Slovak Republic

Jennifer C. Davey
Section of Vascular Surgery, Department of Surgery, Dartmouth Medical School, Lebanon, New Hampshire, USA

Ana Barbosa Marcondes de Mattos
Laboratório de Fisiologia da Nutrição e Desenvolvimento, Departamento de Fisiologia, Universidade do Estado do Rio de Janeiro, Rio de Janeiro, Brazil

Cristiane de Oliveira
Systemic Inflammation Laboratory, Trauma Research, St. Joseph's Hospital and Medical Center, Phoenix, USA

Laetitia Delort
Clermont Université, Université d'Auvergne, UFR Pharmacie, Laboratoire SVFp, 28 place Henri Dunant, and INRA, UMR 1019, ECRIN, CRNH Auvergne, F-63000 Clermont-Ferrand

Marie-Noëlle Dieudonné
Université de Versailles-St-Quentin, Service de Biochimie et Biologie Moléculaire, UPRES-EA 2493, UFR Paris-Ile de France Ouest, PRES Universud Paris, Centre Hospitalier de Poissy-Saint Germain, Poissy Cedex, France

Min Ding
Department of Medicine, and Department of Pharmacology, Yale School of Medicine, New Haven, Connecticut, USA

Debra C. DuBois
Department of Pharmaceutical Sciences, and Department of Biological Sciences, State University of New York at Buffalo, Buffalo, New York, USA

Virginie Dubois
Clermont Université, Université d'Auvergne, UFR Pharmacie, Laboratoire SVFp, 28 place Henri Dunant, and INRA, UMR 1019, ECRIN, CRNH Auvergne, F-63000 Clermont-Ferrand

David J. Dyck
Department of Human Health and Nutritional Sciences, University of Guelph, Guelph, Ontario, Canada

Rina Hemi
Institute of Endocrinology, Sheba Medical Center, Tel-Hashomer, Israel

Kyosuke Hino
Genomic Science Laboratories, Dainippon Sumitomo Pharma Co. Ltd., Konohana-Ku, Osaka, Japan

Thierry Jardé
Clermont Université, Université d'Auvergne, UFR Pharmacie, Laboratoire SVFp, 28 place Henri Dunant, and INRA, UMR 1019, ECRIN, CRNH Auvergne, F-63000 Clermont-Ferrand

Yu Jin
Department of Medicine, and Department of Pharmacology, Yale School of Medicine, New Haven, Connecticut, USA

Robert L. Judd
Department of Anatomy, Physiology and Pharmacology, College of Veterinary Medicine, Auburn University, Auburn, Alabama, USA

Cathleen Juhl
Faculty of Biosciences, Pharmacy and Psychology, Institute of Biochemistry, Universität Leipzig, Leipzig, Germany

William J. Jusko
Department of Pharmaceutical Sciences, and New York State Center of Excellence in Bioinformatics and Life Sciences, State University of New York at Buffalo, Buffalo, New York, USA

Hannah Kanety
Institute of Endocrinology, Sheba Medical Center, Tel-Hashomer, Israel

Atsushi Kumanogoh
Department of Respiratory Medicine, Allergy and Rheumatic Diseases, Osaka University Graduate School of Medicine, Suita, Osaka, Japan

Todd Leff
Department of Pathology, The Cardiovascular Research Institute, Wayne State University School of Medicine, Detroit, Michigan, USA

Li Li
Department of Physiology and Pathophysiology, Peking University Health Science Center and Key Laboratory of Molecular Cardiovascular Sciences, Ministry of Education, Beijing, China

Renjing Liu
Department of Medicine, and Department of Pharmacology, Yale School of Medicine, New Haven, Connecticut, USA

Yaniv Lustig
Institute of Endocrinology, Sheba Medical Center, Tel-Hashomer, Israel

Paolo Magni
Department of Pharmacological and Biomolecular Sciences, Università degli Studi di Milano, and Centro Universitario Dislipidemie, Ospedale Niguarda Cà Granda, Milan, Italy

Kathleen A. Martin
Department of Medicine; Department of Pharmacology, Yale School of Medicine, New Haven, Connecticut, and Section of Vascular Surgery, Department of Surgery, Dartmouth Medical School, Lebanon, New Hampshire, USA

Philippe de Mazancourt
Université de Versailles-St-Quentin, Service de Biochimie et Biologie Moléculaire, UPRES-EA 2493, UFR Paris-Ile de France Ouest, PRES Universud Paris, Centre Hospitalier de Poissy-Saint Germain, Poissy Cedex, France

João Felipe Mota
Faculdade de Nutrição, Universidade Federal de Goiás (FANUT), Goiânia, Brazil

Bruce D. Murphy
Université de Montréal, St-Hyacinthe, Quebec, Canada

Hidetaka Nagata
Genomic Science Laboratories, Dainippon Sumitomo Pharma Co. Ltd., Konohana-Ku, Osaka, Japan

Kaori Nakanishi
Department of Respiratory Medicine, Allergy and Rheumatic Diseases, Osaka University Graduate School of Medicine, Suita, Osaka, Japan

Marie-France Palin
Agriculture and Agri-Food Canada, Sherbrooke, Quebec, Canada

René Pecquery
Université de Versailles-St-Quentin, Service de Biochimie et Biologie Moléculaire, UPRES-EA 2493, UFR Paris-Ile de France Ouest, PRES Universud Paris, Centre Hospitalier de Poissy-Saint Germain, Poissy Cedex, France

Eric P. Plaisance
Laboratory of Nutrient Sensing and Adipocyte Signaling, Pennington Biomedical Research Center, Louisiana State University System, Baton Rouge, Louisiana, USA

Lindsay E. Robinson
Department of Human Health and Nutritional Sciences, University of Guelph, Guelph, Ontario, Canada

Massimiliano Ruscica
Department of Pharmacological and Biomolecular Sciences, Università degli Studi di Milano, and Centro Universitario Dislipidemie, Ospedale Niguarda Cà Granda, Milan, Italy

Eva M. Rzucidlo
Section of Vascular Surgery, Department of Surgery, Dartmouth Medical School, Lebanon, New Hampshire, USA

Esther Dos Santos
Université de Versailles-St-Quentin, Service de Biochimie et Biologie Moléculaire, UPRES-EA 2493, UFR Paris-Ile de France Ouest, PRES Universud Paris, Centre Hospitalier de Poissy-Saint Germain, Poissy Cedex, France

Carolina Biz Rodrigues Silva
Disciplina de Fisiologia da Nutrição, Departamento de Fisiologia, Universidade Federal de São Paulo, São Paulo, Brazil

Lindsey Stavola
Department of Medicine, and Department of Pharmacology, Yale School of Medicine, New Haven, Connecticut, USA

Liliana Steffani
Department of Pharmacological and Biomolecular Sciences, Università degli Studi di Milano, and Centro Universitario Dislipidemie, Ospedale Niguarda Cà Granda, Milan, Italy

Siddharth Sukumaran
Department of Pharmaceutical Sciences, State University of New York at Buffalo, Buffalo, New York, USA

Isao Tachibana
Department of Respiratory Medicine, Allergy and Rheumatic Diseases, Osaka University Graduate School of Medicine, Suita, Osaka, Japan

Yoshito Takeda
Department of Respiratory Medicine, Allergy and Rheumatic Diseases, Osaka University Graduate School of Medicine, Suita, Osaka, Japan

Justine M. Tishinsky
Department of Human Health and Nutritional Sciences, University of Guelph, Guelph, Ontario, Canada

Marie-Paule Vasson
Clermont Université, Université d'Auvergne, UFR Pharmacie, Laboratoire SVFp, 28 place Henri Dunant; INRA, UMR 1019, ECRIN, CRNH Auvergne, and Centre Jean Perrin, Unité de Nutrition, F-63000 Clermont-Ferrand

Desiree Wanders
Department of Anatomy, Physiology and Pharmacology, College of Veterinary Medicine, Auburn University, Auburn, Alabama, USA

Li-Ling Wu
Department of Physiology and Pathophysiology, Peking University Health Science Center and Key Laboratory of Molecular Cardiovascular Sciences, Ministry of Education, Beijing, China

Yi Xie
Department of Medicine, and Department of Pharmacology, Yale School of Medicine, New Haven, Connecticut, USA

Juliane Costa Silva Zemdegs
Disciplina de Fisiologia da Nutrição, Departamento de Fisiologia, Universidade Federal de São Paulo, São Paulo, Brazil

PREFACE

Until recently, the adipocyte or fat cell has been regarded solely as a repository of energy reserves in the form of fat. Newer evidence provides an important role of the adipocyte as a hormone-secreting cell and fat tissue as an endocrine organ in addition to its traditional role. Several hormones are secreted by the adipocyte and some of the chief ones are adiponectin, leptin, and resistin. There is much new evidence in the literature concerning adiponectin and this volume collects and reviews most of that research. Adiponectin heightens the bodily increase in sensitivity to insulin. Together with leptin, adiponectin can normalize insulin action in insulin-resistant experimental animals. The hormones secreted by the fat cell as well as the receptors for these hormones may be critical for the management of obesity, hyperlipidemia, and insulin resistance. In this volume, the focus is on the central hormone of the fat cell, adiponectin.

In the first section, many of the chapters deal with the more basic aspects of adiponectin. Chapter 1 concerns "Lifestyle factors increasing adiponectin synthesis and secretion" by J. M. Tishinsky, D. J. Dyck, and L. E. Robinson. C. Juhl and A. G. Beck-Sickinger report on "Molecular tools to characterize adiponectin activity" in Chapter 2. "Nutritional and hormonal modulation of adiponectin and its receptors adipoR1 and adipoR2" is the subject of C. de Oliveira, A. B. Marcondes de Mattos, C. B. Rodrigues Silva, J. F. Mota, and J. Costa Silva Zemdegs (Chapter 3). Y. Lustig, R. Hemi, and H. Kanety, in Chapter 4, review "Regulation and function of adiponectin receptors in skeletal muscle." "Screening for adiponectin secretion regulators" is the topic provided by K. Hino and H. Nagata in Chapter 5. O. Astapova and T. Leff, in Chapter 6, review "Adiponectin and PPARγ: cooperative and interdependent actions of two key regulators of metabolism." The last chapter in this section reports on "Glucocorticoid effects on adiponectin expression" by S. Sukumaran, D. C. DuBois, W. J. Jusko, and R. R. Almon (Chapter 7).

The next section deals with the broader biological aspects of adiponectin. In this vein, E. Dos Santos, R. Pecquery, P. de Mazancourt, and M.-N. Dieudonne contribute "Adiponectin and reproduction" (Chapter 8). Chapter 9, "Adiponectin and its receptors in preimplantation embryo development," is supplied by S. Cikos. In Chapter 10, M.-F. Palin, V. Bordignon, and B. D. Murphy write on "Adiponectin and the control of female reproductive functions." "Adiponectin in the heart and vascular system" is the topic described by M. Ding, E. M. Rzucidlo, J. C. Davey, Y. Xie, R. Liu, Y. Jin, L. Stavola, and K. A. Martin (Chapter 11). This section concludes with a chapter by M. Ruscica, L. Steffani, and P. Magni entitled "Adiponectin interactions in bone and cartilage biology and disease" (Chapter 12).

The final section deals with disease and the affects of drugs. Of the latter, the subject of "Lipid-lowering drugs and circulating adiponectin" is offered by D. Wanders, E. P. Plaisance, and R. L. Judd (Chapter 13). In relation to disease aspects, L. Li and L.-L. Wu, in Chapter 14, report on "Adiponectin and interleukin-6 in inflammation-associated disease." "New insights into anticarcinogenic properties of adiponectin: a potential therapeutic approach in breast cancer?" is reported by L. Delort, T. Jardé, V. Dubois, M.-P. Vasson, and F. Caldefie-Chézet in Chapter 15. Finally, Y. Takeda, K. Nakanishi, I. Tachibana, and A. Kumanogoh contribute "Adiponectin: a novel link between adipocytes and COPD" (Chapter 16).

As for previous volumes in this series, my thanks go to Sarah Latham, Mary Ann Zimmerman, and Lisa Tickner of Elsevier for their cooperation.

The cover illustration is a reproduction of Figure 3 of Chapter 2 and the three three-dimensional crystal structures show the ribbon structure of globular adiponectin as a monomer and two aspects of the protein as a trimer.

GERALD LITWACK
Toluca Lake, California
26 May, 2012

CHAPTER ONE

Lifestyle Factors Increasing Adiponectin Synthesis and Secretion

Justine M. Tishinsky, David J. Dyck, Lindsay E. Robinson[1]
Department of Human Health and Nutritional Sciences, University of Guelph, Guelph, Ontario, Canada
[1]Corresponding author: e-mail address: lindsay.robinson@uoguelph.ca

Contents

Abstract

Adiponectin is an anti-inflammatory adipokine released from adipose tissue that is known to exert insulin-sensitizing effects in skeletal muscle and liver. Given that the secretion of adiponectin is impaired in obesity and related pathologies, strategies to enhance its synthesis and secretion are of interest. There is evidence that several lifestyle factors, including consumption of dietary long-chain n-3 PUFA, TZD administration, and weight loss can increase adiponectin synthesis and secretion. The effect of chronic exercise, independent of weight loss, is variable and less convincing. Potential mechanisms by which such lifestyle factors exert their favorable effects on adiponectin

Vitamins and Hormones, Volume 90
ISSN 0083-6729
http://dx.doi.org/10.1016/B978-0-12-398313-8.00001-4

include activation of PPARγ and AMPK, regulation of posttranslational modifications, and changes in adipose tissue morphology and macrophage infiltration. As a clear role for adiponectin in mitigating obesity-related impairments in lipid metabolism and insulin sensitivity is evident, further research investigating factors that enhance adiponectin synthesis and secretion is distinctly warranted.

1. INTRODUCTION

1.1. Adiponectin isoforms and its receptors

Adiponectin plays a pivotal role in protecting against the development and progression of obesity, type 2 diabetes, and cardiovascular disease (Berg, Combs, & Scherer, 2002; Hu, Liang, & Spiegelman, 1996; Maeda et al., 1996; Nakano, Tobe, Choi-Miura, Mazda, & Tomita, 1996; Scherer, Williams, Fogliano, Baldini, & Lodish, 1995). Adiponectin circulates in different oligomeric forms of trimeric, hexameric, or high-molecular weight (HMW) (Waki et al., 2003) with the HMW isoform most closely correlated with whole-body insulin sensitivity (Hara et al., 2006). While circulating adiponectin exists predominantly in full-length form, its proteolytic cleavage produces a globular form of adiponectin (gAd) that represents $<1\%$ of total adiponectin (Fruebis et al., 2001) and possesses biological activity, particularly in skeletal muscle (Mullen et al., 2009). Unlike other adipokines, circulating adiponectin concentrations are inversely correlated with increasing adiposity (Arita et al., 1999) and insulin resistance (Hotta et al., 2001).

Two transmembrane proteins, AdipoR1 and AdipoR2, have been identified as receptors for adiponectin (Yamauchi, Kamon, Ito, et al., 2003). AdipoR1 is expressed ubiquitously among tissues and is most highly concentrated in skeletal muscle (Yamauchi, Kamon, Ito, et al., 2003). AdipoR1 has a high affinity for gAd and a lower affinity for full-length adiponectin, suggesting that gAd acts primarily in skeletal muscle (Yamauchi, Kamon, Ito, et al., 2003). In contrast, AdipoR2 is expressed predominantly in the liver and has an intermediate affinity for both gAd and full-length adiponectin, suggesting that full-length adiponectin exerts its functions predominantly in the liver (Yamauchi, Kamon, Ito, et al., 2003). In addition, Hug et al. (2004) have identified a third adiponectin receptor, T-cadherin, that is exclusively expressed in endothelial and smooth muscle cells.

1.2. Adiponectin signaling in liver and muscle

Adiponectin is known to exert insulin-sensitizing effects in both the liver and skeletal muscle, two key tissues involved in glucose homeostasis. Upon binding to its receptor, it is believed that adiponectin signals primarily through AMP-activated protein kinase (AMPK) to inhibit hepatic gluconeogenesis in liver and to promote fatty acid oxidation in skeletal muscle, respectively (Tomas et al., 2002; Yamauchi et al., 2002). An adaptor protein (APPL1) upstream of AMPK seems necessary for adiponectin signaling given that adiponectin-induced AMPK activation in myocytes is enhanced or suppressed when APPL1 is overexpressed or inhibited, respectively (Mao et al., 2006). In contrast, APPL2, an isoform of APPL1, acts as a negative regulator of adiponectin signaling (Wang et al., 2009). In addition to AMPK, adiponectin also activates peroxisome proliferator-activated receptor (PPAR) α (Yamauchi, Kamon, Ito, et al., 2003; Yoon et al., 2006), a nuclear hormone receptor that leads to upregulation of numerous lipid and glucose metabolism genes (Gulick, Cresci, Caira, Moore, & Kelly, 1994). As such, adiponectin treatment of hepatocytes and myocytes stimulates PPARα ligand activity and increases fatty acid oxidation in myocytes (Yamauchi, Kamon, Ito, et al., 2003), while the targeted disruption of AdipoR2 in the liver of diabetic mice decreases expression of PPARα target genes involved in glucose metabolism (Yamauchi et al., 2007).

1.3. Adiponectin effects in liver and muscle

In the first study, to provide evidence for an insulin-sensitizing effect of adiponectin in liver, the administration of a single dose of recombinant, full-length adiponectin led to a significant decrease in circulating glucose levels in wild-type mice and normalized hyperglycemia in three diabetic mouse models (Berg, Combs, Du, Brownlee, & Scherer, 2001). To further elucidate the mechanism by which adiponectin improves whole-body insulin-sensitivity, Combs, Berg, Obici, Scherer, and Rossetti (2001) used a pancreatic insulin clamp technique to determine if the adiponectin-mediated decrease in circulating glucose levels could be attributed to a decrease in hepatic glucose production, an increase in skeletal muscle-mediated glucose clearance, or both. This study found an adiponectin-mediated inhibition of hepatic glucose production without any changes in glucose clearance, glycolysis, or glycogenesis (Combs et al., 2001). While no change in peripheral glucose clearance was found in the aforementioned work, it is important to recognize that those studies utilized full-length adiponectin, known to signal primarily via AdipoR2, which is expressed predominantly at the liver (Yamauchi, Kamon, Ito, et al., 2003). There is considerable evidence to suggest that gAd signals via AdipoR1 to enhance glucose uptake in skeletal muscle (Yamauchi, Kamon, Ito, et al., 2003), further contributing to overall improvements in whole-body insulin sensitivity.

While the administration of full-length adiponectin improves glucose tolerance largely through an increase in hepatic insulin sensitivity and reduced glucose production (Combs et al., 2004), the administration of gAd to lipoatrophic mice (Yamauchi, Kamon, Waki, Terauchi, et al., 2001) and its overexpression in ob/ob mice (Yamauchi, Kamon, Waki, et al., 2003) results in improved peripheral insulin sensitivity, due at least in part to an increase in muscle fatty acid oxidation (Fruebis et al., 2001). Acutely, gAd results in the phosphorylation/deactivation of acetyl CoA carboxylase (ACC) (Mullen et al., 2009) which in turn decreases malonyl CoA, a potent inhibitor of mitochondrial carnitine palmitoyltransferase 1, thereby promoting fatty acid entry into mitochondria. Interestingly, the ability of gAd to stimulate AMPK in skeletal muscle, the upstream kinase that phosphorylates ACC, is somewhat controversial with glycolytic rodent muscle and human muscle clearly showing an acute gAd-induced stimulation of AMPK (Bruce, Mertz, Heigenhauser, & Dyck, 2005; Tomas et al., 2002), while this effect is absent in oxidative rodent muscle (Mullen et al., 2009; Tomas et al., 2002). gAd administration also increases markers of oxidative capacity and mitochondrial content in muscle (Civitarese et al., 2004; Iwabu et al., 2010), which appears to be due to the AdipoR1-mediated calcium activation of AMPK, CaMKII, and SIRT1 (Iwabu et al., 2010). Interestingly, each of these studies also found evidence that adiponectin may reduce oxidative stress, presumably due to either increased mitochondrial content and/or increased oxidative stress-detoxifying enzymes. Overall, there is a good consensus that adiponectin, at least its globular form, can increase the ability of muscle to oxidize fatty acids both acutely and chronically, thereby improving insulin sensitivity.

Obesity and insulin resistance have also been shown to be accompanied by a state of resistance to adiponectin in skeletal muscle. Specifically, the ability of gAd to stimulate fatty acid oxidation is markedly impaired in skeletal muscle isolated from obese humans and rodents fed a high-fat diet (Bruce et al., 2005; Mullen et al., 2009). The mechanisms underlying this impaired response of muscle to adiponectin have not been clearly elucidated, but surprisingly, do not seem to involve an increase in markers of muscle inflammation (Mullen, Tishinsky, Robinson, & Dyck, 2010). Changes in AdipoR1 receptor mRNA in obese/diabetic humans and rodents are conflicting (Chen et al., 2005; Civitarese et al., 2004; Debard et al., 2004; Tsuchida et al., 2004); furthermore, others have not detected any difference in AdipoR1 protein in muscle from lean and obese humans (Kuoppamaa et al., 2008; Thrush, Heigenhauser, Mullen, Wright, & Dyck, 2008) or high-fat-fed rodents (Mullen et al., 2009). Interestingly, the loss of adiponectin-stimulated fatty acid oxidation is induced as rapidly as 3 days by a high-fat diet and precedes lipid accumulation and impaired insulin response (that occurs in 2–4 weeks) (Mullen et al., 2009). It is tempting to interpret these findings as evidence that the loss of adiponectin response in skeletal muscle leads to ectopic lipid deposition and ultimately insulin resistance; however, to date there is no direct evidence to support this. In summary, adiponectin resistance exists in conditions of insulin resistance and can be experimentally induced prior to the onset of metabolic disturbances and insulin resistance; however, its underlying causes and significance requires further study.

2. LIFESTYLE FACTORS THAT INCREASE ADIPONECTIN SYNTHESIS AND SECRETION

Given the known insulin-sensitizing properties of adiponectin, and that circulating adiponectin concentrations are markedly reduced in obesity and type 2 diabetes (36–37), strategies to augment adiponectin synthesis and secretion from adipose tissue are of significance. This review will explore four areas of lifestyle modification: dietary fatty acids, thiazolidinediones (TZDs), weight loss, and exercise that show promise as strategies to increase adiponectin.

2.1. Long-chain n-3 fatty acids

Dietary consumption of the long-chain n-3 polyunsaturated fatty acids (PUFA), eicosapentaenoic acid (EPA; 20:5 n-3), and docosahexaenoic acid (DHA; 22:6 n-3) is associated with a reduced risk of developing type 2 diabetes and cardiovascular disease (Berg et al., 2002). The anti-inflammatory and insulin-sensitizing properties of n-3 PUFA are thought to be at least partly attributable to the regulation of adipose tissue-derived factors, including adiponectin (Puglisi, Hasty, & Saraswathi, 2011).

2.1.1 Evidence from cell culture studies

Several studies using cultured adipocytes have shown an increase in adiponectin gene expression (Oster, Tishinsky, Yuan, & Robinson, 2010), cellular protein content (Banga et al., 2009; Tishinsky, Ma, & Robinson, 2011), and secretion (Banga et al., 2009; Itoh et al., 2007; Johnson, Williams, Wei, & Ranganathan, 2011; Oster et al., 2010; Tishinsky et al., 2011) in response to n-3 PUFA treatment. Work from our laboratory found that treatment of 3T3-L1 murine adipocytes with 125 μM of EPA and DHA for 24 h induced an increase in both adiponectin gene expression and secreted protein concentration in the culture media (Oster et al., 2010). We subsequently confirmed an increase in both cellular and secreted adiponectin protein in response to 48 h of 100 μM of EPA and DHA in isolated primary human adipocytes from overweight donors (Tishinsky et al., 2011). It is also noteworthy that when EPA was combined with the TZD antidiabetic drug, rosiglitazone (see section 2.3), we found an additive increase in adiponectin secretion (Tishinsky et al., 2011), suggesting that n-3 PUFA in combination with a TZD may represent a more potent inducer of adiponectin secretion.

Our findings that long-chain n-3 PUFA increase adiponectin are consistent with those of Banga et al. (2009) who reported increased adiponectin secretion from rat adipocytes treated overnight with 25 μM EPA or DHA. Interestingly, this increase in adiponectin secretion occurred in the absence of any changes in cellular gene or protein expression adiponectin (Banga et al., 2009), indicating that n-3 PUFA may exert a direct stimulatory effect on adiponectin translation. Last, EPA effectively reversed the decrease in adiponectin secretion from 3T3-L1 adipocytes that was induced by a coculture with RAW264 macrophages (Itoh et al., 2007). The restoration of adiponectin secretion occurred in parallel with a decrease in macrophage TNF-α mRNA expression (Itoh et al., 2007), suggesting that n-3 PUFA may elicit changes in adiponectin secretion from adipose tissue by regulating the paracrine interactions between adipocytes and macrophages.

In contrast to the above studies, incubation of rat adipocytes with 100 or 200 μM EPA for 96 h resulted in a decrease in both adiponectin gene expression and secreted protein (Lorente-Cebrian, Perez-Matute, Martinez, Marti, & Moreno-Aliaga, 2006). Similarly, Bueno et al. (2008) reported that treatment of 3T3-L1 adipocytes with 250 μM EPA or DHA for 48 h decreased adiponectin gene expression, citing lipid peroxidation as a possible mechanism for this effect of n-3 PUFA. Last, EPA had no effect on adiponectin gene expression in 3T3-L1 adipocytes after 12 h (Neschen et al., 2006), possibly due to the relatively short fatty acid exposure time.

2.1.2 Evidence from rodent studies

Numerous rodent-based studies also provide evidence for a role of n-3 PUFA in modulating adiponectin secretion. For example, 15 days of dietary fish oil, rich in both EPA and DHA, increased plasma adiponectin levels in a dose-dependent manner in mice compared with those fed an isocaloric safflower oil diet, rich in linoleic acid (18:2 n-6) (Neschen et al., 2006). Interestingly, fish-oil-induced adiponectin gene expression in epididymal adipose tissue, but elicited no change in adiponectin mRNA in subcutaneous fat, suggesting that the effects of n-3 PUFA may vary depending on adipose tissue depot (Neschen et al., 2006). Similar elevations in circulating adiponectin were observed in both genetically obese (ob/ob) mice and high-fat diet-induced obese mice fed EPA (5% of kcal) (Itoh et al., 2007).

Significant uptake of DHA in murine heart, liver, and adipose tissue occurs in as little as 4 days of supplementation; furthermore, plasma concentrations remain stable for a significant period, that is, 16 days after DHA supplementation has stopped (Lefils, Geloen, Vidal, Lagarde, & Bernoud-Hubac, 2010). Although there is a general consensus that n-3 PUFA increase adiponectin in rodent models, there is also some evidence for a negative association between consumption of a fish-oil-enriched diet and adiponectin (Bueno et al., 2008). This negative association was only evident with acute (2 days) but not chronic (60 days) fish oil administration and was accompanied by a reduction in adiponectin gene expression in retroperitoneal, but not epididymal or subcutaneous adipose tissue (Bueno et al., 2008). Thus, it is possible that acute administration of a fish-oil-enriched diet may produce a transient and depot-specific impairment in adiponectin secretion that is subsequently followed by a sustained elevation in adiponectin; however, this requires further study.

Additionally, n-3 PUFA have been demonstrated to counter the reduction in adiponectin secretion induced by high-fat diets (Flachs et al., 2006; Kalupahana et al., 2010; Todoric et al., 2006). For example, incorporation of EPA (16% of kcal) into a high-fat diet resulted in plasma adiponectin concentrations that were similar to those in the low fat control group in wild-type mice (Kalupahana et al., 2010). Similarly, inclusion of n-3 PUFA in a high-fat diet effectively prevented a decrease in adiponectin gene expression and restored serum adiponectin levels in both wild-type (Flachs et al., 2006) and db/db mice (Todoric et al., 2006). Last, dietary fish oil has been demonstrated to restore impairments in adiponectin levels induced by long-term exposure to a high sucrose diet (Rossi et al., 2005). In rats fed a sucrose-rich diet for 7 months, fish oil supplementation restored plasma adiponectin levels and this was also associated with a reversal of whole-body insulin resistance and dyslipidemia (Rossi et al., 2005). These studies strongly suggest a valuable role for n-3 PUFA in normalizing diet-induced impairments in adiponectin, findings that may have important implications for common dietary habits in today's society.

2.1.3 Evidence from human studies

Numerous clinical studies have found that n-3 PUFA consumption is positively correlated with circulating adiponectin concentrations in humans (An et al., 2011; Itoh et al., 2007; Kondo et al., 2010; Krebs et al., 2006; Nomura et al., 2009). EPA supplementation (1.8 g/day) for 3 months significantly increased plasma adiponectin levels in obese men and women with metabolic syndrome (Itoh et al., 2007). Similarly, in overweight insulin resistant women undergoing a 24 week weight loss program, supplementation with 4.2 g/day of EPA and DHA led to an 18% increase in serum adiponectin, while weight loss alone had no effect (Krebs et al., 2006). A positive correlation between plasma adiponectin levels and erythrocyte n-3 PUFA content has also been observed in dialysis patients (An et al., 2011). Furthermore, increases in serum adiponectin concentrations were observed in nonobese Japanese women consuming a minimum of 3 g/day of n-3 PUFA in the form of fish for 8 weeks, suggesting that n-3 PUFA can also elevate circulating adiponectin levels in lean, healthy populations (Kondo et al., 2010). However, this study is limited by a relatively small sample size (17 subjects), as well as variability in the quantity and type of fish consumed. In contrast, in a study that supplemented hyperlipidemic patients with 1.8 g of EPA daily for 6 months, a significant increase in adiponectin levels was observed in only those subjects with type 2 diabetes and initial hypoadiponectinemia (Nomura et al., 2009), implying that response to n-3 PUFA intake may be influenced by baseline adiponectin levels. Thus, additional studies are needed to clarify the role of n-3 PUFA in regulating adiponectin concentrations in healthy individuals.

Although the aforementioned studies provide evidence that dietary n-3 PUFA intake can augment circulating adiponectin in humans, there are also studies that observe no significant association (Kratz et al., 2008; Lara et al., 2007; Vargas, Almario, Buchan, Kim, & Karakas, 2011). For example, moderately obese men and women who consumed 2.9 g/day of n-3 PUFA for 14 weeks had no change in either total plasma or HMW adiponectin (Kratz et al., 2008). Likewise, 3.5 g/day of n-3 PUFA for 6 weeks failed to elicit any change in circulating adiponectin levels in women with polycystic ovary syndrome (Vargas et al., 2011). Finally, no significant changes in plasma adiponectin were found in nonobese healthy men consuming 2.4 g/day of EPA and DHA in the form of salmon for 4 weeks, although a trend ($P=0.086$) for a positive role of n-3 PUFA was evident (Lara et al., 2007).

In summary, despite considerable variability among studies in terms of n-3 PUFA dosage, duration of dietary intake or supplementation, and composition of the background diet, there is evidence that n-3 PUFA increase adiponectin. While most of the work to date has examined the ability of adiponectin to mitigate preexisting hypoadiponectinemia present in obese and/or diabetic models, there does remain a need to further evaluate the role and physiological relevance of n-3 PUFA in modulating adiponectin secretion in lean, healthy populations. Furthermore, the ability of n-3 PUFA to target secretion of higher order multimers of adiponectin warrants examination. Nevertheless, n-3 PUFA represent a potential dietary intervention to increase adiponectin, thereby improving the secretory profile of adipose tissue and associated metabolic outcomes.

2.2. Saturated fatty acids

Consumption of saturated fatty acids (SFA) has been demonstrated to modulate the secretion of adipose tissue-derived factors such as adiponectin (Oller do Nascimento, Ribeiro, & Oyama, 2009), representing a possible mechanistic link between high dietary SFA intake and elevated risk of developing insulin resistance (Vessby et al., 2001) and cardiovascular disease (Yamagishi et al., 2010).

2.2.1 Evidence from cell culture studies

Studies examining SFA-induced modulation of cellular adiponectin content have yielded conflicting results. In 3T3-L1 adipocytes, treatment with 250 μM palmitic acid for 48 h led to a significant decrease in adiponectin gene expression (Bueno et al., 2008). Similarly, exposure of cultured rat adipocytes to 300 μM palmitic acid induced a 43% and 36% decrease in adiponectin mRNA and protein expression, respectively (Xi, Qian, Xu, Zhou, & Sun, 2007). However, neither of these studies measured adiponectin concentration in the media, making it impossible to determine if gene expression changes reflected SFA-induced changes in cellular or secreted adiponectin protein.

In contrast to the aforementioned studies, our laboratory demonstrated that treatment of 3T3-L1 murine adipocytes with 125 μM palmitic acid for 24 h had no effect on either adiponectin gene expression or secreted adiponectin protein measured in the culture media (Oster et al., 2010). Similarly, we found no change in cellular adiponectin protein expression or secreted adiponectin protein from isolated human primary adipocytes cultured with 100 μM palmitic acid for 48 h (Tishinsky et al., 2011). Our results are consistent with reports that short-term (2 h) incubations of primary rat adipocytes with 0.06–0.5 mM concentrations of palmitic acid did not alter secreted adiponectin protein in the media (Szkudelski, Nogowski, & Szkudelska, 2011). Last, in a study that treated 3T3-L1 adipocytes with palmitic and stearic acid for 24 h, both 10 and 100 μM concentrations of these SFA increased secreted adiponectin protein (Schaeffler et al., 2009). These discrepancies among studies may be somewhat attributable to differences in both fatty acid concentration and exposure time, although it is apparent that these contributing factors do not consistently account for differences in findings and more work is required to fully understand the role of SFA in adiponectin synthesis and secretion. There is also a need to evaluate the ability of different types of SFA to modulate adiponectin in cultured adipocytes as most studies to date, including our own work, have focused primarily on palmitic acid while other fatty acids also have direct relevance to current dietary patterns.

2.2.2 Evidence from rodent studies

The majority of *in vivo* studies have reported a negative association between dietary SFA and adiponectin. For example, consumption of a high SFA diet led to a significant decrease in adiponectin gene expression in both diabetic and wild-type mice (Todoric et al., 2006), as well as in male Wistar rats (Saravanan, Haseeb, Ehtesham, & Ghafoorunissa, 2005). In a study by Bueno et al. (2008), mice fed a diet high in lard (17.5% of kcal) had significantly lower serum adiponectin concentrations both acutely (2 days) and chronically (8 weeks). Furthermore, acute ingestion of the lard-enriched diet led to a decrease in adiponectin mRNA expression in retroperitoneal adipose tissue (Bueno et al., 2008). Interestingly, no significant changes in adiponectin gene expression in the epididymal or subcutaneous fat pads were found, suggesting that SFA-induced modulation of adiponectin mRNA may be adipose tissue depot specific. Additionally, our own laboratory has shown that as little as 3 days of a high SFA diet (60% of kcal) induced a significant decrease in plasma adiponectin concentrations in

Sprague–Dawley rats (Mullen et al., 2010). As with the cell culture evidence discussed above, it is noteworthy that most rodent studies in this area, including our own work, have focused primarily on lard-based diets (containing palmitic acid as the most abundant SFA, as well as high levels of the monounsaturated oleic acid, 18:1n-9) and the ability of different types of SFA to modulate adiponectin requires further study.

2.2.3 Evidence from human studies

Plasma concentrations of SFA have been found to be negatively associated ($r = -0.24$) with circulating adiponectin levels in both lean and obese (nondiabetic) humans (Fernandez-Real, Vendrell, & Ricart, 2005). More specifically, percentages (of total fatty acids) of myristic and palmitic acid in the blood were correlated strongly with lower circulating adiponectin (Fernandez-Real et al., 2005). These findings are supported by an intervention study in which consumption of a high SFA (19% of kcal) diet for 8 weeks in overweight subjects at risk for the metabolic syndrome led to decreased adiponectin gene expression in subcutaneous adipose tissue, as well as significantly lower plasma adiponectin levels (van Dijk et al., 2009).

Collectively, these studies suggest that SFA may decrease adiponectin, and this may provide a mechanistic link between SFA consumption and elevated risk of metabolic disease.

2.3. Thiazolidinediones

TZDs are a class of antidiabetic drugs commonly prescribed to type 2 diabetic patients to improve insulin sensitivity (Day, 1999). They were initially discovered in the early 1980s and include the clinically available drugs, rosiglitazone and pioglitazone, as well as troglitazone, which was withdrawn from the market due to risk of hepatotoxicity (Bouskila, Pajvani, & Scherer, 2005). Although the precise mechanisms by which TZDs ameliorate insulin resistance remain somewhat unclear, considerable evidence suggests that an increase in adiponectin secretion is involved. In support of this, rosiglitazone-induced improvements in insulin sensitivity in obese mice are significantly attenuated when these mice concomitantly lack adiponectin (Nawrocki et al., 2006). Additionally, research suggests that genetic variations in the adiponectin gene may actually be predictive of TZD responsiveness in type 2 diabetic patients (Kang et al., 2005), further implicating a role of adiponectin in TZD-mediated improvements in insulin sensitivity.

TZDs have been shown to increase adiponectin secretion from cultured murine 3T3-L1 adipocytes (Combs et al., 2002; Maeda et al., 2001), rat adipocytes (Banga et al., 2009), and isolated primary human adipocytes (Motoshima et al., 2002; Tishinsky et al., 2011), as well as to increase circulating adiponectin in animal models (Combs et al., 2002; Maeda et al., 2001). Elevations in plasma adiponectin concentrations have also been achieved with both short- (14 days) (Combs et al., 2002) and long-term (12 weeks) (Maeda et al., 2001; Phillips et al., 2003) TZD interventions in humans. In an initial study by Maeda et al. (2001), troglitazone led to a significant increase in plasma adiponectin concentrations in glucose intolerant individuals in the absence of any changes in body weight. In the same study, adiponectin expression was upregulated in adipose tissue from both lean and obese mice treated with troglitazone, pioglitazone, or rosiglitazone, and troglitazone increased adiponectin mRNA expression in a dose- and time-dependent manner in 3T3-L1 adipocytes (Maeda et al., 2001). Similarly, both obese and obese diabetic individuals exhibited a twofold increase in circulating adiponectin levels in response to 12 weeks of troglitazone administration (Yu et al., 2002). Interestingly, elevations in circulating adiponectin in response to TZD treatment have also been observed in healthy, nondiabetic individuals. Both Combs et al. (2002) and Yu et al. (2002) reported increased plasma adiponectin levels with TZD treatment in lean, healthy subjects with normal baseline adiponectin concentrations, suggesting that TZD-mediated increases in adiponectin are not limited to obese or diabetic individuals typically characterized by initial hypoadiponectinemia.

In addition to modulating total adiponectin secretion, TZDs are thought to specifically increase higher order multimers of adiponectin (Hammarstedt, Sopasakis, Gogg, Jansson, & Smith, 2005; Pajvani et al., 2004; Phillips et al., 2009). Scherer's group illustrated that rosiglitazone induced a dramatic increase in the ratio of HMW to total adiponectin in both diabetic mice and humans (Pajvani et al., 2004). Interestingly, improvements in insulin sensitivity with TZD treatment correlated with the ratio of HMW adiponectin to total, but not with total adiponectin itself (Pajvani et al., 2004), suggesting that the shift in the secretion of higher order multimers of adiponectin is the prevailing factor by which TZD-mediated increases in adiponectin enhance insulin sensitivity. Phillips et al. (2009) confirmed that TZD treatment increases circulating HMW adiponectin in type 2 diabetic individuals but found no effect of TZD treatment on cellular multimer distribution of adiponectin (Phillips et al., 2009), indicating that TZDs may upregulate secretion of HMW adiponectin without necessarily increasing cellular content.

In short, these studies provide convincing evidence for a TZD-mediated increase in adiponectin, which may serve as an important mechanism by which TZD treatment enhances insulin sensitivity.

2.4. Weight loss

Weight loss is associated with improvements in insulin sensitivity, glucose tolerance, and a reduction in markers of metabolic syndrome. Given the strong evidence for adiponectin's beneficial role for these parameters, it is reasonable to speculate that part of the metabolic benefits of weight loss are conferred by either increases in circulating adiponectinor altered proportions of the different molecular weight isoforms. Weight loss induced by caloric restriction, that is, not exercise, has fairly consistently been reported to cause an increase in total circulating adiponectin concentrations, typically ranging from 10% to 50% (Bobbert et al., 2005; Hotta et al., 2000; Yang et al., 2001). The vast majority of studies have examined overweight or obese adults, although there are also reports of weight loss-induced increases in adiponectin in obese children (Reinehr, Roth, Menke, & Andler, 2004). The pronounced weight loss induced by bariatric surgery results in robust increases in total adiponectin concentrations ranging from 40% to 280% (Coughlin et al., 2007; Engl et al., 2007; Hulver et al., 2002; Ress et al., 2010), greater than those typically reported in studies inducing smaller weight losses through traditional dietary restriction. However, while the main finding of increased total adiponectin is fairly consistent, there are limitations that influence the interpretation. Lack of control of exercise is a common issue, and there is also some disparity in terms of how tightly dietary composition is controlled or monitored. Finally, many studies only report changes in total adiponectin concentration. Studies that have assessed the three major molecular weight isoforms of adiponectin have generally reported changes in the medium and HMW isoforms (Bobbert et al., 2005; Engl et al., 2007), suggesting that weight loss may target increases in the higher order forms of adiponectin.

2.5. Acute exercise

In general, the effects of acute exercise on circulating adiponectin concentrations have been inconsistent. This inconsistency may be due to various factors including subject status (trained vs. untrained; lean vs. overweight), mode of exercise (cycling, running, rowing), and the recovery period over which potential changes in adiponectin are measured. Two studies, using highly trained subjects, have reported small but significant increases in total adiponectin within the first 30 min of recovery following intense aerobic exercise (Jurimae, Purge, & Jurimae, 2006; Kraemer et al., 2003). However, the majority of studies report either no change in total adiponectin or a slight decrease (Ferguson et al., 2004; Jamurtas et al., 2006; Jurimae, Hofmann, et al., 2006; Yamauchi, Kamon, Waki, et al., 2003). Given these limited findings and the variability in subjects and exercise modality, firm conclusions are not possible. In addition, only two reported studies accounted for changes in plasma volume that may occur during acute exercise which could alter or mask changes in adiponectin concentration. Nonetheless, notwithstanding these issues, it is interesting to note that the only identified increases in adiponectin following acute exercise (Jurimae, Purge, et al., 2006; Kraemer et al., 2003) appear to be characterized by (i) highly trained subjects, (ii) intense exercise, and (iii) sampling within the initial 30 min of recovery. However, further better controlled research is required to confirm these findings. At this time, the overall consensus appears to be that acute exercise has little effect on circulating adiponectin.

2.6. Chronic exercise

The effect of chronic exercise on circulating adiponectin is also variable. Part of this variability may be due to the nature of the exercise program (modality, intensity, duration), but also due to whether or not weight loss also occurred. Although some studies report significant increases in total adiponectin post training (Bluher et al., 2006; Hsieh & Wang, 2005), the majority do not (Hulver et al., 2002; Jurimae, Hofmann, et al., 2006, Jurimae, Purge, et al., 2006; Nassis et al., 2005; O'Leary et al., 2006; Yatagai, Nagasaka, et al., 2003). It should also be recognized that weight loss, rather than the exercise itself, may account for the majority of the change in adiponectin (Hsieh & Wang, 2005). While weight loss no doubt represents a confounding factor in determining whether exercise has an independent effect on adiponectin concentrations, it is interesting to note that of the 15 exercise intervention studies that reported information on both adiponectin concentrations and body weight changes, a linear regression analysis did not demonstrate a significant relationship between changes in body weight and circulating total adiponectin (Simpson & Singh, 2008). Thus, the independent effects of exercise and weight loss remain somewhat unclear.

3. MECHANISMS FOR LIFESTYLE-MEDIATED EFFECTS ON ADIPONECTIN

Various mechanisms have been proposed for the role of lifestyle factors in modulating adiponectin in rodents and humans. The following section will discuss the role of PPARγ and AMPK activation, posttranslational modifications and changes in adipose tissue morphology, macrophage infiltration, and inflammation as potential underlying mechanisms by which lifestyle factors influence adiponectin synthesis and secretion.

3.1. Peroxisome proliferator-activated receptor γ

PPARγ represents a nuclear receptor protein involved in the regulation of adipogenesis and carbohydrate and lipid metabolism (Lehmann et al., 1995). PPARγ is a ligand-activated transcription factor known to heterodimerize with retinoid × receptor and bind to PPAR-responsive elements (PPRE), thereby upregulating expression of its target genes (Schoonjans, Staels, & Auwerx, 1996). Iwaki et al. (2003) identified the existence of a PPRE in the promoter region of the adiponectin gene, and research strongly suggests that PPARγ agonists serve to upregulate the expression of adiponectin (Gustafson, Jack, Cushman, & Smith, 2003; Neschen et al., 2006; Oster et al., 2010; Tishinsky et al., 2011).

Several lines of research implicate n-3 PUFA and their derivatives as activators of PPARγ (Yamamoto et al., 2005). Recently, both *in vitro* and *in vivo* studies have signified the necessity of PPARγ in the n-3 PUFA-mediated upregulation of adiponectin (Bajaj et al., 2004; Neschen et al., 2006; Oster et al., 2010; Tishinsky et al., 2011). Work from our laboratory has demonstrated that antagonism of PPARγ significantly attenuates the n-3 PUFA-mediated induction of adiponectin secretion from both mouse (Oster et al., 2010) and human (Tishinsky et al., 2011) adipocytes. These results are in agreement with findings in rat adipocytes (Bajaj et al., 2004). Interestingly, whereas PPARγ antagonism completely abolished the DHA-mediated increase in secreted adiponectin from human adipocytes, it only partially attenuated the EPA-mediated response (Tishinsky et al., 2011). Thus, while DHA appears to enhance adiponectin exclusively via a PPARγ-dependent mechanism, EPA likely acts via a combination of PPARγ activation as well as an additional unknown mechanism, signifying a need to further evaluate the independent effects of these two long-chain n-3 PUFA on adiponectin.

Perhaps the strongest *in vivo* evidence for a role of PPARγ in mediating n-3 PUFA-induced increases in adiponectin comes from a study in which the feeding of a 27% (of kcal) fish oil diet for 15 days led to a two- to three-fold increase in plasma adiponectin concentrations in mice (Neschen et al., 2006). However, feeding the fish oil diet with the PPARγ antagonist, BADGE completely abolished the n-3 PUFA-mediated rise in circulating adiponectin (Neschen et al., 2006). It is also noteworthy that when PPARα-null mice received the same fish oil diet, adiponectin levels increased to similar levels observed in the wild-type mice, suggesting that n-3 PUFA mediate adiponectin specifically via the gamma isoform of PPAR (Neschen et al., 2006). Finally, in contrast to n-3 PUFA, evidence suggests that SFA may actually downregulate PPARγ expression (van Dijk et al., 2009). In overweight subjects at risk for metabolic syndrome, consumption of a high SFA diet for 8 weeks led to a significant decrease in PPARγ expression in subcutaneous adipose tissue (van Dijk et al., 2009), which may serve as an underlying mechanism for the dietary SFA-induced impairment in adiponectin, but this requires further study.

Similar to n-3 PUFA, PPARγ is also the primary molecular target of TZDs (Lehmann et al., 1995). Data from an adiponectin knockout mouse model demonstrated that adiponectin is necessary for TZD-mediated improvements in insulin sensitivity (Nawrocki et al., 2006), implicating a role for PPARγ agonism in the upregulation of adiponectin. Indeed, other work has shown that pioglitazone directly stimulates expression of PPARγ and this is required for TZD-mediated increases in adiponectin secretion from cultured adipocytes (Gustafson et al., 2003). Additionally, rosiglitazone treatment failed to increase adiponectin levels in patients with a dominant-negative PPARγ mutation, further emphasizing that TZDs induce adiponectin transcription via a PPARγ-dependent mechanism at the whole body level (Combs et al., 2002). Given that PPARγ is well-established as the primary receptor protein for TZDs (Lehmann et al., 1995), it is logical to largely accredit the enhancement of adiponectin secretion by TZD treatment to a PPARγ-dependent mechanism.

In summary, substantial evidence exists for a mechanistic role of PPARγ in the n-3 PUFA- and TZD-mediated upregulation of adiponectin.

3.2. AMP-activated protein kinase

A more limited body of evidence suggests that lifestyle factors, including dietary n-3 PUFA, may also increase adiponectin via stimulation of AMPK. Pharmacological activation of AMPK by AICAR has been demonstrated to increase adiponectin gene expression in a dose-dependent manner in adipose tissue explants from normal and overweight individuals (Lihn, Jessen, Pedersen, Lund, & Richelsen, 2004). Although the direct AMPK-mediated regulation of adiponectin secretion by n-3 PUFA has yet to be examined, several groups have reported that n-3 PUFA can effectively stimulate AMPK (Gonzalez-Periz et al., 2009; Lorente-Cebrian, Bustos, Marti, Martinez, & Moreno-Aliaga, 2009; Motawi, Hashem, Rashed, & El-Razek, 2009). Incubation of both mature murine adipocytes and preadipocytes with 100–200 μM EPA for 24 h induced a significant increase in the phosphorylation of AMPK (Lorente-Cebrian et al., 2009). Furthermore, dietary n-3 PUFA increased adiponectin mRNA expression in liver and skeletal muscle of high-fat-fed rats (Motawi et al., 2009) and induced AMPK phosphorylation in liver, skeletal muscle, and adipose tissue of ob/ob mice (Gonzalez-Periz et al., 2009).

In contrast to the abovementioned studies, Dobrzyn, Dobrzyn, Miyazaki, and Ntambi (2005) observed no effect of 14 days of dietary fish oil (5% of kcal) on AMPK phosphorylation and protein content in skeletal muscle, liver, and heart in wild-type mice. In addition to using a healthy animal model, the n-3 PUFA feeding duration was relatively short (14 days vs. 5–6 weeks in preceding studies), which may provide an explanation for their conflicting results. Given that n-3 PUFA can successfully stimulate AMPK (Gonzalez-Periz et al., 2009; Lorente-Cebrian et al., 2009; Motawi et al., 2009), and AMPK activation is known to increase adiponectin secretion (Lihn et al., 2004), investigating the role of AMPK as a potential mechanism by which n-3 PUFA increase adiponectin is warranted.

Additionally, while TZDs are classically accepted as PPARγ agonists, a single dose of troglitazone has also been shown to phosphorylate AMPK in skeletal muscle, liver, and adipose tissue in Sprague–Dawley rats (LeBrasseur et al., 2006), suggesting that TZDs may also regulate adiponectin via AMPK although this requires further study. Finally, exercise has been demonstrated to activate AMPK in adipose tissue (Park et al., 2002), which may provide a possible underlying mechanism by which exercise training has been shown to increase circulating adiponectin concentrations in some studies.

3.3. Posttranslational modification of adiponectin

Adiponectin is known to undergo considerable posttranslational modifications prior to its release from adipocytes (Wang, Xu, Knight, Xu, & Cooper, 2002). Recently, Wang et al. (2007) discovered that adiponectin is bound to ERp44, an endoplasmic reticulum (ER) chaperone protein that retains adiponectin intracellularly. Ero1-Lα, a second ER chaperone, competes with ERp44 to bind adiponectin and stimulate its release from the adipocyte (Wang et al., 2007). Interestingly, activation of PPARγ by troglitazone has been demonstrated to increase expression of Ero1-Lα in 3T3-L1 adipocytes, and this was associated with an increase in adiponectin secretion (Wang et al., 2007). From this, it stands to reason that TZDs may also increase adiponectin by facilitating the posttranslational modifications necessary for its release from adipocytes. Similarly, rosiglitazone has been shown to stimulate the expression of DsbA-L, an adiponectin-interactive protein that promotes adiponectin multimerization and secretion (Liu et al., 2008), further implicating a role for TZDs in the posttranslational modulation of adiponectin. To the best of our knowledge, the ability of n-3 PUFA to modulate the expression and function of the aforementioned chaperone proteins has not yet been examined. As TZDs and n-3 PUFA have been shown to have other commonalities such as increasing adiponectin via similar PPARγ-dependent mechanisms, investigating the role of n-3 PUFA in regulating posttranslational modifications of adiponectin is warranted.

3.4. Adipose tissue morphology

In addition to increasing adiponectin expression and secretion directly, n-3 PUFA may also stimulate adiponectin by regulating adipogenesis and inflicting a change in adipocyte morphology. Since adipocyte size represents a known determinant of the adipokine secretory profile, with smaller adipocytes synthesizing and secreting more adiponectin than larger adipocytes (Yamauchi, Kamon, Waki, Murakami, et al., 2001), such a change may provide another mechanism for lifestyle-induced modifications of adiponectin (Fig. 1.1). Manickam, Sinclair, and Cameron-Smith (2010) found that incubation of 3T3-L1 adipocytes with 100 μM EPA exerted a suppressive effect on lipid droplet formation, and this was associated with an elevation in adiponectin gene expression (Manickam et al., 2010). Similarly, 25–200 μM DHA inhibited adipocyte differentiation, induced apoptosis, and promoted lipolysis in postconfluent preadipocytes (Kim, Della-Fera, Lin, & Baile, 2006). Evidence from these *in vitro* studies is complemented by a study examining the n-3 PUFA-mediated regulation of adipocyte proliferation and differentiation in mice with diet-induced obesity (Hensler et al., 2011). After consuming a high-fat (35% of energy) corn oil-based diet for 8 weeks, the replacement of 15% of the dietary fat with an n-3 PUFA concentrate

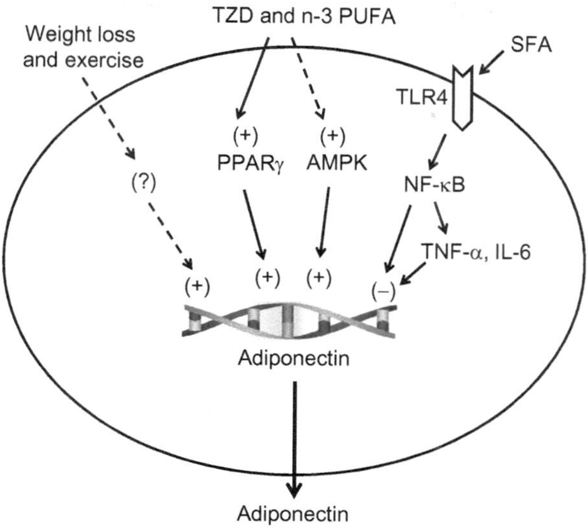

Figure 1.1 Proposed mechanisms for lifestyle-induced modulation of adiponectin secretion from adipocytes. TZDs and n-3 PUFA activate PPARγ, and possibly AMPK, to upregulate adiponectin secretion. SFA bind to TLR-4 to activate inflammatory pathways involving NF-κB to downregulate adiponectin secretion. Weight loss and exercise also act to increase adiponectin secretion through an unknown mechanism. Thiazolidinediones (TZDs), n-3 polyunsaturated fatty acids (PUFA), peroxisome proliferator-activated receptor (PPAR)γ, AMP-activated protein kinase (AMPK), saturated fatty acids (SFA), toll-like receptor (TLR)-4, nuclear factor (NF)-κB.

resulted in an increase in plasma adiponectin and a corresponding decrease in epididymal fat pad weight as well as adipocyte size (Hensler et al., 2011).

Likewise, TZD administration has also been shown to regulate adipose tissue mass and morphology. More specifically, 15 days of troglitazone treatment resulted in a fourfold increase in the number of small adipocytes and induced apoptosis of larger adipocytes, without any change in total adipose tissue weight, in obese Zucker rats (Okuno et al., 1998). Similarly, rosiglitazone treatment led to a significant decrease in adipocyte size in Sprague–Dawley rats fed either a low- or high-fat diet (Johnson, Trasino, Ferrante, & Vasselli, 2007). Given that the secretion of adipokines, including adiponectin, from adipose tissue is highly dependent on both adiposity (Arita et al., 1999) and adipocyte size (Yamauchi, Kamon, Waki, Terauchi, et al., 2001), it stands to reason that n-3 PUFA and TZD may upregulate adiponectin, in part, by reducing adipocyte hyperplasia and hypertrophy.

Furthermore, n-3 PUFA and TZD administration may also alleviate impairments in adiponectin secretion induced by mitochondrial dysfunction in adipocytes. In 3T3-L1 adipocytes, adiponectin synthesis was increased following induction of mitochondrial biogenesis by nuclear respiratory factor-1 overexpression, whereas impairments in mitochondrial function induced by JNK activation significantly impaired adiponectin transcription (Koh et al., 2007). As such, mitochondrial integrity appears to be required for adiponectin synthesis. As both n-3 PUFA (Flachs et al., 2005) and TZDs (Wilson-Fritch et al., 2003) are known to improve the mitochondrial biogenesis and morphology in adipocytes, it is possible that they may serve to improve cellular adiponectin synthesis. It is also interesting to note that the chaperone protein, DsbA-L involved in the assembly and secretion of adiponectin (see section 3.3) is largely localized to mitochondria (Wang & Scherer, 2008), presenting a potential link between mitochondrial biogenesis and the posttranslational modifications of adiponectin involved in its secretion.

3.5. Macrophage infiltration and inflammation (Todoric et al., 2006)

Changes in the release of inflammatory adipokines from adipose tissue in obesity are largely accredited to an increase in macrophage infiltration (Weisberg et al., 2003). Macrophages have been shown to inhibit adiponectin secretion and promote the release of inflammatory cytokines, including tumor necrosis factor (TNF)-α and interleukin (IL)-6 (Suganami, Nishida, & Ogawa, 2005). Furthermore, there is considerable evidence to suggest that the release of proinflammatory adipokines from adipose tissue-derived macrophages can exert paracrine effects on adipocytes, thereby inhibiting the synthesis and secretion of adiponectin. Treatment of 3T3-L1 adipocytes with IL-6 decreased the expression and secretion of adiponectin by 50% and 75%, respectively (Fasshauer et al., 2003). Similar impairments in adiponectin transcription are reported with TNF-α treatment in adipocytes (Zappala & Rechler, 2009). Recently, data has emerged to suggest that n-3 PUFA may act to prevent impairments in adiponectin secretion induced by inflammatory cytokines (Huber et al., 2007; Itoh et al., 2007; Perez-Matute, Perez-Echarri, Martinez, Marti, & Moreno-Aliaga, 2007; Todoric et al., 2006). In a study by Itoh et al. (2007) using a coculture system of 3T3-L1 adipocytes and RAW264 macrophages, 100–200 μM EPA was shown to mitigate the coculture-induced increase in TNF-α gene expression in macrophages. Furthermore, 5 weeks of

EPA supplementation (1 g/kg/day) in high-fat-fed Wistar rats was shown to effectively prevent the induction of TNF-α expression in adipose tissue (Perez-Matute et al., 2007). From these two studies, it seems likely that n-3 PUFA may alleviate reductions in adiponectin by downregulating the expression of proinflammatory adipocytes that interfere with its secretion.

Moreover, n-3 PUFA may also prevent the decrease in adiponectin that is induced by proinflammatory cytokines through inhibition of macrophage infiltration into the adipose tissue itself. Todoric et al. (2006) found that feeding a high SFA diet (30% of kcal from lard) for 6 weeks induced a marked increase in macrophage infiltration in obese diabetic mice and this was paralleled by an increase in inflammatory gene expression (Todoric et al., 2006). However, replacement of 12% of the lard with a high n-3 PUFA fish oil concentrate completely prevented the exacerbated state of adipose tissue macrophage infiltration and upregulated inflammatory gene expression that was triggered by the high SFA diet (Todoric et al., 2006). Intriguingly, n-3 PUFA have also been shown to downregulate the expression of matrix metalloproteinases required for macrophage infiltration into adipose tissue (Huber et al., 2007), supporting a role for n-3 PUFA in preventing inflammation by contributing to a remodeling of adipose tissue.

In addition to increases in macrophage infiltration in obesity, the activation of inflammatory pathways may also contribute to a downregulation of adiponectin secretion by proinflammatory cytokines. SFA are known ligands for toll-like receptor (TLR)-4 and, upon binding, initiate an inflammatory cascade that culminates in the activation of nuclear factor (NF)-κB and expression of inflammatory cytokines (Milanski et al., 2009). Shi et al. (2006) demonstrated that SFA treatment induced a TLR-4 dependent activation of NF-κB in macrophages and this resulted in increased IL-6 expression. Interestingly, research suggests that n-3 PUFA may aid in downregulating the SFA-mediated activation of TLR-4 and NF-κB, thereby preventing increases in proinflammatory adipokines that can inhibit adiponectin secretion. For example, 5 μM EPA or DHA prevented lauric acid-induced activation of NF-κB in RAW264 macrophages (Lee, Sohn, Rhee, & Hwang, 2001), while EPA effectively suppressed induction of NF-κB activity by palmitic acid in RAW264 macrophages (Itoh et al., 2007). Taken together, these studies suggest that n-3 PUFA may downregulate inflammatory signaling pathways induced by SFA, thereby alleviating the suppression of adiponectin by proinflammatory cytokines.

4. CONCLUSIONS

Adiponectin synthesis and secretion can be modulated by lifestyle factors. Long-chain n-3 PUFA can increase adiponectin by direct activation of PPARγ and AMPK, as well as indirectly via changes in adipose tissue morphology, macrophage infiltration and inflammation. TZDs are known activators of PPARγ, and thereby also elicit an increase in adiponectin. n-3 PUFA and TZDs are also thought to play a role in mediating posttranslational modifications that are required for the release of adiponectin from adipocytes. Finally, weight loss has been shown to exert modest increases in circulating adiponectin levels, although the underlying mechanism by which this occurs is currently unclear. The effects of exercise, independent of weight loss, are not consistent and require additional study. Given that adiponectin represents an anti-inflammatory and insulin-sensitizing adipokine that is classically downregulated in obesity and related diseases, the use of lifestyle strategies to increase the synthesis and secretion of adiponectin represents a promising approach to ameliorate these pathologies.

REFERENCES

An, W. S., Son, Y. K., Kim, S. E., Kim, K. H., Bae, H. R., Lee, S., et al. (2011). Association of adiponectin and leptin with serum lipids and erythrocyte omega-3 and omega-6 fatty acids in dialysis patients. *Clinical Nephrology*, *75*, 195–203.

Arita, Y., Kihara, S., Ouchi, N., Takahashi, M., Maeda, K., Miyagawa, J., et al. (1999). Paradoxical decrease of an adipose-specific protein, adiponectin, in obesity. *Biochemical and Biophysical Research Communications*, *257*, 79–83.

Bajaj, M., Suraamornkul, S., Piper, P., Hardies, L. J., Glass, L., Cersosimo, E., et al. (2004). Decreased plasma adiponectin concentrations are closely related to hepatic fat content and hepatic insulin resistance in pioglitazone-treated type 2 diabetic patients. *The Journal of Clinical Endocrinology and Metabolism*, *89*, 200–206.

Banga, A., Unal, R., Tripathi, P., Pokrovskaya, I., Owens, R. J., Kern, P. A., et al. (2009). Adiponectin translation is increased by the PPARgamma agonists pioglitazone and omega-3 fatty acids. *American Journal of Physiology. Endocrinology and Metabolism*, *296*, E480–E489.

Berg, A. H., Combs, T. P., Du, X., Brownlee, M., & Scherer, P. E. (2001). The adipocyte-secreted protein acrp30 enhances hepatic insulin action. *Nature Medicine*, *7*, 947–953.

Berg, A. H., Combs, T. P., & Scherer, P. E. (2002). ACRP30/adiponectin: An adipokine regulating glucose and lipid metabolism. *Trends in Endocrinology and Metabolism*, *13*, 84–89.

Bluher, M., Bullen, J. W., Jr., Lee, J. H., Kralisch, S., Fasshauer, M., Kloting, N., et al. (2006). Circulating adiponectin and expression of adiponectin receptors in human skeletal muscle: Associations with metabolic parameters and insulin resistance and regulation by physical training. *The Journal of Clinical Endocrinology and Metabolism*, *91*, 2310–2316.

Bobbert, T., Rochlitz, H., Wegewitz, U., Akpulat, S., Mai, K., Weickert, M. O., et al. (2005). Changes of adiponectin oligomer composition by moderate weight reduction. *Diabetes*, *54*, 2712–2719.

Bouskila, M., Pajvani, U. B., & Scherer, P. E. (2005). Adiponectin: A relevant player in PPARgamma-agonist-mediated improvements in hepatic insulin sensitivity? *International Journal of Obesity, 29*(Suppl. 1), S17–S23.

Bruce, C. R., Mertz, V. A., Heigenhauser, G. J., & Dyck, D. J. (2005). The stimulatory effect of globular adiponectin on insulin-stimulated glucose uptake and fatty acid oxidation is impaired in skeletal muscle from obese subjects. *Diabetes, 54,* 3154–3160.

Bueno, A. A., Oyama, L. M., de Oliveira, C., Pisani, L. P., Ribeiro, E. B., Silveira, V. L., et al. (2008). Effects of different fatty acids and dietary lipids on adiponectin gene expression in 3T3-L1 cells and C57BL/6J mice adipose tissue. *Pflügers Archiv, 455,* 701–709.

Chen, M. B., McAinch, A. J., Macaulay, S. L., Castelli, L. A., O'brien, P. E., Dixon, J. B., et al. (2005). Impaired activation of AMP-kinase and fatty acid oxidation by globular adiponectin in cultured human skeletal muscle of obese type 2 diabetics. *The Journal of Clinical Endocrinology and Metabolism, 90,* 3665–3672.

Civitarese, A. E., Jenkinson, C. P., Richardson, D., Bajaj, M., Cusi, K., Kashyap, S., et al. (2004). Adiponectin receptors gene expression and insulin sensitivity in non-diabetic Mexican Americans with or without a family history of type 2 diabetes. *Diabetologia, 47,* 816–820.

Combs, T. P., Berg, A. H., Obici, S., Scherer, P. E., & Rossetti, L. (2001). Endogenous glucose production is inhibited by the adipose-derived protein Acrp30. *The Journal of Clinical Investigation, 108,* 1875–1881.

Combs, T. P., Pajvani, U. B., Berg, A. H., Lin, Y., Jelicks, L. A., Laplante, M., et al. (2004). A transgenic mouse with a deletion in the collagenous domain of adiponectin displays elevated circulating adiponectin and improved insulin sensitivity. *Endocrinology, 145,* 367–383.

Combs, T. P., Wagner, J. A., Berger, J., Doebber, T., Wang, W. J., Zhang, B. B., et al. (2002). Induction of adipocyte complement-related protein of 30 kilodaltons by PPARgamma agonists: A potential mechanism of insulin sensitization. *Endocrinology, 143,* 998–1007.

Coughlin, C. C., Finck, B. N., Eagon, J. C., Halpin, V. J., Magkos, F., Mohammed, B. S., et al. (2007). Effect of marked weight loss on adiponectin gene expression and plasma concentrations. *Obesity (Silver Spring), 15,* 640–645.

Day, C. (1999). Thiazolidinediones: A new class of antidiabetic drugs. *Diabetic Medicine, 16,* 179–192.

Debard, C., Laville, M., Berbe, V., Loizon, E., Guillet, C., Morio-Liondore, B., et al. (2004). Expression of key genes of fatty acid oxidation, including adiponectin receptors, in skeletal muscle of type 2 diabetic patients. *Diabetologia, 47,* 917–925.

Dobrzyn, A., Dobrzyn, P., Miyazaki, M., & Ntambi, J. M. (2005). Polyunsaturated fatty acids do not activate AMP-activated protein kinase in mouse tissues. *Biochemical and Biophysical Research Communications, 332,* 892–896.

Engl, J., Bobbert, T., Ciardi, C., Laimer, M., Tatarczyk, T., Kaser, S., et al. (2007). Effects of pronounced weight loss on adiponectin oligomer composition and metabolic parameters. *Obesity (Silver Spring), 15,* 1172–1178.

Fasshauer, M., Kralisch, S., Klier, M., Lossner, U., Bluher, M., Klein, J., et al. (2003). Adiponectin gene expression and secretion is inhibited by interleukin-6 in 3T3-L1 adipocytes. *Biochemical and Biophysical Research Communications, 301,* 1045–1050.

Ferguson, M. A., White, L. J., McCoy, S., Kim, H. W., Petty, T., & Wilsey, J. (2004). Plasma adiponectin response to acute exercise in healthy subjects. *European Journal of Applied Physiology, 91,* 324–329.

Fernandez-Real, J. M., Vendrell, J., & Ricart, W. (2005). Circulating adiponectin and plasma fatty acid profile. *Clinical Chemistry, 51,* 603–609.

Flachs, P., Horakova, O., Brauner, P., Rossmeisl, M., Pecina, P., Franssen-van Hal, N., et al. (2005). Polyunsaturated fatty acids of marine origin upregulate mitochondrial biogenesis and induce beta-oxidation in white fat. *Diabetologia, 48,* 2365–2375.

Flachs, P., Mohamed-Ali, V., Horakova, O., Rossmeisl, M., Hosseinzadeh-Attar, M. J., Hensler, M., et al. (2006). Polyunsaturated fatty acids of marine origin induce adiponectin in mice fed a high-fat diet. *Diabetologia, 49*, 394–397.

Fruebis, J., Tsao, T. S., Javorschi, S., Ebbets-Reed, D., Erickson, M. R., Yen, F. T., et al. (2001). Proteolytic cleavage product of 30-kDa adipocyte complement-related protein increases fatty acid oxidation in muscle and causes weight loss in mice. *Proceedings of the National Academy of Sciences of the United States of America, 98*, 2005–2010.

Gonzalez-Periz, A., Horrillo, R., Ferre, N., Gronert, K., Dong, B., Moran-Salvador, E., et al. (2009). Obesity-induced insulin resistance and hepatic steatosis are alleviated by omega-3 fatty acids: A role for resolvins and protectins. *The FASEB Journal, 23*, 1946–1957.

Gulick, T., Cresci, S., Caira, T., Moore, D. D., & Kelly, D. P. (1994). The peroxisome proliferator-activated receptor regulates mitochondrial fatty acid oxidative enzyme gene expression. *Proceedings of the National Academy of Sciences of the United States of America, 91*, 11012–11016.

Gustafson, B., Jack, M. M., Cushman, S. W., & Smith, U. (2003). Adiponectin gene activation by thiazolidinediones requires PPAR gamma 2, but not C/EBP alpha-evidence for differential regulation of the aP2 and adiponectin genes. *Biochemical and Biophysical Research Communications, 308*, 933–939.

Hammarstedt, A., Sopasakis, V. R., Gogg, S., Jansson, P. A., & Smith, U. (2005). Improved insulin sensitivity and adipose tissue dysregulation after short-term treatment with pioglitazone in non-diabetic, insulin-resistant subjects. *Diabetologia, 48*, 96–104.

Hara, K., Horikoshi, M., Yamauchi, T., Yago, H., Miyazaki, O., Ebinuma, H., et al. (2006). Measurement of the high-molecular weight form of adiponectin in plasma is useful for the prediction of insulin resistance and metabolic syndrome. *Diabetes Care, 29*, 1357–1362.

Hensler, M., Bardova, K., Jilkova, Z. M., Wahli, W., Meztger, D., Chambon, P., et al. (2011). The inhibition of fat cell proliferation by n-3 fatty acids in dietary obese mice. *Lipids in Health and Disease, 10*, 128.

Hotta, K., Funahashi, T., Arita, Y., Takahashi, M., Matsuda, M., Okamoto, Y., et al. (2000). Plasma concentrations of a novel, adipose-specific protein, adiponectin, in type 2 diabetic patients. *Arteriosclerosis, Thrombosis, and Vascular Biology, 20*, 1595–1599.

Hotta, K., Funahashi, T., Bodkin, N. L., Ortmeyer, H. K., Arita, Y., Hansen, B. C., et al. (2001). Circulating concentrations of the adipocyte protein adiponectin are decreased in parallel with reduced insulin sensitivity during the progression to type 2 diabetes in rhesus monkeys. *Diabetes, 50*, 1126–1133.

Hsieh, C. J., & Wang, P. W. (2005). Effectiveness of weight loss in the elderly with type 2 diabetes mellitus. *Journal of Endocrinological Investigation, 28*, 973–977.

Hu, E., Liang, P., & Spiegelman, B. M. (1996). AdipoQ is a novel adipose-specific gene dysregulated in obesity. *The Journal of Biological Chemistry, 271*, 10697–10703.

Huber, J., Loffler, M., Bilban, M., Reimers, M., Kadl, A., Todoric, J., et al. (2007). Prevention of high-fat diet-induced adipose tissue remodeling in obese diabetic mice by n-3 polyunsaturated fatty acids. *International Journal of Obesity, 31*, 1004–1013.

Hug, C., Wang, J., Ahmad, N. S., Bogan, J. S., Tsao, T. S., & Lodish, H. F. (2004). T-cadherin is a receptor for hexameric and high-molecular-weight forms of Acrp30/adiponectin. *Proceedings of the National Academy of Sciences of the United States of America, 101*, 10308–10313.

Hulver, M. W., Zheng, D., Tanner, C. J., Houmard, J. A., Kraus, W. E., Slentz, C. A., et al. (2002). Adiponectin is not altered with exercise training despite enhanced insulin action. *American Journal of Physiology. Endocrinology and Metabolism, 283*, E861–E865.

Itoh, M., Suganami, T., Satoh, N., Tanimoto-Koyama, K., Yuan, X., Tanaka, M., et al. (2007). Increased adiponectin secretion by highly purified eicosapentaenoic acid in rodent models of obesity and human obese subjects. *Arteriosclerosis, Thrombosis, and Vascular Biology, 27*, 1918–1925.

Iwabu, M., Yamauchi, T., Okada-Iwabu, M., Sato, K., Nakagawa, T., Funata, M., et al. (2010). Adiponectin and AdipoR1 regulate PGC-1alpha and mitochondria by Ca(2+) and AMPK/SIRT1. *Nature, 464*, 1313–1319.

Iwaki, M., Matsuda, M., Maeda, N., Funahashi, T., Matsuzawa, Y., Makishima, M., et al. (2003). Induction of adiponectin, a fat-derived antidiabetic and antiatherogenic factor, by nuclear receptors. *Diabetes, 52*, 1655–1663.

Jamurtas, A. Z., Theocharis, V., Koukoulis, G., Stakias, N., Fatouros, I. G., Kouretas, D., et al. (2006). The effects of acute exercise on serum adiponectin and resistin levels and their relation to insulin sensitivity in overweight males. *European Journal of Applied Physiology, 97*, 122–126.

Johnson, J. A., Trasino, S. E., Ferrante, A. W., Jr., & Vasselli, J. R. (2007). Prolonged decrease of adipocyte size after rosiglitazone treatment in high- and low-fat-fed rats. *Obesity (Silver Spring), 15*, 2653–2663.

Johnson, C., Williams, R., Wei, J. Y., & Ranganathan, G. (2011). Regulation of serum response factor and adiponectin by PPARgamma agonist docosahexaenoic acid. *Journal of Lipids, 2011*, 670479.

Jurimae, J., Hofmann, P., Jurimae, T., Maestu, J., Purge, P., Wonisch, M., et al. (2006). Plasma adiponectin response to sculling exercise at individual anaerobic threshold in college level male rowers. *International Journal of Sports Medicine, 27*, 272–277.

Jurimae, J., Purge, P., & Jurimae, T. (2006). Adiponectin and stress hormone responses to maximal sculling after volume-extended training season in elite rowers. *Metabolism, 55*, 13–19.

Kalupahana, N. S., Claycombe, K., Newman, S. J., Stewart, T., Siriwardhana, N., Matthan, N., et al. (2010). Eicosapentaenoic acid prevents and reverses insulin resistance in high-fat diet-induced obese mice via modulation of adipose tissue inflammation. *The Journal of Nutrition, 140*, 1915–1922.

Kang, E. S., Park, S. Y., Kim, H. J., Ahn, C. W., Nam, M., Cha, B. S., et al. (2005). The influence of adiponectin gene polymorphism on the rosiglitazone response in patients with type 2 diabetes. *Diabetes Care, 28*, 1139–1144.

Kim, H. K., Della-Fera, M., Lin, J., & Baile, C. A. (2006). Docosahexaenoic acid inhibits adipocyte differentiation and induces apoptosis in 3T3-L1 preadipocytes. *The Journal of Nutrition, 136*, 2965–2969.

Koh, E. H., Park, J. Y., Park, H. S., Jeon, M. J., Ryu, J. W., Kim, M., et al. (2007). Essential role of mitochondrial function in adiponectin synthesis in adipocytes. *Diabetes, 56*, 2973–2981.

Kondo, K., Morino, K., Nishio, Y., Kondo, M., Fuke, T., Ugi, S., et al. (2010). Effects of a fish-based diet on the serum adiponectin concentration in young, non-obese, healthy Japanese subjects. *Journal of Atherosclerosis and Thrombosis, 17*, 628–637.

Kraemer, R. R., Aboudehen, K. S., Carruth, A. K., Durand, R. T., Acevedo, E. O., Hebert, E. P., et al. (2003). Adiponectin responses to continuous and progressively intense intermittent exercise. *Medicine & Science in Sports & Exercise, 35*, 1320–1325.

Kratz, M., Swarbrick, M. M., Callahan, H. S., Matthys, C. C., Havel, P. J., & Weigle, D. S. (2008). Effect of dietary n-3 polyunsaturated fatty acids on plasma total and high-molecular-weight adiponectin concentrations in overweight to moderately obese men and women. *The American Journal of Clinical Nutrition, 87*, 347–353.

Krebs, J. D., Browning, L. M., McLean, N. K., Rothwell, J. L., Mishra, G. D., Moore, C. S., et al. (2006). Additive benefits of long-chain n-3 polyunsaturated fatty acids and weight-loss in the management of cardiovascular disease risk in overweight hyperinsulinaemic women. *International Journal of Obesity, 30*, 1535–1544.

Kuoppamaa, H., Skrobuk, P., Sihvo, M., Hiukka, A., Chibalin, A. V., Zierath, J. R., et al. (2008). Globular adiponectin stimulates glucose transport in type 2 diabetic muscle. *Diabetes/Metabolism Research and Reviews*, *24*, 554–562.

Lara, J. J., Economou, M., Wallace, A. M., Rumley, A., Lowe, G., Slater, C., et al. (2007). Benefits of salmon eating on traditional and novel vascular risk factors in young, non-obese healthy subjects. *Atherosclerosis*, *193*, 213–221.

LeBrasseur, N. K., Kelly, M., Tsao, T. S., Farmer, S. R., Saha, A. K., Ruderman, N. B., et al. (2006). Thiazolidinediones can rapidly activate AMP-activated protein kinase in mammalian tissues. *American Journal of Physiology. Endocrinology and Metabolism*, *291*, E175–E181.

Lee, J. Y., Sohn, K. H., Rhee, S. H., & Hwang, D. (2001). Saturated fatty acids, but not unsaturated fatty acids, induce the expression of cyclooxygenase-2 mediated through toll-like receptor 4. *The Journal of Biological Chemistry*, *276*, 16683–16689.

Lefils, J., Geloen, A., Vidal, H., Lagarde, M., & Bernoud-Hubac, N. (2010). Dietary DHA: Time course of tissue uptake and effects on cytokine secretion in mice. *The British Journal of Nutrition*, *104*, 1304–1312.

Lehmann, J. M., Moore, L. B., Smith-Oliver, T. A., Wilkison, W. O., Willson, T. M., & Kliewer, S. A. (1995). An antidiabetic thiazolidinedione is a high affinity ligand for peroxisome proliferator-activated receptor gamma (PPAR gamma). *The Journal of Biological Chemistry*, *270*, 12953–12956.

Lihn, A. S., Jessen, N., Pedersen, S. B., Lund, S., & Richelsen, B. (2004). AICAR stimulates adiponectin and inhibits cytokines in adipose tissue. *Biochemical and Biophysical Research Communications*, *316*, 853–858.

Liu, M., Zhou, L., Xu, A., Lam, K. S., Wetzel, M. D., Xiang, R., et al. (2008). A disulfide-bond A oxidoreductase-like protein (DsbA-L) regulates adiponectin multimerization. *Proceedings of the National Academy of Sciences of the United States of America*, *105*, 18302–18307.

Lorente-Cebrian, S., Bustos, M., Marti, A., Martinez, J. A., & Moreno-Aliaga, M. J. (2009). Eicosapentaenoic acid stimulates AMP-activated protein kinase and increases visfatin secretion in cultured murine adipocytes. *Clinical Science (London, England)*, *117*, 243–249.

Lorente-Cebrian, S., Perez-Matute, P., Martinez, J. A., Marti, A., & Moreno-Aliaga, M. J. (2006). Effects of eicosapentaenoic acid (EPA) on adiponectin gene expression and secretion in primary cultured rat adipocytes. *Journal of Physiology and Biochemistry*, *62*, 61–69.

Maeda, K., Okubo, K., Shimomura, I., Funahashi, T., Matsuzawa, Y., & Matsubara, K. (1996). CDNA cloning and expression of a novel adipose specific collagen-like factor, apM1 (AdiPose most abundant gene transcript 1). *Biochemical and Biophysical Research Communications*, *221*, 286–289.

Maeda, N., Takahashi, M., Funahashi, T., Kihara, S., Nishizawa, H., Kishida, K., et al. (2001). PPARgamma ligands increase expression and plasma concentrations of adiponectin, an adipose-derived protein. *Diabetes*, *50*, 2094–2099.

Manickam, E., Sinclair, A. J., & Cameron-Smith, D. (2010). Suppressive actions of eicosapentaenoic acid on lipid droplet formation in 3T3-L1 adipocytes. *Lipids in Health and Disease*, *9*, 57.

Mao, X., Kikani, C. K., Riojas, R. A., Langlais, P., Wang, L., Ramos, F. J., et al. (2006). APPL1 binds to adiponectin receptors and mediates adiponectin signalling and function. *Nature Cell Biology*, *8*, 516–523.

Milanski, M., Degasperi, G., Coope, A., Morari, J., Denis, R., Cintra, D. E., et al. (2009). Saturated fatty acids produce an inflammatory response predominantly through the activation of TLR4 signaling in hypothalamus: Implications for the pathogenesis of obesity. *The Journal of Neuroscience*, *29*, 359–370.

Motawi, T. M., Hashem, R. M., Rashed, L. A., & El-Razek, S. M. (2009). Comparative study between the effect of the peroxisome proliferator activated receptor-alpha ligands fenofibrate and n-3 polyunsaturated fatty acids on activation of 5'-AMP-activated protein kinase-alpha1 in high-fat fed rats. *The Journal of Pharmacy and Pharmacology*, *61*, 1339–1346.

Motoshima, H., Wu, X., Sinha, M. K., Hardy, V. E., Rosato, E. L., Barbot, D. J., et al. (2002). Differential regulation of adiponectin secretion from cultured human omental and subcutaneous adipocytes: Effects of insulin and rosiglitazone. *The Journal of Clinical Endocrinology and Metabolism*, *87*, 5662–5667.

Mullen, K. L., Pritchard, J., Ritchie, I., Snook, L. A., Chabowski, A., Bonen, A., et al. (2009). Adiponectin resistance precedes the accumulation of skeletal muscle lipids and insulin resistance in high-fat-fed rats. *American Journal of Physiology. Regulatory, Integrative and Comparative Physiology*, *296*, R243–R251.

Mullen, K. L., Tishinsky, J. M., Robinson, L. E., & Dyck, D. J. (2010). Skeletal muscle inflammation is not responsible for the rapid impairment in adiponectin response with high-fat feeding in rats. *American Journal of Physiology. Regulatory, Integrative and Comparative Physiology*, *299*, R500–R508.

Nakano, Y., Tobe, T., Choi-Miura, N. H., Mazda, T., & Tomita, M. (1996). Isolation and characterization of GBP28, a novel gelatin-binding protein purified from human plasma. *Journal of Biochemistry*, *120*, 803–812.

Nassis, G. P., Papantakou, K., Skenderi, K., Triandafillopoulou, M., Kavouras, S. A., Yannakoulia, M., et al. (2005). Aerobic exercise training improves insulin sensitivity without changes in body weight, body fat, adiponectin, and inflammatory markers in overweight and obese girls. *Metabolism*, *54*, 1472–1479.

Nawrocki, A. R., Rajala, M. W., Tomas, E., Pajvani, U. B., Saha, A. K., Trumbauer, M. E., et al. (2006). Mice lacking adiponectin show decreased hepatic insulin sensitivity and reduced responsiveness to peroxisome proliferator-activated receptor gamma agonists. *The Journal of Biological Chemistry*, *281*, 2654–2660.

Neschen, S., Morino, K., Rossbacher, J. C., Pongratz, R. L., Cline, G. W., Sono, S., et al. (2006). Fish oil regulates adiponectin secretion by a peroxisome proliferator-activated receptor-gamma-dependent mechanism in mice. *Diabetes*, *55*, 924–928.

Nomura, S., Shouzu, A., Omoto, S., Inami, N., Ueba, T., Urase, F., et al. (2009). Effects of eicosapentaenoic acid on endothelial cell-derived microparticles, angiopoietins and adiponectin in patients with type 2 diabetes. *Journal of Atherosclerosis and Thrombosis*, *16*, 83–90.

Okuno, A., Tamemoto, H., Tobe, K., Ueki, K., Mori, Y., Iwamoto, K., et al. (1998). Troglitazone increases the number of small adipocytes without the change of white adipose tissue mass in obese Zucker rats. *The Journal of Clinical Investigation*, *101*, 1354–1361.

O'Leary, V. B., Marchetti, C. M., Krishnan, R. K., Stetzer, B. P., Gonzalez, F., & Kirwan, J. P. (2006). Exercise-induced reversal of insulin resistance in obese elderly is associated with reduced visceral fat. *Journal of Applied Physiology*, *100*, 1584–1589.

Oller do Nascimento, C. M., Ribeiro, E. B., & Oyama, L. M. (2009). Metabolism and secretory function of white adipose tissue: Effect of dietary fat. *Anais da Academia Brasileira de Ciências*, *81*, 453–466.

Oster, R. T., Tishinsky, J. M., Yuan, Z., & Robinson, L. E. (2010). Docosahexaenoic acid increases cellular adiponectin mRNA and secreted adiponectin protein, as well as PPARgamma mRNA, in 3T3-L1 adipocytes. *Applied Physiology, Nutrition, and Metabolism*, *35*, 783–789.

Pajvani, U. B., Hawkins, M., Combs, T. P., Rajala, M. W., Doebber, T., Berger, J. P., et al. (2004). Complex distribution, not absolute amount of adiponectin, correlates with thiazolidinedione-mediated improvement in insulin sensitivity. *The Journal of Biological Chemistry*, *279*, 12152–12162.

Park, H., Kaushik, V. K., Constant, S., Prentki, M., Przybytkowski, E., Ruderman, N. B., et al. (2002). Coordinate regulation of Malonyl-CoA decarboxylase, sn-glycerol-3-phosphate acyltransferase, and acetyl-CoA carboxylase by AMP-activated protein kinase in rat tissues in response to exercise. *The Journal of Biological Chemistry*, *277*, 32571–32577.

Perez-Matute, P., Perez-Echarri, N., Martinez, J. A., Marti, A., & Moreno-Aliaga, M. J. (2007). Eicosapentaenoic acid actions on adiposity and insulin resistance in control and high-fat-fed rats: Role of apoptosis, adiponectin and tumour necrosis factor-alpha. *The British Journal of Nutrition*, *97*, 389–398.

Phillips, S. A., Ciaraldi, T. P., Kong, A. P., Bandukwala, R., Aroda, V., Carter, L., et al. (2003). Modulation of circulating and adipose tissue adiponectin levels by antidiabetic therapy. *Diabetes*, *52*, 667–674.

Phillips, S. A., Kung, J., Ciaraldi, T. P., Choe, C., Christiansen, L., Mudaliar, S., et al. (2009). Selective regulation of cellular and secreted multimeric adiponectin by antidiabetic therapies in humans. *American Journal of Physiology. Endocrinology and Metabolism*, *297*, E767–E773.

Puglisi, M. J., Hasty, A. H., & Saraswathi, V. (2011). The role of adipose tissue in mediating the beneficial effects of dietary fish oil. *The Journal of Nutritional Biochemistry*, *22*, 101–108.

Reinehr, T., Roth, C., Menke, T., & Andler, W. (2004). Adiponectin before and after weight loss in obese children. *The Journal of Clinical Endocrinology and Metabolism*, *89*, 3790–3794.

Ress, C., Tschoner, A., Engl, J., Klaus, A., Tilg, H., Ebenbichler, C. F., et al. (2010). Effect of bariatric surgery on circulating chemerin levels. *European Journal of Clinical Investigation*, *40*, 277–280.

Rossi, A. S., Lombardo, Y. B., Lacorte, J. M., Chicco, A. G., Rouault, C., Slama, G., et al. (2005). Dietary fish oil positively regulates plasma leptin and adiponectin levels in sucrose-fed, insulin-resistant rats. *American Journal of Physiology. Regulatory, Integrative and Comparative Physiology*, *289*, R486–R494.

Saravanan, N., Haseeb, A., Ehtesham, N. Z., & Ghafoorunissa, (2005). Differential effects of dietary saturated and trans-fatty acids on expression of genes associated with insulin sensitivity in rat adipose tissue. *European Journal of Endocrinology*, *153*, 159–165.

Schaeffler, A., Gross, P., Buettner, R., Bollheimer, C., Buechler, C., Neumeier, M., et al. (2009). Fatty acid-induced induction of toll-like receptor-4/nuclear factor-kappaB pathway in adipocytes links nutritional signalling with innate immunity. *Immunology*, *126*, 233–245.

Scherer, P. E., Williams, S., Fogliano, M., Baldini, G., & Lodish, H. F. (1995). A novel serum protein similar to C1q, produced exclusively in adipocytes. *The Journal of Biological Chemistry*, *270*, 26746–26749.

Schoonjans, K., Staels, B., & Auwerx, J. (1996). The peroxisome proliferator activated receptors (PPARS) and their effects on lipid metabolism and adipocyte differentiation. *Biochimica et Biophysica Acta*, *1302*, 93–109.

Shi, H., Kokoeva, M. V., Inouye, K., Tzameli, I., Yin, H., & Flier, J. S. (2006). TLR4 links innate immunity and fatty acid-induced insulin resistance. *The Journal of Clinical Investigation*, *116*, 3015–3025.

Simpson, K. A., & Singh, M. A. (2008). Effects of exercise on adiponectin: A systematic review. *Obesity (Silver Spring)*, *16*, 241–256.

Suganami, T., Nishida, J., & Ogawa, Y. (2005). A paracrine loop between adipocytes and macrophages aggravates inflammatory changes: Role of free fatty acids and tumor necrosis factor alpha. *Arteriosclerosis, Thrombosis, and Vascular Biology*, *25*, 2062–2068.

Szkudelski, T., Nogowski, L., & Szkudelska, K. (2011). Short-term regulation of adiponectin secretion in rat adipocytes. *Physiological Research*, *60*, 521–530.

Thrush, A. B., Heigenhauser, G. J., Mullen, K. L., Wright, D. C., & Dyck, D. J. (2008). Palmitate acutely induces insulin resistance in isolated muscle from obese but not lean humans. *American Journal of Physiology. Regulatory, Integrative and Comparative Physiology, 294*, R1205–R1212.

Tishinsky, J. M., Ma, D. W., & Robinson, L. E. (2011). Eicosapentaenoic acid and rosiglitazone increase adiponectin in an additive and PPARgamma-dependent manner in human adipocytes. *Obesity (Silver Spring), 19*, 262–268.

Todoric, J., Loffler, M., Huber, J., Bilban, M., Reimers, M., Kadl, A., et al. (2006). Adipose tissue inflammation induced by high-fat diet in obese diabetic mice is prevented by n-3 polyunsaturated fatty acids. *Diabetologia, 49*, 2109–2119.

Tomas, E., Tsao, T. S., Saha, A. K., Murrey, H. E., Zhang Cc, C., Itani, S. I., et al. (2002). Enhanced muscle fat oxidation and glucose transport by ACRP30 globular domain: Acetyl-CoA carboxylase inhibition and AMP-activated protein kinase activation. *Proceedings of the National Academy of Sciences of the United States of America, 99*, 16309–16313.

Tsuchida, A., Yamauchi, T., Ito, Y., Hada, Y., Maki, T., Takekawa, S., et al. (2004). Insulin/Foxo1 pathway regulates expression levels of adiponectin receptors and adiponectin sensitivity. *The Journal of Biological Chemistry, 279*, 30817–30822.

van Dijk, S. J., Feskens, E. J., Bos, M. B., Hoelen, D. W., Heijligenberg, R., Bromhaar, M. G., et al. (2009). A saturated fatty acid-rich diet induces an obesity-linked proinflammatory gene expression profile in adipose tissue of subjects at risk of metabolic syndrome. *The American Journal of Clinical Nutrition, 90*, 1656–1664.

Vargas, M. L., Almario, R. U., Buchan, W., Kim, K., & Karakas, S. E. (2011). Metabolic and endocrine effects of long-chain versus essential omega-3 polyunsaturated fatty acids in polycystic ovary syndrome. *Metabolism, 60*, 1711–1718.

Vessby, B., Uusitupa, M., Hermansen, K., Riccardi, G., Rivellese, A. A., Tapsell, L. C., et al. (2001). Substituting dietary saturated for monounsaturated fat impairs insulin sensitivity in healthy men and women: The KANWU Study. *Diabetologia, 44*, 312–319.

Waki, H., Yamauchi, T., Kamon, J., Ito, Y., Uchida, S., Kita, S., et al. (2003). Impaired multimerization of human adiponectin mutants associated with diabetes. Molecular structure and multimer formation of adiponectin. *The Journal of Biological Chemistry, 278*, 40352–40363.

Wang, Z. V., & Scherer, P. E. (2008). DsbA-L is a versatile player in adiponectin secretion. *Proceedings of the National Academy of Sciences of the United States of America, 105*, 18077–18078.

Wang, Z. V., Schraw, T. D., Kim, J. Y., Khan, T., Rajala, M. W., Follenzi, A., et al. (2007). Secretion of the adipocyte-specific secretory protein adiponectin critically depends on thiol-mediated protein retention. *Molecular and Cellular Biology, 27*, 3716–3731.

Wang, C., Xin, X., Xiang, R., Ramos, F. J., Liu, M., Lee, H. J., et al. (2009). Yin-Yang regulation of adiponectin signaling by APPL isoforms in muscle cells. *The Journal of Biological Chemistry, 284*, 31608–31615.

Wang, Y., Xu, A., Knight, C., Xu, L. Y., & Cooper, G. J. (2002). Hydroxylation and glycosylation of the four conserved lysine residues in the collagenous domain of adiponectin. Potential role in the modulation of its insulin-sensitizing activity. *The Journal of Biological Chemistry, 277*, 19521–19529.

Weisberg, S. P., McCann, D., Desai, M., Rosenbaum, M., Leibel, R. L., & Ferrante, A. W., Jr. (2003). Obesity is associated with macrophage accumulation in adipose tissue. *The Journal of Clinical Investigation, 112*, 1796–1808.

Wilson-Fritch, L., Burkart, A., Bell, G., Mendelson, K., Leszyk, J., Nicoloro, S., et al. (2003). Mitochondrial biogenesis and remodeling during adipogenesis and in response to the insulin sensitizer rosiglitazone. *Molecular and Cellular Biology, 23*, 1085–1094.

Xi, L., Qian, Z., Xu, G., Zhou, C., & Sun, S. (2007). Crocetin attenuates palmitate-induced insulin insensitivity and disordered tumor necrosis factor-alpha and adiponectin expression in rat adipocytes. *British Journal of Pharmacology, 151*, 610–617.

Yamagishi, K., Iso, H., Yatsuya, H., Tanabe, N., Date, C., Kikuchi, S., et al. (2010). Dietary intake of saturated fatty acids and mortality from cardiovascular disease in Japanese: The Japan Collaborative Cohort Study for evaluation of cancer risk (JACC) study. *The American Journal of Clinical Nutrition, 92*, 759–765.

Yamamoto, K., Itoh, T., Abe, D., Shimizu, M., Kanda, T., Koyama, T., et al. (2005). Identification of putative metabolites of docosahexaenoic acid as potent PPARgamma agonists and antidiabetic agents. *Bioorganic & Medicinal Chemistry Letters, 15*, 517–522.

Yamauchi, T., Kamon, J., Ito, Y., Tsuchida, A., Yokomizo, T., Kita, S., et al. (2003). Cloning of adiponectin receptors that mediate antidiabetic metabolic effects. *Nature, 423*, 762–769.

Yamauchi, T., Kamon, J., Minokoshi, Y., Ito, Y., Waki, H., Uchida, S., et al. (2002). Adiponectin stimulates glucose utilization and fatty-acid oxidation by activating AMP-activated protein kinase. *Nature Medicine, 8*, 1288–1295.

Yamauchi, T., Kamon, J., Waki, H., Imai, Y., Shimozawa, N., Hioki, K., et al. (2003). Globular adiponectin protected ob/ob mice from diabetes and ApoE-deficient mice from atherosclerosis. *The Journal of Biological Chemistry, 278*, 2461–2468.

Yamauchi, T., Kamon, J., Waki, H., Murakami, K., Motojima, K., Komeda, K., et al. (2001). The mechanisms by which both heterozygous peroxisome proliferator-activated receptor gamma (PPARgamma) deficiency and PPARgamma agonist improve insulin resistance. *The Journal of Biological Chemistry, 276*, 41245–41254.

Yamauchi, T., Kamon, J., Waki, H., Terauchi, Y., Kubota, N., Hara, K., et al. (2001). The fat-derived hormone adiponectin reverses insulin resistance associated with both lipoatrophy and obesity. *Nature Medicine, 7*, 941–946.

Yamauchi, T., Nio, Y., Maki, T., Kobayashi, M., Takazawa, T., Iwabu, M., et al. (2007). Targeted disruption of AdipoR1 and AdipoR2 causes abrogation of adiponectin binding and metabolic actions. *Nature Medicine, 13*, 332–339.

Yang, W. S., Lee, W. J., Funahashi, T., Tanaka, S., Matsuzawa, Y., Chao, C. L., et al. (2001). Weight reduction increases plasma levels of an adipose-derived anti-inflammatory protein, adiponectin. *The Journal of Clinical Endocrinology and Metabolism, 86*, 3815–3819.

Yatagai, T., Nagasaka, S., Taniguchi, A., Fukushima, M., Nakamura, T., Kuroe, A., et al. (2003). Hypoadiponectinemia is associated with visceral fat accumulation and insulin resistance in Japanese men with type 2 diabetes mellitus. *Metabolism, 52*, 1274–1278.

Yatagai, T., Nishida, Y., Nagasaka, S., Nakamura, T., Tokuyama, K., Shindo, M., et al. (2003). Relationship between exercise training-induced increase in insulin sensitivity and adiponectinemia in healthy men. *Endocrine Journal, 50*, 233–238.

Yoon, M. J., Lee, G. Y., Chung, J. J., Ahn, Y. H., Hong, S. H., & Kim, J. B. (2006). Adiponectin increases fatty acid oxidation in skeletal muscle cells by sequential activation of AMP-activated protein kinase, p38 mitogen-activated protein kinase, and peroxisome proliferator-activated receptor alpha. *Diabetes, 55*, 2562–2570.

Yu, J. G., Javorschi, S., Hevener, A. L., Kruszynska, Y. T., Norman, R. A., Sinha, M., et al. (2002). The effect of thiazolidinediones on plasma adiponectin levels in normal, obese, and type 2 diabetic subjects. *Diabetes, 51*, 2968–2974.

Zappala, G., & Rechler, M. M. (2009). IGFBP-3, hypoxia and TNF-alpha inhibit adiponectin transcription. *Biochemical and Biophysical Research Communications, 382*, 785–789.

CHAPTER TWO

Molecular Tools to Characterize Adiponectin Activity

Cathleen Juhl, Annette G. Beck-Sickinger[1]
Faculty of Biosciences, Pharmacy and Psychology, Institute of Biochemistry, Universität Leipzig, Leipzig, Germany
[1]Corresponding author: e-mail address: beck-sickinger@uni-leipzig.de

Contents

Abstract

Within the past years, numerous hormones were found to be secreted by adipose tissue. As these adipokines exert different physiological effects with great importance in obesity, they provide new strategies for the treatment of obesity associated disorders. Adiponectin is one of the most promising targets due to its protective properties in glucose and lipid metabolism, which are mediated by the adiponectin receptor 1 and 2. Within the past decades, substantial progress in understanding the molecular function of this unique ligand–receptor system could be achieved. This review summarizes the most important approaches for the investigation of adiponectin activity. Even though many insights into adipokine function could be achieved, clarification of the

Vitamins and Hormones, Volume 90
ISSN 0083-6729
http://dx.doi.org/10.1016/B978-0-12-398313-8.00002-6

detailed mode of action is still challenging. For this reason, this review gives an overview of frequently used methods, which led to the molecular characterization of adiponectin and might help to get more detailed insights into the broad aspects of obesity.

ABBREVIATIONS

ACRP30 adipocyte complement-related protein of 30 kDa
AdipoR adiponectin receptor
AMPK AMP-activated protein kinase
apM1 adipose most abundant gene transcript 1
APPL1 adaptor protein-containing pleckstrin homology domain, phosphotyrosine-binding domain, and leucine zipper motif
BiFC bimolecular fluorescence complementation
CD circular dichroism
CK2 casein kinase 2
ERp endoplasmic reticulum protein
eYFP enhanced yellow fluorescent protein
fAd full-length adiponectin
gAd globular adiponectin
GBP28 gelatin-binding protein of 28 kDa
GFP green fluorescent protein
GLUT glucose transporter
GPCR G-protein coupled receptor
HFD high-fat diet
HMW high-molecular weight
LTBR lymphotoxin-b receptor
MKK mitogen-activated protein kinase
PAQR progesterone and AdipoQ receptors
PGC peroxisome proliferator-activated receptor-coactivator
p38MAPK p38 mitogen-activated protein kinase
PPAR peroxisome proliferator-activated receptor
RACK1 receptor for activated C-kinase 1
S1P sphingosine 1-phosphate
TAK transforming growth factor-β-activated kinase
T2D type 2 diabetes
Y2H yeast two-hybrid

1. INTRODUCTION

Overweight is one of the main health problems worldwide, leading to the death of 2.8 million adults every year. In 2008, 1.5 billion adults were overweight including about 500 million obese individuals. By now, even 43 million children under the age of five are overweight (WHO, 2011). Obesity is strongly correlated with disorders such as type 2 diabetes (T2D), hypertension, cardiovascular diseases, cancer, inflammation, and fatty liver disease (Blüher, 2009). A combination of obesity, dyslipidemia, hypertension, and insulin resistance is designated as the metabolic syndrome, which furthermore increases the risk to develop T2D and cardiovascular diseases (Gaddam, Ventura, & Lavie, 2011). These data strongly demonstrate the relevance to understand the molecular aspects of obesity.

Within the past years, adipose tissue could be identified as an endocrine organ secreting numerous important hormones. These adipokines show huge differences in expression levels and functions and play a major role in obesity and its correlated disorders (Singla, Bardoloi, & Parkash, 2010; Tesauro et al., 2011). Whereas leptin, resistin, tumor necrosis factor α, interleukin 6, and nicotinamide phosphoribosyltransferase are proinflammatory, adiponectin is an important anti-inflammatory adipokine (Ouchi, Parker, Lugus, & Walsh, 2011).

Adiponectin is one of the most abundant adipokines with plasma concentrations from 0.5 to 30 μg/ml and was found in a full-length (fAd) and a globular form (gAd). The latter is generated by proteolytic cleavage of fAd (Fruebis et al., 2001). Adiponectin was discovered first by Scherer, Williams, Fogliano, Baldini, and Lodish (1995) and named adipocyte complement-related protein of 30 kDa (ACRP30) due to its similarity to the complement factor C1q. Three other groups were also able to identify adiponectin independently, which led to its numerous designations as AdipoQ (Hu, Liang, & Spiegelman, 1996), gelatin-binding protein of 28 kDa (GBP28; Nakano, Tobe, Choi-Miura, Mazda, & Tomita, 1996), and adipose most abundant gene transcript 1 (apM1) (Maeda et al., 1996). Adiponectin is mainly secreted by white adipose tissue. However, adiponectin expression was also found in skeletal muscle, brown adipose tissue, and liver (Heiker, Kosel, & Beck-Sickinger, 2010).

Adiponectin is characterized by its insulin-sensitizing, antiatherogenic, and anti-inflammatory properties. On the one hand, adiponectin increases fatty acid oxidation resulting in lower triglyceride tissue concentrations. Furthermore, adiponectin promotes the glucose uptake in skeletal muscles and decreases gluconeogenesis in the liver. These effects finally lead to its insulin-sensitizing function (Ceddia et al., 2005; Miller et al., 2011; Yamauchi et al., 2002). On the other hand, adiponectin was shown to counteract atherosclerosis by inhibition of adhesion molecule activation, formation of foam cells but increasing nitric oxide production (Chen, Montagnani, Funahashi, Shimomura, & Quon, 2003; Hui, Lam, Vanhoutte, & Xu, 2011). In addition, adiponectin shows anti-inflammatory function by stimulating anti-inflammatory and decreasing proinflammatory cytokine secretion (Okamoto, 2011). It also seems to play a protective role in carcinogenesis (Jarde, Perrier, Vasson, & Caldefie-Chezet, 2011; Man et al., 2010). In 2003, Yamauchi et al. described the cloning of the adiponectin receptor (AdipoR) 1 and 2 which are expressed ubiquitary but mainly in skeletal muscle and liver (Yamauchi et al., 2003). This new type of seven transmembrane receptors was shown to mediate adiponectin-derived effects (Yamauchi et al., 2003) and now belongs to the group of progesterone and AdipoQ receptors (PAQR; Tang et al., 2005).

Since adiponectin exerts numerous protective effects, it is obvious that this adipokine is a promising target for obesity treatment. However, serum levels of adiponectin are negatively correlated with body mass index (BMI) (Matsubara, Maruoka, & Katayose, 2002), T2D (Weyer et al., 2001), and cardiovascular diseases (Pischon et al., 2004). AdipoR expression levels were shown to be downregulated in obese humans as well (Rasmussen et al., 2006). This indicates two possibilities for medical application: on the one hand, administration of adiponectin and on the other hand, upregulation of the endogenous adiponectin level. Several drugs, such as statins, angiotensin-converting enzyme inhibitors and receptor blocker, and thiazolidinediones, were shown to increase adiponectin expression (Phillips & Kung, 2010). Until now, obesity treatment is insufficient, as detailed molecular mechanisms remain unclear. Therefore, the development of biochemical, -medical, and -physical tools is essential for a complete understanding of adipokine function.

Adipokine function can be divided into four major parts: (1) the cellular biology, which includes the intracellular trafficking to the membrane as well as internalization processes from the membrane; (2) the structural biology to get detailed insights into the molecular constitution; (3) protein–protein interactions; and (4) signaling cascades that are induced upon binding of adiponectin to its receptors.

Several techniques have been developed to study these main parts of adiponectin function (Fig. 2.1). This review focuses on the most important approaches, which built up our current understanding from its cellular biology, structural properties to protein–protein interactions, and induced signaling cascades.

2. ANALYSIS OF THE CELLULAR BIOLOGY

After biosynthesis on ribosomes, proteins need to be delivered to their target sites. Membrane and extracellular proteins pass through the secretory pathway among the endoplasmic reticulum and Golgi apparatus for

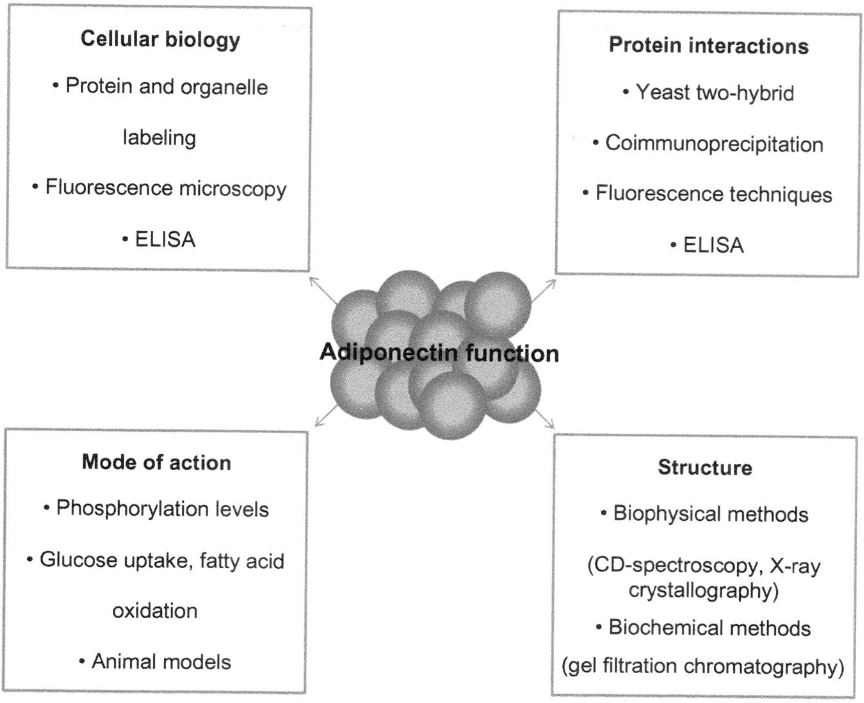

Figure 2.1 Overview of current methods to study adiponectin function including cell biology, interaction partner, mode of action, and structure. (For color version of this figure, the reader is referred to the online version of this chapter.)

modifications and are finally transferred to the cell membrane. Subsequent to the activation of signaling cascades, ligands and receptors can be internalized for either degradation or recycling to the membrane. The investigation of these processes apparently requires the determination of protein translocation and localization within the cell. Primarily, fluorescence techniques are used to monitor proteins within living or fixed cells. Therefore, the following section briefly describes protein-labeling strategies and the subsequent analysis by fluorescence microscopy and ELISA. These techniques revealed first insights into the intracellular trafficking of adiponectin and its receptors.

2.1. Protein labeling

In order to visualize transport processes within the cell, proteins need to be fluorescently labeled. Therefore, autofluorescent proteins, whose powerful properties became accessible with the discovery of the green fluorescent

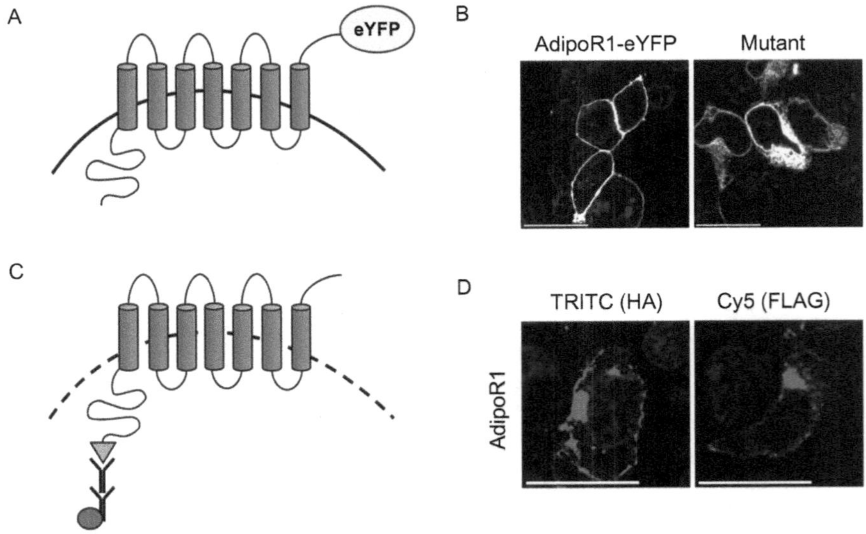

Figure 2.2 Visualization of AdipoR. (A) AdipoR fusion to eYFP enables direct monitoring in living cells. (B) Live-cell imaging of the AdipoR1 and a receptor mutant which is located intracellularly. (C) Immunostaining of the AdipoR using fluorophore-conjugated antibodies. (D) Immunostaining of HA-AdipoR1 and Flag-AdipoR1 using TRITC- and Cy5-conjugated antibodies, respectively. (See Color Insert.)

protein (GFP) from jellyfish *Aequorea victoria*, are highly suitable (Fig. 2.2A). Furthermore, the red-emitting protein DsRed was isolated from the coral *Discosoma*, which allowed reduction of the autofluorescence of cells and tissues (Wiedenmann, Oswald, & Nienhaus, 2009). Until now, many GFP-like derivatives, that facilitate a broad color spectrum from blue to red emitting proteins, have been generated. However, most autofluorescent proteins form oligomers. Mutagenesis of GFP analogs revealed numerous variants with improved brightness, folding, photostability, and decreased oligomer formation (Wang, Shyy, & Chien, 2008). One of the most frequently used fluorophores is the enhanced yellow fluorescent protein (eYFP), which emits at 527 nm. Fusion of eYFP to the AdipoR1 provided detailed insights into its trafficking among the endoplasmic reticulum and Golgi to the cell membrane and revealed the regulation of this process by the N-terminal domain (Fig. 2.2B; Juhl, Kosel, & Beck-Sickinger, 2012).

Not only autofluorescent proteins can be used but also fluorescent dyes allow protein imaging within the cell. Within the past decades, a huge number of small organic dyes have been described that show high photostability, aqueous solubility, and pH tolerance (Böhme & Beck-Sickinger, 2009).

The covalent linkage is mainly performed via amine and thiol groups, but specific labeling of proteins is rarely possible due to the numerous functional groups in one protein. However, the unspecific labeling of adiponectin via its amino groups using Alexa Fluor 555 succinimidylester allowed the visualization of a specific adiponectin internalization for the first time (Ding, Wang, & Chen, 2009).

In contrast to live cell imaging, immunostaining of proteins with fluorescently labeled antibodies will be another possibility if direct labeling strategies are not applicable or fail (Fig. 2.2C/D). To visualize intracellular proteins, immunostaining requires fixation and permeabilization of the cells. Usually, primary antibodies that recognize the protein of interest or small tags and fluorophore-conjugated secondary antibodies are used (Giepmans, Adams, Ellisman, & Tsien, 2006). The combination of an anti-myc tag primary antibody and Alexa 546-conjugated secondary antibody revealed a ligand-independent internalization of the AdipoR1-myc construct. However, the stimulation with fluorescently labeled adiponectin showed a colocalization of adiponectin and AdipoR1 suggesting that the receptor internalizes together with its ligand (Ding et al., 2009).

Since AdipoR was shown to be different from G-protein coupled receptor (GPCR) (Yamauchi et al., 2003), the analysis of the receptor orientation within the cell membrane was of great importance. Therefore, AdipoR constructs were cloned that carry an N-terminal hemagglutinin (HA) tag, while YFP was fused to the C-terminus. This approach revealed that the HA tag could only be recognized by the respective antibody after permeabilization of the cells. These studies confirmed the opposite orientation of AdipoR with an intracellular N-terminus and an extracellular C-terminus (Deckert, Heiker, & Beck-Sickinger, 2006).

Recently, a fourth labeling strategy for gAd was described by Almer et al. (2011) using fluorescently labeled nanoparticles. They generated gAd-coated Atto655-labeled stealth liposomes as well as AF488-labeled particles which allowed either staining of the surface or the cytoplasm of atherosclerotic plaques, respectively.

2.2. Fluorescence microscopy

Fluorescently labeled proteins and their cellular distribution are most commonly analyzed by fluorescence microscopy. This approach is based on the excitation with the respective wavelengths leading to fluorophore emission. Thereby, huge datasets can quickly be obtained, but resolutions are limited and photobleaching is a main problem. Within the past years, improved techniques have been developed to overcome these limitations. Confocal microscopy was shown to reduce background signals due to the installation of a pinhole. Hence, only the fluorescence of a single spot can be measured leading to higher resolutions, but slow data-acquisition rates (Stephens & Allan, 2003). To further increase the resolution and to decrease photobleaching, stimulated emission depletion microscopy is the method of choice. This technique implies a second beam which stimulates the fluorescence reversion into the ground state at the frame of a visualized spot. Thereby, this method overcomes the diffraction limit and high resolutions can be obtained (Schermelleh, Heintzmann, & Leonhardt, 2010).

In summary, it is a challenge to choose the appropriate microscopy technique. Thus, the scope of work and required sensitivities and acquisition rates should always be taken into consideration.

2.3. Cell surface ELISA

A second method for the analysis of fluorescently labeled proteins is the cell surface ELISA. This approach includes the relation of cell surface to total receptor expression levels and, therewith, reveals whether receptors are correctly integrated within the membrane or are located intracellularly. This can be realized by blocking fixed cells with or without detergents. The addition of detergents leads to the permeabilization of the cell membrane and reflects the amount of total receptor expression. Subsequent detection is performed with primary antibodies against the protein and enzyme-conjugated secondary antibodies. This allows the spectrophotometrical analysis, as these enzymes catalyze color reactions after addition of the respective substrates (Lai, Feldman, & Clark, 2005).

On the basis of this approach, the cellular distribution of AdipoR1-eYFP constructs using an anti-GFP specific antibody was studied. As N-terminal truncated receptor analogs showed decreased membrane expression compared to the wild type, a great impact of the AdipoR1 N-terminus in the proper membrane localization could be detected (Juhl et al., 2012).

Whereas the intracellular trafficking of GPCR is already well understood, our knowledge on these processes is still insufficient for adiponectin and its receptors. Until now, most of these questions still need to be answered. How are these processes regulated? Which proteins are involved in the transport? What happens after the ligand and receptor internalization? Fluorescence-based techniques might help to answer all these questions.

However, the specific labeling of proteins is challenging due to various functional groups, and suitable methods need to be developed. Affecting the intracellular transport might provide interesting possibilities for therapeutic intervention, as obesity and its correlated disorders are characterized by deregulated ligand and receptor concentrations.

3. PROTEIN STRUCTURES

Protein structures are the basis of every cellular process as signaling cascades and protein interactions. Hence, it is essential to investigate the structural biology of recently discovered proteins in order to completely understand all molecular mechanisms. Biophysical methods such as CD (circular dichroism) spectroscopy and X-ray crystallography are most important to determine the secondary and tertiary structure, respectively. In addition, several biochemical approaches such as electrophoresis, size-exclusion chromatography, and density gradient centrifugation are highly suitable, particularly, for the analysis of quaternary structures.

3.1. Biophysical methods

3.1.1 Circular dichroism spectroscopy

CD spectroscopy is one of the fastest methods to gain first structural information and is based on different absorptions of left and right circularly polarized light of asymmetric molecules. The measurement is usually performed between 190 and 250 nm, as peptide bonds absorb within this wavelength range, and it mainly reveals insights into the secondary structure of proteins. These structural elements are characterized by defined absorption values. Alpha helices show two minima at 222 and 208 nm and one maximum at 193 nm, whereas β sheets have only one minimum at 218 nm and a maximum at 195 nm. Proteins with disordered structures show a negative band at 195 nm (Ranjbar & Gill, 2009). However, the main limitation is that only low salt buffers can be used, as high salt concentrations and detergents lead to strong background absorptions. Furthermore, residue-specific information can rarely be achieved. On the other hand, CD spectroscopy provides fast structural information using small amounts of proteins and has broad applications in monitoring folding processes, interactions, and thermal stability (Dodero, Quirolo, & Sequeira, 2011). Conformational changes as response to different buffer conditions such as pH value, ionic strength, or organic solvents can also be investigated. CD spectroscopy of the AdipoR1 N-terminal domain revealed a mainly alpha helical structure which is depending on pH values (C. Juhl, T. Leimer, A.G. Beck-Sickinger, unpublished data). Additionally, CD spectroscopy was also shown to confirm the β sheet structure of recombinantly expressed globular adiponectin (Heiker, Klöting, Blüher, & Beck-Sickinger, 2010).

3.1.2 X-ray crystallography

X-ray crystallography is the method of choice for the determination of high-resolution three-dimensional structures. Atoms in a particular arrangement that are exposed to X-rays scatter the beam by their electrons. These diffractions depend on the sample shape and symmetry and are characterized by their intensity, wavelength of the X-ray source, and phase. Phases cannot directly be measured but can be determined by several methods. After identification of all parameters, an electron density map is obtained and finally reveals the atom positions within the molecule. The protein structure can be received by fine tuning of the atomic model, which includes the setting of each amino acid into the electron density map (Wlodawer, Minor, Dauter, & Jaskolski, 2008). However, many drawbacks have to be overcome in order to gain high-resolution structures. First of all, proteins need to be produced in high amounts and purities, which is a main challenge particularly for membrane proteins. Additionally, many crystallization conditions have to be screened and diffractions of various proteins cannot be gained in required resolutions due to disordered and high solvent-containing protein domains (Acharya & Lloyd, 2005).

Shapiro and Scherer (1998) were able to solve the three-dimensional structure of the murine globular adiponectin trimer. Crystallization of the adiponectin trimer was performed by hanging drop method, and a resolution of 2.1 Å was achieved, which provided detailed insights into the structural biology of adiponectin. The monomers showed a 10-β-strand jellyroll folding topology, whereas trimers are bell-shaped through mainly hydrophobic interactions near the base. These data strongly demonstrated a high homology of adiponectin to the tumor necrosis factor (TNF) family (Fig. 2.3).

3.2. Biochemical approaches

Besides biophysical methods, several biochemical approaches as native electrophoresis and density gradient centrifugation allow the estimation of molecular weights and quaternary structures. Furthermore, gel filtration chromatography is one of the most important methods to study an oligomer formation. The stationary phase is composed of resin with defined pores and separates proteins with different sizes and shape. Large molecules do not

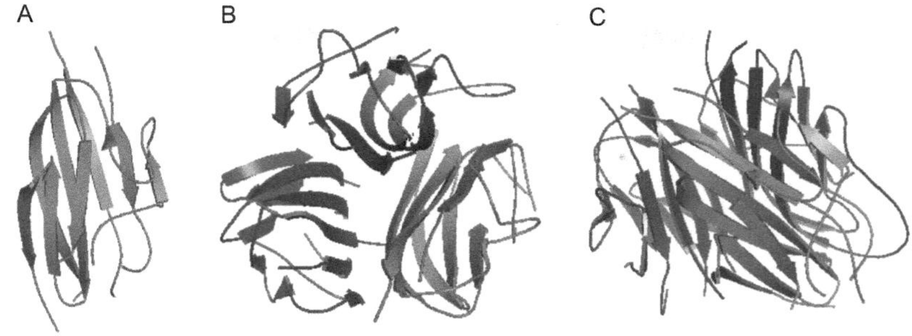

Figure 2.3 Structure of the murine gAd based on PDB entry 1C28. (A) Ribbon model of the gAd monomer. (B) Top view of the bell-shaped gAd trimer. The chains a, b, and c are colored in green, blue, and red, respectively. (C) Side view of the gAd trimer. All structures were created with the program PyMOL. (See Color Insert.)

enter the pores, whereas small molecules pass through them resulting in later retention times (Kunji, Harding, Butler, & Akamine, 2008). Unfortunately, molecules with similar shape or hydrodynamic volume cannot be separated, which leads to low resolutions. Phosphate buffers are used as mobile phase, but addition of salts, amino acids, detergents, or organic solvents is frequently required to decrease nonspecific interactions with the resin (Arakawa, Ejima, Li, & Philo, 2010). However, gel filtration chromatography enables a quick and easy separation of proteins with different sizes under native conditions. The analysis of adiponectin samples by gel filtration chromatography revealed the formation of trimeric, hexameric, and high-molecular weight (HMW) complexes (Wang et al., 2006). Using this approach, Wang et al. furthermore identified adiponectin modifications to be important for the oligomer formation. They could show that the exchange of four lysines in the collagenous domain to arginine prevented adiponectin assembly into HMW oligomers suggesting a role of lysine hydroxylations and glycosylations in the HMW formation. The dependency of disulfide bonds in adiponectin oligomerization could also be shown as mutation of cysteine 39 to serine led to detection of only trimers (Pajvani et al., 2003).

A high-resolution gel filtration assay was developed by Schraw, Wang, Halberg, Hawkins, and Scherer (2008) and allowed a correlation of adiponectin oligomer levels with obesity and its associated disorders. While HMW adiponectin levels were decreased with high BMI and body weight, trimeric adiponectin was positively correlated with diabetes (Hamilton et al., 2011).

Finally, protein structures are of great interest as they give insights into the molecular constitution of protein–protein interactions and signaling events. Furthermore, the application of proteins as lead structures in the drug development requires a detailed understanding of the structure–function relationship. Until now, X-ray crystallography is the method of choice as it reveals high-resolution three-dimensional structures. However, the analysis of huge proteins like seven transmembrane receptors is still a main challenge, and other approaches were found for structural investigations. Thereby, the investigation of single protein fragments can also give first evidence for the structural configuration. So far, the three-dimensional structure of globular adiponectin could be solved and much is known about its posttranslational modifications. With the exception of the AdipoR1 N-terminus secondary structure, unfortunately, no structural information for AdipoR is available. This needs to be elucidated in the future, as this is also essential to understand the detailed binding mode of adiponectin to its receptors.

4. SEARCHING FOR NEW INTERACTION PARTNERS

Subsequent to correct integration of AdipoR into the cell membrane, binding of adiponectin induces important downstream events. The identification of protein–protein interactions can reveal first insights into these pathways, as they have great biological importance in signaling cascades, transport, and folding processes. Hence, the identification of interacting proteins was one of the main questions when studying adiponectin function. Several methods have been developed to address this issue including genetic methods (yeast two-hybrid, Y2H), fluorescence techniques (FRET, BRET, BiFC (bimolecular fluorescence complementation)), and biochemical approaches (coimmunoprecipitations) (Fig. 2.4).

4.1. Yeast two-hybrid

As a first step in the identification of AdipoR interacting proteins, Y2H approaches are most often described, as huge cDNA libraries can be screened in living cells. Preferably, *Saccharomyces cerevisiae* strains are used, as they are simple eukaryotes and are characterized by rapid cell growth. To analyze an interaction, the protein of interest X (bait) is fused to the N-terminal DNA-binding domain of Gal4 transcription factor, while the putative interaction partner Y (prey) is fused to the C-terminal activation domain. Interaction of both proteins leads to the formation of an active Gal4 transcription factor (Fig. 2.4A). The subsequent expression of reporter genes enables the selection of yeast cells either by β-galactosidase catalyzed color reaction or by growth on minimal media lacking histidine, leucine, or tryptophan (Bruckner, Polge, Lentze, Auerbach, & Schlattner, 2009). Finally, the

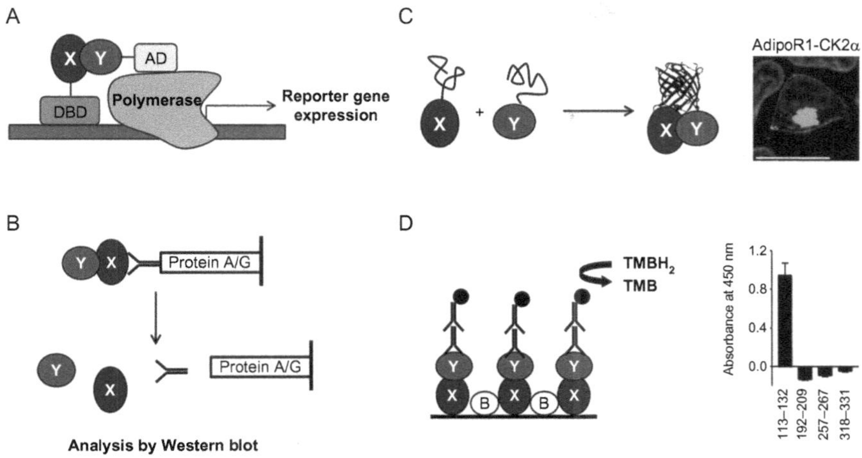

Figure 2.4 Most common approaches to detect AdipoR interactions (X AdipoR, Y interacting protein). (A) Y2H systems allow the identification of novel interaction partners (DBD DNA-binding domain, AD activation domain). (B) Coimmunoprecipitation of AdipoR complexes. (C) BiFC experiment and visualization of the AdipoR1 interaction with CK2α (Juhl, Mörl, & Beck-Sickinger, 2011). (D) ELISA and identification of the interaction site between the AdipoR1 and CK2β (Heiker et al., 2009). (See Color Insert.)

interaction partners can be identified by simple DNA sequencing. On the one hand, Y2H screens can be performed easily. They allow high-throughput analysis, and transient interactions can be detected (Petschnigg, Moe, & Stagljar, 2011). On the other hand, Y2H is characterized by a high number of false positives due to unspecific interactions of proteins. Therefore, the results obtained by Y2H always require the confirmation by a second method. Another limitation is that the protein complex needs to relocate into the nucleus, which complicates the analysis of membrane proteins. However, one possibility to study the interactome of membrane proteins is the fusion of truncated proteins to the DNA-binding domain, which was successfully shown for the AdipoR1. Screening the cDNA of the AdipoR1 N-terminal domain against a human testis cDNA library led to the identification of the regulatory subunit of protein kinase CK2 (CK2β) (Heiker et al., 2009), whereas screening against a human fetal brain cDNA library revealed adaptor protein-containing pleckstrin homology domain, phosphotyrosine-binding domain, and leucine zipper motif (APPL1) as AdipoR1-binding proteins (Mao et al., 2006). Mao et al. furthermore demonstrated Y2H as a tool to determine interaction sites as they found the APPL1 PTB domain to be essential for the interaction by screening different domains of APPL1. Interestingly, Xu et al. were able to identify RACK1 (receptor for activated C-kinase 1) and LTBR (lymphotoxin-b receptor) using the complete AdipoR1 cDNA as bait and a human liver cDNA library as prey (Xu, Wang, Ling, Li, & Gao, 2009, 2011; Xu et al., 2011). Y2H was also applied for the screening of adiponectin-binding proteins. Liu et al. found GST-κ/disulfide-bond A oxidoreductase-like protein as an adiponectin-interacting protein (Liu et al., 2008). The binding of adiponectin to the AdipoR1 C-terminus could also be detected by Y2H screen (Mao et al., 2006).

4.2. Coimmunoprecipitation

To confirm protein–protein interactions, coimmunoprecipitation experiments are highly suitable and include the isolation of protein complexes using specific antibodies (Fig. 2.4B). As initial step, cell lysates are incubated with the respective antibody and the subsequent binding to immobilized protein A or G enables the separation of interacting proteins by affinity chromatography (Berggard et al., 2007). Several immobilization strategies for protein A/G have been developed. Heiker et al. (2009) showed that the regulatory subunit of protein kinase CK2 interacts with the AdipoR1 using protein A magnetic microbeads. Another possibility is the conjugation of protein A/G to agarose, which was used to verify the AdipoR1 RACK1 interaction (Xu et al., 2009). The analysis of immunoprecipitates is usually carried out by mass spectrometry or immunoblotting, but to confirm the specificity of binding, each experiment should be performed with adequate control IgG.

Coimmunoprecipitations are highly specific, but choosing the right antibody is one of the main challenges. To improve the specificity of the antibody binding, proteins can be tagged, which allow the application of tag-specific antibodies. Not only small peptide sequences as FLAG- or HA tags are conceivable, but also huge proteins such as GFP can be added. These approaches led to the confirmation of numerous AdipoR1-binding proteins such as endoplasmic reticulum protein (ERp) 46 (Charlton et al., 2010) and both subunits of CK2, respectively (Heiker et al., 2009; Juhl et al., 2011). On the other hand, immunoprecipitations can be carried out conversely using antibodies against the putative interaction partner. This was shown by Xu et al. (2011), as they performed immunoprecipitations using FLAG-tagged LTBR and isolated the AdipoR1–LTBR complex with an anti-FLAG antibody.

As stringent wash conditions during protein complex purification may disturb the interaction, cross-linking of cells is frequently required to stabilize weak or transient protein–protein interactions. Heiker et al. (2009)

showed that only cross-linking with dithiobis(succinimidyl-propionate) leads to successful coimmunoprecipitation of CK2β and AdipoR1.

4.3. Fluorescence techniques

4.3.1 Bimolecular fluorescence complementation

Besides the visualization of protein transport processes (discussed in Section 2), several fluorescence techniques have been recently developed for the detection of protein–protein interactions in living cells. Apart from bioluminescence and fluorescence resolution energy transfer, BiFC is a powerful tool to study protein complexes. Therefore, fluorescent proteins such as GFP, YFP, or improved mutants such as Venus are splitted into two nonfluorescent halves and fused to the proteins of interest. If proteins interact, both halves will come in a spatial proximity and the fluorophore will be formed (Fig. 2.4C; Kerppola, 2009). Accordingly, this enables the detection of transient interactions. However, the formation of the fluorophore is not reversible, which requires a thorough validation of the interaction specificity (Rose et al., 2010).

Using this approach, an interaction of the catalytic subunit of CK2 and AdipoR1 at the cell membrane could be detected (Fig. 2.4C) (Juhl et al., 2011). As the Venus fragments carry FLAG- and HA tags, the fusion protein expression could be shown by immunostaining using specific antibodies against these tags. Besides studying the interaction with cytosolic proteins, BiFC is also highly applicable for the detection of receptor dimerization, which was shown for the AdipoR1 homodimerization by Kosel et al. (2010). Furthermore, fluorescence-activated cell sorting revealed that the dimerization is inhibited by fAd and collagen-like domain-derived peptides and thereby enabled quantification of this interaction. However, the stimulation was needed to be directly carried out after cotransfection of the fusion proteins, as the formation of the Venus protein is irreversible.

4.3.2 Colocalization studies

In addition, colocalization studies can also reveal important information about the regulation of protein–protein interactions. Therefore, the proteins are fused to different fluorophores whose overlapping can be determined by fluorescence microscopy. Consequently, protein translocations within the cell can be monitored and assigned to special compartments (Berggard et al., 2007; Giepmans et al., 2006).

Heiker et al. (2009) could visualize a CK2β recruitment to the cell membrane leading to an increased interaction with the AdipoR1 as response to adiponectin stimulation.

Interestingly, colocalization studies of the AdipoR1 and ERp46 revealed that this ERp also translocates to the cell membrane for AdipoR1 interaction (Charlton et al., 2010). Hence, the fusion of different fluorophores to the proteins of interest using either autofluorescent proteins or fluorophore-conjugated antibodies (discussed in Section 2.1) allows determination of interaction alterations and the assignment of these complexes to special cell compartments.

4.4. ELISA

After the identification and confirmation of protein–protein interactions, the quantification and determination of interaction sites can be easily performed by ELISA approaches. Immobilization and blocking of free binding sites are similar to the cell surface ELISA (discussed in Section 2.3) but is followed by incubation with the interaction partner. Finally, the detection is performed using specific antibodies against the interacting protein (Fig. 2.4D).

Masaie et al. (2007) performed such ELISA experiments and found the chemokines stromal cell-derived factor 1 and the murine macrophage inflammatory protein-1 gamma to interact with adiponectin. The immobilization of fAd and gAd revealed that the globular domain is sufficient for chemokine binding. Additionally, the incubation with truncated chemokine analogs detected the binding site in the middle domain of both chemokines.

A different approach was used by Heiker et al. (2009), as they could show the region between amino acids 121 and 132 of the AdipoR1 as the interaction site for CK2β (Fig. 2.4D). Therefore, small peptide fragments of intracellular receptor domains were immobilized and incubated with CK2β. The analysis of analogs of the putative binding site indicated that basic amino acids are essential for the interaction and therewith provided residue-specific information of the interaction site.

Within the past years, improved strategies for the identification of protein–protein interactions have been developed. However, the analysis cannot be restricted to one method only, as each technique can reveal both false positive and negative data. The search for new interaction partners usually starts with the screening of huge cDNA libraries and validation can be performed by coimmunoprecipitations. Furthermore, fluorescence-based approaches can define protein complexes to particular cell compartments, and ELISA studies can reveal additional information about interaction sites and the quantity of binding. However, the detection of weak or transient interactions is still challenging, though there has been recent progress in developing applicable methods.

Until now, the analysis of the adiponectin and AdipoR interactome allowed detailed insights into cellular processes as most functions are exerted by protein complexes. This mainly includes transport and folding processes as well as signaling pathways and physiological effects. Finally, the identification of protein complexes might support an improved drug development.

5. SIGNALING CASCADES AND PHYSIOLOGICAL EFFECT

Subsequent to the identification of the adiponectin–AdipoR system, the most extensive challenge was the analysis of the induced signaling cascades. The characterization of binding proteins can give first insights into these downstream events. However, this unique ligand–receptor system shows major differences to conventional GPCR. Until now, there is no evidence that AdipoR couple to G-proteins. Hence, many established methods such as inositol triphosphate accumulation assays cannot be used.

The most important readout system for the adiponectin signaling is the determination of protein phosphorylation levels. Since adiponectin has great importance in glucose and lipid metabolism, the measurement of these physiological effects can also be used to analyze the impact of proteins that may be involved in adiponectin-induced pathways. Moreover, numerous animal models have been generated which allow the investigation of adiponectin function in obesity. Within the following section, the most important techniques are summarized that have been used to characterize adiponectin signal transduction.

5.1. Phosphorylation

Protein phosphorylations are key processes in many signaling cascades and are, therewith, an important tool for the analysis of adiponectin downstream events. The determination of phosphorylation levels can be easily performed by Western blotting using phosphospecific antibodies. After cell lysis or homogenization of tissue extracts, proteins are separated by SDS-PAGE and afterward transferred to PVDF or nitrocellulose membranes. Prior to antibody incubation, free binding sites are blocked to avoid an unspecific binding. Subsequent incubation with phosphospecific antibodies and horseradish peroxidase-conjugated secondary antibodies allows the detection by chemiluminescence (MacPhee, 2010). Finally, the quantification is performed by calculating the ratio of phosphorylated to unphosphorylated protein levels.

Adiponectin signaling in muscle cells was shown to be mainly mediated by phosphorylation of AMPK (AMP-activated protein kinase) or p38MAPK (p38 mitogen-activated protein kinase; Yamauchi et al., 2003). Activation of these kinases led to increased fatty acid oxidation and glucose uptake. Consequently, the measurement of AMPK or p38MAPK phosphorylation levels as response to protein stimulations or suppressions can reveal the impact of novel proteins on adiponectin-induced signaling pathways. Xin et al. (2011) analyzed the adiponectin-induced fatty acid oxidation in C2C12 myoblasts. The measurement of p38MAPK phosphorylation levels identified transforming growth factor-β-activated kinase (TAK) 1 and mitogen-activated protein kinase (MKK)3 as effector molecules. Stimulation with adiponectin, while suppressing TAK1 and MKK3, resulted in significantly decreased p38MAPK phosphorylation levels. Interestingly, the knockdown of AMPKα2 had no effect on p38MAPK phosphorylation, indicating that p38MAPK activation is independent of AMPKα2. Furthermore, the comparison of TAK1, MKK3, and p38MAPK phosphorylation at different time points after adiponectin stimulation revealed detailed insights into the chronology of these signaling events.

5.2. Measurement of physiological effects

5.2.1 Glucose uptake

A different approach for the analysis of adiponectin signaling is the measurement of its physiological effects. Since adiponectin was shown to stimulate glucose transporter (GLUT) 4 translocation (Ceddia et al., 2005; Mao et al., 2006; Yamauchi et al., 2007), glucose uptake assays are frequently used for the investigation of adiponectin-induced pathways. Therefore, cells are stimulated with adiponectin prior to incubation with 2-deoxy-D-glucose. This glucose analog was shown to be taken up but not to be metabolized due to the exchange of the 2-hydroxylgroup to hydrogen. 2-Deoxy-D-glucose is commonly labeled with radioactive isotopes such as ^{3}H or ^{14}C, which allows direct measurement of radioactivity in the cell lysates (Fang et al., 2005; Yea et al., 2009). Furthermore, the assays are performed in the presence of cytochalasin B, as this substance was shown to inhibit glucose uptake and, thus, reflects the unspecific transport (Jijakli et al., 2002).

Using this approach, AMPK was identified as a key regulator in adiponectin-derived antidiabetic effects. Inhibition of AMPK activity, while stimulating with gAd or fAd, led to extenuated glucose levels in C2C12 myocytes (Yamauchi et al., 2002). The impact of the AdipoR-binding protein APPL1 was also analyzed by glucose uptake assays and revealed an important role of APPL1 in these effects. It was shown that knockdown of APPL1 led to diminished glucose transport rates (Mao et al., 2006).

To overcome the drawbacks of radioactivity, fluorescence-based approaches can also be applied. Xu et al. described the adiponectin-induced uptake of 2-(*N*-(7-nitrobenz-2-oxa-1,3-diazol-4-yl)amino)-2-deoxyglucose. These studies also revealed that RACK1, a second AdipoR1 interaction partner, stimulates adiponectin-derived effects in glucose metabolism (Xu et al., 2009).

5.2.2 Fatty acid oxidation

In addition to its protective function in the carbohydrate metabolism, adiponectin plays a major role in fatty acid oxidation (Kadowaki & Yamauchi, 2005; Tomas et al., 2002). For the determination of the oxidation rates, fatty acids need to be labeled. Therefore, ^{3}H- or ^{14}C-labeled palmitate or oleic acids are most frequently used. Subsequent addition of strong acids as perchloric acid leads to the release of radioactive carbon dioxide, which can be collected by NaOH saturated filter paper locking the reaction tube (Fang et al., 2010). This approach revealed that APPL1 regulates not only the glucose transport but also the fatty acid degradation as Fang et al. (2010) could show that APPL1 knockdown, while stimulating with adiponectin, led to a decrease of fatty acid oxidation. Additionally, the oxidation rate can be determined by measuring $^{3}H_2O$, as this radioactive product is also being produced in the cell media (Yoon et al., 2006). Yoon et al. could demonstrate that adiponectin activates fatty acid oxidation by stimulating the AMPK, p38MAPK, and peroxisome proliferator-activated receptor (PPAR) α-signaling cascade. They could show that the inhibition of p38MAPK and AMPK reduces the PPARα activity and fatty acid oxidation in C2C12 myotubes (Yoon et al., 2006).

5.3. Animal models

Apart from the developed *in vitro* methods, numerous genetic and diet-induced animal models have been generated. Even though animal models cannot fully reflect the phenotypes of human disorders, these models provide a powerful tool to study protein function *in vivo*.

ob/ob and *db/db* mice are most frequently used and are generated by either leptin or leptin receptor knockout, respectively. Their phenotypes are characterized by obesity, insulin resistance, and T2D (Kanasaki & Koya, 2011). Besides these genetically induced animals, high-fat diet (HFD) was also shown to cause obesity, impaired glucose tolerance, and dyslipidemia. Furthermore, their phenotype highly resembles human disorders in the metabolic syndrome. Commonly, both types of obese mice are combined leading to different degrees of insulin resistance (Bergheim et al., 2011).

Interestingly, several adiponectin-specific models were generated that are characterized by either knockout or overexpression of adiponectin. Adiponectin KO mice showed glucose intolerance, insulin resistance, hyperlipidemia, and atherosclerosis. However, no significant differences in the body weight, compared to wild-type mice, were found (Kubota et al., 2002). Transgenic overexpression of adiponectin in *ob/ob* mice reversed the insulin resistance and decreased diabetes, but the animals were still obese. In contrast, overexpression of adiponectin in the liver of diet-induced obesity models led to insulin sensitivity and decreased body weight and adipocyte size (Bauche et al., 2007). These models revealed further information about how adiponectin exerts its anti-inflammatory function. Mice overexpressing adiponectin in white adipose tissue extenuated the secretion of proinflammatory factors, whereas this expression pattern was opposite in adiponectin KO mice (Ge et al., 2011).

Since adiponectin was shown to mediate its function via two receptors, several receptor KO mice were generated. AdipoR1 and AdipoR2 KO mice showed normal body weight and food intake but elevated insulin levels. Furthermore, AdipoR1 KO led to impaired glucose tolerance, while AdipoR2 KO mice were glucose tolerant (Yamauchi et al., 2007).

As ceramide levels correlate with insulin resistance, Holland et al. (2011) recently analyzed the role of adiponectin in the ceramide metabolism using different insulin-resistant mouse models. They could proof that adiponectin led to decreased ceramide content in *ob/ob* and HFD-induced obese mice. Animals that transgenically overexpressed adiponectin had the lowest ceramide levels after HFD, whereas adiponectin-deficient animals showed the highest concentrations. They furthermore investigated the role of sphingosine 1-phosphate (S1P), as S1P is a degradation product of ceramide and inhibits apoptosis while stimulating proliferation. These studies revealed decreased S1P levels in adiponectin null mice and increased ceramidase activity after HFD in mice that overexpressed AdipoR1 or 2. In summary, these animal studies revealed the sphingolipid metabolism as one key event in adiponectin-induced effects (Holland et al., 2011).

A second animal study was performed by Iwabu et al. (2010) who investigated the impact of muscle AdipoR1 on adiponectin signaling. Therefore, they generated insulin-resistant muscle AdipoR1 KO mice and found out that adiponectin stimulates peroxisome proliferator-activated receptor-coactivator (PGC)-1α deacetylation and activation via AdipoR1, AMPK, and SIRT1. PGC-1α is involved in the regulation of mitochondria and fatty acid oxidation, and its expression was found to be decreased in muscle AdipoR1 KO mice, which led to increased triglyceride content in skeletal muscles. Furthermore, they could show that adiponectin increases Ca^{2+} influx via AdipoR1 and subsequently activates calmodulin-dependent protein kinase β leading to increased PGC-1a expression (Iwabu et al., 2010).

In conclusion, a more detailed understanding of adiponectin signaling could be achieved within the past decade. Therefore, a broad method spectrum was used comprising *in vitro* methods such as the determination of phosphorylation levels and the measurement of adiponectin-induced physiological effects in different cell lines. Additionally, important animal models were generated, which allowed numerous insights into adiponectin function *in vivo*. However, whether all downstream signals and tissues are truly linked is still unknown.

6. CONCLUSION

Adiponectin is one of the most important adipokines, as it shows protective function in glucose and lipid metabolism, atherosclerosis, inflammation, and cancer. These effects are mediated by the seven transmembrane receptors AdipoR1 and 2. This unique ligand–receptor system shows numerous structural and functional differences to conventional GPCR. AdipoR are characterized by an opposite orientation with an intracellular N-terminus and an extracellular C-terminus and seem not to be coupled to G-proteins. Additionally, adiponectin consists of 244 amino acids and thus exceeds the size of common peptide hormones involved in obesity as gastric inhibitory polypeptide, pancreatic polypeptide, peptide YY, or ghrelin. These characteristics complicate the analysis of adiponectin function. Much information about the molecular mechanisms of adiponectin could be obtained, as important tools have been developed. This review summarizes recent techniques that were used to analyze adiponectin function. This includes fluorophore-labeling strategies to visualize the intracellular transport as well as methods to determine the structural biology. Additionally, the analysis by Y2H, coimmunoprecipitations, and fluorescence-based approaches revealed information about the interactome of adiponectin and AdipoR. Detailed mechanisms of adiponectin-derived pathways could be obtained using a broad method spectrum from *in vitro* approaches to *in vivo* models. Within the past decade, enormous progress in understanding the physiological effects could be achieved. However, numerous questions are still open including the regulation of the intracellular transport, the binding mode of adiponectin to its receptors, the molecular cause for the opposite AdipoR membrane orientation, and a complete correlation of all identified molecules that are involved in the adiponectin signaling. Only few pharmaceuticals are currently in medical application, although much is already known about adipokines. As adiponectin is one of the most promising targets for the development of effective antiobesity drugs, all of its molecular facets have to be understood. Hence, this review provides an overview of the most important techniques and might help to further characterize adiponectin. As the described methods can be easily transferred to other adipokines with so far unknown function, it might also support the analysis of the broad aspects of obesity in general.

REFERENCES

Acharya, K. R., & Lloyd, M. D. (2005). The advantages and limitations of protein crystal structures. *Trends in Pharmacological Sciences*, *26*, 10–14.

Almer, G., Wernig, K., Saba-Lepek, M., Haj-Yahya, S., Rattenberger, J., Wagner, J., et al. (2011). Adiponectin-coated nanoparticles for enhanced imaging of atherosclerotic plaques. *International Journal of Nanomedicine*, *6*, 1279–1290.

Arakawa, T., Ejima, D., Li, T., & Philo, J. S. (2010). The critical role of mobile phase composition in size exclusion chromatography of protein pharmaceuticals. *Journal of Pharmaceutical Sciences*, *99*, 1674–1692.

Bauche, I. B., El Mkadem, S. A., Pottier, A. M., Senou, M., Many, M. C., Rezsohazy, R., et al. (2007). Overexpression of adiponectin targeted to adipose tissue in transgenic mice: Impaired adipocyte differentiation. *Endocrinology*, *148*, 1539–1549.

Berggard, T., Linse, S., & James, P. (2007). Methods for the detection and analysis of protein-protein interactions. *Proteomics*, *7*, 2833–2842.

Bergheim, I., Eagon, P. K., Dooley, S., & Breitkopf-Heinlein, K. (2011). Alcoholic liver disease and exacerbation by malnutrition and infections: What animal models are currently available? *Annals of the New York Academy of Sciences*, *1216*, 41–49.

Blüher, M. (2009). Adipose tissue dysfunction in obesity. *Experimental and Clinical Endocrinology & Diabetes*, *117*, 241–250.

Böhme, I., & Beck-Sickinger, A. G. (2009). Illuminating the life of GPCRs. *Cell Communication and Signaling*, *7*, 16.

Bruckner, A., Polge, C., Lentze, N., Auerbach, D., & Schlattner, U. (2009). Yeast two-hybrid, a powerful tool for systems biology. *International Journal of Molecular Sciences*, *10*, 2763–2788.

Ceddia, R. B., Somwar, R., Maida, A., Fang, X., Bikopoulos, G., & Sweeney, G. (2005). Globular adiponectin increases GLUT4 translocation and glucose uptake but reduces glycogen synthesis in rat skeletal muscle cells. *Diabetologia*, *48*, 132–139.

Charlton, H. K., Webster, J., Kruger, S., Simpson, F., Richards, A. A., & Whitehead, J. P. (2010). ERp46 binds to AdipoR1, but not AdipoR2, and modulates adiponectin signalling. *Biochemical and Biophysical Research Communications, 392*, 234–239.
Chen, H., Montagnani, M., Funahashi, T., Shimomura, I., & Quon, M. J. (2003). Adiponectin stimulates production of nitric oxide in vascular endothelial cells. *Journal of Biological Chemistry, 278*, 45021–45026.
Deckert, C. M., Heiker, J. T., & Beck-Sickinger, A. G. (2006). Localization of novel adiponectin receptor constructs. *Journal of Receptor and Signal Transduction Research, 26*, 647–657.
Ding, Q., Wang, Z., & Chen, Y. (2009). Endocytosis of adiponectin receptor 1 through a clathrin- and Rab5-dependent pathway. *Cell Research, 19*, 317–327.
Dodero, V. I., Quirolo, Z. B., & Sequeira, M. A. (2011). Biomolecular studies by circular dichroism. *Frontiers in Bioscience, 16*, 61–73.
Fang, X., Palanivel, R., Cresser, J., Schram, K., Ganguly, R., Thong, F. S., et al. (2010). An APPL1-AMPK signaling axis mediates beneficial metabolic effects of adiponectin in the heart. *American Journal of Physiology, Endocrinology and Metabolism, 299*, E721–E729.
Fang, X., Palanivel, R., Zhou, X., Liu, Y., Xu, A., Wang, Y., et al. (2005). Hyperglycemia- and hyperinsulinemia-induced alteration of adiponectin receptor expression and adiponectin effects in L6 myoblasts. *Journal of Molecular Endocrinology, 35*, 465–476.
Fruebis, J., Tsao, T. S., Javorschi, S., Ebbets-Reed, D., Erickson, M. R., Yen, F. T., et al. (2001). Proteolytic cleavage product of 30-kDa adipocyte complement-related protein increases fatty acid oxidation in muscle and causes weight loss in mice. *Proceedings of the National Academy of Sciences of the United States of America, 98*, 2005–2010.
Gaddam, K. K., Ventura, H. O., & Lavie, C. J. (2011). Metabolic syndrome and heart failure—The risk, paradox, and treatment. *Current Hypertension Reports, 13*, 142–148.
Ge, Q., Rycken, L., Noel, L., Maury, E., & Brichard, S. M. (2011). Adipokines identified as new downstream targets for adiponectin: Lessons from adiponectin-overexpressing or deficient-mice. *American Journal of Physiology, Endocrinology and Metabolism, 301*, E326–E335.
Giepmans, B. N., Adams, S. R., Ellisman, M. H., & Tsien, R. Y. (2006). The fluorescent toolbox for assessing protein location and function. *Science, 312*, 217–224.
Hamilton, M. P., Gore, M. O., Ayers, C. R., Xinyu, W., McGuire, D. K., & Scherer, P. E. (2011). Adiponectin and cardiovascular risk profile in patients with type 2 diabetes mellitus: Parameters associated with adiponectin complex distribution. *Diabetes & Vascular Disease Research, 8*, 190–194.
Heiker, J. T., Kloting, N., Blüher, M., & Beck-Sickinger, A. G. (2010). Access to gram scale amounts of functional globular adiponectin from E. coli inclusion bodies by alkaline-shock solubilization. *Biochemical and Biophysical Research Communications, 398*, 32–37.
Heiker, J. T., Kosel, D., & Beck-Sickinger, A. G. (2010). Molecular mechanisms of signal transduction via adiponectin and adiponectin receptors. *Biological Chemistry, 391*, 1005–1018.
Heiker, J. T., Wottawah, C. M., Juhl, C., Kosel, D., Mörl, K., & Beck-Sickinger, A. G. (2009). Protein kinase CK2 interacts with adiponectin receptor 1 and participates in adiponectin signaling. *Cellular Signalling, 21*, 936–942.
Holland, W. L., Miller, R. A., Wang, Z. V., Sun, K., Barth, B. M., Bui, H. H., et al. (2011). Receptor-mediated activation of ceramidase activity initiates the pleiotropic actions of adiponectin. *Nature Medicine, 17*, 55–63.
Hu, E., Liang, P., & Spiegelman, B. M. (1996). AdipoQ is a novel adipose-specific gene dysregulated in obesity. *Journal of Biological Chemistry, 271*, 10697–10703.
Hui, X., Lam, K. S., Vanhoutte, P. M., & Xu, A. (2011). Adiponectin and cardiovascular health: An update. *British Journal of Pharmacology, 165*, 574–590.

Iwabu, M., Yamauchi, T., Okada-Iwabu, M., Sato, K., Nakagawa, T., Funata, M., et al. (2010). Adiponectin and AdipoR1 regulate PGC-1alpha and mitochondria by Ca(2+) and AMPK/SIRT1. *Nature, 464*, 1313–1319.

Jarde, T., Perrier, S., Vasson, M. P., & Caldefie-Chezet, F. (2011). Molecular mechanisms of leptin and adiponectin in breast cancer. *European Journal of Cancer, 47*, 33–43.

Jijakli, H., Zhang, H. X., Dura, E., Ramirez, R., Sener, A., & Malaisse, W. J. (2002). Effects of cytochalasin B and D upon insulin release and pancreatic islet cell metabolism. *International Journal of Molecular Medicine, 9*, 165–172.

Juhl, C., Mörl, K., & Beck-Sickinger, A. G. (2011). Adiponectin receptor 1 interacts with both subunits of protein kinase CK2. *Molecular and Cellular Biochemistry, 356*, 185–189.

Juhl, C., Kosel, D., & Beck-Sickinger, A. G. (2012). Two motifs with different function regulate the anterograde transport of the adiponectin receptor 1. *Cellular Signalling, 24*, 1762–1769.

Kadowaki, T., & Yamauchi, T. (2005). Adiponectin and adiponectin receptors. *Endocrine Reviews, 26*, 439–451.

Kanasaki, K., & Koya, D. (2011). Biology of obesity: Lessons from animal models of obesity. *Journal of Biomedicine and Biotechnology, 2011*, 197636.

Kerppola, T. K. (2009). Visualization of molecular interactions using bimolecular fluorescence complementation analysis: Characteristics of protein fragment complementation. *Chemical Society Reviews, 38*, 2876–2886.

Kosel, D., Heiker, J. T., Juhl, C., Wottawah, C. M., Blüher, M., Mörl, K., et al. (2010). Dimerization of adiponectin receptor 1 is inhibited by adiponectin. *Journal of Cell Science, 123*, 1320–1328.

Kubota, N., Terauchi, Y., Yamauchi, T., Kubota, T., Moroi, M., Matsui, J., et al. (2002). Disruption of adiponectin causes insulin resistance and neointimal formation. *Journal of Biological Chemistry, 277*, 25863–25866.

Kunji, E. R., Harding, M., Butler, P. J., & Akamine, P. (2008). Determination of the molecular mass and dimensions of membrane proteins by size exclusion chromatography. *Methods, 46*, 62–72.

Lai, Y., Feldman, K. L., & Clark, R. S. (2005). Enzyme-linked immunosorbent assays (ELISAs). *Critical Care Medicine, 33*, S433–S434.

Liu, M., Zhou, L., Xu, A., Lam, K. S., Wetzel, M. D., Xiang, R., et al. (2008). A disulfide-bond A oxidoreductase-like protein (DsbA-L) regulates adiponectin multimerization. *Proceedings of the National Academy of Sciences of the United States of America, 105*, 18302–18307.

MacPhee, D. J. (2010). Methodological considerations for improving Western blot analysis. *Journal of Pharmacological and Toxicological Methods, 61*, 171–177.

Maeda, K., Okubo, K., Shimomura, I., Funahashi, T., Matsuzawa, Y., & Matsubara, K. (1996). cDNA cloning and expression of a novel adipose specific collagen-like factor, apM1 (AdiPose Most abundant Gene transcript 1). *Biochemical and Biophysical Research Communications, 221*, 286–289.

Man, K., Ng, K. T., Xu, A., Cheng, Q., Lo, C. M., Xiao, J. W., et al. (2010). Suppression of liver tumor growth and metastasis by adiponectin in nude mice through inhibition of tumor angiogenesis and downregulation of Rho kinase/IFN-inducible protein 10/matrix metalloproteinase 9 signaling. *Clinical Cancer Research, 16*, 967–977.

Mao, X., Kikani, C. K., Riojas, R. A., Langlais, P., Wang, L., Ramos, F. J., et al. (2006). APPL1 binds to adiponectin receptors and mediates adiponectin signalling and function. *Nature Cell Biology, 8*, 516–523.

Masaie, H., Oritani, K., Yokota, T., Takahashi, I., Shirogane, T., Ujiie, H., et al. (2007). Adiponectin binds to chemokines via the globular head and modulates interactions between chemokines and heparan sulfates. *Experimental Hematology, 35*, 947–956.

Matsubara, M., Maruoka, S., & Katayose, S. (2002). Inverse relationship between plasma adiponectin and leptin concentrations in normal-weight and obese women. *European Journal of Endocrinology*, *147*, 173–180.

Miller, R. A., Chu, Q., Le Lay, J., Scherer, P. E., Ahima, R. S., Kaestner, K. H., et al. (2011). Adiponectin suppresses gluconeogenic gene expression in mouse hepatocytes independent of LKB1-AMPK signaling. *The Journal of Clinical Investigation*, *121*, 2518–2528.

Nakano, Y., Tobe, T., Choi-Miura, N. H., Mazda, T., & Tomita, M. (1996). Isolation and characterization of GBP28, a novel gelatin-binding protein purified from human plasma. *Journal of Biochemistry*, *120*, 803–812.

Okamoto, Y. (2011). Adiponectin provides cardiovascular protection in metabolic syndrome. *Cardiology Research and Practice*, *2011*, 313179.

Ouchi, N., Parker, J. L., Lugus, J. J., & Walsh, K. (2011). Adipokines in inflammation and metabolic disease. *Nature Reviews Immunology*, *11*, 85–97.

Pajvani, U. B., Du, X., Combs, T. P., Berg, A. H., Rajala, M. W., Schulthess, T., et al. (2003). Structure-function studies of the adipocyte-secreted hormone Acrp30/ adiponectin. Implications fpr metabolic regulation and bioactivity. *Journal of Biological Chemistry*, *278*, 9073–9085.

Petschnigg, J., Moe, O. W., & Stagljar, I. (2011). Using yeast as a model to study membrane proteins. *Current Opinion in Nephrology and Hypertension*, *20*, 425–432.

Phillips, S. A., & Kung, J. T. (2010). Mechanisms of adiponectin regulation and use as a pharmacological target. *Current Opinion in Pharmacology*, *10*, 676–683.

Pischon, T., Girman, C. J., Hotamisligil, G. S., Rifai, N., Hu, F. B., & Rimm, E. B. (2004). Plasma adiponectin levels and risk of myocardial infarction in men. *Journal of the American Medical Association*, *291*, 1730–1737.

Ranjbar, B., & Gill, P. (2009). Circular dichroism techniques: biomolecular and nanostructural analyses—A review. *Chemical Biology & Drug Design*, *74*, 101–120.

Rasmussen, M. S., Lihn, A. S., Pedersen, S. B., Bruun, J. M., Rasmussen, M., & Richelsen, B. (2006). Adiponectin receptors in human adipose tissue: Effects of obesity, weight loss, and fat depots. *Obesity (Silver Spring)*, *14*, 28–35.

Rose, R. H., Briddon, S. J., & Holliday, N. D. (2010). Bimolecular fluorescence complementation: Lighting up seven transmembrane domain receptor signalling networks. *British Journal of Pharmacology*, *159*, 738–750.

Scherer, P. E., Williams, S., Fogliano, M., Baldini, G., & Lodish, H. F. (1995). A novel serum protein similar to C1q, produced exclusively in adipocytes. *Journal of Biological Chemistry*, *270*, 26746–26749.

Schermelleh, L., Heintzmann, R., & Leonhardt, H. (2010). A guide to super-resolution fluorescence microscopy. *The Journal of Cell Biology*, *190*, 165–175.

Schraw, T., Wang, Z. V., Halberg, N., Hawkins, M., & Scherer, P. E. (2008). Plasma adiponectin complexes have distinct biochemical characteristics. *Endocrinology*, *149*, 2270–2282.

Shapiro, L., & Scherer, P. E. (1998). The crystal structure of a complement-1q family protein suggests an evolutionary link to tumor necrosis factor. *Current Biology*, *8*, 335–338.

Singla, P., Bardoloi, A., & Parkash, A. A. (2010). Metabolic effects of obesity: A review. *The World Journal of Diabetes*, *1*, 76–88.

Stephens, D. J., & Allan, V. J. (2003). Light microscopy techniques for live cell imaging. *Science*, *300*, 82–86.

Tang, Y. T., Hu, T., Arterburn, M., Boyle, B., Bright, J. M., Emtage, P. C., et al. (2005). PAQR proteins: A novel membrane receptor family defined by an ancient 7-transmembrane pass motif. *Journal of Molecular Evolution*, *61*, 372–380.

Tesauro, M., Canale, M. P., Rodia, G., Di Daniele, N., Lauro, D., Scuteri, A., et al. (2011). Metabolic syndrome, chronic kidney, and cardiovascular diseases: Role of adipokines. *Cardiology Research and Practice*, *2011*, 653182.

Tomas, E., Tsao, T. S., Saha, A. K., Murrey, H. E., Zhang Cc, C., Itani, S. I., et al. (2002). Enhanced muscle fat oxidation and glucose transport by ACRP30 globular domain: Acetyl-CoA carboxylase inhibition and AMP-activated protein kinase activation. *Proceedings of the National Academy of Sciences of the United States of America, 99*, 16309–16313.

Wang, Y., Lam, K. S., Chan, L., Chan, K. W., Lam, J. B., Lam, M. C., et al. (2006). Posttranslational modifications of the four conserved lysine residues within the collagenous domain of adiponectin are required for the formation of its high molecular weight oligomeric complex. *Journal of Biological Chemistry, 281*, 16391–16400.

Wang, Y., Shyy, J. Y., & Chien, S. (2008). Fluorescence proteins, live-cell imaging, and mechanobiology: Seeing is believing. *Annual Review of Biomedical Engineering, 10*, 1–38.

Weyer, C., Funahashi, T., Tanaka, S., Hotta, K., Matsuzawa, Y., Pratley, R. E., et al. (2001). Hypoadiponectinemia in obesity and type 2 diabetes: close association with insulin resistance and hyperinsulinemia. *Journal of Clinical Endocrinology and Metabolism, 86*, 1930–1935.

WHO. (2011). Obesity and overweight. WHO, Fact sheet N°311, May 2012. http://www.who.int/mediacentre/factsheets/fs311/en/index.html.

Wiedenmann, J., Oswald, F., & Nienhaus, G. U. (2009). Fluorescent proteins for live cell imaging: Opportunities, limitations, and challenges. *IUBMB Life, 61*, 1029–1042.

Wlodawer, A., Minor, W., Dauter, Z., & Jaskolski, M. (2008). Protein crystallography for non-crystallographers, or how to get the best (but not more) from published macromolecular structures. *FEBS Journal, 275*, 1–21.

Xin, X., Zhou, L., Reyes, C. M., Liu, F., & Dong, L. Q. (2011). APPL1 mediates adiponectin-stimulated p38 MAPK activation by scaffolding the TAK1-MKK3-p38 MAPK pathway. *American Journal of Physiology, Endocrinology and Metabolism, 300*, E103–E110.

Xu, Y., Wang, N., Ling, F., Li, P., & Gao, Y. (2009). Receptor for activated C-kinase 1, a novel binding partner of adiponectin receptor 1. *Biochemical and Biophysical Research Communications, 378*, 95–98.

Xu, Y., Zhang, C., Wang, N., Ling, F., Li, P., Gao, Y., et al. (2011). Adiponectin inhibits lymphotoxin-beta receptor-mediated NF-kappaB signaling in human umbilical vein endothelial cells. *Biochemical and Biophysical Research Communications, 404*, 1060–1064.

Yamauchi, T., Kamon, J., Ito, Y., Tsuchida, A., Yokomizo, T., Kita, S., et al. (2003). Cloning of adiponectin receptors that mediate antidiabetic metabolic effects. *Nature, 423*, 762–769.

Yamauchi, T., Kamon, J., Minokoshi, Y., Ito, Y., Waki, H., Uchida, S., et al. (2002). Adiponectin stimulates glucose utilization and fatty-acid oxidation by activating AMP-activated protein kinase. *Nature Medicine, 8*, 1288–1295.

Yamauchi, T., Nio, Y., Maki, T., Kobayashi, M., Takazawa, T., Iwabu, M., et al. (2007). Targeted disruption of AdipoR1 and AdipoR2 causes abrogation of adiponectin binding and metabolic actions. *Nature Medicine, 13*, 332–339.

Yea, K., Kim, J., Lim, S., Kwon, T., Park, H. S., Park, K. S., et al. (2009). Lysophosphatidylserine regulates blood glucose by enhancing glucose transport in myotubes and adipocytes. *Biochemical and Biophysical Research Communications, 378*, 783–788.

Yoon, M. J., Lee, G. Y., Chung, J. J., Ahn, Y. H., Hong, S. H., & Kim, J. B. (2006). Adiponectin increases fatty acid oxidation in skeletal muscle cells by sequential activation of AMP-activated protein kinase, p38 mitogen-activated protein kinase, and peroxisome proliferator-activated receptor alpha. *Diabetes, 55*, 2562–2570.

CHAPTER THREE

Nutritional and Hormonal Modulation of Adiponectin and its Receptors adipoR1 and adipoR2

Cristiane de Oliveira[*,1], Ana Barbosa Marcondes de Mattos[†], Carolina Biz Rodrigues Silva[‡], João Felipe Mota[§], Juliane Costa Silva Zemdegs[‡]

[*]Systemic Inflammation Laboratory, Trauma Research, St. Joseph's Hospital and Medical Center, Phoenix, USA
[†]Laboratório de Fisiologia da Nutrição e Desenvolvimento, Departamento de Fisiologia, Universidade do Estado do Rio de Janeiro, Rio de Janeiro, Brazil
[‡]Disciplina de Fisiologia da Nutrição, Departamento de Fisiologia, Universidade Federal de São Paulo, São Paulo, Brazil
[§]Faculdade de Nutrição, Universidade Federal de Goiás (FANUT), Goiânia, Brazil
[1]Corresponding author: e-mail address: crioliva@uol.com.br

Contents

Abstract

Adiponectin is the most abundant plasma protein synthesized mostly by adipose tissue and is an insulin-sensitive hormone, playing a central role in glucose and lipid metabolism. Adiponectin effects are mediated via two receptors, adipoR1 and adipoR2. Several hormones and diet components that are involved in insulin resistance may impair insulin sensitivity at least in part by decreasing adiponectin and adiponectin receptors. Adiponectin expression and serum levels are associated with the amount and type of fatty acids and carbohydrate consumed. Other food items, such as vitamins, alcohol, sodium, green tea, and coffee, have been reported to modify

Vitamins and Hormones, Volume 90
ISSN 0083-6729
http://dx.doi.org/10.1016/B978-0-12-398313-8.00003-8

adiponectin levels. Several hormones, including testosterone, estrogen, prolactin, glucocorticoids, catecholamines, and growth hormone, have been shown to inhibit adiponectin production, but the studies are still controversial. Even so, adiponectin is a potential therapeutic target in the treatment of diabetes mellitus and other diseases associated with hypoadiponectinemia.

ABBREVIATIONS

adipoR1 adiponectin receptor 1
adipoR2 adiponectin receptor 2
AMPK adenosine monophosphate-activated kinase
fT3 free triiodothyronine
fT4 free thyroxin
GH growth hormone
HF high-fat diet
PPAR-α peroxisome proliferator-activated receptors alpha
PUFA polyunsaturated fatty acids
T3 triiodothyronine
T4 thyroxin
WAT white adipose tissue

1. INTRODUCTION

Adiponectin is an abundant plasma protein with concentrations higher than those of other hormones corresponding to 0.01–0.05% of the total serum proteins (Arita et al., 1999; Nakano, Tobe, Choi-Miura, Mazda, & Tonita, 1996). It is primarily synthesized by adipocytes but has also been detected in skeletal muscle (Delaigle, Jonas, Bauche, Cornu, & Brichard, 2004), cardiomyocytes (Pineiro et al., 2005), osteoblasts (Berner et al., 2004), lymphocytes (Crawford, Peake, Price, Morris, & Irvine, 2010), adrenal gland (Paschke et al., 2010), placenta (Caminos et al., 2005), testis (Caminos et al., 2008), ovary (Chabrolle, Tosca, & Dupont, 2007), pituitary gland (Rodriguez-Pacheco et al., 2007), and liver tissue (Ma et al., 2009). Secreted adiponectin has different multimeric forms. Globular adiponectin exists as a trimer and full-length adiponectin exists in three forms: a low-molecular weight trimer (LMW), a medium-molecular weight (MMW) hexamer, and a high-molecular weight oligomer (HMW) (Hada et al., 2007). HMW is considered the most metabolically active form and a reduction in this form is related to insulin resistance, diabetes, and obesity (Lara-Castro, Luo, Wallace, Klein, & Garvey, 2006; Wang, Lam, Yau, & Xu, 2008).

Unlike other adipokines produced by adipose tissue, plasma levels of adiponectin are inversely correlated with body mass index, visceral adiposity, plasma levels of triglycerides, and low-density lipoprotein (Arita et al., 1999; Lihn et al., 2004; Maeda et al., 1996; Yamamoto et al., 2002).

The inverse correlation between plasma levels of adiponectin and obesity is probably a result of altered gene expression induced by oxidative stress and hypoxia (Soares et al., 2005; Trayhurn, Wang, & Wood, 2008). Adiponectin is an insulin-sensitive hormone and plays a central role in lipid and glucose metabolism (Lihn, Pedersen, & Richelsen, 2005). It augments lipid oxidation in skeletal muscle and myocytes (Fruebis et al., 2001; Yamauchi et al., 2001) and reduces hepatic glucose production in liver and hepatocytes (Berg, Combs, Du, Brownlee, & Scherer, 2001; Combs, Berg, Obici, Scherer, & Rossetti, 2001). In addition, adiponectin possesses antiatherogenic activities and anti-inflammatory effects (Lihn et al., 2005; Ouchi et al., 2001).

All these metabolic processes are mediated via two adiponectin receptors, adipoR1 and adipoR2, that activate its downstream signaling pathways mainly by activating adenosine monophosphate-activated kinase (AMPK) and peroxisome proliferator-activated receptors alpha (PPAR-α). In mouse, it was observed that adipoR1 receptor acts via AMPK pathway, whereas adipoR2 through PPAR-α signaling (Yamauchi et al., 2007). AdipoR1 has a strong affinity for globular adiponectin and full-length adiponectin and is mainly expressed in the liver (Yamauchi et al., 2003). AdipoR2 exhibits an intermediate affinity for globular and full-length adiponectin and is mainly expressed in muscle (Beylot, Pinteur, & Peroni, 2006). Adiponectin receptors have also been detected in pancreatic beta cells (Kharroubi, Rasschaert, Eizirik, & Cnop, 2003), macrophages (Chinetti, Zawadski, Fruchart, & Staels, 2004), osteoblasts (Berner et al., 2004), adrenal gland (Paschke et al., 2010), human endometrium (Takemura et al., 2006), testis (Caminos et al., 2008), ovary (Chabrolle et al., 2007), hypothalamus (Qi et al., 2004), endothelial cells (Motoshima, Wu, Mahadev, & Goldstein, 2004), placenta (Caminos et al., 2005), and in adipose tissue (Rasmussen et al., 2006), with adipoR1 being expressed 10–15-fold higher than adipoR2 (Beylot et al., 2006). T-cadherin, a tumor-suppressor protein, was identified as the third adiponectin receptor by its ability to specifically bind the HMW isoforms of adiponectin. This receptor is abundantly expressed in the cardiovascular system and also in muscle cells (Hug et al., 2004).

Several hormones and diet components that are involved in insulin resistance may impair insulin sensitivity at least in part by decreasing adiponectin and adiponectin receptors expression and/or secretion. In this chapter, we review current studies on the effect of some of these factors on adiponectin system.

2. NUTRITIONAL REGULATION OF ADIPONECTIN AND ADIPONECTIN RECEPTORS

2.1. Fatty acids

A high level of dietary fat in the diet is a major factor in promoting weight gain, obesity, and insulin resistance. Different studies have shown that the excess of fat on diet decreases adiponectin gene transcription (Berk et al., 2005; Barnea, Shamay, Stark, & Madar, 2006; Bullen, Bluher, Kelesidis, & Mantzoros, 2007; Boullu-Ciocca, Achard, Tassistro, Dutour, & Grino, 2008; Mullen, Tishinsky, Robinson, & Dyck, 2010; Zhang, Chen, Aravindakshan, & Ram Sairam, 2007). This reduction may play a causal role in the development of insulin resistance. Further, decreasing the expression of adipoR1 and adipoR2 leads to a reduction in adiponectin sensitivity, which in turn leads to insulin resistance (Kadowaki & Yamauchi, 2005). In this sense, some studies have investigated the effect of diet composition on the expression of adiponectin and its receptors in different tissues.

High-fat diet (HF) has a tissue-specific effect on adiponectin expression as demonstrated by Barnea et al. (2006). Mice fed with a 22% soybean HF for 4 months had weight gain, and the insulin resistance was significantly increased. In addition, there was a significant reduction in adiponectin mRNA expression and protein levels in epididymal adipose tissue; also, reduction in adipoR1 and adipoR2 mRNA expression in hepatic tissue when compared to control diet group. In the muscle, mRNA expression of adipoR1 and adipoR2 increased in the HF group combined with a reduction in pAMPK levels. In this study, 4 months of HF was sufficient to affect insulin sensitivity and this was caused, at least in part via impairment of adiponectin action in adipose and hepatic tissue. Consistent with this finding, de Oliveira, Mattos, et al. (2011) demonstrated that rats treated for only 3 weeks with 20% soybean HF promoted hyperglicemia state, but not weight gain compared to the control diet group, with a significant reduction in adiponectin gene expression in subcutaneous adipose tissue, with no alteration in other fat depots or in serum adiponectin levels. In addition, the HF increased adipoR2 mRNA expression in retroperitonal adipose tissue, muscle and liver, but not in other fat depots. AdipoR1 protein secretion was slightly increased in muscle.

In relation to HF, an important point is checking which one, the energy intake, adiposity, or dietary fat content, is affecting adiponectin. The results from de Oliveira study showed that an HF *per se*, and not the change in body weight, could modulate adiponectin system. Qiao, Lee, Kinney, Yoo, and Shao (2011) showed that total energy intake, but not dietary fat, plays an important role in controlling adiponectin gene expression and circulating levels. Also, energy intake had higher effect on circulating adiponectin levels than adiposity. The results from Barnea group and de Oliveira group showed that adiponectin modulation varies between different adipose tissue depots and seems to be a compensatory effect in order to maintain serum adiponectin levels in the high-fat induced adiponectin dysbalance.

It has been documented that adiponectin expression in muscle promotes an increase in insulin sensitivity (Ingelson et al., 2009), while in liver exerts its protective actions against lipid accumulation in liver (Civitarese et al., 2007; Xu et al., 2003). Thus, adiponectin signaling may be a cross-link between HF and hepatic inflammation (Peng et al., 2009). In fact, they found that HF-induced liver steatosis was associated with a reduction in serum adiponectin levels and reduction in hepatic adipoR2 expression. The expression levels of adipoR1, adipoR2, AMPK $\alpha 1/\alpha 2$ (which inhibits fatty acid synthesis), and PPAR-α (which activates fatty acid oxidation) decreased after 8 weeks of HF in liver of obese Zucker rats (Matsunami et al., 2011). Dysregulation of hepatic AMPK signaling represents a crucial mechanism for development of alcoholic fatty liver disease (Ha, Kim, & Chae, 2011). In animal model, studies have demonstrated that diets enriched in saturated fatty acids protect against alcoholic liver injury, while diets enriched in polyunsaturated fatty acids (PUFA) promote liver injury (Mezey, 1998; Nanji et al., 1995). A high-saturated fat diet (40% of calories cocoa butter) attenuated hepatic steatosis caused by chronic ethanol consumption in mice and elevated circulating adiponectin levels. Further, treatment of 3T3-L1 adipocytes with saturated fatty acids (palmitic or stearic acids) in the presence of ethanol increased secretion of adiponectin (You, Considine, Leone, Kelly, & Crabb, 2005). Recently, in addition to these results, Liang, Hu, Rogers, Shen, and You (2011) observed that a high-saturated fat diet selectively upregulated hepatic adipoR2 in mice treated with ethanol, a result that repeated also in hepatic culture cells treated with palmitic acid (major saturated fatty acid) in the presence of ethanol significantly increased adipoR2 mRNA expression. These results show that the protective effect of dietary saturated fatty acids against the development of alcoholic liver disease is partially mediated through induction of adiponectin system.

A low proportion of long-chain unsaturated fatty acids and a high proportion of saturated fatty acids in the diet have been associated with impaired insulin action (Hernandez-Morante et al., 2007; Vessby, 2003). Also, adiponectin expression and serum levels are associated type of fatty acids consumed: saturated fatty acids appearing as negative regulator for adiponectin, and monounsaturated fatty acids (MUFA) and PUFA as positive ones (Peres de Heredia et al., 2009). In mice fed for 2 or 60 days, Bueno et al. (2008) evaluate the effects of four different HFs enriched with soybean oil (ω-6 PUFA), fish oil (ω-3 PUFA), coconut oil (saturated fatty acid rich in medium-chain fatty acids), and lard (saturated fatty acids) on adiponectin expression by white adipose tissue (WAT). The soybean-enriched diet and the saturated coconut and lard-enriched diets reduced serum adiponectin levels. Acute treatment with all diets decreased adiponectin gene expression in retroperitoneal WAT. Chronic treatment with soybean and coconut diets decreased adiponectin gene expression in epididymal WAT. In 3T3-L1 adipocytes, 48 h of incubation with the fatty acids, palmitic, linoleic, eicosapentaenoic (EPA), and docosahexaenoic (DHA), also decreased adiponectin gene expression.

HF rich in ω-3 PUFA leads to an increase in adiponectin concentration in mice combined with attenuation of diet-induced obesity state (Fukumitsu et al., 2008; Shirai & Suzuki, 2008). Similarly, mice fed with diet containing 27% fish oil demonstrated a significant increase in plasma adiponectin compared with a diet high in safflower oil, and this effect was blocked by the treatment with a PPARγ antagonist (Itoh et al., 2007; Iwaki et al., 2003; Neschen et al., 2006; 2007). Moreover, the partial replacement of vegetable oils in a 35% HF by EPA (6%) and DHA (51%) increased mRNA adiponectin expression and secretion in epididymal adipose tissue, leading to increased plasma adiponectin levels by mechanisms independent of adiposity (Flachs et al., 2006). Likewise, Mullen et al. (2010) found that adiponectin resistance in soleus muscle of rats caused by a saturated HF is prevented when saturated fat is replaced by polyunsaturated fatty acids. In addition, treatment with DHA or EPA upregulated adiponectin by activating PPAR-γ, and this was blocked by the addition of PPAR-γ inhibitor (Banga et al., 2009).

In healthy humans, a transversal study observed that the proportion of ω-3 fatty acids in plasma was significantly associated with circulating adiponectin concentrations (Fernández-Real, Vendrell, & Ricart, 2005).

In general, several studies indicate that saturated fatty acids have negative association with adiponectin levels while PUFA positive.

2.2. Carbohydrates

An increase in sugar intake, especially in the form of fructose, is positively correlated with development of insulin resistance, hyperinsulinemia, dyslipidemia, and hypertension. Higher sugar intake is likely one of the major causes of increased incidence of both obesity type 2 diabetes and metabolic syndrome (Elliott, Keim, Stern, Teff, & Havel, 2002; Tappy, Lê, Tran, & Paquot, 2010).

Also in animal models, a high-fructose diet can induce insulin resistance, hyperinsulinemia, and hypertension, and some research groups reported that chronic consumption of a high-fructose diet decreases plasma level of adiponectin in animals (Meeprom, Sompong, Suwannaphet, Yibchok-anun, & Adisakwattana, 2011; Qin, Polansky, & Anderson, 2010). However, the opposite effect was found by Alzamendi et al. (2009): rats fed a fructose diet (addition of 10%) for 3 weeks had increased serum adiponectin, along with increased levels of circulating glucose, triglyceride, and free fat acids, all features implicated in the development of metabolic syndrome. In accordance with this last result, Carvalho et al. (2010) observed an increase in adiponectin mRNA level in adipose culture cells 3T3-L1 after treatment with glucose and fructose.

In the study by Bonnard, Durand, Vidal, and Rieusset (2008), while serum adiponectin levels were increased in mice fed a high-fat, high-sucrose diet for 16 weeks, compared to the control diet group, adiponectin mRNA and protein were decreased in adipose tissue but increased in skeletal muscle, probably due to increase in intramuscular adipose tissue content. These changes in adiponectin were accompanied by a decrease in adipoR1 mRNA in skeletal muscle and adipoR2 in liver. This same research group treated mice for 4 weeks with the high-fat, high-sucrose diet; however, despite their obesity, there were no changes in circulating adiponectin, adipose tissue adiponectin, or adiponectin receptor expression. These results suggest that obesity induced by high-fat, high-sucrose diet, per se, is not responsible for adiponectin modification. A decrease in adiponectin gene expression, in adipose tissue of rats, due to high-fructose diet intake was also found by Xi et al. (2007). Meeprom et al. (2011) confirmed decrease in adipoR1 in skeletal muscle of rats treated with high-fructose diet.

Few clinical studies have evaluated the effect of carbohydrate on adiponectin activity. In this context, a randomized crossover trial comprising insulin-resistant patients documented that a low-fat, carbohydrate-enriched diet increased central fat depot, decreased fat mass deposited in periphery, and decreased postprandial adiponectin expression; compared to monounsaturated fatty acids and saturated-rich isocaloric diets (Paniagua et al., 2007). A cross-sectional study revealed that a diet which induced high glycemic load was associated with decreased adiponectin concentration (Pischon, Girman, Rifai, Hotamisligil, & Rimm, 2005). Likewise, refined grain intake was negatively correlated with adiponectin level, but fruit intake was positively correlated with adiponectin level (Yannakoulia et al., 2008).

When considering low carbohydrate intake, and thus restriction of this macronutrient, the opposite effect in adiponectin expression was not evident. Various studies have reported no improvement in adiponectin activity when carbohydrate was restricted, and no weight loss was observed (Cardillo, Seshadri, & Iqbal, 2006; de Luis et al., 2007; Keogh, Brinkworth, & Clifton, 2007, Keogh et al., 2008). In negative energy balance status, there was an increase in adiponectin expression (Al-Sarraj, Saadi, Calle, Volek, & Fernandez, 2009).

Therefore, from the few studies that examined the effect of carbohydrate on adiponectin expression in humans, we can conclude that the regulatory action of carbohydrate depends on the type of carbohydrate consumed. But it is not clear whether fructose or glucose promotes obesity by modifying either adiponectin or other adipocytokines. It is also unclear whether these hexoses act directly within adipocytes to modify adiponectin gene expression.

2.3. Vitamins

Vitamin E is a fat-soluble vitamin with antioxidant properties. The α-tocopherol and γ-tocopherol are the two predominant forms found in the diet and in plasma (Gray, Swick, & Ronnenberg, 2011). The absorption and bioavailability of vitamin E are greater in foods rich in lipid. Paolisso et al. (1993) reported that vitamin E supplementation improves insulin sensitivity. They verified that diabetic patients supplemented with 900 mg of vitamin E had lower fasting plasma glucose and hemoglobin A1c (HbA1c) than control patients. Other studies also showed beneficial effects of vitamin E on insulin sensitivity with a reduction in type 2 diabetes risk (Costacou, Ma, King, & Mayer-Davis, 2008; Montonen, Knekt, Järvinen, & Reunanen, 2004).

By now, most of these effects have been related to the antioxidant capacity of vitamin E, especially to its ability to scavenge free radicals that could have a negative impact on insulin sensitivity and type 2 diabetes (Landrier et al., 2009). Moreover, vitamin E also modulates gene expression; tocopherols can regulate several genes involved in cell cycle regulation, inflammation and cell adhesion, cell signaling, lipid uptake, and atherosclerosis (Azzi et al., 2004; Gray et al., 2011; Landrier et al., 2009).

The insulin-sensitivity activity of vitamin E appears to be associated with adiponectin modulation. Vitamin E supplementation increased adiponectin mRNA and protein expression in diet-induced obese rats (Shen, Tang, Huang, & Cai, 2010). Landrier et al. (2009) hypothesized that tocopherols were able to induce adiponectin via PPARγ expression. *In vivo*, mice were fed with vitamin E equated to a human "supranutritional" dose and their plasma adiponectin level and adiponectin mRNA expression in epididymal fat depot increased after 4 days of treatment. *In vitro*, 3T3-L1 preadipocytes treated with α- and γ-tocopherol had dose-dependently increased in adiponectin mRNA expression. *N*-Acetylcysteine and Trolox, known antioxidants, were used to verify if the effect on adiponectin were related to antioxidant capacity. These antioxidants were not effective to induce adiponectin expression, which suggest that induction of adiponectin expression by α- and γ-tocopherol was likely independent of their antioxidant capacities. The reduction in oxidative stress and an increase in adiponectin serum levels by antioxidants, including vitamin E, was also observed in humans (Vincent et al., 2009).

Vitamin D, a fat-soluble vitamin, is converted in the body to 25-hydroxyvitamin (25(OH)D) also known as calcidiol; the active circulant form in the organism and in 1,25-dihydroxyvitamin D3 (1,25(OH)2D3) also known as calcitriol. Plasma 25(OH)D was positively associated with adiponectin concentrations (Gannagé-Yared et al., 2009; Nimitphong, Chanprasertyothin, Jongjaroenprasert, & Ongphiphadhanakul, 2009). Liu et al. (2009) observed an increase in adiponectin concentrations among individuals in the higher tertile category of plasma 25(OH)D compared with those in the lowest.

Studies indicate different reasons for these effects: vitamin D receptor has been identified in preadipocytes, and it is possible that 1,25(OH)2D3 may regulate adiponectin gene expression (Lee, Lee, Choi, & Lee, 2005). As an hormone, 1,25(OH)2D3 is thought to play a role in regulating the production of TNF-α (Cantorna & Mahon, 2005), which can affect adiponectin synthesis (Ruan & Lodish, 2003).

In extremely obese subjects, 25(OH)D serum concentrations were inversely associated with several biomarkers of systemic inflammation, regardless of the total quantity of fat mass. However, the researchers did not find any significant linear association between 25(OH)D levels and adiponectin (Bellia et al., 2011).

Nimitphong et al. (2009) demonstrated an association between insufficient vitamin D status and lower circulating adiponectin in subjects with abnormal glucose tolerance independent of adiposity, which may indicate the role of adiponectin as a link between vitamin D status and insulin resistance. The authors suggested that the mechanisms are related to a reduction in inflammation state. 1,25(OH)2D3 binds to C-terminal domain of vitamin D receptors and enhances osteocalcin gene transcription (Carvallo et al., 2007). Uncarboxylated form of osteocalcin, one of the osteoblast hormones, regulates glucose homeostasis (Lee et al., 2007) and also adiponectin expression (Sayinalp, Gedik, & Koray, 1995).

Vitamin C or ascorbic acid is a water-soluble vitamin used as an antioxidant. An inverse relationship between plasma vitamin C and adiposity has been showed (Canoy et al., 2005), but no association with plasma total adiponectin was found by Johnston, Beezhold, Mostow, and Swan (2007). Rose et al. (2010) found that treatment of different adipocyte cell type with vitamin C increased secretion of HMW adiponectin without altering adiponectin mRNA or total adiponectin secretion and that ascorbic acid supplementation together with thiazolidinedione (TZD) treatment, an insulin-sensitizing drug that increases HMW adiponectin levels, had synergistic effect on secretion of HMW adiponectin. Differences in ascorbic levels in patients may influence the efficacy of TZD treatment in humans.

Knowledge about the regulation of adiponectin by vitamins is still a new theme, and the mechanisms still need to be elucidated.

2.4. Other food items

Other food items, beverages, and dietary constituents have been reported to modify adiponectin levels.

2.4.1 Alcohol

Studies have shown protective and harmful effects of alcohol on insulin sensitivity (Joosten, Beulens, Kersten, & Hendriks, 2008; Kawamoto et al., 2009). Alcohol has been shown to affect adiponectin levels: chronic ethanol consumption decreased serum adiponectin levels in rodents (Chen, Sebastian, & Nagy, 2007; Song, Zhou, Deaciuc, Chen, & McClain, 2008; Xu et al., 2003). As shown by Chen et al. (2007), a 4-week liquid diet containing ethanol (36% of total caloric value) decreased adiponectin mRNA had no effect on adiponectin protein in subcutaneous adipose tissue and decreased adiponectin secretion by isolated adipocytes from these rats. Similar suppression of adiponectin secretion by isolated adipocytes was found by Tang et al. (2012). Repunte-Canonigo et al. (2010) showed that alcohol administration in rats increased adipoR2 mRNA levels in the anterior cingulate cortex (a brain region implicated in alcohol and drug addiction) and suggested that adipoR2 receptor is involved in the cellular actions of alcohol and regulation of alcohol intake.

Contradicting the results in animals and cell culture, human observation reports that short-term (3 weeks), but not acute, alcohol consumption was positively associated with circulating adiponectin levels (Beulens, Rimm, Hu, Hendriks, & Mukamal, 2008; Imhof, Plamper, Maier, Trischler, & Koenig, 2009; Joosten et al., 2011; Sierksma et al., 2004; Thamer, Haap, Fritsche, Haering, & Stumvoll, 2004). In a recent meta-analysis of interventional studies about the effects of alcohol consumption on biological markers associated with the risk of coronary heart disease, alcohol consumption was shown to increase adiponectin levels (Brien, Ronksley, Turner, Mukamal, & Ghali, 2011). This increase in adiponectin can partially explain the association between moderate alcohol consumption and reduced risk for type 2 diabetes mellitus (Koppes, Dekker, Hendriks, Bouter, & Heine, 2005).

2.4.2 High-salt diet

High-sodium intake is associated with high blood pressure and insulin resistance (Donovan, Solomon, Seely, Williams, & Simonson, 1993; Fuenmayor, Moreira, & Cubeddu, 1998). Kamari et al. (2010) showed that high-salt diet intake (3% Na) increased plasma adiponectin levels in rats, independent of an increase in blood pressure. Similarly, consumption of a high-sodium diet (200 mmol/d) by healthy humans increased plasma adiponectin level (Lely et al., 2007), whereas dietary sodium restriction decreased HMW-adiponectin (Krikken, Dallinga-Thie, Navis, & Dullaart, 2012). These results would suggest that adiponectin may have a protective role in the pathogenesis of hypertension. However, in salt-resistant rats and in salt-sensitive hypertensive rats, Usukura et al. (2009) showed that a high-salt diet (7%) did not change serum adiponectin concentration. Likewise, dietary salt restriction by human did not modify plasma adiponectin (Nakandakare et al., 2008).

2.4.3 Green tea

The health benefits of green tea have been attributed mainly to the catechins such as (−)-epicatechin (EC), (−)-epigallocatechin (EGC), (−)-epicatechin gallate (ECG), and (−)-epigallocatechin gallate (EGCG). The EGCG, the main catechin in green tea, has antiobesity effects and insulin-sensitive action (Iso, Date, Wakai, Fukui, & Tamakoshi, 2006; Lee, Kim, & Kim, 2009). In obese and insulin-resistant animals and in nonobese diabetic animals, ECGC increased plasma adiponectin level, while decreased triacylclycerol and ameliorated hypertension (Li, Douglas, Maiyoh, Adeli, & Theriault, 2006; Potenza et al., 2007; Shimada, Mochizuki, Sakurai, & Goda, 2007). These results suggest that the effect of ECGC in improving insulin resistance is partly due to an increase in adiponectin levels. In another study by Cho et al. (2007), (-)-catechin increased expression and secretion of adiponectin in murine adipocytes 3T3-L1.

In one cross-sectional study of humans, adiponectin concentration was positively associated with green tea intake (Wu, Yu, Stanczyk, Tseng, & Pike, 2011). In contrast, other human studies showed no association between green tea intake and adiponectin levels (Basu et al., 2010; Imatoh et al., 2011).

2.4.4 Coffee

Several coffee constituents have a possible role in metabolism, including nonphenolic compounds (caffeine, trigonelline, and nicotinic acid), phenolic compounds (chlorogenic acid and caffeic acid), and the minerals magnesium and potassium (Higdon & Frei, 2006).

In rats fed for 3 months with a HF (58% of energy from fat, 24% from carbohydrates, and 5.6 kcal/g), adipo-R2 gene and protein expressions in liver were significantly reduced, compared with the control group (5% of energy from fat, 77% from carbohydrates, and 3.3 kcal/g). The association of HF with decaffeinated coffee or coffee polyphenols and melanoidins treatment counteracted this downregulation. This finding suggests that coffee consumption protects the liver from damage caused by a HF, and this protection was partially mediated by a reduction in liver inflammation, through increases of adipoR2 and anti-inflammatory cytokines IL-4 and IL-10 (Vitaglione et al., 2010).

In a prospective cohort study, Williams et al. (2008) showed that caffeine-containing coffee, but not decaffeinated coffee, increased plasma adiponectin concentration in women compared to the group that consumed lower amounts. A similar result was found by Wedick et al. (2011). This group suggested that increased adiponectin levels can explain the beneficial effects of regular coffee consumption on low-grade systemic inflammation and type 2 diabetes incidence. Habitual coffee consumption has been associated with a lower risk of type 2 diabetes, as showed in a systematic review by Van Dam and Hu (2005). Other studies also positively correlated coffee consumption with adiponectin levels (Imatoh et al., 2011; Kempf et al., 2010), but contradictory results by Kotani, Fujiwara, Hamada, Tsuzaki, and Sakane (2008) found no association between coffee consumption and adiponectin levels in healthy men and women.

3. HORMONAL REGULATION OF ADIPONECTIN AND ADIPONECTIN RECEPTORS

3.1. Adrenal hormones

The adrenal steroid hormones, glucocorticoids, are involved in glucose and lipid metabolism, and insulin secretion, and contribute to the development of obesity and insulin resistance (Andrews & Walker, 1999; Müssig et al., 2010).

Studies *in vitro* and *in vivo* showed inhibition of adiponectin expression and reduction on plasma adiponectin levels by glucocorticoids (Degawa-Yamauchi et al., 2005; Fallo et al., 2004; Fasshauer, Klein, Neumann, Eszlinger, & Paschke, 2002; Halleux et al., 2001; Jang, Inder, Obeyesekere, & Alford, 2008). Viengchareun, Zennaro, Pascual-Le Tallec, and Lombes (2002) also demonstrated reduction in adiponectin expression in differentiated brown adipocytes T37i cells by treatment with dexamethasone, a synthetic glucocorticoid with a higher affinity for the glucocorticoid receptor. Gavrila, Chan, et al. (2003) and Gavrila, Peng, et al. (2003) studying if serum adiponectin levels exhibit diurnal or ultradian rhythms in healthy men found that serum adiponectin levels have a diurnal variation with a significant decline at night and changed in the same direction and behind cortisol variations by a few hours, suggesting adiponectin regulation by cortisol in humans.

Makimura, Mizuno, Bergen, and Mobbs (2002) noticed that adrenalectomy normalized expression and increased adiponectin levels in obese ob/ob mice and in several models of obesity reversed hyperglycemia (Blair, Caterson, & Cooney, 1996; Freedman, Stern, Reaven, & Mondon, 1986). In a recently study, de Oliveira, Iwanaga-Carvalho, et al. (2011) showed that the excess of glucocorticoids due to dexamethasone therapy decreased serum adiponectin concentrations together with increased insulin levels, whereas adrenalectomy decreased the mRNA expression of adiponectin and adipoR2 in epididymal adipose tissue but not in other fat depots and increased adipoR2 gene expression in skeletal muscle compared to control group sham-operated. Degawa-Yamauchi et al. (2005) also showed a tissue-specific regulation of glucocorticoids: dexamethasone had an inhibitory effect on adiponectin release from subcutaneous but not in omental adipocytes. In humans, adrenalectomy has been shown to increase adiponectin levels in patients with Cushing's syndrome (Ashizawa, Takagi, Seto, Suzuki, & Yano, 2007).

On the other hand, Libe et al. (2005) suggest that cortisol excess did not directly affect adiponectin levels in patients with Cushing's disease, as surgical cure of Cushing's disease does not alter serum adiponectin, despite a decrease in cortisol and insulin resistance indices. To confirm this, Lewandowski, Szosland, and Lewinski (2006) studied the effect of oral dexamethasone suppression in human subjects and did not found any significant effect of short-term dexamethasone administration on serum adiponectin. Similar result was found by Patel et al. (2006): a short-term dexamethasone treatment in healthy men did not significantly change levels of adiponectin, despite a hyperinsulinaemic response. In opposition to all the previous studies, Jang et al. (2008) observed an increase in serum adiponectin levels after dexamethasone treatment in nondiabetic subjects and Rieth et al. (2009) showed that short-term therapeutic administration of prednisolone for 1 week significantly increased adiponectin circulating levels in healthy men.

In the study by de Oliveira, Iwanaga-Carvalho, et al. (2011), as dexamethasone treatment did not reverse the effects of adrenalectomy on adiponectin (decrease in adiponectin and adipoR2 mRNA and increase in adipoR2 mRNA in muscle) and treatment with glucocorticoid receptor blockade did not reproduce the effects of adrenalectomy in rats, they hypothesized that the observed changes in adiponectin and adipoR2 could be due to the lack of other adrenal hormones also decreased with adrenalectomy such as mineralocorticoids and catecholamines. In 3T3-L1 adipocytes culture cells, the same group observed that adrenaline increased the expression of adipoR2, thus suggesting a stimulatory effect of catecholamines on the adiponectin system. Another study by Fu et al. (2007) showed that a β-adrenoceptor agonist increases the expression of adipoR2 in 3T3-L1 adipocytes, and Oana et al. (2005) showed increased plasma adiponectin levels and adiponectin mRNA in the EPI WAT of mice. However, Fasshauer, Klein, Neumann, Eszlinger, and Paschke (2001) showed that catecholamines inhibit adiponectin mRNA expression in 3T3-L1 adipocytes, and Fujioka et al. (2006) showed inhibition of adipoR1 and adipoR2 expression in cultured cardiomyocytes.

Few and contradictory studies analyzed the effects of adrenal glucocorticoids, catecholamines, and other adrenal hormones on the expression of adiponectin receptors. de Oliveira, Iwanaga-Carvalho, et al. (2011) showed that dexamethasone and adrenaline increased the expression of adipoR2 in 3T3-L1 adipocytes. In hepatocytes, dexamethasone has been shown to stimulate the adipoR2 promoter, an effect that can be abolished by treatment with the glucocorticoid receptor antagonist, RU486 (Sun, Han, Wang, & Chen, 2006). On the other hand, dexamethasone has been shown to decrease adipoR2 mRNA expression in human skeletal muscle (Jang et al., 2008).

Despite the fact that adiponectin, adipoR1, and adipoR2 are known to be expressed in the mouse adrenal gland (Li et al., 2009; Paschke et al., 2010), no direct effects of the mineralocorticoid aldosterone were observed on adiponectin system. Furuhashi et al. (2003) observed that rennin–angiotensin–aldosterone system blockade increased serum adiponectin concentrations with improvement in insulin sensitivity in patients with hypertension. Karbowska & Kochan (2005) and Pérez-de-Heredia et al. (2008) showed that exogenous administration of dehydroepiandrosterone (DHEA) stimulated adiponectin expression and secretion in rat adipose tissue, but we did not find studies about the effects of DHEA on adiponectin receptors.

3.2. Thyroid hormones

The thyroid hormones, thyroxin (T4) and triiodothyronine (T3), are critical in regulating the growth and differentiation of many tissues and organs, as well as energy homeostasis. Thyroid hormones have important effects on WAT, such as on cell proliferation and differentiation, and also affect adipocyte function, which influences lipid and carbohydrate metabolism, and can induce insulin resistance (Chidakel, Mentuccia, & Celi, 2005; Obregon, 2008; Pucci, Chiovato, & Pinchera, 2000).

Patients with hyperthyroidism are associated with weight loss, fat, and muscle mass reduction. Disturbances in carbohydrate metabolism are also frequent in these patients. Thyroid hormone excess is known to induce insulin resistance in liver and peripheral tissues (Dimitriadis & Raptis, 2001; Duntas, 2002). Hypothyroidism is also associated with metabolic alterations and changes in body composition. Patients with hypothyroidism often show decreased appetite with a slight weight gain (Pucci et al., 2000).

Both adiponectin and thyroid hormones are involved in energy homeostasis and weight changes (Ahima, Qi, Singhal, Jackson, & Scherer, 2006; Iglesias & Díez, 2007). The first evidence for a role of thyroid hormones in the regulation of adiponectin comes from a study showing increased adiponectin levels in mice exposed to cold. Yoda et al. (2001) showed that adiponectin could regulate body temperature and basal metabolic rate in response to change in environmental conditions, suggesting that adiponectin may have an important role in adaptive thermogenesis. They proposed that this interaction between thyroid hormones and adiponectin could be direct, through stimulating the synthesis of hormones, or indirectly, by improving insulin sensitivity during exposure to cold.

The animal studies show conflicting results. Aragão, Souza, Cabanelas, Oliveira, and Pazos-Moura (2007) showed in rats that circulating levels of adiponectin were higher in experimental hyperthyroidism but not changed in experimental hypothyroidism. The authors suggest that one possible mechanism for this increase could be related to reduction in visceral adipose tissue depot present in hyperthyroid rats. Another possible explanation suggested by the authors was associated with increased brown adipose tissue (BAT) present in the hyperthyroid rats. Fujimoto et al. (2005) showed in brown adipocytes culture that incubation with T4 increases adiponectin secretion and expression. Thus, BAT could contribute to the increase in serum adiponectin in hyperthyroid rats.

Kokkinos et al. (2007) using almost the same protocol to make hyper or hypothyroid rats found opposites results to Aragão group study: no change in plasma adiponectin concentration of hyperthyroid rats and increase in hypothyroid rats. The authors suggested that this increase may be an adaptive response to changes in body weight associated with thyroid hormone, as hypothyroid rats showed a reduction in body weight during the treatment.

In 2002, Fasshauer et al. showed that treatment of 3T3-L1 adipocytes with T3 for 16 h was not able to change adiponectin gene expression. In isolated explants from subcutaneous or visceral WAT of normal rats, Cabanelas et al. (2010) observed that T3 treatment was unable to modify the secretion of adiponectin in both depots. Nevertheless, T3 treatment had an inhibitory effect on gene expression of adiponectin in subcutaneous adipose tissue explants. This model highlights the fact that the regulation of adiponectin production may be dependent on the type of fat deposit. In a recent work, Luvizotto et al. (2011) showed that supraphysiological doses of T3 decreased serum adiponectin and adiponectin gene expression in adipose tissue of obese rats on caloric restriction.

Fernández-Real, López-Bermejo, Casamitjana, and Ricart (2003) found an association between the circulating concentration of adiponectin and free thyroxin (fT4) in humans, showing that fT4 was significant predictors of adiponectin concentrations. This association could be indirect through the modulation of insulin sensitivity, as thyroid hormones increase the gene expression of several genes involved in insulin production (Chidakel et al., 2005). These authors also proposed that circulating adiponectin could have some effect on thyroid hormone synthesis. Since then, some authors have investigated the influence of thyroid hormones on adiponectin, but with conflicting results.

Iglesias, Alvarez Fidalgo, Codoceo, and Díez (2003) evaluated the serum concentrations of adiponectin in patients with hypothyroidism and hyperthyroidism in both before therapy and after normalization of thyroid dysfunction, and found no significant differences in adiponectin circulating levels between thyroid dysfunctions group and control subjects. Nagasaki et al. (2005) also demonstrated that patients with hypothyroidism have similar levels of serum adiponectin, and these levels remained normal and unchanged after hormone replacement therapy. Similar results, no alteration on plasma adiponectin in patients with thyroid disfuntion, were found by others studies (Altinova et al., 2006; Caixàs et al., 2009; Chu et al., 2011; Loustis et al., 2006; Santini et al., 2004; Siemińska et al., 2010). Several studies also showed that patients with hyperthyroidism presented insulin resistance. However, no correlations between adiponectin, glucose, insulin, and HOMA-R were found (Caixàs et al., 2009; Chu et al., 2011; Iglesias et al., 2003). These observations suggest that insulin resistance associated with hyperthyroidism may not be due to reduction in adiponectin levels.

Unlike the previous studies, Yaturu, Prado, and Grimes (2004) and Hsieh and Wang (2008) evaluated patients with hyperthyroidism before and after antithyroid drugs or radioactive iodine therapy, to make these patients with a hypothyroid function. These authors observed decrease in adiponectin levels in the "hypothyroidism state" in relation to the "hyperthyroidism state." Also, adiponectin levels correlated positively with both fT4 and free triiodothyronine (fT3). Hyperthyroid patients showed elevated levels of insulin with increased values of the HOMA-R index. Insulin resistance index by HOMA-R did not correlate with adiponectin but correlated with fT4 and fT3 during "hyperthyroid state." But in these studies, no control group with normal thyroid function was used to compare. These results showed that thyroid hormones may influence the concentrations of serum adiponectin during the reestablishment of normal levels of T3 and T4 in patients with hyperthyroidism and perhaps this relationship is associated with weight gain or loss during treatment.

Other authors investigated patients with hyperthyroidism and euthyroid subjects. Serum adiponectin levels were higher in patients with hyperthyroidism compared with euthyroid (Bossowski et al., 2010; Saito et al., 2005; Sieminska et al., 2008; Yu et al., 2006), and adiponectin levels were closely associated with serum thyroid hormone and thyroid receptor antibodies (TRAb) (Saito et al., 2005; Sieminska et al., 2008). In contradiction to previous studies, these findings indicate that hyperthyroidism induces the production of adiponectin in adipose tissue.

Another approach to study the relationship between thyroid hormones and adiponectin is the evaluation of this interaction in patients with thyroid cancer. Botella-Carretero, Alvarez-Blasco, Sancho, and Escobar-Morreale (2006) and Lin, Huang, and Sheu (2010) studied patients with differentiated thyroid carcinoma, but they did not find any difference in serum adiponectin levels in these patients during T4 withdrawal (hypothyroid state), when compared to the values observed in control patients. However, Mitsiades et al. (2011) showed that patients with thyroid carcinoma had decreased plasma adiponectin levels compared with controls. They also found, using immunohistochemistry, that both adiponectin receptors (adipoR1 and adipoR2) are expressed in human cells of thyroid cancer. They also confirmed the expression of both adiponectin receptors in cell lines of thyroid cancer (BHP7 and SW579). These results suggest that the inverse association between circulating adiponectin levels and risk of thyroid cancer in humans could be attributed to indirect effects of adiponectin, as the treatment with recombinant adiponectin had no major effect on cell proliferation or cell death *in vitro*.

The studies here discussed are controversial and no definitive conclusions can be made. Because most thyroid disorders are autoimmune diseases, many other factors may be offsetting or modifying the levels of adiponectin. Further, differences between patient characteristics, duration of thyroid dysfunction, changes in energy expenditure, and nutrients may also explain these controversial results in humans.

3.3. Growth hormone

Growth hormone (GH) is an important anabolic and lipolytic hormone is associated with insulin resistance and hyperinsulinemia; this hormone antagonizes insulin action on insulin-sensitive tissues such as muscle, fat, and liver (Dominici et al., 2005; Frank, 2001; Hew et al., 1996). Adipose tissue is one of the major targets of GH, where GH receptors are expressed (Nam & Lobie, 2000). As with adiponectin, obesity is associated with reduced levels of GH (Nam & Marcus, 2000) and reduction in body weight recovers GH to normal levels (Tanaka et al., 1990).

Adiponectin and adiponectin receptors were found to be expressed in the anterior pituitary gland of rats and humans (Psilopanagioti, Papadaki, Kranioti, Alexandrides, & Varakis, 2009; Rodriguez-Pacheco et al., 2007; Steyn et al., 2009). Steyn et al. (2009) demonstrated that adiponectin stimulates GH secretion in pituitary cells of rats. However, the relationship between GH, adiponectin levels, and insulin sensitivity is not clear. In 2002, Fasshauer et al. showed that GH treatment did not change adiponectin mRNA levels in 3T3-L1 adipocytes culture cell. The effects *in vitro* GH on adiponectin levels have conflicting results: GH increased gene expression (Xu et al., 2004) and stimulated adiponectin secretion in 3T3-L1 cells (Wölfing et al., 2008); GH suppressed adiponectin secretion in cultured human adipose tissue as well as adiponectin secretion by adipose tissue of mice (Nilsson et al., 2005).

In humans, the results are also conflicting: GH replacement therapy in patients with growth hormone deficiency (GHD) did not change circulating adiponectin levels (Edén Engström, Burman, Holdstock, & Karlsson, 2003; Hana et al., 2004); acromegalic patients, with excess of GH, have higher adiponectin levels when compared to control subjects (Silha et al., 2003), and GHD individuals have lower adiponectin levels when compared to normal subjects (Joaquin et al., 2008; Lanes et al., 2006); patients with acromegaly were shown to have lower serum adiponectin levels, which was reversible with GH-lowering therapy (Lam et al., 2004).

With regard to the adiponectin receptors, Fasshauer et al. (2004) showed that GH treatment upregulated adipoR2 mRNA in culture cells of 3T3-L1 adipocytes. Qin and Tian (2010) showed in mice that chronic exogenous GH treatment upregulated hepatic adipoR2 but not adipoR1 in skeletal muscle and raised the serum HMW adiponectin and the ratio of HMW to total adiponectin. In these last cases, the increase in adipoR2 could be a compensatory mechanism to increase adiponectin action and restore insulin sensitivity, as the increase in adiponectin may be insufficient to counterbalance the decrease in body insulin sensitivity induced by GH. But Nilsson et al. (2005) showed decrease in adipoR2 mRNA expression following GH treatment in human subcutaneous adipose tissue culture.

If adiponectin levels may, in part, be correlated with the insulin-resistance related to GH is not yet clear. This relationship may not be direct but rather mediated by other factors related to GH and adiponectin, like adiposity. Chronic GH treatment decreased the body fat percentage and increased body weight in the study by Qin and Tian (2010), so the rise in serum adiponectin level may also be due directly to the change in body composition.

3.4. Sex hormones

The main sex steroids are androgens and estrogens. Androgens include dihydrotestosterone (DHT), DHEA, and DHEA sulfate (DHEA-S) and the two principal, testosterone and androstenedione. In women, androgens are converted into the female hormones estrogens in the ovaries, adrenal glands, and fat cells.

Adiponectin concentration in rodents and humans is higher in females compared with males possibly as a result of inhibition of adiponectin production by androgens (Combs et al., 2003; Nishizawa et al., 2002). Also, adiponectin as well as adipoR1 and adipoR2 receptors are expressed in the rat testis (Caminos et al., 2008) and ovary (Chabrolle et al., 2007) suggesting a direct effect of adiponectin in the gonads.

In animal models, adiponectin levels dynamically changed with sexual maturation and were regulated by estrogens or testosterone: surgical castration induced a rise in plasma adiponectin and testosterone injections reduced plasma adiponectin (Combs et al., 2003; Nishizawa et al., 2002). Nishizawa et al. (2002) observed that in mice and in 3T3-L1 adipocytes, testosterone treatment reduced the production of adiponectin in plasma or media without changing adiponectin mRNA and protein levels of in adipose cells, probably due to intracellular retention of the protein adiponectin (Xu et al., 2005). In adult female mice, ovariectomy induced an increase in plasma adiponectin levels above sham-group, and the use of estrogen implant had a suppressive effect on adiponectin levels. In old female mice, when estrogen level is low due to ovarian cessation, circulant adiponectin was significantly higher than in the young mice (Combs et al., 2003). Also, addition of estrogen in adipocyte culture cell media suppressed adiponectin mRNA and protein expression, and this effect was blocked by treatment with estrogen antagonist (Combs et al., 2003). According to Tworoger, Mantzoros, and Hankinson (2007), between the estrogens, the free estradiol is the one with strong inverse association with adiponectin levels. Contradicting these results, Nishizawa et al. (2002) found no significant difference in adiponectin plasma concentrations between ovariectomized and sham-operated, indicating no effect of estrogen on plasma adiponectin.

In humans, adiponectin levels decreased with the puberty progression in boys, and this decrease was related to testosterone and DHEA-S serum concentrations and may account for the gender differences seen in adults (Bottner et al., 2004). Tan et al. (2006) showed that human subcutaneous adipocytes treated with testosterone and the estrogen 17β-oestradiol had increase in adipoR1 and adipoR2 gene expression. While Chalvatzas et al. (2009) showed no effect of estrogen treatment on plasma adiponectin of women after ovariectomy.

During the menstrual cycle, sex hormonal changes are accompanied by changes of insulin sensitivity (Bruns & Kemnitz, 2004), and the hormonal fluctuation during the menstrual cycle could affect serum adiponectin levels that in turn could modulate insulin sensitivity. Kleiblová, Springer, and Haluzík (2006) and Dafopoulos, Sourlas, Kallitsaris, Pournaras, and Messinis (2009) observed that in women, total circulating adiponectin levels were stable throughout the menstrual cycle, not being significantly affected by hormonal changes during the cycle and despite the fact that both androgens and estrogens were showed able to modify circulating adiponectin levels. In this sense, in human fat cells SGBS, incubation with male and female serum led to a downregulation of adiponectin expression, with stronger inhibitory effect of the male serum, but the incubation of these cells with testosterone or estradiol did not influence adiponectin mRNA expression and secretion (Horenburg, Fischer-Posovszky, Debatin, & Wabitsch, 2008). This group suggested the existence of a serum factor that is differently regulated by sex steroids and subsequently causes the sex dimorphism in circulating adiponectin levels.

In women with polycystic ovary syndrome (PCOS), characterized by menstrual dysregulation and excess of androgen production, adiponectin concentrations levels are controversial: some studies report no difference in serum adiponectin concentration between PCOS women and controls (Spranger et al., 2004; Tan et al., 2006), while others report circulating adiponectin levels lower in women with PCOS compared with non-PCOS (Panidis et al., 2003; Wang et al., 2011; Yilmaz et al., 2009). The reduction of adiponectin levels, due to the direct or indirect effect of higher testosterone levels, could be the explanation for the insulin resistance frequently associated with PCOS. The group of Tan et al. (2006) observed an upregulation of adipoR1 and adipoR2 in subcutaneous and omental adipose tissue depots of women with PCOS compared to non-PCOS women. An upregulation in adiponectin receptors may sensitize adipocytes to the effects of a probably low circulanting adiponectin levels. But controversial results were found by Seow et al. (2009): adipoR1 and adipoR2 mRNA levels in adipose tissue of women with PCOS were significantly lower than in controls. Due to the negative correlation between adiponectin receptors' mRNA levels and insulin levels present in women with PCOS, this group suggested that adiponectin receptors expression was downregulated by hyperinsulinemia in PCOS, resulting in adiponectin resistance.

In the postmenopause, when endogenous estrogen is decreased, some studies showed women with higher plasma adiponectin levels compared with premenopausal women (Chalvatzas et al., 2009; Gavrila, Chan, et al., 2003, Gavrila, Peng, et al., 2003; Jürimäe and Jürimäe, 2007; Tamakoshi et al., 2007). This increase in adiponectin can be due to decrease in the estrogens levels. But no change in adiponectin levels between pre- and postmenopausal women was found by Ryan et al. (2003) and Hong et al. (2007).

Androgens decrease insulin sensitivity (Shoupe & Lobo, 1984), but it is not clear if this is due to the inhibitory effect of these hormones on adiponectin levels, as the results are so conflicting. The decrease in adiponectin levels, caused by androgens, may explain in part the high risks of diabetes and atherosclerosis in men compared to women.

3.5. Prolactin

Prolactin, hormone essential for lactation, secreted by pituitary gland and also by human adipose tissue (Brandebourg, Hugo, & Ben-Jonathan, 2007) was shown with suppressive effect on circulating adiponectin levels (Asai-Sato et al., 2006; Combs et al., 2003; Nilsson et al., 2005). In *in vitro* cultured human adipose tissue, this hormone increased adipoR1 mRNA expression (Nilsson et al., 2005). In mice with a dominant mutation in the collagenous domain of adiponectin that elevated circulating adiponectin values, the serum prolactin value was found increased (Combs et al., 2004).

4. CONCLUSION

In this chapter, we reviewed the effects of some diet components and hormonal factors involved in the regulation of adiponectin and adiponectin receptors in both experimental and human models. The conflicting results about the effects of food items on adiponectin levels are possibly due to differences in the amount and frequency of the food item consumed, the dietary context or nutritional status and the explanation for the discrepancies regarding the effects of hormones on adiponectin levels can be explained by interspecies differences in hormone action and secretion, different time of sample collecting, and changes in body composition. Also, adiponectin circulates in plasma as distinct forms and a reduction in the ratio of HMW to total adiponectin is related to insulin-sensitizing effects of adiponectin (Katsuki et al., 2006). So, measurement of total plasma adiponectin or HMW cannot reflect properly the changes in adiponectin. In addition, the circulating levels of adiponectin do not always correlate with their gene or protein expression in WAT, possibly due to a posttranscriptional mechanism. So, increase in mRNA or protein levels does not always mean increase in circulating and active adiponectin. Alteration in adiponectin receptors is also possible of modulation and a potential target for diet and hormones. In this review, it is clear that this field was not largely studied.

Even so, these results about nutritional and hormonal regulation of adiponectin suggest a potential therapeutic role for adiponectin in the treatment of diabetes mellitus, cardiovascular diseases, and other diseases. Potential strategies to treat disorders associated with adiponectin activity impairment must consider nutritional factors as a toll to modulate adiponectin and its receptors.

REFERENCES

Ahima, R. S., Qi, Y., Singhal, N. S., Jackson, M. B., & Scherer, P. E. (2006). Brain adipocytokine action and metabolic regulation. *Diabetes*, *55*(Suppl. 2), 145–554.

Al-Sarraj, T., Saadi, H., Calle, M. C., Volek, J. S., & Fernandez, M. L. (2009). Carbohydrate restriction, as a first-line dietary intervention, effectively reduces biomarkers of metabolic syndrome in Emiratiadults. *The Journal of Nutrition*, *139*, 1667–1676.

Altinova, A. E., Törüner, F. B., Aktürk, M., Bukan, N., Cakir, N., Ayvaz, G., et al. (2006). Adiponectin levels and cardiovascular risk factors in hypothyroidism and hyperthyroidism. *Clinical Endocrinology*, *65*(4), 530–535.

Alzamendi, A., Giovambattista, A., Raschia, A., Madrid, V., Gaillard, R. C., Rebolledo, O., et al. (2009). Fructose-rich diet-induced abdominal adipose tissue endocrine dysfunction in normal male rats. *Endocrine, 35*, 227–232.

Andrews, R. C., & Walker, B. R. (1999). Glucocorticoids and insulin resistance: Old hormones, new targets. *Clinical Science (London, England), 96*(5), 513–523.

Aragão, C. N., Souza, L. L., Cabanelas, A., Oliveira, K. J., & Pazos-Moura, C. C. (2007). Effect of experimental hypo and hyperthyroidism on serum adiponectin. *Metabolism, 56*(1), 6–11.

Arita, Y., Kihara, S., Ouchi, N., Takahashi, M., Maeda, K., Miyagawa, J., et al. (1999). Paradoxical decrease of an adipose-specific protein, adiponectin, in obesity. *Biochemical and Biophysical Research Communications, 257*(1), 79–83.

Asai-Sato, M., Okamoto, M., Endo, M., Yoshida, H., Murase, M., Ikeda, M., et al. (2006). Hypoadiponectinemia in lean lactating women: Prolactin inhibits adiponectin secretion from human adipocytes. *Endocrine Journal, 53*(4), 555–562.

Ashizawa, N., Takagi, M., Seto, S., Suzuki, S., & Yano, K. (2007). Serum adiponectin and leptin in a patient with Cushing's syndrome before and after adrenalectomy. *Internal Medicine, 46*, 383–385.

Azzi, A., Gysin, R., Kempná, P., Munteanu, A., Villacorta, L., Visarius, T., et al. (2004). Regulation of gene expression by alpha-tocopherol. *Biological Chemistry, 385*(7), 585–591.

Banga, A., Unal, R., Tripathi, P., Pokrovskaya, I., Owens, R. J., Kern, P. A., et al. (2009). Adiponectin translation is increased by the PPAR agonists pioglitazone and ω-3 fatty acids. *American Journal of Physiology. Endocrinology and Metabolism, 296*, E480–E489.

Barnea, M., Shamay, A., Stark, A. H., & Madar, Z. (2006). A high-fat diet has a tissue-specific effect on adiponectin and related enzyme expression. *Obesity, 14*, 2145–2153.

Basu, A., Du, M., Sanchez, K., Leyva, M. J., Betts, N. M., Blevins, S., et al. (2010). Green tea minimally affects biomarkers of inflammation in obese subjects with metabolic syndrome. *Nutrition, 27*, 206–213.

Bellia, A., Garcovich, C., D'Adamo, M., Lombardo, M., Tesauro, M., Donadel, G., et al. (2011). Serum 25-hydroxyvitamin D levels are inversely associated with systemic inflammation in severe obese subjects. *Internal and Emergency Medicine*, 2011 Mar 25.

Berg, A. H., Combs, T. P., Du, X., Brownlee, M., & Scherer, P. E. (2001). The adipocyte-secreted protein Acrp30 enhances hepatic insulin action. *Nature Medicine*, 7(8), 947–953.

Berk, E. S., Kovera, A. J., Boozer, C. N., Pi-Sunyer, F. X., Johnson, J. A., & Albu, J. B. (2005). Adiponectin levels during low- and high-fat eucaloric diets in lean and obese women. *Obesity Research, 13*, 1566–1571.

Berner, H. S., Lyngstadaas, S. P., Spahr, A., Monjo, M., Thommesen, L., Drevon, C. A., et al. (2004). Adiponectin and its receptors are expressed in bone-forming cells. *Bone, 35*, 842–849.

Beulens, J. W., Rimm, E. B., Hu, F. B., Hendriks, H. F., & Mukamal, K. J. (2008). Alcohol consumption, mediating biomarkers, and risk of type 2 diabetes among middle-aged women. *Diabetes Care, 31*, 2050–2055.

Beylot, M., Pinteur, C., & Peroni, O. (2006). Expression of the adiponectin receptors AdipoR1 and AdipoR2 in lean rats and in obese Zucker rats. *Metabolism, 55*(3), 396–401.

Blair, S. C., Caterson, I. D., & Cooney, G. J. (1996). Glucose tolerance and insulin secretion after adrenalectomy in mice made obese with gold thioglucose. *The Journal of Endocrinology, 148*, 391–398.

Bonnard, C., Durand, A., Vidal, H., & Rieusset, J. (2008). Changes in adiponectin, its receptors and AMPK activity in tissues of diet-induced diabetic mice. *Diabetes & Metabolism, 34*(1), 52–61.

Bossowski, A., Sawicka, B., Szalecki, M., Koput, A., Wysocka, J., & Zelazowska-Rutkowska, B. (2010). Analysis of serum adiponectin, resistin and leptin levels in children and adolescents with autoimmune thyroid disorders. *Journal of Pediatric Endocrinology & Metabolism, 23*(4), 369–377.

Botella-Carretero, J. I., Alvarez-Blasco, F., Sancho, J., & Escobar-Morreale, H. F. (2006). Effects of thyroid hormones on serum levels of adipokines as studied in patients with differentiated thyroid carcinoma during thyroxine withdrawal. *Thyroid, 16*(4), 397–402.

Bottner, A., Kratzsch, J., Muller, G., Kapellen, T. M., Bluher, S., Keller, E., et al. (2004). Gender differences of adiponectin levels develop during the progression of puberty and are related to serum androgen levels. *The Journal of Clinical Endocrinology and Metabolism, 89*, 4053–4061.

Boullu-Ciocca, S., Achard, V., Tassistro, V., Dutour, A., & Grino, M. (2008). Postnatal programming of glucocorticoid metabolism in rats modulates high-fat diet-induced regulation of visceral adipose tissue glucocorticoid exposure and sensitivity and adiponectin and proinflammatory adipokines gene expression in adulthood. *Diabetes, 57*, 669–677.

Brandebourg, T., Hugo, E., & Ben-Jonathan, N. (2007). Adipocyte prolactin: Regulation of release and putative functions. *Diabetes, Obesity & Metabolism, 9*(4), 464–476.

Brien, S. E., Ronksley, P. E., Turner, B. J., Mukamal, K. J., & Ghali, W. A. (2011). Effect of alcohol consumption on biological markers associated with risk of coronary heart disease: Systematic review and meta-analysis of interventional studies. *British Medical Journal, 342*, d636. http://dx.doi.org/10.1136/bmj.d636.

Bruns, C. M., & Kemnitz, J. W. (2004). Sex hormones, insulin sensitivity, and diabetes mellitus. *ILAR Journal, 45*(2), 160–169.

Bueno, A. A., Oyama, L. M., de Oliveira, C., Pisani, L. P., Ribeiro, E. B., Silveira, V. L., et al. (2008). Effects of different fatty acids and dietary lipids on adiponectin gene expression in 3T3-L1 cells and C57BL/6J mice adipose tissue. *Pflügers Archiv, 455*, 701–709.

Bullen, J. W., Jr., Bluher, S., Kelesidis, T., & Mantzoros, C. S. (2007). Regulation of adiponectin and its receptors in response to development of diet-induced obesity in mice. *American Journal of Physiology. Endocrinology and Metabolism, 292*, 1079–1086.

Cabanelas, A., Cordeiro, A., Santos Almeida, N. A., Monteiro de Paula, G. S., Coelho, V. M., Ortiga-Carvalho, T. M., et al. (2010). Effect of triiodothyronine on adiponectin expression and leptin release by white adipose tissue of normal rats. *Hormone and Metabolic Research, 42*(4), 254–260.

Caixàs, A., Tirado, R., Vendrell, J., Gallart, L., Megía, A., Simón, I., et al. (2009). Plasma visfatin concentrations increase in both hyper and hypothyroid subjects after normalization of thyroid function and are not related to insulin resistance, anthropometric or inflammatory parameters. *Clinical Endocrinology, 71*(5), 733–738.

Caminos, J. E., Nogueiras, R., Gallego, R., Bravo, S., Tovar, S., Garcia-Caballero, T., et al. (2005). Expression and regulation of adiponectin and receptor in human and rat placenta. *The Journal of Clinical Endocrinology and Metabolism, 90*, 4276–4286.

Caminos, J. E., Nogueiras, R., Gaytán, F., Pineda, R., González, C. R., Barreiro, M. L., et al. (2008). Novel expression and direct effects of adiponectin in the rat testis. *Endocrinology, 149*(7), 3390–3402.

Canoy, D., Wareham, N., Welch, A., Bingham, S., Luben, R., Day, N., et al. (2005). Plasma ascorbic acid concentrations and fat distribution in 19,068 British men and women in the European Prospective Investigation into Cancer and Nutrition Norfolk cohort study. *The American Journal of Clinical Nutrition, 82*, 1203–1209.

Cantorna, M. T., & Mahon, B. D. (2005). D-hormone and the immune system. *The Journal of Rheumatology, 76*, 11–20.

Cardillo, S., Seshadri, P., & Iqbal, N. (2006). The effects of a low carbohydrate versus low-fat diet on adipocytokines in severely obese adults: Three-year follow-up of a randomized trial. *European Review for Medical and Pharmacological Sciences*, *10*, 99–106.

Carvalho, C. R., Bueno, A. A., Mattos, A. M., Biz, C., de Oliveira, C., Pisani, L. P., et al. (2010). Fructose alters adiponectin, haptoglobin and angiotensinogen gene expression in 3T3-L1 adipocytes. *Nutrition Research*, *30*(9), 644–649.

Carvallo, L., Henriquez, B., Olate, J., van Wijnen, A. J., Lian, J. B., Stein, G. S., et al. (2007). The 1alpha,25-dihydroxy Vitamin D3 receptor preferentially recruits the coactivator SRC-1 during up-regulation of the osteocalcin gene. *The Journal of Steroid Biochemistry and Molecular Biology*, *103*(3–5), 420–424.

Chabrolle, C., Tosca, L., & Dupont, J. (2007). Regulation of adiponectin and its receptors in rat ovary by human chorionic gonadotrophin treatment and potential involvement of adiponectin in granulosa cell steroidogenesis. *Reproduction*, *133*, 719–731.

Chalvatzas, N., Dafopoulos, K., Kosmas, G., Kallitsaris, A., Pournaras, S., & Messinis, I. E. (2009). Effect of ovarian hormones on serum adiponectin and resistin concentrations. *Fertility and Sterility*, *91*(4), 1189–1194.

Chen, X., Sebastian, B. M., & Nagy, L. E. (2007). Chronic ethanol feeding to rats decreases adiponectin secretion by subcutaneous adipocytes. *American Journal of Physiology. Endocrinology and Metabolism*, *292*, 621–628.

Chidakel, A., Mentuccia, D., & Celi, F. S. (2005). Peripheral metabolism of thyroid hormone and glucose homeostasis. *Thyroid*, *15*(8), 899–903.

Chinetti, G., Zawadski, C., Fruchart, J. C., & Staels, B. (2004). Expression of adiponectin receptors in human macrophages and regulation by agonists of the nuclear receptors PPARa, PPARg, and LXR. *Biochemical and Biophysical Research Communications*, *314*, 151–158.

Cho, S. Y., Park, P. J., Shin, H. J., Kim, Y. K., Shin, D. W., Shin, E. S., et al. (2007). (-)-Catechin suppresses expression of Kruppel-like factor 7 and increases expression and secretion of adiponectin protein in 3T3-L1 cells. *American Journal of Physiology. Endocrinology and Metabolism*, *292*(4), 1166–1172.

Chu, C. H., Lam, H. C., Lee, J. K., Lu, C. C., Sun, C. C., Wang, M. C., et al. (2011). Hyperthyroidism-associated insulin resistance is not mediated by adiponectin levels. *Journal of Thyroid Research*, *2011*, 194721. http://dx.doi.org/10.4061/2011/194721.

Civitarese, A. E., Carling, S., Heilbronn, L. K., Hulver, M. H., Ukropcova, B., Deutsch, W. A., et al. (2007). Calorie restriction increases muscle mitochondrial biogenesis in healthy humans. *PLoS Medicine*, *4*, e76.

Combs, T. P., Berg, A. H., Obici, S., Scherer, P. E., & Rossetti, L. (2001). Endogenous glucose production is inhibited by the adipose-derived protein Acrp30. *The Journal of Clinical Investigation*, *108*, 1875–1881.

Combs, T. P., Berg, A. H., Rajala, M. W., Klebanov, S., Iyengar, P., Jimenez-Chillaron, J. C., et al. (2003). Sexual differentiation, pregnancy, calorie restriction, and aging affect the adipocyte-specific secretory protein adiponectin. *Diabetes*, *52* (2), 268–276.

Combs, T. P., Pajvani, U. B., Berg, A. H., Lin, Y., Jelicks, L. A., Laplante, M., et al. (2004). A transgenic mouse with a deletion in the collagenous domain of adiponectin displays elevated circulating adiponectin and improved insulin sensitivity. *Endocrinology*, *145*(1), 367–383.

Costacou, T., Ma, B., King, I. B., & Mayer-Davis, E. J. (2008). Plasma and dietary vitamin E in relation to insulin secretion and sensitivity. *Diabetes, Obesity & Metabolism*, *10*(3), 223–228.

Crawford, L. J., Peake, R., Price, S., Morris, T. C., & Irvine, A. E. (2010). Adiponectin is produced by lymphocytes and is a negative regulator of granulopoiesis. *Journal of Leukocyte Biology*, *88*(4), 807–811.

Dafopoulos, K., Sourlas, D., Kallitsaris, A., Pournaras, S., & Messinis, I. E. (2009). Blood ghrelin, resistin, and adiponectin concentrations during the normal menstrual cycle. *Fertility and Sterility*, *92*(4), 1389–1394.

de Luis, D. A., Aller, R., Izaola, O., Gonzalez-Sagrado, M., Bellioo, D., & Conde, R. (2007). Effects of a low-fat versus a low-carbohydrate diet on adipocytokines in obese adults. *Hormone Research*, *67*, 296–300.

de Oliveira, C., Iwanaga-Carvalho, C., Mota, J. F., Oyama, L. M., Ribeiro, E. B., & Oller do Nascimento, C. M. (2011). Effects of adrenal hormones on the expression of adiponectin and adiponectin receptors in adipose tissue, muscle and liver. *Steroids*, *76*(12), 1260–1267.

de Oliveira, C., Mattos, A. B. M., Biz, C., Oyama, L. M., Ribeiro, E. B., & Nascimento, C. M. (2011). High-fat diet and glucocorticoid treatment cause hyperglycemia associated with adiponectin receptor alterations. *Lipids in Health and Disease*, *10*, 11.

Degawa-Yamauchi, M., Moss, K. A., Bovenkerk, J. E., Shankar, S. S., Morrison, C. L., Lelliott, C. J., et al. (2005). Regulation of adiponectin expression in human adipocytes: Effects of adiposity, glucocorticoids, and tumor necrosis factor alpha. *Obesity Research*, *13*(4), 662–669.

Delaigle, A. M., Jonas, J. C., Bauche, I. B., Cornu, O., & Brichard, S. M. (2004). Induction of adiponectin in skeletal muscle by inflammatory cytokines: in vivo and in vitro studies. *Endocrinology*, *145*, 5589–5597.

Dimitriadis, G. D., & Raptis, S. A. (2001). Thyroid hormone excess and glucose intolerance. *Experimental and Clinical Endocrinology & Diabetes*, *109*(Suppl. 2), 225–239.

Dominici, F. P., Argentino, D. P., Muñoz, M. C., Miquet, J. G., Sotelo, A. I., & Turyn, D. (2005). Influence of the crosstalk between growth hormone and insulin signalling on the modulation of insulin sensitivity. *Growth Hormone & IGF Research*, *15*(5), 324–336.

Donovan, D. S., Solomon, C. G., Seely, E. W., Williams, G. H., & Simonson, D. C. (1993). Effect of sodium intake on insulin sensitivity. *The American Journal of Physiology*, *264*(★★★5 Pt 1), 730–734.

Duntas, L. H. (2002). Thyroid disease and lipids. *Thyroid*, *12*(4), 287–293.

Edén Engström, B., Burman, P., Holdstock, C., & Karlsson, F. A. (2003). Effects of growth hormone (GH) on ghrelin, leptin, and adiponectin in GH-deficient patients. *The Journal of Clinical Endocrinology and Metabolism*, *88*, 5193–5198.

Elliott, S. S., Keim, N. L., Stern, J. S., Teff, K., & Havel, P. J. (2002). Fructose, weight gain, and the insulin resistance syndrome. *The American Journal of Clinical Nutrition*, *76*, 911–922.

Fallo, F., Scarda, A., Sonino, N., Paoletta, A., Boscaro, M., Pagano, C., et al. (2004). Effect of glucocorticoids on adiponectin: A study in healthy subjects and in Cushing's syndrome. *European Journal of Endocrinology*, *150*(3), 339–344.

Fasshauer, M., Klein, J., Kralisch, S., Klier, M., Lössner, U., Blüher, M., et al. (2004). Growth hormone is a positive regulator of adiponectin receptor 2 in 3T3-L1 adipocytes. *FEBS Letters*, *558*(1–3), 27–32.

Fasshauer, M., Klein, J., Neumann, S., Eszlinger, M., & Paschke, R. (2001). Adiponectin gene expression is inhibited by β-adrenergic stimulation via protein kinase A in 3T3-L1 adipocytes. *FEBS Letters*, *507*, 142–146.

Fasshauer, M., Klein, J., Neumann, S., Eszlinger, M., & Paschke, R. (2002). Hormonal regulation of adiponectin gene expression in 3T3-L1 adipocytes. *Biochemical and Biophysical Research Communications*, *290*(3), 1084–1089.

Fernández-Real, J. M., López-Bermejo, A., Casamitjana, R., & Ricart, W. (2003). Novel interactions of adiponectin with the endocrine system and inflammatory parameters. *The Journal of Clinical Endocrinology and Metabolism*, *88*(6), 2714–2718.

Fernández-Real, J. M., Vendrell, J., & Ricart, W. (2005). Circulating adiponectin and plasma fatty acid profile. *Clinical Chemistry*, *51*, 603–609.

Flachs, P., Mohamed-Ali, V., Horakova, O., Rossmeisl, M., Hosseinzadeh-Attar, M. J., Hensler, M., et al. (2006). Polyunsaturated fatty acids of marine origin induce adiponectin in mice fed a high-fat diet. *Diabetologia, 49*(2), 394–397.

Frank, S. J. (2001). Growth hormone signalling and its regulation: preventing too much of a good thing. *Growth Horm IGF Res, 11*(4), 201–212.

Freedman, M. R., Stern, J. S., Reaven, G. M., & Mondon, C. E. (1986). Effect of adrenalectomy on in vivo glucose metabolism in insulin resistant Zucker obese rats. *Hormone and Metabolic Research, 18*, 296–298.

Fruebis, J., Tsao, T. S., Javorschi, S., Ebbets-Reed, D., Erickson, M. R., Yen, F. T., et al. (2001). Proteolytic cleavage product of 30-kDa adipocyte complement-related protein increases fatty acid oxidation in muscle and causes weight loss in mice. *Proceedings of the National Academy of Sciences of the United States of America, 98*(4), 2005–2010.

Fu, L., Isobe, K., Zeng, Q., Suzukawa, K., Takekoshi, K., & Kawakami, Y. (2007). Beta adrenoceptor agonists downregulate adiponectin, but upregulate adiponectin receptor 2 and tumor necrosis factor-alpha expression in adipocytes. *European Journal of Pharmacology, 569*(1–2), 155–162.

Fuenmayor, N., Moreira, E., & Cubeddu, L. X. (1998). Salt sensitivity is associated with insulin resistance in essential hypertension. *American Journal of Hypertension, 11*(4 Pt 1), 397–402.

Fujimoto, N., Matsuo, N., Sumiyoshi, H., Yamaguchi, K., Saikawa, T., Yoshimatsu, H., et al. (2005). Adiponectin is expressed in the brown adipose tissue and surrounding immature tissues in mouse embryos. *Biochimica et Biophysica Acta, 1731*(1), 1–12.

Fujioka, D., Kawabata, K., Saito, Y., Kobayashi, T., Nakamura, T., Kodama, Y., et al. (2006). Role of adiponectin receptors in endothelin-induced cellular hypertrophy in cultured cardiomyocytes and their expression in infarcted heart. *American Journal of Physiology. Heart and Circulatory Physiology, 290*(6), 2409–2416.

Fukumitsu, S., Aida, K., Ueno, N., Ozawa, S., Takahashi, Y., & Kobori, M. (2008). Flaxseed lignan attenuates high-fat diet-induced fat accumulation and induces adiponectin expression in mice. *The British Journal of Nutrition, 100*, 669–676.

Furuhashi, M., Ura, N., Higashiura, K., Murakami, H., Tanaka, M., Moniwa, N., et al. (2003). Blockade of the renin-angiotensin system increases adiponectin concentrations in patients with essential hypertension. *Hypertension, 42*(1), 76–81.

Gannagé-Yared, M. H., Chedid, R., Khalife, S., Azzi, E., Zoghbi, F., & Halaby, G. (2009). Vitamin D in relation to metabolic risk factors, insulin sensitivity and adiponectin in a young Middle-Eastern population. *European Journal of Endocrinology, 160*(6), 965–971.

Gavrila, A., Chan, J. L., Yiannakouris, N., Kontogianni, M., Miller, L. C., Orlova, C., et al. (2003). Serum adiponectin levels are inversely associated with overall and central fat distribution but are not directly regulated by acute fasting or leptin administration in humans: Cross-sectional and interventional studies. *The Journal of Clinical Endocrinology and Metabolism, 88*, 4823–4831.

Gavrila, A., Peng, C. K., Chan, J. L., Mietus, J. E., Goldberger, A. L., & Mantzoros, C. S. (2003). Diurnal and ultradian dynamics of serum adiponectin in healthy men: Comparison with leptin, circulating soluble leptin receptor, and cortisol patterns. *Journal of Clinical Endocrinology and Metabolism, 88*, 2838–2843.

Gray, B., Swick, J., & Ronnenberg, A. G. (2011). Vitamin E and adiponectin: Proposed mechanism for vitamin E-induced improvement in insulin sensitivity. *Nutrition Reviews, 69*(3), 155–161.

Ha, S. K., Kim, J., & Chae, C. (2011). Role of AMP-activated protein kinase and adiponectin during development of hepatic steatosis in high-fat diet-induced obesity in rats. *Journal of Comparative Pathology, 145*(1), 88–94.

Hada, Y., Yamauchi, T., Waki, H., Tsuchida, A., Hara, K., Yago, H., et al. (2007). Selective purification and characterization of adiponectin multimer species from human plasma. *Biochemical and Biophysical Research Communications, 356*, 487–493.

Halleux, C. M., Takahashi, M., Delporte, M. L., Detry, R., Funahashi, T., Matsuzawa, Y., et al. (2001). Secretion of adiponectin and regulation of apM1 gene expression in human visceral adipose tissue. *Biochemical and Biophysical Research Communications, 288*(5), 1102–1107.

Hana, V., Silha, J. V., Justova, V., Lacinova, Z., Stepan, J. J., & Murphy, L. J. (2004). The effects of GH replacement in adult GH-deficient patients: Changes in body composition without concomitant changes in the adipokines and insulin resistance. *Clinical Endocrinology, 60*(4), 442–450.

Hernandez-Morante, J. J., Milagro, F. I., Larque, E., Lujan, J., Martinez, J. A., Zamora, S., et al. (2007). Relationship among adiponectin, adiponectin gene expression and fatty acids composition in morbidly obese patients. *Obesity Surgery, 17*, 516–524.

Hew, F. L., Koschmann, M., Christopher, M., Rantzau, C., Vaag, A., Ward, G., et al. (1996). Insulin resistance in growth hormone-deficient adults: Defects in glucose utilization and glycogen synthase activity. *The Journal of Clinical Endocrinology and Metabolism, 81*, 555–564.

Higdon, J. V., & Frei, B. (2006). Coffee and health: A review of recent human research. *Critical Reviews in Food Science and Nutrition, 46*(2), 101–123.

Hong, S. C., Yoo, S. W., Cho, G. J., Kim, T., Hur, J. Y., Park, Y. K., et al. (2007). Correlation between estrogens and serum adipocytokines in premenopausal and postmenopausal women. *Menopause, 14*(5), 835–840.

Horenburg, S., Fischer-Posovszky, P., Debatin, K. M., & Wabitsch, M. (2008). Influence of sex hormones on adiponectin expression in human adipocytes. *Hormone and Metabolic Research, 40*(11), 779–786.

Hsieh, C. J., & Wang, P. W. (2008). Serum concentrations of adiponectin in patients with hyperthyroidism before and after control of thyroid function. *Endocrine Journal, 55*(3), 489–494.

Hug, C., Wang, J., Ahmad, N. S., Bogan, J. S., Tsao, T. S., & Lodish, H. F. (2004). T-cadherin is a receptor for hexameric and high-molecular-weight forms of Acrp30/adiponectin. *Proceedings of the National Academy of Sciences of the United States of America, 101*(28), 10308–10313.

Iglesias, P., Alvarez Fidalgo, P., Codoceo, R., & Díez, J. J. (2003). Serum concentrations of adipocytokines in patients with hyperthyroidism and hypothyroidism before and after control of thyroid function. *Clinical Endocrinology, 59*(5), 621–629.

Iglesias, P., & Díez, J. J. (2007). Influence of thyroid dysfunction on serum concentrations of adipocytokines. *Cytokine, 40*(2), 61–70.

Imatoh, T., Tanihara, S., Miyazaki, M., Momose, Y., Uryu, Y., & Une, H. (2011). Coffee consumption but not green tea consumption is associated with adiponectin levels in Japanese males. *European Journal of Nutrition, 50*(4), 279–284.

Imhof, A., Plamper, I., Maier, S., Trischler, G., & Koenig, W. I. (2009). Effect of drinking on adiponectin in healthy men and women: A randomized intervention study of water, ethanol, red wine, and beer with or without alcohol. *Diabetes Care, 32*, 1101–1103.

Ingelson, E., Arnlov, J., Zethelius, B., Vasan, R. S., Flyvbjerg, A., Frystyk, J., et al. (2009). Associations of serum adiponectin with skeletal muscle morphology and insulin sensitivity. *Endocrinology and Metabolism, 94*, 953–957.

Iso, H., Date, C., Wakai, K., Fukui, M., & Tamakoshi, A. (2006). The relationship between green tea and total caffeine intake and risk for self-reported type 2 diabetes among Japanese adults. *Annals of Internal Medicine, 144*(8), 554–562.

Itoh, M., Suganami, T., Satoh, N., Tanimoto-Koyama, K., Yuan, X., Tanaka, M., et al. (2007). Increased adiponectin secretion by highly purified eicosapentaenoic acid in rodent models of obesity and human obese subjects. *Arteriosclerosis, Thrombosis, and Vascular Biology, 27*, 1918–1925.

Iwaki, M., Matsuda, M., Maeda, N., Funahashi, T., Matsuzawa, Y., Makishima, M., et al. (2003). Induction of adiponectin, a fat-derived antidiabetic and antiatherogenic factor, by nuclear receptors. *Diabetes, 52*(7), 1655–1663.

Jang, C., Inder, W. J., Obeyesekere, V. R., & Alford, F. P. (2008). Adiponectin, skeletal muscle adiponectin receptor expression and insulin resistance following dexamethasone. *Clinical Endocrinology, 69*(5), 745–750.

Joaquin, C., Aguilera, E., Granada, M. L., Pastor, M. C., Salinas, I., Alonso, N., et al. (2008). Effects of GH treatment in GH-deficient adults on adiponectin, leptin and pregnancy-associated plasma protein-A. *European Journal of Endocrinology, 158*(4), 483–490.

Johnston, C. S., Beezhold, B. L., Mostow, B., & Swan, P. D. (2007). Plasma vitamin C is inversely related to body mass index and waist circumference but not to plasma adiponectin in nonsmoking adults. *The Journal of Nutrition, 137*(7), 1757–1762.

Joosten, M. M., Beulens, J. W., Kersten, S., & Hendriks, H. F. (2008). Moderate alcohol consumption increases insulin sensitivity and ADIPOQ expression in postmenopausal women: A randomised, crossover trial. *Diabetologia, 51*, 1375–1381.

Joosten, M. M., Chiuve, S. E., Mukamal, K. J., Hu, F. B., Hendriks, H. F., & Rimm, E. B. (2011). Changes in alcohol consumption and subsequent risk of type 2 diabetes in men. *Diabetes, 60*, 74–79.

Jürimäe, J., & Jürimäe, T. (2007). Plasma adiponectin concentration in healthy pre and post-menopausal women: Relationship with body composition, bone mineral, and metabolic variables. *American Journal of Physiology. Endocrinology and Metabolism, 293*(1), 42–47.

Kadowaki, T., & Yamauchi, T. (2005). Adiponectin and adiponectin receptors. *Endocrine Reviews, 26*, 439–451.

Kamari, Y., Shimoni, N., Koren, F., Peleg, E., Sharabi, Y., & Grossman, E. (2010). High-salt diet increases plasma adiponectin levels independent of blood pressure in hypertensive rats: The role of the renin-angiotensin-aldosterone system. *Journal of Hypertension, 28*(1), 95–101.

Karbowska, J., & Kochan, Z. (2005). Effect of DHEA on endocrine functions of adipose tissue, the involvement of PPAR gamma. *Biochemical Pharmacology, 70*(2), 249–257.

Katsuki, A., Suematsu, M., Gabazza, E. C., Murashima, S., Nakatani, K., Togashi, K., et al. (2006). Decreased high-molecular weight adiponectin-to-total adiponectin ratio in sera is associated with insulin resistance in Japanese metabolically obese, normal-weight men with normal glucose tolerance. *Diabetes Care, 29*(10), 2327–2328.

Kawamoto, R., Kohara, K., Tabara, Y., Miki, T., Ohtsuka, N., Kusunoki, T., et al. (2009). Alcohol consumption is associated with decreased insulin resistance independent of body mass index in Japanese community-dwelling men. *The Tohoku Journal of Experimental Medicine, 218*, 331–337.

Kempf, K., Herder, C., Erlund, I., Kolb, H., Martin, S., Carstensen, M., et al. (2010). Effects of coffee consumption on subclinical inflammation and other risk factors for type 2 diabetes: A clinical trial. *The American Journal of Clinical Nutrition, 91*(4), 950–957.

Keogh, J. B., Brinkworth, G. D., & Clifton, P. M. (2007). Effects of weight loss on a low carbohydrate diet on flow-mediated dilatation, adhesion molecules and adiponectin. *The British Journal of Nutrition, 98*, 852–859.

Keogh, J. B., Brinkworth, G. D., Noakes, M., Belobrajdic, D. P., Buckley, J. D., & Clifton, P. M. (2008). Effects of weight loss from a very-lowcarbohydrate diet on endothelial function and markers of cardiovascular disease risk in subjects with abdominal obesity. *The American Journal of Clinical Nutrition, 87*, 567–576.

Kharroubi, I., Rasschaert, J., Eizirik, D. L., & Cnop, M. (2003). Expression of adiponectin receptors in pancreatic beta cells. *Biochemical and Biophysical Research Communications, 312*, 1118–1122.

Kleiblová, P., Springer, D., & Haluzík, M. (2006). The influence of hormonal changes during menstrual cycle on serum adiponectin concentrations in healthy women. *Physiological Research, 55*(6), 661–666.

Kokkinos, A., Mourouzis, I., Kyriaki, D., Pantos, C., Katsilambros, N., & Cokkinos, D. V. (2007). Possible implications of leptin, adiponectin and ghrelin in the regulation of energy homeostasis by thyroid hormone. *Endocrine, 32*(1), 30–32.

Koppes, L. L., Dekker, J. M., Hendriks, H. F., Bouter, L. M., & Heine, R. J. (2005). Moderate alcohol consumption lowers the risk of type 2 diabetes: A meta-analysis of prospective observational studies. *Diabetes Care, 28*, 719–725.

Kotani, K., Fujiwara, S., Hamada, T., Tsuzaki, K., & Sakane, N. (2008). Coffee consumption is associated with higher plasma adiponectin concentrations in women with or without type 2 diabetes: Response to Williams et al. *Diabetes Care, 31*(5), 46–47.

Krikken, J. A., Dallinga-Thie, G. M., Navis, G., & Dullaart, R. P. (2012). Short term dietary sodium restriction decreases HDL cholesterol, apolipoprotein A-I and high molecular weight adiponectin in healthy young men: Relationships with renal hemodynamics and RAAS activation. *Nutrition, Metabolism, and Cardiovascular Diseases, 22*, 35–41.

Lam, K. S., Xu, A., Tan, K. C., Wong, L. C., Tiu, S. C., & Tam, S. (2004). Serum adiponectin is reduced in acromegaly and normalized after correction of growth hormone excess. *The Journal of Clinical Endocrinology and Metabolism, 89*(11), 5448–5453.

Landrier, J. F., Gouranton, E., El Yazidi, C., Malezet, C., Balaguer, P., Borel, P., et al. (2009). Adiponectin expression is induced by vitamin E via a peroxisome proliferator-activated receptor gamma-dependent mechanism. *Endocrinology, 150*(12), 5318–5325.

Lanes, R., Soros, A., Gunczler, P., Paoli, M., Carrillo, E., Villaroel, O., et al. (2006). Growth hormone deficiency, low levels of adiponectin, and unfavorable plasma lipid and lipoproteins. *The Journal of Pediatrics, 149*(3), 324–329.

Lara-Castro, C., Luo, N., Wallace, P., Klein, R. L., & Garvey, W. T. (2006). Adiponectin multimeric complexes and the metabolic syndrome trait cluster. *Diabetes, 55*, 249–259.

Lee, M. S., Kim, C. T., & Kim, Y. (2009). Green tea (-)-epigallocatechin-3-gallate reduces body weight with regulation of multiple genes expression in adipose tissue of diet-induced obese mice. *Annals of Nutrition & Metabolism, 54*(2), 151–157.

Lee, S., Lee, D. K., Choi, E., & Lee, J. W. (2005). Identification of a functional vitamin D response element in themurine Insig-2 promoter and its potential role in the differentiation of 3T3-L1 preadipocytes. *Molecular Endocrinology, 19*, 399–408.

Lee, N. K., Sowa, H., Hinoi, E., Ferron, M., Ahn, J. D., Confavreux, C., et al. (2007). Endocrine regulation of energy metabolism by the skeleton. *Cell, 130*(3), 456–469.

Lely, A. T., Krikken, J. A., Bakker, S. J., Boomsma, F., Dullaart, R. P., Wolffenbuttel, B. H., et al. (2007). Low dietary sodium and exogenous angiotensin II infusion decrease plasma adiponectin concentrations in healthy men. *The Journal of Clinical Endocrinology and Metabolism, 92*(5), 1821–1826.

Lewandowski, K., Szosland, K., & Lewinski, A. (2006). Short-term dexamethasone administration does not alter serum adiponectin or resistin concentrations in overweight and obese subjects despite an increase in insulin resistance. *Clinical Endocrinology, 65*, 551–552.

Li, R. W., Douglas, T. D., Maiyoh, G. K., Adeli, K., & Theriault, A. G. (2006). Green tea leaf extract improves lipid and glucose homeostasis in a fructose-fed insulin-resistant hamster model. *Journal of Ethnopharmacology, 104*(1–2), 24–31.

Li, P., Sun, F., Cao, H. M., Ma, Q. Y., Pan, C. M., Ma, J. H., et al. (2009). Expression of adiponectin receptors in mouse adrenal glands and the adrenocortical Y-1 cell line: Adiponectin regulates steroidogenesis. *Biochemical and Biophysical Research Communications, 390*(4), 1208–1213.

Liang, X., Hu, M., Rogers, C. Q., Shen, Z., & You, M. (2011). Role of SIRT1-FoxO1 signaling in dietary saturated fat-dependent upregulation of liver adiponectin receptor 2 in ethanol-administered mice. *Antioxidants & Redox Signaling, 15*(2), 425–435.

Libe, R., Morpurgo, P. S., Cappiello, V., Maffini, A., Bondioni, S., Locatelli, M., et al. (2005). Ghrelin and adiponectin in patients with Cushing's disease before and after transsphenoidal surgery. *Clinical Endocrinology, 62*, 30–36.

Lihn, A. S., Bruun, J. M., He, G., Pedersen, S. B., Jensen, P. F., & Richelsen, B. (2004). Lower expression of adiponectin mRNA in visceral adipose tissue in lean and obese subjects. *Molecular and Cellular Endocrinology, 219*, 9–15.

Lihn, A. S., Pedersen, S. B., & Richelsen, B. (2005). Adiponectin: Action, regulation and association to insulin sensitivity. *Obesity Reviews, 6*, 13–21.

Lin, S. Y., Huang, S. C., & Sheu, W. H. (2010). Circulating adiponectin concentrations were related to free thyroxine levels in thyroid cancer patients after thyroid hormone withdrawal. *Metabolism, 59*(2), 195–199.

Liu, E., Meigs, J. B., Pittas, A. G., McKeown, N. M., Economos, C. D., Booth, S. L., et al. (2009). Plasma 25-hydroxyvitamin d is associated with markers of the insulin resistant phenotype in nondiabetic adults. *The Journal of Nutrition, 139*(2), 329–334.

Loustis, K., Pontikides, N., Koliakos, G., Constantinidis, T. H., Papadopoulou, F., & Krassas, G. E. (2006). Serum cytokines profile in Graves' hyperthyroidism before and after restoration of thyroid function: Correlation with different anthropometric parameters. In: *Proceedings of the 31st European thyroid association meeting, Naples* P143, 96.

Luvizotto, R. A., Síbio, M. T., Olímpio, R. M., Nascimento, A. F., Lima-Leopoldo, A. P., Leopoldo, A. S., et al. (2011). Supraphysiological triiodothyronine doses diminish leptin and adiponectin gene expression, but do not alter resistin expression in calorie restricted obese rats. *Hormone and Metabolic Research, 43*(7), 452–457.

Ma, H., Gomez, V., Lu, L., Yang, X., Wu, X., & Xiao, S. Y. (2009). Expression of adiponectin and its receptors in livers of morbidly obese patients with non-alcoholic fatty liver disease. *Journal of Gastroenterology and Hepatology, 24*(2), 233–237.

Maeda, K., Okubo, K., Shimomura, I., Funahashi, T., Matsuzawa, Y., & Matsubara, K. (1996). cDNA cloning and expression of a novel adipose specific collagen-like factor, apM1 (AdiPose Most abundant Gene transcript 1). *Biochemical and Biophysical Research Communications, 221*(2), 286–289.

Makimura, H., Mizuno, T. M., Bergen, H., & Mobbs, C. V. (2002). Adiponectin is stimulated by adrenalectomy in ob/ob mice and is highly correlated with resistin mRNA. *American Journal of Physiology. Endocrinology and Metabolism, 283*, E1266–E1271.

Matsunami, T., Sato, Y., Ariga, S., Sato, T., Shimomura, T., Kashimura, H., et al. (2011). Regulation of synthesis and oxidation of fatty acids by adiponectin receptors (AdipoR1/R2) and insulin receptor substrate isoforms (IRS-1/-2) of the liver in a nonalcoholic steatohepatitis animal model. *Metabolism, 60*, 805–814.

Meeprom, A., Sompong, W., Suwannaphet, W., Yibchok-anun, S., & Adisakwattana, S. (2011). Grape seed extract supplementation prevents high-fructose diet-induced insulin resistance in rats by improving insulin and adiponectin signalling pathways. *The British Journal of Nutrition, 106*, 1173–1181.

Mezey, E. (1998). Dietary fat and alcoholic liver disease. *Hepatology, 28*, 901–905.

Mitsiades, N., Pazaitou-Panayiotou, K., Aronis, K. N., Moon, H. S., Chamberland, J. P., Liu, X., et al. (2011). Circulating adiponectin is inversely associated with risk of thyroid cancer: in vivo and in vitro studies. *The Journal of Clinical Endocrinology and Metabolism, 96*, E2023–E2028. http://dx.doi.org/10.1210/jc.2010-1908.

Montonen, J., Knekt, P., Järvinen, R., & Reunanen, A. (2004). Dietary antioxidant intake and risk of type 2 diabetes. *Diabetes Care, 27*(2), 362–366.

Motoshima, H., Wu, X., Mahadev, K., & Goldstein, B. J. (2004). Adiponectin suppresses proliferation and superoxide generation and enhances eNOS activity in endothelial cells treated with oxidized LDL. *Biochemical and Biophysical Research Communications, 315*, 264–271.

Mullen, K. L., Tishinsky, J. M., Robinson, L. E., & Dyck, D. J. (2010). Skeletal muscle inflammation is not responsible for the rapid impairment in adiponectin response with high-fat feeding in rats. *American Journal of Physiology. Regulatory, Integrative and Comparative Physiology, 299*, 500–508.

Müssig, K., Remer, T., & Maser-Gluth, C. (2010). Brief review: Glucocorticoid excretion in obesity. *The Journal of Steroid Biochemistry and Molecular Biology*, *121*(3–5), 589–593.

Nagasaki, T., Inaba, M., Hiura, Y., Tahara, H., Kumeda, Y., Shirakawa, K., et al. (2005). Plasma levels of adiponectin and soluble thrombomodulin in hypothyroid patients with normal thyroid function following levothyroxine replacement therapy. *Biomedicine & Pharmacotherapy*, *59*(10), 571–577.

Nakandakare, E. R., Charf, A. M., Santos, F. C., Nunes, V. S., Ortega, K., Lottenberg, A. M., et al. (2008). Dietary salt restriction increases plasma lipoprotein and inflammatory marker concentrations in hypertensive patients. *Atherosclerosis*, *200* (2), 410–416.

Nakano, Y., Tobe, T., Choi-Miura, N. H., Mazda, T., & Tonita, M. (1996). Isolation and characterization of GBP28, a novel gelatine-binding protein purified from human plasma. *Journal of Biochemistry*, *120*, 803–812.

Nam, S. Y., & Lobie, P. E. (2000). The mechanism of effect of growth hormone on preadipocyte and adipocyte function. *Obesity Reviews*, *1*(2), 73–86.

Nam, S. Y., & Marcus, C. (2000). Growth hormone and adipocyte function in obesity. *Hormone Research*, *53*(Suppl. 1), 87–97.

Nanji, A. A., Sadrzadeh, S. M., Yang, E. K., Fogt, F., Meydani, M., & Dannenberg, A. J. (1995). Dietary saturated fatty acids: A novel treatment for alcoholic liver disease. *Gastroenterology*, *109*, 547–554.

Neschen, S., Morino, K., Dong, J., Wang-Fischer, Y., Cline, G. W., Romanelli, A. J., et al. (2007). ω-3 fatty acids preserve insulin sensitivity in vivo in a peroxisome proliferator-activated receptor-alpha-dependent manner. *Diabetes*, *56*, 1034–1041.

Neschen, S., Morino, K., Rossbacher, J. C., Pongratz, R. L., Cline, G. W., Sono, S., et al. (2006). Fish oil regulates adiponectin secretion by a peroxisome proliferator-activated receptor-gamma-dependent mechanism in mice. *Diabetes*, *55*, 924–928.

Nilsson, L., Binart, N., Bohlooly-Y, M., Bramnert, M., Egecioglu, E., Kindblom, J., et al. (2005). Prolactin and growth hormone regulate adiponectin secretion and receptor expression in adipose tissue. *Biochemical and Biophysical Research Communications*, *331*(4), 1120–1126.

Nimitphong, H., Chanprasertyothin, S., Jongjaroenprasert, W., & Ongphiphadhanakul, B. (2009). The association between vitamin D status and circulating adiponectin independent of adiposity in subjects with abnormal glucose tolerance. *Endocrine*, *36*(2), 205–210.

Nishizawa, H., Shimomura, I., Kishida, K., Maeda, N., Kuriyama, H., Nagaretani, H., et al. (2002). Androgens decrease plasma adiponectin, an insulin-sensitizing adipocyte-derived protein. *Diabetes*, *51*(9), 2734–2741.

Oana, F., Takeda, H., Matsuzawa, A., Akahane, S., Isaji, M., & Akahane, M. (2005). Adiponectin receptor 2 expression in liver and insulin resistance in db/db mice given a beta3-adrenoceptor agonist. *European Journal of Pharmacology*, *518*(1), 71–76.

Obregon, M. J. (2008). Thyroid hormone and adipocyte differentiation. *Thyroid*, *18*(2), 185–195.

Ouchi, N., Kihara, S., Arita, Y., Nishida, M., Matsuyama, A., Okamoto, Y., et al. (2001). Adipocyte-derived plasma protein, adiponectin, suppresses lipid accumulation and class A scavenger receptor expression inhuman monocyte-derived macrophages. *Circulation*, *103*, 1057–1063.

Paniagua, J. A., Gallego de la Sacristana, A., Romero, I., Vidal-Puig, A., Latre, J. M., Sanchez, E., et al. (2007). Monounsaturated fat-rich diet prevents central body fat distribution and decreases postprandial adiponectin expression induced by a carbohydrate-rich diet in insulin-resistant subjects. *Diabetes Care*, *30*, 1717–1723.

Panidis, D., Kourtis, A., Farmakiotis, D., Mouslech, T., Rousso, D., & Koliakos, G. (2003). Serum adiponectin levels in women with polycystic ovary syndrome. *Human Reproduction*, *18*(9), 1790–1796.

Paolisso, G., D'Amore, A., Giugliano, D., Ceriello, A., Varricchio, M., & D'Onofrio, F. (1993). Pharmacologic doses of vitamin E improve insulin action in healthy subjects and non-insulin-dependent diabetic patients. *The American Journal of Clinical Nutrition*, *57*(5), 650–656.

Paschke, L., Zemleduch, T., Rucinski, M., Ziolkowska, A., Szyszka, M., & Malendowicz, L. K. (2010). Adiponectin and adiponectin receptor system in the rat adrenal gland: Ontogenetic and physiologic regulation, and its involvement in regulating adrenocortical growth and steroidogenesis. *Peptides*, *31*(9), 1715–1724.

Patel, J. V., Cummings, D. E., Girod, J. P., Mascarenhas, A. V., Hughes, E. A., Gupta, M., et al. (2006). Role of metabolically active hormones in the insulin resistance associated with short-term glucocorticoid treatment. *Journal of Negative Results in Biomedicine*, *5*, 14.

Peng, Y., Rideout, D., Rakita, S., Sajan, M., Farese, R., You, M., et al. (2009). Down-regulation of adiponectin/adipoR2 is associated with steatohepatitis in obese mice. *Journal of Gastrointestinal Surgery*, *13*, 2043–2049.

Peres de Heredia, F., Sanchez, J., Priego, T., Larque, E., del Puy Portillo, M., Palou, A., et al. (2009). Adiponectin is associated with serum and adipose tissue fatty acid composition in rats. *Journal of Endocrinological Investigation*, *32*, 659–665.

Pérez-de-Heredia, F., Sánchez, J., Priego, T., Nicolás, F., Portillo Mdel, P., Palou, A., et al. (2008). Adiponectin is involved in the protective effect of DHEA against metabolic risk in aged rats. *Steroids*, *73*(11), 1128–1136.

Pineiro, R., Iglesias, M. J., Gallego, R., Raghay, K., Eiras, S., Rubio, J., et al. (2005). Adiponectin is synthesized and secreted by human and murine cardiomyocytes. *FEBS Letters*, *579*, 5163–5169.

Pischon, T., Girman, C. J., Rifai, N., Hotamisligil, G. S., & Rimm, E. B. (2005). Association between dietary factors and plasma adiponectin concentrations in men. *The American Journal of Clinical Nutrition*, *81*, 780–786.

Potenza, M. A., Marasciulo, F. L., Tarquinio, M., Tiravanti, E., Colantuono, G., Federici, A., et al. (2007). EGCG, a green tea polyphenol, improves endothelial function and insulin sensitivity, reduces blood pressure, and protects against myocardial I/R injury in SHR. *American Journal of Physiology. Endocrinology and Metabolism*, *292*(5), 1378–1387.

Psilopanagioti, A., Papadaki, H., Kranioti, E. F., Alexandrides, T. K., & Varakis, J. N. (2009). Expression of adiponectin and adiponectin receptors in human pituitary gland and brain. *Neuroendocrinology*, *89*, 38–47.

Pucci, E., Chiovato, L., & Pinchera, A. (2000). Thyroid and lipid metabolism. *International Journal of Obesity and Related Metabolic Disorders*, *24*(Suppl. 2), S109–S112.

Qi, Y., Takahashi, N., Hileman, S. M., Patel, H. R., Berg, A. H., Pajvani, U. B., et al. (2004). Adiponectin acts in the brain to decrease body weight. *Nature Medicine*, *10*, 524–529.

Qiao, L., Lee, B., Kinney, B., Yoo, H. S., & Shao, J. (2011). Energy intake and adiponectin gene expression. *American Journal of Physiology. Endocrinology and Metabolism*, *300*(5), 809–816.

Qin, B., Polansky, M. M., & Anderson, R. A. (2010). Cinnamon extract regulates plasma levels of adipose-derived factors and expression of multiple genes related to carbohydrate metabolism and lipogenesis in adipose tissue of fructose fed rats. *Hormone and Metabolic Research*, *42*, 187–193.

Qin, Y., & Tian, Y. P. (2010). Hepatic adiponectin receptor R2 expression is up-regulated in normal adult male mice by chronic exogenous growth hormone levels. *Molecular Medicine Reports*, *3*(3), 525–530.

Rasmussen, M. S., Lihn, A. S., Pedersen, S. B., Bruun, J. M., Rasmussen, M., & Richelsen, B. (2006). Adiponectin receptors in human adipose tissue: Effects of obesity, weight loss, and fat depots. *Obesity (Silver Spring)*, *14*, 28–35.

Repunte-Canonigo, V., Berton, F., Cottone, P., Reifel-Miller, A., Roberts, A. J., Morales, M., et al. (2010). A potential role for adiponectin receptor 2 (AdipoR2) in the regulation of alcohol intake. *Brain Research*, *1339*, 11–17.

Rieth, N., Jollin, L., Le Panse, B., Lecoq, A. M., Arlettaz, A., De Ceaurriz, J., et al. (2009). Effects of short-term corticoid ingestion on food intake and adipokines in healthy recreationally trained men. *European Journal of Applied Physiology*, *105*(2), 309–313.

Rodriguez-Pacheco, F., Martinez-Fuentes, A. J., Tovar, S., Pinilla, L., Tena-Sempere, M., Dieguez, C., et al. (2007). Regulation of pituitary cell function by adiponectin. *Endocrinology*, *148*, 401–410.

Rose, F. J., Webster, J., Barry, J. B., Phillips, L. K., Richards, A. A., & Whitehead, J. P. (2010). Synergistic effects of ascorbic acid and thiazolidinedione on secretion of high molecular weight adiponectin from human adipocytes. *Diabetes, Obesity & Metabolism*, *12*, 1084–1089.

Ruan, H., & Lodish, H. F. (2003). Insulin resistance in adipose tissue: Direct and indirect effects of tumor necrosis factor-alpha. *Cytokine & Growth Factor Reviews*, *14*, 447–455.

Ryan, A. S., Berman, D. M., Nicklas, B. J., Sinha, M., Gingerich, R. L., Meneilly, G. S., et al. (2003). Plasma adiponectin and leptin levels, body composition, and glucose utilization in adult women with wide ranges of age and obesity. *Diabetes Care*, *26*, 2383–2388.

Saito, T., Kawano, T., Saito, T., Ikoma, A., Namai, K., Tamemoto, H., et al. (2005). Elevation of serum adiponectin levels in Basedow disease. *Metabolism*, *54*(11), 1461–1466.

Santini, F., Marsili, A., Mammoli, C., Valeriano, R., Scartabelli, G., Pelosini, C., et al. (2004). Serum concentrations of adiponectin and leptin in patients with thyroid dysfunctions. *Journal of Endocrinological Investigation*, *27*(2), RC5–RC7.

Sayinalp, S., Gedik, O., & Koray, Z. (1995). Increasing serum osteocalcin after glycemic control in diabetic men. *Calcified Tissue International*, *57*(6), 422–425.

Seow, K. M., Tsai, Y. L., Juan, C. C., Lin, Y. H., Hwang, J. L., & Ho, L. T. (2009). Omental fat expression of adiponectin and adiponectin receptors in non-obese women with PCOS: A preliminary study. *Reproductive Biomedicine Online*, *19*(4), 577–582.

Shen, X. H., Tang, Q. Y., Huang, J., & Cai, W. (2010). Vitamin E regulates adipocytokine expression in a rat model of dietary-induced obesity. *Experimental Biology and Medicine (Maywood, N.J.)*, *235*(1), 47–51.

Shimada, M., Mochizuki, K., Sakurai, N., & Goda, T. (2007). Dietary supplementation with epigallocatechin gallate elevates levels of circulating adiponectin in non-obese type-2 diabetic Goto-Kakizaki rats. *Bioscience, Biotechnology, and Biochemistry*, *71*(8), 2079–2082.

Shirai, N., & Suzuki, H. (2008). Effects of simultaneous intakes of fish oil and green tea extracts on plasma, glucose, insulin, C-peptide, and adiponectin and on liver lipid concentrations in mice fed low- and high-fat diets. *Annals of Nutrition & Metabolism*, *52*(3), 241–249.

Shoupe, D., & Lobo, R. A. (1984). The influence of androgens on insulin resistance. *Fertility and Sterility*, *41*, 385–388.

Sieminska, L., Niedziolka, D., Pillich, A., Kos-Kudla, B., Marek, B., Nowak, M., et al. (2008). Serum concentrations of adiponectin and resistin in hyperthyroid Graves' disease patients. *Journal of Endocrinological Investigation*, *31*(9), 745–749.

Siemińska, L., Wojciechowska, C., Kos-Kudła, B., Marek, B., Kajdaniuk, D., Nowak, M., et al. (2010). Serum concentrations of leptin, adiponectin, and interleukin-6 in postmenopausal women with Hashimoto's thyroiditis. *Endokrynologia Polska*, *61*(1), 112–116.

Sierksma, A., Patel, H., Ouchi, N., Kihara, S., Funahashi, T., Heine, R. J., et al. (2004). Effect of moderate alcohol consumption on adiponectin, TNF-alpha, and insulin sensitivity. *Diabetes Care*, *27*, 184–189.

Silha, J. V., Krsek, M., Hana, V., Marek, J., Jezkova, J., Weiss, V., et al. (2003). Perturbations in adiponectin, leptin and resistin levels in acromegaly: Lack of correlation with insulin resistance. *Clinical Endocrinology*, *58*(6), 736–742.

Soares, A. F., Guichardant, M., Cozzone, D., Bernoud-Hubac, N., Bouzaidi-Tiali, N., Lagarde, M., et al. (2005). Effects of oxidative stress on adiponectin secretion and lactate production in 3T3-L1 adipocytes. *Free Radical Biology & Medicine*, *38*, 882–889.

Song, Z., Zhou, Z., Deaciuc, I., Chen, T., & McClain, C. J. (2008). Inhibition of adiponectin production by homocysteine: A potential mechanism for alcoholic liver disease. *Hepatology*, *47*, 867–879.

Spranger, J., Möhlig, M., Wegewitz, U., Ristow, M., Pfeiffer, A. F., Schill, T., et al. (2004). Adiponectin is independently associated with insulin sensitivity in women with polycystic ovary syndrome. *Clinical Endocrinology*, *61*(6), 738–746.

Steyn, F. J., Boehme, F., Vargas, E., Wang, K., Parkington, H. C., Rao, J. R., et al. (2009). Adiponectin regulate growth hormone secretion via adiponectin receptor mediated Ca(2+) signalling in rat somatotrophs in vitro. *Journal of Neuroendocrinology*, *21*(8), 698–704.

Sun, X., Han, R., Wang, Z., & Chen, Y. (2006). Regulation of adiponectin receptors in hepatocytes by the peroxisome proliferator-activated receptor-gamma agonist rosiglitazone. *Diabetologia*, *49*(6), 1303–1310.

Takemura, Y., Osuga, Y., Yamauchi, T., Kobayashi, M., Harada, M., Hirata, T., et al. (2006). Expression of adiponectin receptors and its possible implication in the human endometrium. *Endocrinology*, *147*(7), 3203–3210.

Tamakoshi, K., Yatsuya, H., Wada, K., Matsushita, K., Otsuka, R., Yang, P. O., et al. (2007). The transition to menopause reinforces adiponectin production and its contribution to improvement of insulin-resistant state. *Clinical Endocrinology*, *66*, 65–71.

Tan, B. K., Chen, J., Digby, J. E., Keay, S. D., Kennedy, C. R., & Randeva, H. S. (2006). Upregulation of adiponectin receptor 1 and 2 mRNA and protein in adipose tissue and adipocytes in insulin-resistant women with polycystic ovary syndrome. *Diabetologia*, *49* (11), 2723–2728.

Tanaka, K., Inoue, S., Numata, K., Okazaki, H., Nakamura, S., & Takamura, Y. (1990). Very low-calorie diet-induced weight reduction reverses impaired growth hormone secretion response to growth hormone-releasing hormone, arginine, and L-Dopa in obesity. *Metabolism*, *39*, 892–896.

Tang, H., Sebastian, B. M., Axhemi, A., Chen, X., Hillian, A. D., Jacobsen, D. W., et al. (2012). Ethanol-induced oxidative stress via the CYP2E1 pathway disrupts adiponectin secretion from adipocytes. *Alcoholism, Clinical and Experimental Research*, *36*, 214–222. http://dx.doi.org/10.1111/j.1530-0277.2011.01607.x.

Tappy, L., Lê, K. A., Tran, C., & Paquot, N. (2010). Fructose and metabolic diseases: New findings, new questions. *Nutrition*, *26*(11–12), 1044–1049.

Thamer, C., Haap, M., Fritsche, A., Haering, H., & Stumvoll, M. (2004). Relationship between moderate alcohol consumption and adiponectin and insulin sensitivity in a large heterogeneous population. *Diabetes Care*, *27*(5), 1240.

Trayhurn, P., Wang, B., & Wood, I. S. (2008). Hypoxia and the endocrine and signalling role of white adipose tissue. *Archives of Physiology and Biochemistry*, *114*(4), 267–276.

Tworoger, S. S., Mantzoros, C., & Hankinson, S. E. (2007). Relationship of plasma adiponectin with sex hormone and insulin-like growth factor levels. *Obesity (Silver Spring)*, *15*(9), 2217–2224.

Usukura, M., Zhu, A., Yoneda, T., Karashima, S., Yagi, K., Yamagishi, M., et al. (2009). Effects of a high-salt diet on adipocyte glucocorticoid receptor and 11-beta hydroxysteroid dehydrogenase 1 in salt-sensitive hypertensive rats. *Steroids*, *74*(12), 978–982.

Van Dam, R. M., & Hu, F. B. (2005). Coffee consumption and risk of type 2 diabetes: A systematic review. *JAMA: The Journal of the American Medical Association*, *294*(1), 97–104.

Vessby, B. (2003). Dietary fat, fatty acid composition in plasma and the metabolic syndrome. *Current Opinion in Lipidology*, *14*, 15–19.
Viengchareun, S., Zennaro, M. C., Pascual-Le Tallec, L., & Lombes, M. (2002). Brown adipocytes are novel sites of expression and regulation of adiponectin and resistin. *FEBS Letters*, *532*(3), 345–350.
Vincent, H. K., Bourguignon, C. M., Weltman, A. L., Vincent, K. R., Barrett, E., Innes, K. E., et al. (2009). Effects of antioxidant supplementation on insulin sensitivity, endothelial adhesion molecules, and oxidative stress in normal-weight and overweight young adults. *Metabolism*, *58*(2), 254–262.
Vitaglione, P., Morisco, F., Mazzone, G., Amoruso, D. C., Ribecco, M. T., Romano, A., et al. (2010). Coffee reduces liver damage in a rat model of steatohepatitis: The underlying mechanisms and the role of polyphenols and melanoidins. *Hepatology*, *52*(5), 1652–1661.
Wang, Q., Guo, T., Tao, Y., Wang, Q., Song, Y., & Huang, W. (2011). Association between serum adipocyte factor level and insulin resistance in polycystic ovarian syndrome. *Gynecological Endocrinology*, *27*(11), 931–934.
Wang, Y., Lam, K., Yau, M., & Xu, A. (2008). Post-translational modifications of adiponectin: Mechanisms and functional implications. *The Biochemical Journal*, *409*, 623–633.
Wedick, N. M., Brennan, A. M., Sun, Q., Hu, F. B., Mantzoros, C. S., & van Dam, R. M. (2011). Effects of caffeinated and decaffeinated coffee on biological risk factors for type 2 diabetes: A randomized controlled trial. *Nutrition Journal*, *10*, 93.
Williams, C. J., Fargnoli, J. L., Hwang, J. J., van Dam, R. M., Blackburn, G. L., Hu, F. B., et al. (2008). Coffee consumption is associated with higher plasma adiponectin concentrations in women with or without type 2 diabetes: A prospective cohort study. *Diabetes Care*, *31*(3), 504–507.
Wölfing, B., Neumeier, M., Buechler, C., Aslanidis, C., Schölmerich, J., & Schäffler, A. (2008). Interfering effects of insulin, growth hormone and glucose on adipokine secretion. *Experimental and Clinical Endocrinology & Diabetes*, *116*(1), 47–52.
Wu, A. H., Yu, M. C., Stanczyk, F., Tseng, C. C., & Pike, M. C. (2011). Anthropometric, dietary and hormonal correlates of serum adiponectin in Asian American women. *Nutrition and Cancer*, *63*(4), 549–557.
Xi, L., Qian, Z., Xu, G., Zheng, S., Sun, S., Wen, N., et al. (2007). Beneficial impact of crocetin, a carotenoid from saffron, on insulin sensitivity in fructose-fed rats. *The Journal of Nutritional Biochemistry*, *18*(1), 64–72.
Xu, A., Chan, K. W., Hoo, R. L., Wang, Y., Tan, K. C., Zhang, J., et al. (2005). Testosterone selectively reduces the high molecular weight form of adiponectin by inhibiting its secretion from adipocytes. *The Journal of Biological Chemistry*, *280*, 18073–18080.
Xu, A., Wang, Y., Keshaw, H., Xu, L. Y., Lam, K. S., & Cooper, G. J. (2003). The fat-derived hormone adiponectin alleviates alcoholic and nonalcoholic fatty liver diseases in mice. *The Journal of Clinical Investigation*, *112*, 91–100.
Xu, A., Wong, L. C., Wang, Y., Xu, J. Y., Cooper, G. J., & Lam, K. S. (2004). Chronic treatment with growth hormone stimulates adiponectin gene expression in 3T3-L1 adipocytes. *FEBS Letters*, *572*(1–3), 129–134.
Yamamoto, Y., Hirose, H., Saito, I., Tomita, M., Taniyama, M., Matsubara, K., et al. (2002). Correlation of the adipocyte-derived protein adiponectin with insulin resistance index and serum high-density lipoprotein-cholesterol, independent of body mass index, in the Japanese population. *Clinical Science (London, England)*, *103*, 137–142.
Yamauchi, T., Kamon, J., Ito, Y., Tsuchida, A., Yokomizo, T., Kita, S., et al. (2003). Cloning of adiponectin receptors that mediate antidiabetic metabolic effects. *Nature*, *423*, 762–769.
Yamauchi, T., Kamon, J., Waki, H., Murakami, K., Motojima, K., Komeda, K., et al. (2001). The mechanisms by which both heterozygous peroxisome proliferator-activated

receptor gamma (PPARgamma) deficiency and PPARgamma agonist improve insulin resistance. *The Journal of Biological Chemistry*, *276*(44), 41245–41254.

Yamauchi, T., Nio, Y., Maki, T., Kobayashi, M., Takazawa, T., Iwabu, M., et al. (2007). Targeted disruption of AdipoR1 and AdipoR2 causes abrogation of adiponectin binding and metabolic actions. *Nature Medicine*, *13*(3), 332–339.

Yannakoulia, M., Yiannakouris, N., Melistas, L., Fappa, E., Vidra, N., Kontogianni, M. D., et al. (2008). Dietary factors associated with plasma high molecular weight and total adiponectin levels in apparently healthy women. *European Journal of Endocrinology*, *159* (4), 5–10.

Yaturu, S., Prado, S., & Grimes, S. R. (2004). Changes in adipocyte hormones leptin, resistin, and adiponectin in thyroid dysfunction. *Journal of Cellular Biochemistry*, *93*(3), 491–496.

Yilmaz, M., Bukan, N., Demirci, H., Oztürk, C., Kan, E., Ayvaz, G., et al. (2009). Serum resistin and adiponectin levels in women with polycystic ovary syndrome. *Gynecological Endocrinology*, *25*(4), 246–252.

Yoda, M., Nakano, Y., Tobe, T., Shioda, S., Choi-Miura, N. H., & Tomita, M. (2001). Characterization of mouse GBP28 and its induction by exposure to cold. *International Journal of Obesity and Related Metabolic Disorders*, *25*(1), 75–83.

You, M., Considine, R. V., Leone, T. C., Kelly, D. P., & Crabb, D. W. (2005). Role of adiponectin in the protective action of dietary saturated fat against alcoholic fatty liver in mice. *Hepatology*, *42*(3), 568–577.

Yu, H., Yang, Y., Zhang, M., Lu, H., Zhang, J., Wang, H., et al. (2006). Thyroid status influence on adiponectin, acylation stimulating protein ASP and complement C3 in hyperthyroid and hypothyroid subjects. *Nutrition & Metabolism (London)*, *3*, 13.

Zhang, H., Chen, X., Aravindakshan, J., & Ram Sairam, M. (2007). Changes in adiponectin and inflammatory genes in response to hormonal imbalances in female mice and exacerbation of depot selective visceral adiposity by high-fat diet: Implications for insulin resistance. *Endocrinology*, *148*, 5667–5679.

CHAPTER FOUR

Regulation and Function of Adiponectin Receptors in Skeletal Muscle

Yaniv Lustig, Rina Hemi, Hannah Kanety[1]
Institute of Endocrinology, Sheba Medical Center, Tel-Hashomer, Israel
[1]Corresponding author: e-mail address: hkanety@sheba.health.gov.il

Contents

Abstract

Obesity-induced insulin resistance is a primary contributing factor in the pathogenesis of type 2 diabetes. Adiponectin, an adipocyte-derived abundant plasma protein, has profound effects on systemic insulin sensitivity through direct action of the hormone on liver and muscle. The biological responses to adiponectin are mediated by two distinct receptors, AdipoR1 and AdipoR2, which differ in their affinities for adiponectin isoforms and exhibit cell type-specific effects. Disruption of AdipoR1 expression in muscle revealed a pivotal role of adiponectin/AdipoR1 in the regulation of mitochondrial biogenesis and insulin resistance. Here, we review the recent progress regarding adiponectin/AdipoRs signaling and function in skeletal muscle and summarize a range of physiological and pathophysiological conditions, as well as transcriptional and posttranscriptional mechanisms, controlling muscle AdipoR1 mRNA, and

Vitamins and Hormones, Volume 90
ISSN 0083-6729
http://dx.doi.org/10.1016/B978-0-12-398313-8.00004-X

protein levels. Comprehensive understanding of the pathways that regulate AdipoRs expression in muscle is critical to benefit from the full therapeutic potential of the adiponectin–AdipoR system.

1. INTRODUCTION

The prevalence of obesity has become a worldwide health problem due to association with many metabolic disorders such as hypertension, dyslipidemia, and glucose intolerance resulting in atherosclerotic cardiovascular diseases and type 2 diabetes (T2D). One of the potential mechanisms underlying these metabolic disorders is altered endocrine effects of adipose-derived hormones. Adiponectin is the most abundantly secreted adipokine and has profound effects on systemic insulin sensitivity through direct action of the hormone on liver and muscle. Adiponectin biological effects depend not only on the relative circulating concentrations of the hormone but also on the expression level and function of its receptors, adiponectin receptor (AdipoR) 1 and 2. Skeletal muscle plays a primary role in determining whole-body insulin sensitivity and glucose disposal and is an important target of adiponectin and its receptors in the regulation of energy metabolism. In this chapter, we have summarized the function of AdipoRs in skeletal muscle and the regulatory mechanisms that control their expression under physiological and pathophysiological conditions.

2. ADIPONECTIN AND ADIPONECTIN RECEPTORS

Adiponectin is an adipocyte-derived abundant plasma protein (Hu, Liang, & Spiegelman, 1996; Maeda et al., 1996; Nakano, Tobe, Choi-Miura, Mazda, & Tomita, 1996; Scherer, Williams, Fogliano, Baldini, & Lodish, 1995) with insulin-sensitizing, anti-inflammatory, and antiatherogenic properties (Brochu-Gaudreau et al., 2010; Kishida, Funahashi, & Shimomura, 2012; Shetty, Kusminski, & Scherer, 2009; Yamauchi & Kadowaki, 2008). Intriguingly, the circulating levels of adiponectin are decreased in obesity, insulin resistance, and T2D (Arita et al., 1999; Hotta et al., 2000; Kadowaki et al., 2006; Trujillo and Scherer, 2005). Adiponectin knockout mice exhibit insulin resistance and diabetes (Kubota et al., 2002; Maeda et al., 2002; Nawrocki et al., 2006), while administration of adiponectin has been shown to elicit glucose lowering effects and ameliorate insulin resistance, primarily through its action in liver and muscle (Berg et al., 2001; Combs et al., 2001; Satoh et al., 2005; Yamauchi et al., 2001). The beneficial effects of adiponectin are mediated primarily by stimulation of AMP-activated protein kinase (AMPK) (Yamauchi et al., 2002) and peroxisome proliferator-activated receptor α (PPARα) (Yamauchi, Kamon, Waki, et al., 2003), and through its effect on sphingolipid metabolism (Holland et al., 2011).

The 30-kDa adiponectin protein consists of an amino-terminal signal sequence followed by a collagenous domain and a carboxyl-terminal globular domain. Circulating adiponectin exists predominantly as hexamers and high-molecular weight (HMW) oligomers that are believed to possess differing biological activities (Pajvani et al., 2003; Tsao et al., 2002). In addition, the full-length adiponectin (fAd) may be cleaved to a fragment containing the C-terminal globular domain (gAd), which exhibits potent metabolic effects, particularly in skeletal muscle; however, its circulating presence remains controversial (Fruebis et al., 2001; Waki et al., 2005).

The pleiotropic actions of adiponectin are mediated by two receptors, termed as AdipoR1 and AdipoR2, which share approximately 67% sequence identity (Yamauchi, Kamon, Ito, et al., 2003). AdipoR1 is ubiquitously expressed and is the predominant form expressed in skeletal muscle, whereas AdipoR2 is most abundantly expressed in the liver (Fang et al., 2005; Yamauchi, Kamon, Ito, et al., 2003). Using AdipoR1 and AdipoR2 knockout mice, several studies have shown that AdipoR1 and AdipoR2 act as the major receptors for adiponectin *in vivo* and have important roles in the regulation of glucose metabolism and insulin sensitivity (Bjursell et al., 2007; Iwabu et al., 2010; Liu et al., 2007; Yamauchi et al., 2007).

Both AdipoR1 and AdipoR2 are seven-transmembrane domain (7TM) receptors; however, they are structurally and functionally distinct from the G-protein-coupled 7TM receptors (e.g., the GPCR), due to their extracellular C-terminus and cytosolic N-terminus (Deckert et al., 2006; Yamauchi, Kamon, Ito, et al., 2003) as well as their downstream signaling mechanisms (Buechler et al., 2010; Heiker et al., 2010; Holland et al., 2011; Mao et al., 2006; Yamauchi, Kamon, Ito, et al., 2003). Based on their distinct topology, AdipoR1 and AdipoR2 have been included in a novel 7TM receptor family, termed the progestin and AdipoQ receptor family, which currently consists of 11 mammalian proteins (Tang et al., 2005).

AdipoR1 and AdipoR2 may form both homo- and heteromultimers (Kosel et al., 2010; Yamauchi, Kamon, Ito, et al., 2003), and recent findings provide evidence that AdipoR1 homodimerization depends on a specific GxxxG motif in its fifth transmembrane domain (Kosel et al., 2010). Several studies indicate that AdipoR1 is a high-affinity receptor for gAd and a low-affinity receptor for fAd, whereas AdipoR2 is an intermediate-affinity receptor for both fAd and gAd (Yamauchi, Kamon, Ito, et al., 2003; Yamauchi et al., 2002).

Another putative receptor for adiponectin is T-cadherin, which can only bind to the hexameric and HMW oligomers of adiponectin, but not to its trimeric and globular forms (Hug et al., 2004). T-cadherin is abundantly expressed in skeletal muscle (Ordelheide et al., 2011); however, its function remains elusive in this tissue. As T-cadherin lacks an intracellular domain, it may be considered as an adiponectin binding protein that associates with or works in concert with other transmembrane proteins, including AdipoR1 and AdipoR2, to transmit adiponectin signal to intracellular pathways (Denzel et al., 2010; Hug et al., 2004).

3. ADIPONECTIN SIGNALING AND FUNCTION IN SKELETAL MUSCLE

Skeletal muscle is an important target of adiponectin in the regulation of energy metabolism. Early studies suggested that adiponectin might participate in energy homeostasis because its mRNA was decreased in adipose tissue of obese mice and humans (Hu et al., 1996). In addition, a negative correlation between plasma adiponectin levels and both insulin resistance and obesity has been described (Arita et al., 1999; Hotta et al., 2000). Animal models and cell-based *in vitro* studies demonstrated that adiponectin regulates both lipid metabolism and fatty acid oxidation in skeletal muscle (Combs et al., 2004; Maeda et al., 2002; Tomas et al., 2002; Yamauchi, Kamon, Waki, et al., 2003; Yamauchi et al., 2001; Yoon et al., 2006). By administrating gAd to C2C12 myotubes, isolated hindlimb muscle *ex vivo*, and mice *in vivo*, Lodish and colleagues have first shown that adiponectin regulates energy balance by stimulating muscle free fatty acid (FFA) oxidation by mitochondria (Fruebis et al., 2001). In addition, adiponectin has a major impact on glucose homeostasis and induces the stimulation of glucose uptake in muscle cells via enhanced GLUT4 translocation to the plasma membrane (Ceddia et al., 2005; Fang, Fetros, Dadson, Xu and Sweeney, 2009, Fang et al., 2005; Mao et al., 2006). Adiponectin is also viewed as having anti-inflammatory properties in skeletal muscle, as inflammatory cytokines and proinflammatory conditions were shown to upregulate adiponectin in human and rodent myotubes (Delaigle et al., 2004). As a support for this hypothesis, adiponectin was recently shown to enhance skeletal muscle insulin sensitivity by inhibition of nuclear factor (NF)-κB-inducing kinase-mediated skeletal muscle inflammation (Choudhary et al., 2011).

Mechanistically, AMPK, p38MAPK, and PPARα have all been shown to mediate adiponectin biological activities in skeletal muscle (Yamauchi, Kamon, Ito, et al., 2003; Yamauchi et al., 2002; Yoon et al., 2006). As a response to adiponectin, AMPK is activated by phosphorylation on thr^{172} in its α subunit and inhibits acetyl CoA carboxylase (ACC) by phosphorylation of Ser^{79} (Tomas et al., 2002; Yamauchi, Kamon, Ito, et al., 2003). This leads to reduced malonyl CoA concentration in the cytoplasm and therefore more active carnitine palmitoyltransferase 1 that imports to the mitochondria FFAs for β oxidation (Li, Wu, Wang, Liu & Zhao, 2007; Tomas et al., 2002). Adiponectin administration also activates p38MAPK which consequently upregulates the activity of PPARα, a transcription factor that regulates fatty acid transport, synthesis, and combustion (Yamauchi, Kamon, Ito, et al., 2003; Yoon et al., 2006). Adiponectin-dependent activation of PPARα results in reduced triglycerides (TG) and an increased combustion of FFA due to increased expression of CD36, acetyl CoA oxidase, and uncoupling protein 2 (Yamauchi et al., 2001). These data correlate well with animal studies showing that overexpression of adiponectin decreases serum TG and FFA and successfully corrects high-fat diet (HFD) induced insulin resistance in skeletal muscle (Combs et al., 2004; Maeda et al., 2002; Yamauchi, Kamon, Waki, et al., 2003; Yamauchi et al., 2001). Using adiponectin gene knockout and transgenic mouse models, studies have demonstrated that adiponectin increases mitochondrial biogenesis and oxidative capacity in skeletal muscle through activated AMPK and increased peroxisome-proliferator-activated receptor gamma coactivator-1 α (PGC-1α) expression (Civitarese et al., 2006; Kadowaki et al., 2006). In addition, it was shown that the MAPK phosphatase-1 (MKP1) plays an important role in mediating adiponectin-enhanced mitochondrial biogenesis and oxidative metabolism and that suppression of MKP1 by adiponectin leads to p38MAPK activation and increased PGC-1α gene expression (Qiao et al., 2012).

In the past few years, growing evidence suggests that in addition to adipose tissue, skeletal muscle and several other tissues express adiponectin that can mediate functionally important autocrine and paracrine effects (Amin et al., 2010; Delaigle et al., 2004, 2006; Jortay et al., 2010; Krause et al., 2008; Liu et al., 2009; Van Berendoncks et al., 2010). Therefore, adiponectin is now described also as a myokine which could be expressed and secreted by skeletal muscle (Pedersen and Febbraio, 2008). Adiponectin's autocrine signaling in skeletal muscle exerts functional metabolic effects including improvement of insulin sensitivity and protection from diet-induced insulin resistance (Amin et al., 2010; Liu et al., 2009). It was suggested that local adiponectin production and secretion occurs as a response to stress conditions such as inflammation and obesity in order to enable local anti-inflammatory and antioxidative protection (Delaigle et al., 2004, 2006; Jortay et al., 2010). Recently, Fiaschi et al. (2009) showed for the first time that gAd is able to induce muscle gene expression and cell differentiation in C2C12 cells. gAd induced differentiation via activation of p38MAPK, AMPK, and Akt pathways culminating in increased expression of muscle differentiation markers such as myosin heavy chain and caveolin-3. Moreover, adiponectin was shown to be expressed in differentiated cells, suggesting a novel function of adiponectin directly coordinating the myogenic differentiation program and serving an autocrine function during skeletal myogenesis (Fiaschi et al., 2009).

Adiponectin has also been hypothesized to be involved in the regulation of skeletal muscle growth. Yamauchi et al. (2001) demonstrated that adiponectin is potentially able to enhance protein synthesis and inhibit protein degradation via stimulation of the insulin signaling pathway by activating insulin receptor substrate 1, which in turn activates the phosphorinositide 3-kinase (PI3K)–Akt cascade. Moreover, recent studies suggest that nutrients such as FFAs and amino acids may play an important role in protein metabolism regulation by adiponectin (Du et al., 2007; Zhou et al., 2007).

Both AdipoR1 and AdipoR2 mediate adiponectin beneficial metabolic effects (Yamauchi et al., 2007). A recent study has added to our understanding of the mechanism of action of adiponectin and AdipoR1 in skeletal muscle. Kadowaki and colleagues (Iwabu et al., 2010) have demonstrated that specific deletion of the *AdipoR1* gene in mice skeletal muscle (muscle-R1KO) significantly decreased glucose disposal rate and glucose infusion rate, indicating decreased insulin sensitivity in muscle. In addition, muscle endurance during treadmill running was significantly lower, and type 1 oxidative fibers were significantly reduced in soleus muscle of muscle-R1KO mice. Importantly, this study revealed that adiponectin, through AdipoR1, can increase mitochondrial mass and oxidative capacity at least in part via inducing extracellular Ca^{2+} influx and subsequently activating the Ca^{2+}/

calmodulin-dependent protein kinase kinase β (CaMKKβ), AMPK, and Sirt1 pathway leading to increased PGC-1α expression and activation (Iwabu et al., 2010).

To date, only a few intracellular proteins were found to bind AdipoRs and mediate their downstream effects in skeletal muscle. Adaptor protein containing the pleckstrin homology domain, phosphotyrosine domain, and leucine zipper motif (APPL1) binds directly to both AdipoR1 and AdipoR2 in skeletal muscle and liver (Mao et al., 2006). Suppression of APPL1 in C2C12 myotubes significantly attenuated adiponectin phosphorylation of AMPK, p38MAPK, ACC, and fatty acid oxidation, suggesting an important role of APPL1 in mediating adiponectin regulated lipid metabolism in skeletal muscle cells (Wang et al., 2009). An additional protein that was found to interact with AdipoR1 in C2C12 cells is protein kinase CK2 β subunit (CK2β) (Heiker et al., 2009). When CK2 activity is blocked by the specific inhibitor 2-dimethylamino-4,5,6,7-tetrabromo-1H-benz-imidazole, adiponectin-induced phosphorylation of ACC is reduced in C2C12 cells.

4. TRANSCRIPTIONAL REGULATION OF ADIPORS IN SKELETAL MUSCLE UNDER DIFFERENT PHYSIOLOGICAL AND PATHOPHYSIOLOGICAL CONDITIONS

Due to the important role of adiponectin in skeletal muscle, the gene expression of AdipoR1 and AdipoR2 has been assessed in rodent models of obesity and diabetes as well as in muscle biopsies obtained from individuals with impaired glucose tolerance (IGT) or T2D. While several studies have revealed decreased AdipoRs mRNA expression in obesity/diabetes (Civitarese et al., 2004; Inukai et al., 2005; Tsuchida et al., 2004), other studies have shown upregulation or no significant change in AdipoR1 gene expression (Bluher et al., 2006; Holmes et al., 2011; Metais et al., 2008). The following sections and Table 4.1 summarize the recent data in this area according to different physiological and pathophysiological conditions.

4.1. Nutritional conditions

The expression of AdipoR1 and AdipoR2 in skeletal muscle appears to be inversely regulated by insulin during fasting and refeeding conditions. Tsuchida et al. (2004) demonstrated that expression of both AdipoR1 and AdipoR2 in skeletal muscle is significantly increased in fasted mice and decreased in refed mice. Moreover, streptozotocin (STZ) treatment

Table 4.1 AdipoRs mRNA expression in skeletal muscle under different physiological and pathophysiological conditions in humans and rodents

Condition			Effects on AdipoR1/R2 expression	References
Nutrition	Mice	Fasting	↑AdipoR1, ↑/↔AdipoR	Inukai et al. (2005), Tsuchida et al. (2004)
		Refeeding	↓AdipoR1, ↓AdipoR2	Tsuchida et al. (2004)
		STZ treatment	↑AdipoR1, ↑AdipoR2	Tsuchida et al. (2004)
	Rats	Undernutrition	↑AdipoR1	Prior et al. (2008)
		Short-time lipid diet	↑AdipoR1	Anavi et al. (2010)
		Fasting	↔AdipoR1, ↔AdipoR2	Beylot et al. (2006)
		Short high-fat diet	↔AdipoR1, ↔AdipoR2	Beylot et al. (2006), Mullen et al. (2010)
	Humans	High-fat meal	↓AdipoR1, ↓AdipoR2	Heilbronn et al. (2007)
Obesity/ diabetes	Mice	4 months high-fat diet	↑AdipoR1, ↑AdipoR2	Barnea et al. (2006), Bullen et al. (2007), de Oliveira et al. (2011)
		16 weeks HFHS diet	↓AdipoR1	Bonnard et al. (2008)
		ob/ob	↓AdipoR1, ↓AdipoR2	Tsuchida et al. (2004)
		db/db	↓AdipoR1	Inukai et al. (2005)
		KKAy	↓AdipoR1	Huang et al. (2006)
	Rats	Obese Zucker	↓/↔AdipoR1, ↔AdipoR2	Beylot et al. (2006), Chang et al. (2006)

		ZDF	↔AdipoR1, ↔AdipoR2	Metais et al. (2008)
		SHR	↑AdipoR1, ↑AdipoR2	Rodriguez et al. (2008)
		CLA diet in Zucker rats	↑AdipoR1	Inoue et al. (2006)
	Humans	NGT subjects with family history of T2D	↓AdipoR1, ↓AdipoR2	Civitarese et al. (2004)
		Obese T2D and IGT subjects	↑AdipoR1, ↑AdipoR2	Bluher et al. (2006)
		T2D subjects	↔AdipoR1, ↔AdipoR2	Debard et al. (2004), Holmes et al. (2011)
		3 h hyperinsulinemic clamp of T2D subjects	↑AdipoR1, ↔AdipoR2	Debard et al. (2004)
		Obesity	↑AdipoR1	Holmes et al. (2011)
		Weight loss after bariatric surgery	↑AdipoR1, ↑AdipoR2	Holmes et al. (2011)
Physical activity	Rodents	Prolonged exercise in obese and diabetic models	↑AdipoR1, ↔AdipoR2	Chang et al. (2006), Huang et al. (2006), Zeng et al. (2007)
	Humans	Prolonged physical training in insulin resistance or T2D patients	↑AdipoR1, ↑AdipoR2	Bluher et al. (2006), Christiansen et al. (2010), O'Leary et al. (2007), Sixt et al. (2010), Van Berendoncks et al. (2011)

Continued

Table 4.1 AdipoRs mRNA expression in skeletal muscle under different physiological and pathophysiological conditions in humans and rodents—cont'd

Condition			Effects on AdipoR1/R2 expression	References
PPAR-γ agonists	Mice	Pioglitazone for 7 days in db/db mice	↔AdipoR1	Inukai et al. (2005)
		Rosiglitazone for 7 weeks in diabetic mice	↔AdipoR1	Yao et al. (2005)
	Rats	Pioglitazone or rosiglitazone for 6 weeks in Zucker fatty rats	↑AdipoR1	Pita et al. (2012)
	Humans	Rosiglitazone for 12 weeks in T2D patients	↓AdipoR1, ↔AdipoR2	Tan et al. (2005)
		Pioglitazone for 21 days	↔AdipoR1, ↔AdipoR2	Li, Tonelli, et al. (2007)
		Pioglitazone for 6 months in T2D patients	↑AdipoR1, ↑AdipoR2	Coletta et al. (2009)
	In vitro	20 h of Rosiglitazone or troglitazone in differentiated myotubes	↔AdipoR1, ↔AdipoR2	Kaltenbach et al. (2005)

↑ designates upregulation, ↓ downregulation, and ↔ no change in AdipoRs mRNA expression.

in mice, which diminished plasma insulin, caused a marked increase in AdipoR1 and AdipoR2 mRNA levels, whereas replenishment of insulin into STZ-treated mice reduced the levels to those observed in mice without STZ treatment (Tsuchida et al., 2004). A similar study in mice which investigated the expression levels of AdipoRs after 48 h of starvation detected significantly higher levels of AdipoR1 compared to the control group; however, no change was observed in the mRNA levels of AdipoR2 (Inukai et al., 2005). In agreement with these results, a recent study showed that undernutrition confined to the suckling period in rats, elevated both circulating adiponectin and AdipoR1 mRNA expression in skeletal muscle (Prior et al., 2008). Another study in rainbow trout demonstrated that the expression of AdipoR1 and AdipoR2 increased in response to fasting while insulin injection decreased AdipoR1 expression in white and red muscle (Sanchez-Gurmaches et al., 2012). In humans, the mRNA expression of AdipoRs has decreased in response to a single high-fat meal with a more significant decrease in AdipoR1 compared to AdipoR2, indicating that AdipoR1 may be more sensitive than AdipoR2 to dietary fat (Heilbronn et al., 2007).

While these results in mice, humans, and rainbow trout suggest that nutritional conditions such as fasting/refeeding and HFD affect the expression levels of AdipoRs in skeletal muscle, studies in rats found no or opposing differences in AdipoRs levels. A study investigating the impact of lipid oversupply on adiponectin levels and AMPK pathway in Wistar rats found increased expression of AdipoR1 mRNA following a short-time (6 days) infusion of lipid emulsion (Anavi et al., 2010). Another study found that both fasting and HFD in Wistar rats did not modify the expression levels of AdipoR1 and AdipoR2 in skeletal muscle (Beylot et al., 2006). In addition, no change was observed in AdipoR1 expression after 3 days of high saturated or high polysaturated fat feeding (Mullen et al., 2010), suggesting that the expression of muscle AdipoRs in rats is poorly responsive to changes in nutritional conditions. Similarly, there was no change in AdipoR1 and AdipoR2 mRNA levels in pigs after a 24-h fast (Liu et al., 2008). These discrepancies in the regulation of AdipoRs levels following changes in nutritional conditions in various species may be explained either by differences in the sampling time between fed and fasted animals or in the variation between species.

4.2. Obesity and diabetes

Most of the expression studies of AdipoR1 and AdipoR2 under pathophysiological conditions were performed in conditions of obesity and diabetes. Studies in mice demonstrated that prolonged HFD feeding significantly increased the expression of both AdipoR1 and AdipoR2 in skeletal muscle (Barnea et al., 2006; de Oliveira et al., 2011). In addition, Mantzoros group showed that HFD feeding increased AdipoR1 mRNA in 18-week-old diet-induced obesity (DIO)-prone C57BL/6J and DIO-resistant A/J mice (Bullen et al., 2007). In contrast to these results, skeletal muscle AdipoR1 mRNA levels decreased in C57CL/6 mice fed for 16 weeks with high-fat and high-sucrose diet suggesting decreased adiponectin sensitivity in the muscle of diabetic mice (Bonnard et al., 2008). In skeletal muscle of ob/ob mice, which have a mutation in the leptin gene and serve as a model of insulin resistance linked to obesity, both AdipoR1 and AdipoR2 mRNA expression levels were significantly decreased (Tsuchida et al., 2004). A significant suppression of AdipoR1 mRNA levels in skeletal muscle was also observed in db/db mice, which lack a functional leptin receptor (Inukai et al., 2005) as well as in KKAy mice, a model of T2D possessing insulin resistance (Huang et al., 2006). Since plasma adiponectin levels were shown to be downregulated in ob/ob mice (Yamauchi, Kamon, Waki, et al., 2003), these data collectively raise the interesting possibility that altered expression of AdipoRs and plasma adiponectin levels may play a casual role in the regulation of insulin sensitivity (Kadowaki et al., 2006).

AdipoRs expression under pathophysiological conditions was also investigated in rats. In agreement with mice models of obesity and diabetes, one study showed reduced AdipoR1 expression in soleus muscle of a genetic model of insulin resistance, the obese Zucker rat (Chang et al., 2006). A few other studies, however, demonstrated that compared to wild-type animals, AdipoR1 and AdipoR2 expression was unchanged in muscle of obese Zucker rats (Beylot et al., 2006), as well as in obese, insulin resistance ZDF (Zucker Diabetic Fatty) rats (Metais et al., 2008). These results do not support a role for decreased expression of AdipoRs in the development of insulin resistance. In contrast, both AdipoR1 and AdipoR2 expression levels were increased in skeletal muscle of spontaneously hypersensitive rats (SHR), which show overweight, dyslipidemia, glucose intolerance, and insulin resistance (Rodriguez et al., 2008). Furthermore, a significant upregulation of AdipoR1 was measured in skeletal muscle of Zucker rats fed a conjugated linoleic acid (CLA) diet (Inoue et al., 2006). Therefore, and based on the results obtained from mice and rats, a role for decreased expression of AdipoRs in the pathogenesis of obesity and diabetes remains controversial.

A controversial relationship between insulin sensitivity and AdipoRs mRNA expression in muscle was also observed in humans. Civitarese et al. (2004)

showed that both AdipoR1 and AdipoR2 expression levels were significantly lower in normal glucose tolerant (NGT) individuals with a strong family history of T2D (which are at high risk of developing the disease) than in volunteers with no family history. In contrast to this result, Bluher et al. (2006) investigated AdipoR1 and AdipoR2 mRNA expression in human skeletal muscle in a cross-sectional study of 140 subjects with NGT, IGT, or T2D and found that both AdipoR1 and AdipoR2 levels were positively associated with obesity. Even though this result seems to imply that obesity increases AdipoRs mRNA levels, the NGT subjects in the study were significantly younger than IGT and T2D subjects, a factor that may have an opposite impact on AdipoRs mRNA expression. In an additional study that compared AdipoRs mRNA in skeletal muscle of type 2 diabetic patients with insulin-sensitive lean individuals, no alterations were found in the expression of AdipoR1 and AdipoR2 (Debard et al., 2004). However, mRNA expression of AdipoR1 but not AdipoR2 was significantly induced during a 3 h of hyperinsulinemic–euglycemic clamp of both lean and type 2 diabetic subjects, further supporting a role of AdipoR1 in glucose transport and fatty acid oxidation pathways (Debard et al., 2004). Recently, the expression of AdipoR1 in skeletal muscle was shown to be significantly higher in obese compared with lean controls but was not altered in type 2 diabetic patients (Holmes et al., 2011). In addition, AdipoR1 and AdipoR2 mRNA expression was determined following weight loss after bariatric surgery in morbidly obese, nondiabetic participants with the metabolic syndrome. Following weight loss, both AdipoR1 and AdipoR2 mRNA levels showed a significant twofold increase (Holmes et al., 2011). The data from these experiments imply that AdipoRs mRNA expression may be differentially regulated by obesity and T2D through different mechanisms and can partially explain the contradicting and diverse AdipoRs expression levels obtained from several studies in different species.

4.3. Physical activity

Skeletal muscle plays a primary role in determining whole-body insulin sensitivity and glucose disposal. Under hyperinsulinemic–euglycemic conditions, around 80% of total body glucose uptake occurs in skeletal muscle (Wasserman, 2009). In individuals with IGT or T2D, glucose disposal into muscle is reduced by as much as 50% (DeFronzo et al., 1985). Skeletal muscle is comprised of heterogeneous muscle fibers that differ in their contractile and metabolic characteristics including slow-twitch (type I) and fast-twitch (type II) fibers. Type I fibers are predominantly oxidative, contain more mitochondria, and are more responsive to insulin compared with type II fibers. In addition, whole-body glucose uptake and muscle glucose transport are positively associated with a higher percentage of type I slow-twitch fibers and thus oxidative capacity (Crunkhorn et al., 2007), and patients who present with insulin resistance and T2D have a decreased proportion of type I fibers.

Exercise training is known to improve insulin sensitivity and oxidative capacity (Andersen et al., 2003; Sigal et al., 2006) and to prevent or delay the development of T2D mellitus in subjects with IGT (Burr et al., 2010; Eriksson and Lindgarde, 1991; Knowler et al., 2002; Pan et al., 1997; Tuomilehto et al., 2001). Endurance exercise training was shown to promote mitochondrial biogenesis in skeletal muscle and enhance muscle oxidative capacity, through activation of calcium (Ca^{2+}), AMPK, SIRT1, and PGC1-α (Henstridge and Febbraio, 2010; Iwabu et al., 2010). Recently, it was reported that adiponectin, similar to exercise, induces calcium (Ca^{2+}) influx into myotubes and thereby activates CaMKKβ, AMPK, SIRT1, and PGC-1α to subsequently increase mitochondrial biogenesis (Iwabu et al., 2010). Importantly, these effects were mediated and dependent on AdipoR1, as muscle-specific AdipoR1 disruption in mice suppressed these exercise mimetic effects, resulting in insulin resistance and decreased exercise endurance capacity.

The similar metabolic effects of adiponectin and exercise provided the rationale for studies that examined the potential role of adiponectin in mediating the insulin-sensitizing action of exercise. These studies assessed in both rodents and humans whether physical training alters levels of adiponectin and its receptors *in vivo*. From work published to date, alterations in plasma adiponectin concentrations do not appear to be a major contributing factor in the enhanced insulin action observed upon acute or prolong exercise (Ando et al., 2009; Magkos et al., 2010; Numao et al., 2008; Vu et al., 2007), although increases in total and/or HMW-adiponectin levels have been observed in insulin resistant or obese hypoadiponectinemic individuals (Bluher et al., 2006; Kelly et al., 2012; O'Leary et al., 2007). On the other hand, the insulin-sensitizing effects of exercise appear to be associated with increased AdipoRs expression in skeletal muscle. In obese and diabetic rodent models, exercise has been found to upregulate the mRNA expression of AdipoR1, but not AdipoR2, in skeletal muscle (Chang et al., 2006; Huang et al., 2006; Zeng et al., 2007). Similarly, studies in insulin resistant or diabetic individuals have demonstrated a significant increase in skeletal muscle AdipoR1 and AdipoR2 expression upon prolonged physical training

(Bluher et al., 2006; Christiansen et al., 2010; O'Leary et al., 2007; Sixt et al., 2010; Van Berendoncks et al., 2011). Furthermore, the exercise-induced improvements in insulin resistance were associated with the increase in AdipoR1 and AdipoR2 expression (Bluher et al., 2006). These data suggest that part of the improvement in insulin sensitivity following exercise may be due to enhanced AdipoRs mRNA expression. The latter is dependent on the intensity and duration of the exercise program (a change in receptors expression is only observed in longer duration training programs) as well as the baseline characteristics of the subjects studied.

4.4. Regulation by PPAR-γ agonists

Thiazolidinediones (TZDs), drugs with PPAR-γ agonist action, are known to improve whole-body insulin sensitivity and lipid profile, and to reduce several cardiovascular risk factors such as arterial hypertension, and procoagulant and proinflammatory factors. It is now well established that circulating adiponectin levels are increased by TZDs in concert with their insulin-sensitizing effects (Maeda et al., 2001; Pajvani et al., 2004). Several studies have also assessed whether TZDs regulate muscle AdipoRs gene expression in animal models of diabetes and insulin resistance and in subjects with T2D. Despite a significant improvement in insulin sensitivity, associated with a marked increase in circulating adiponectin levels, muscle AdipoR1 expression was not altered in genetically obese db/db mice fed with a standard rodent chow containing 0.01% (w/w) pioglitazone for 7 days (Inukai et al., 2005) or in diabetic rats treated for 7 weeks with rosiglitazone (2 mg/kg every day) (Yao et al., 2005). Conversely, treatment of Zucker fatty rats with pioglitazone or rosiglitazone (3 mg/kg every day) for 6 weeks resulted in improved systemic insulin sensitivity, increased muscle insulin-stimulated glucose transport, and enhanced muscle AdipoR1 mRNA expression, suggesting that the insulin-sensitizing action of the TZDs may be mediated, at least in part, by their influence on AdipoR1 expression (Pita et al., 2012). The effects of the insulin-sensitizing agents, rosiglitazone, and pioglitazone on muscle AdipoRs expression were also assessed in subjects with T2D. AdipoR1 mRNA expression was downregulated, while AdipoR2 expression was unchanged in skeletal muscle of type 2 diabetic patients following 12 weeks treatment with rosiglitazone (Tan et al., 2005). This decrease in AdipoR1 expression was unrelated to the TZDs-mediated change in muscle glucose uptake. In another study, pioglitazone treatment for 21 days did not affect AdipoR1 and AdipoR2 expression in muscle, and AdipoRs expression did not correlate with baseline or TZD-enhanced insulin action (Li, Tonelli, et al., 2007). By contrast, treatment of individuals with T2D with pioglitazone for 6 months resulted in increased mRNA levels for AdipoR1 and AdipoR2 in skeletal muscle biopsies, associated with increased whole-body insulin sensitivity (Coletta et al., 2009). The differences found on TZDs effect on AdipoRs expression in the different studies mentioned above could be due to the doses of drugs and/or duration of therapy.

Currently, the mechanisms by which TZDs treatment promote the expression of AdipoRs in skeletal muscle are largely unknown. A direct effect of the PPAR-γ agonists on muscle is unlikely as *in vitro* differentiated human myotubes treated with troglitazone or rosiglitazone for 20 h, showed no significant changes in AdipoR1 and AdipoR2 mRNA expression (Kaltenbach et al., 2005). The increase in AdipoRs expression observed after prolonged TZDs treatment may result from a decrease in insulin levels, as insulin has been shown to negatively regulate AdipoRs expression (as discussed in section 5).

5. MOLECULAR MECHANISMS REGULATING MUSCLE ADIPORS TRANSCRIPTION

In light of the accumulating data showing the significance of AdipoRs in the regulation of adiponectin biology, it is necessary to identify the molecular pathways that control AdipoRs gene expression. So far, some aspects of AdipoRs regulation in muscle have been identified. In 2004, Kadowaki and colleagues discovered that insulin negatively regulates the expression levels of AdipoRs, most likely via the PI3K/Foxo1 pathway (Tsuchida et al., 2004). This finding was supported by Inukai et al. (2005) who demonstrated that PI3K is required for the regulation of AdipoR1 expression upon insulin signaling in skeletal muscle cells. Recently, the promoters of human AdipoR1 and AdipoR2 were characterized, and the transcriptional regulation of AdipoRs by insulin was analyzed in C2C12 myoblasts. In this study, insulin repressed the promoter activity of AdipoR1 but not AdipoR2 via the PI3K and Foxo1 pathway. Molecular analysis of AdipoR1 promoter has characterized a putative insulin responsive region which does not contain classical insulin responsive sequences or a Foxo1 binding site, but a strong repressor element termed nuclear inhibitory protein (NIP).

Furthermore, the NIP element was demonstrated to be required for the inhibitory effect of insulin on AdipoR1 transcription (Sun et al., 2008). The effects of insulin and glucose on AdipoRs expression were characterized also in L6 rat skeletal muscle cells. Interestingly, while AdipoR1 expression was shown to decrease by both hyperinsulinemia and hyperglycemia, AdipoR2 expression was induced by hyperinsulinemia. These changes were correlated with alterations in the functional effects of gAd and fAd on glucose and fatty acid uptake and metabolism (Fang et al., 2005). In contrast to the results implicating insulin and glucose as regulators of AdipoRs expression in skeletal muscle, another study in C2C12 and human differentiated myotubes did not detect any significant effect of different concentrations of insulin on AdipoR1 mRNA expression (Staiger et al., 2004), suggesting that a more complicated mechanism for AdipoRs regulation may exist. Recently, the ER stress-inducible factor activating transcription factor 3 (ATF3) was shown to bind to an ATF3 responsive region in human AdipoR1 promoter and negatively regulate AdipoR1 expression in C2C12 and HepG2 cells (Park et al., 2010). This result raises the interesting possibility that ATF3 acts as a transcriptional repressor in the regulation of AdipoR1 which may impair adiponectin signaling under obesity and diabetic conditions.

6. POSTTRANSCRIPTIONAL REGULATION OF ADIPORS IN SKELETAL MUSCLE

Gene expression is a dynamic process and is controlled at many levels, including transcription, mRNA splicing, mRNA stability, translation, and posttranslational events. As summarized in section 4, the transcriptional regulation of AdipoRs, primarily AdipoR1, in skeletal muscle of rodents and humans has been studied extensively (Table 4.1). However, it is not yet known whether the changes in AdipoR1 mRNA levels, observed in these studies under diverse physiological and pathophysiological conditions, are reflected in parallel alterations in AdipoR1 protein levels. In this section, we discuss recent data addressing the importance of posttranscriptional regulation on AdipoR1 protein levels in skeletal muscle.

6.1. Translational control of AdipoR1

To date, the effects of posttranscriptional regulatory mechanisms on AdipoR1 protein levels are largely unknown. To address this issue, we have recently compared the mRNA and protein levels of AdipoR1 during muscle cells differentiation (Ashwal et al., 2011). Our experiments in cultured cells revealed high expression of AdipoR1 protein in differentiated C2C12 myotubes compared to undifferentiated cells. However, despite the robust increase in AdipoR1 protein, no significant change was detected in AdipoR1 mRNA levels during myoblast–myotube differentiation. These findings support an important role for posttranscriptional mechanisms in the regulation of muscle AdipoR1 protein expression.

One of the major posttranscriptional mechanisms affecting the levels of expressed proteins is translational control. A growing body of evidence supports a key role of translational control in metabolic regulation and implicates translational mechanisms in the pathogenesis of metabolic disorders such as T2D and the metabolic syndrome. Translational control is involved in many components of T2D including insulin biosynthesis, hepatic and peripheral insulin sensitivity, and diabetic complications (Adeli, 2011)

To assess the role of translational control in the regulation of AdipoR1 expression during myogenesis, we cloned the human AdipoR1 3′untranslated (UTR) region downstream of the luciferase coding region of a pGL3 reporter vector and performed luciferase assays in C2C12 myoblasts and myotubes. These experiments demonstrated a considerably higher luciferase activity in myotubes, suggesting that AdipoR1 3′UTR may have a profound role in the significant increase of AdipoR1 protein expression observed during skeletal muscle cells differentiation (Ashwal et al., 2011). Since the mRNA translational process is modulated efficiently by both RNA-binding proteins (RBPs) and microRNAs, by interacting with specific elements in the 3′UTR (Adeli, 2011; Bartel, 2009), our findings suggest that the translational regulation of AdipoR1 3′UTR may be linked to alterations in specific microRNAs or RBPs that occur during myoblast–myotube differentiation (Fig. 4.1).

6.2. Alternative mRNA splicing of AdipoR1

Another important mechanism for generating posttranscriptional modulations is alternative mRNA splicing, a process that occurs in at least 90% of human genes (Wang et al., 2008). Alternative mRNA splicing is one of the major sources for structural and functional diversity of proteins. Additionally, it can alter mRNA stability and translation efficiency and thereby affect protein levels. Alternative mRNA splicing is tissue and developmentally specific, is regulated by hormonal and metabolic stimuli, and is associated with physiological and pathological regulation including

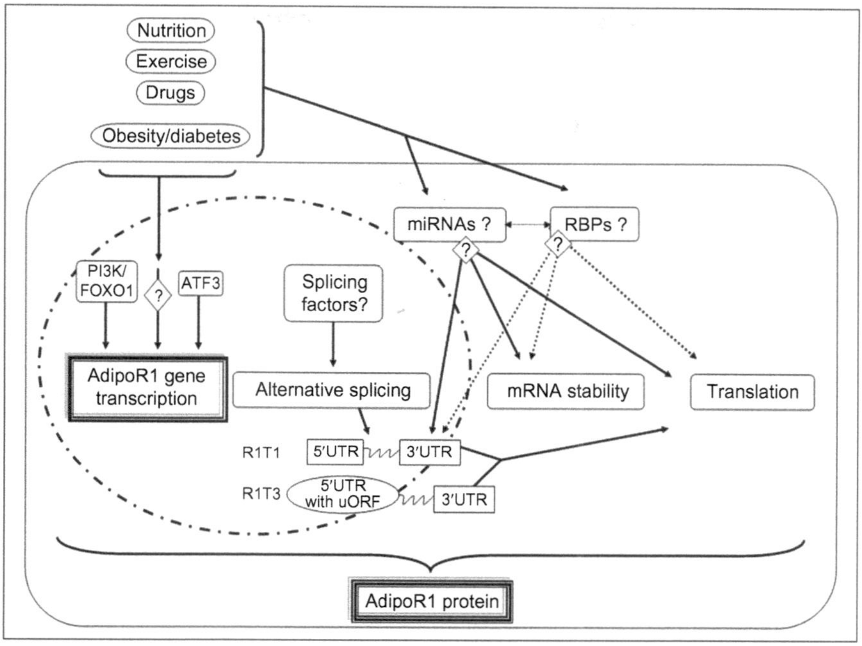

Figure 4.1 Molecular mechanisms regulating AdipoR1 protein levels in skeletal muscle. AdipoR1 protein expression is differentially controlled by several mechanisms depending on the upstream signaling pathways activated by physiological and pathophysiological conditions. At the transcriptional level, insulin has been suggested to repress AdipoR1 transcription via the PI3K and FOXO1 pathway (Inukai et al., 2005; Tsuchida et al., 2004). This may contribute to the decrease in muscle AdipoR1 mRNA expression, observed in diabetes, and to the increase in AdipoR1 gene expression following fasting, prolonged physical training, or prolonged TZDs therapy (summarized in Table 4.1). The ER stress-inducible transcription factor ATF3 negatively regulates AdipoR1 transcription via binding to an ATF3-responsive region in the AdipoR1 promoter; however, the significance of ER stress and its role in regulating skeletal muscle energy metabolism is controversial. At the posttranscriptional level alternative splicing of human AdipoR1 produces a novel muscle-specific transcript, termed R1T3 (Ashwal et al., 2011). This transcript is increased during fetal development and myogenesis, is associated with insulin sensitivity, and is repressed in diabetes. Translational control emerges as an important posttranscriptional mechanism in the regulation of muscle AdipoR1 protein expression (Ashwal et al., 2011). Future studies will be required in order to decipher the role of microRNAs (miRNAs) and RNA-binding proteins (RBPs) in AdipoR1 translation.

obesity and insulin resistance (Ghosh et al., 2007; Patel et al., 2005; Pihlajamaki et al., 2011).

We have recently reported the identification of previously unrecognized splice variants that encode the human AdipoR1 (Ashwal et al., 2011). Screening of potential AdipoR1 5′UTR splice variants revealed a novel transcript (termed R1T3), which is derived from inclusion of a short exon located in intron 1 at the 5′UTR of the human AdipoR1 gene. The novel transcript was found to be abundantly and predominantly expressed in adult human muscle, contrary to the previously described human AdipoR1 mRNA transcript R1T1 which is ubiquitously expressed in adult human tissues including muscle (Yamauchi, Kamon, Ito, et al., 2003). The two distinct transcripts are driven by the same promoter, and both encode AdipoR1 receptor; however, R1T3, unlike R1T1, is subjected to developmental regulation, and its expression is significantly increased during fetal development and human myoblast–myotube differentiation, in parallel to a significant increase in AdipoR1 protein expression. Importantly, R1T3-specific siRNA decreased significantly the expression of AdipoR1 protein in human skeletal muscle cells.

Intriguingly, the expression of AdipoR1 transcripts was affected by physiological and pathophysiological conditions. Our findings demonstrated a marked reduction in muscle expression of both R1T1 and R1T3 in individuals with T2D compared to subjects with NGT, paralleled with a decreased expression of the differentiation marker myogenin. Moreover, a significant decrease in R1T3-to-R1T1 ratio was found in the type 2 diabetic group compared with the NGT group indicating that AdipoR1 transcripts expression is not regulated solely by transcriptional mechanisms but also by a diabetes-specific repression of the alternative splicing process. Among NGT subjects, R1T3 expression was positively correlated with insulin sensitivity, and R1T1 and R1T3 were negatively associated with insulin levels.

Further analysis of the distinct 5′UTR regions of R1T1 and R1T3 revealed that the 5′UTR of R1T3 contains upstream ORFs (uORFs) that attenuate the translation of downstream coding sequences compared with the 5′UTR of R1T1. Conversely, AdipoR1 3′UTR (identical in both transcripts) was associated with enhanced translation efficiency during myoblast–myotube differentiation, as discussed in the previous section. These findings suggest that the introduction of uORFs, which is upregulated by alternative splicing during muscle differentiation, may serve as a negative control mechanism to attenuate the concomitant-marked increase in AdipoR1 protein translation mediated by the 3′UTR during differentiation (Fig. 4.1). Furthermore, the significant decrease in R1T3-to-R1T1 ratio found in diabetes raises the possibility that R1T3 repression in diabetes may represent a compensatory mechanism to enhance AdipoR1 biosynthesis when AdipoR1 gene transcription is significantly reduced.

7. CONCLUDING REMARKS

Much progress has been made in the past few years toward understanding the physiological actions and regulation of adiponectin and its receptors, AdipoR1 and AdipoR2. It is now well documented that adiponectin, through AdipoRs, exerts antidiabetic, antiatherogenic, and anti-inflammatory effects and influences a wide range of biological functions including glucose and lipid metabolism, inflammation, and skeletal muscle growth. In particular, skeletal muscle emerged as an important target of adiponectin and AdipoRs in the regulation of energy metabolism. AdipoR1, specifically, has a crucial role in mitochondrial dysfunction and insulin resistance in muscle and thus may be highly important therapeutic target.

Future research in this field will have to focus on several open questions regarding AdipoR1 signaling in order to fully comprehend its function in muscle. For example, what is the mechanism by which AdipoR1 increase calcium influx upon activation? Are there more proteins that directly bind AdipoR1? Is T-cadherin involved in adiponectin and AdipoR1 functions in muscle? Is AdipoR1 biosynthesis and/or functions impaired in obesity or metabolic diseases related to obesity? Furthermore, data concerning the molecular pathways regulating AdipoR1 protein levels are scarce, and conflicting data about its mRNA expression under physiological and pathophysiological conditions suggest that mRNA levels alone cannot fully account for AdipoR1 protein expression. We have recently demonstrated for the first time an important role for posttranscriptional mechanisms, including alternative splicing and translational control, in the regulation of AdipoR1 protein levels in skeletal muscle (Fig. 4.1), which may explain, at least in part, the contradicting mRNA results. Further studies will be required to elucidate the specific factors that influence these posttranscriptional mechanisms and AdipoR1 biosynthesis and to understand their effects on adiponectin biological action in muscle. Collectively, answers to these questions will provide better tools to enhance muscle AdipoR1 levels and function which may improve insulin resistance, and T2D linked to obesity.

ACKNOWLEDGMENTS

Related work in the authors' laboratory has been supported by the D-Cure Foundation for Diabetes Care in Israel and the Israel Association for the Study of Diabetes. Y. L. is supported by a postdoctoral fellowship from the International Human Frontier Science Program Organization. Unfortunately, it was not possible to quote all relevant literature, and we apologize for any omissions of pertinent research papers.

REFERENCES

Adeli, K. (2011). Translational control mechanisms in metabolic regulation: Critical role of RNA binding proteins, microRNAs, and cytoplasmic RNA granules. *American Journal of Physiology. Endocrinology and Metabolism, 301*, E1051–E1064.

Amin, R. H., Mathews, S. T., Camp, H. S., Ding, L., & Leff, T. (2010). Selective activation of PPARgamma in skeletal muscle induces endogenous production of adiponectin and protects mice from diet-induced insulin resistance. *American Journal of Physiology. Endocrinology and Metabolism, 298*, E28–E37.

Anavi, S., Ilan, E., Tirosh, O., & Madar, Z. (2010). Infusion of a lipid emulsion modulates AMPK and related proteins in rat liver, muscle, and adipose tissues. *Obesity (Silver Spring), 18*, 1108–1115.

Andersen, J. L., Schjerling, P., Andersen, L. L., & Dela, F. (2003). Resistance training and insulin action in humans: Effects of de-training. *The Journal of Physiology, 551*, 1049–1058.

Ando, D., Hosaka, Y., Suzuki, K., & Yamagata, Z. (2009). Effects of exercise training on circulating high molecular weight adiponectin and adiponectin oligomer composition: A randomized controlled trial. *Journal of Atherosclerosis and Thrombosis, 16*, 733–739.

Arita, Y., Kihara, S., Ouchi, N., Takahashi, M., Maeda, K., Miyagawa, J., et al. (1999). Paradoxical decrease of an adipose-specific protein, adiponectin, in obesity. *Biochemical and Biophysical Research Communications, 257*, 79–83.

Ashwal, R., Hemi, R., Tirosh, A., Gordin, R., Yissachar, E., Cohen-Dayag, A., et al. (2011). Differential expression of novel adiponectin receptor-1 transcripts in skeletal muscle of subjects with normal glucose tolerance and type 2 diabetes. *Diabetes, 60*, 936–946.

Barnea, M., Shamay, A., Stark, A. H., & Madar, Z. (2006). A high-fat diet has a tissue-specific effect on adiponectin and related enzyme expression. *Obesity (Silver Spring), 14*, 2145–2153.

Bartel, D. P. (2009). MicroRNAs: Target recognition and regulatory functions. *Cell, 136*, 215–233.

Berg, A. H., Combs, T. P., Du, X., Brownlee, M., & Scherer, P. E. (2001). The adipocyte-secreted protein Acrp30 enhances hepatic insulin action. *Nature Medicine, 7*, 947–953.

Beylot, M., Pinteur, C., & Peroni, O. (2006). Expression of the adiponectin receptors AdipoR1 and AdipoR2 in lean rats and in obese Zucker rats. *Metabolism, 55*, 396–401.

Bjursell, M., Ahnmark, A., Bohlooly, Y., William-Olsson, L., Rhedin, M., Peng, X. R., et al. (2007). Opposing effects of adiponectin receptors 1 and 2 on energy metabolism. *Diabetes, 56*, 583–593.

Bluher, M., Bullen, J. W., Jr., Lee, J. H., Kralisch, S., Fasshauer, M., Kloting, N., et al. (2006). Circulating adiponectin and expression of adiponectin receptors in human skeletal muscle: Associations with metabolic parameters and insulin resistance and regulation by physical training. *The Journal of Clinical Endocrinology and Metabolism, 91*, 2310–2316.

Bonnard, C., Durand, A., Vidal, H., & Rieusset, J. (2008). Changes in adiponectin, its receptors and AMPK activity in tissues of diet-induced diabetic mice. *Diabetes & Metabolism, 34*, 52–61.

Brochu-Gaudreau, K., Rehfeldt, C., Blouin, R., Bordignon, V., Murphy, B. D., & Palin, M. F. (2010). Adiponectin action from head to toe. *Endocrine, 37*, 11–32.

Buechler, C., Wanninger, J., & Neumeier, M. (2010). Adiponectin receptor binding proteins—Recent advances in elucidating adiponectin signalling pathways. *FEBS Letters, 584*, 4280–4286.

Bullen, J. W., Jr., Bluher, S., Kelesidis, T., & Mantzoros, C. S. (2007). Regulation of adiponectin and its receptors in response to development of diet-induced obesity in mice. *American Journal of Physiology. Endocrinology and Metabolism, 292*, E1079–E1086.

Burr, J. F., Rowan, C. P., Jamnik, V. K., & Riddell, M. C. (2010). The role of physical activity in type 2 diabetes prevention: Physiological and practical perspectives. *The Physician and Sportsmedicine, 38*, 72–82.

Ceddia, R. B., Somwar, R., Maida, A., Fang, X., Bikopoulos, G., & Sweeney, G. (2005). Globular adiponectin increases GLUT4 translocation and glucose uptake but reduces glycogen synthesis in rat skeletal muscle cells. *Diabetologia, 48*, 132–139.

Chang, S. P., Chen, Y. H., Chang, W. C., Liu, I. M., & Cheng, J. T. (2006). Increase of adiponectin receptor gene expression by physical exercise in soleus muscle of obese Zucker rats. *European Journal of Applied Physiology, 97*, 189–195.

Choudhary, S., Sinha, S., Zhao, Y., Banerjee, S., Sathyanarayana, P., Shahani, S., et al. (2011). NF-kappaB-inducing kinase (NIK) mediates skeletal muscle insulin resistance: Blockade by adiponectin. *Endocrinology, 152*, 3622–3627.

Christiansen, T., Paulsen, S. K., Bruun, J. M., Ploug, T., Pedersen, S. B., & Richelsen, B. (2010). Diet-induced weight loss and exercise alone and in combination enhance the expression of adiponectin receptors in adipose tissue and skeletal muscle, but only diet-induced weight loss enhanced circulating adiponectin. *The Journal of Clinical Endocrinology and Metabolism, 95*, 911–919.

Civitarese, A. E., Jenkinson, C. P., Richardson, D., Bajaj, M., Cusi, K., Kashyap, S., et al. (2004). Adiponectin receptors gene expression and insulin sensitivity in non-diabetic Mexican Americans with or without a family history of Type 2 diabetes. *Diabetologia, 47*, 816–820.

Civitarese, A. E., Ukropcova, B., Carling, S., Hulver, M., DeFronzo, R. A., Mandarino, L., et al. (2006). Role of adiponectin in human skeletal muscle bioenergetics. *Cell Metabolism, 4*, 75–87.

Coletta, D. K., Sriwijitkamol, A., Wajcberg, E., Tantiwong, P., Li, M., Prentki, M., et al. (2009). Pioglitazone stimulates AMP-activated protein kinase signalling and increases the expression of genes involved in adiponectin signalling, mitochondrial function and fat oxidation in human skeletal muscle in vivo: A randomised trial. *Diabetologia, 52*, 723–732.

Combs, T. P., Berg, A. H., Obici, S., Scherer, P. E., & Rossetti, L. (2001). Endogenous glucose production is inhibited by the adipose-derived protein Acrp30. *The Journal of Clinical Investigation, 108*, 1875–1881.

Combs, T. P., Pajvani, U. B., Berg, A. H., Lin, Y., Jelicks, L. A., Laplante, M., et al. (2004). A transgenic mouse with a deletion in the collagenous domain of adiponectin displays elevated circulating adiponectin and improved insulin sensitivity. *Endocrinology, 145*, 367–383.

Crunkhorn, S., Dearie, F., Mantzoros, C., Gami, H., da Silva, W. S., Espinoza, D., et al. (2007). Peroxisome proliferator activator receptor gamma coactivator-1 expression is reduced in obesity: Potential pathogenic role of saturated fatty acids and p38 mitogen-activated protein kinase activation. *The Journal of Biological Chemistry, 282*, 15439–15450.

de Oliveira, C., de Mattos, A. B., Biz, C., Oyama, L. M., Ribeiro, E. B., & do Nascimento, C. M. (2011). High-fat diet and glucocorticoid treatment cause hyperglycemia associated with adiponectin receptor alterations. *Lipids in Health and Disease, 10*, 11.

Debard, C., Laville, M., Berbe, V., Loizon, E., Guillet, C., Morio-Liondore, B., et al. (2004). Expression of key genes of fatty acid oxidation, including adiponectin receptors, in skeletal muscle of Type 2 diabetic patients. *Diabetologia, 47*, 917–925.

Deckert, C. M., Heiker, J. T., & Beck-Sickinger, A. G. (2006). Localization of novel adiponectin receptor constructs. *Journal of Receptor and Signal Transduction Research, 26*, 647–657.

DeFronzo, R. A., Gunnarsson, R., Bjorkman, O., Olsson, M., & Wahren, J. (1985). Effects of insulin on peripheral and splanchnic glucose metabolism in noninsulin-dependent (type II) diabetes mellitus. *The Journal of Clinical Investigation, 76*, 149–155.

Delaigle, A. M., Jonas, J. C., Bauche, I. B., Cornu, O., & Brichard, S. M. (2004). Induction of adiponectin in skeletal muscle by inflammatory cytokines: in vivo and in vitro studies. *Endocrinology, 145*, 5589–5597.

Delaigle, A. M., Senou, M., Guiot, Y., Many, M. C., & Brichard, S. M. (2006). Induction of adiponectin in skeletal muscle of type 2 diabetic mice: in vivo and in vitro studies. *Diabetologia, 49*, 1311–1323.

Denzel, M. S., Scimia, M. C., Zumstein, P. M., Walsh, K., Ruiz-Lozano, P., & Ranscht, B. (2010). T-cadherin is critical for adiponectin-mediated cardioprotection in mice. *The Journal of Clinical Investigation, 120*, 4342–4352.

Du, M., Shen, Q. W., Zhu, M. J., & Ford, S. P. (2007). Leucine stimulates mammalian target of rapamycin signaling in C2C12 myoblasts in part through inhibition of adenosine monophosphate-activated protein kinase. *Journal of Animal Science, 85*, 919–927.

Eriksson, K. F., & Lindgarde, F. (1991). Prevention of type 2 (non-insulin-dependent) diabetes mellitus by diet and physical exercise. The 6-year Malmo feasibility study. *Diabetologia, 34*, 891–898.

Fang, X., Fetros, J., Dadson, K. E., Xu, A., & Sweeney, G. (2009). Leptin prevents the metabolic effects of adiponectin in L6 myotubes. *Diabetologia, 52*, 2190–2200.

Fang, X., Palanivel, R., Zhou, X., Liu, Y., Xu, A., Wang, Y., et al. (2005). Hyperglycemia- and hyperinsulinemia-induced alteration of adiponectin receptor expression and adiponectin effects in L6 myoblasts. *Journal of Molecular Endocrinology, 35*, 465–476.

Fiaschi, T., Cirelli, D., Comito, G., Gelmini, S., Ramponi, G., Serio, M., et al. (2009). Globular adiponectin induces differentiation and fusion of skeletal muscle cells. *Cell Research, 19*, 584–597.

Fruebis, J., Tsao, T. S., Javorschi, S., Ebbets-Reed, D., Erickson, M. R., Yen, F. T., et al. (2001). Proteolytic cleavage product of 30-kDa adipocyte complement-related protein increases fatty acid oxidation in muscle and causes weight loss in mice. *Proceedings of the National Academy of Sciences of the United States of America, 98*, 2005–2010.

Ghosh, N., Patel, N., Jiang, K., Watson, J. E., Cheng, J., Chalfant, C. E., et al. (2007). Ceramide-activated protein phosphatase involvement in insulin resistance via Akt, serine/arginine-rich protein 40, and ribonucleic acid splicing in L6 skeletal muscle cells. *Endocrinology, 148*, 1359–1366.

Heiker, J. T., Kosel, D., & Beck-Sickinger, A. G. (2010). Molecular mechanisms of signal transduction via adiponectin and adiponectin receptors. *Biological Chemistry, 391*, 1005–1018.

Heiker, J. T., Wottawah, C. M., Juhl, C., Kosel, D., Morl, K., & Beck-Sickinger, A. G. (2009). Protein kinase CK2 interacts with adiponectin receptor 1 and participates in adiponectin signaling. *Cellular Signalling, 21*, 936–942.

Heilbronn, L. K., Gregersen, S., Shirkhedkar, D., Hu, D., & Campbell, L. V. (2007). Impaired fat oxidation after a single high-fat meal in insulin-sensitive nondiabetic individuals with a family history of type 2 diabetes. *Diabetes, 56*, 2046–2053.

Henstridge, D. C., & Febbraio, M. A. (2010). Adiponectin sparks an interest in calcium. *Cell Metabolism, 11*, 447–449.

Holland, W. L., Miller, R. A., Wang, Z. V., Sun, K., Barth, B. M., Bui, H. H., et al. (2011). Receptor-mediated activation of ceramidase activity initiates the pleiotropic actions of adiponectin. *Nature Medicine, 17*, 55–63.

Holmes, R. M., Yi, Z., De Filippis, E., Berria, R., Shahani, S., Sathyanarayana, P., et al. (2011). Increased abundance of the adaptor protein containing pleckstrin homology domain, phosphotyrosine binding domain and leucine zipper motif (APPL1) in patients with obesity and type 2 diabetes: Evidence for altered adiponectin signalling. *Diabetologia, 54*, 2122–2131.

Hotta, K., Funahashi, T., Arita, Y., Takahashi, M., Matsuda, M., Okamoto, Y., et al. (2000). Plasma concentrations of a novel, adipose-specific protein, adiponectin, in type 2 diabetic patients. *Arteriosclerosis, Thrombosis, and Vascular Biology, 20*, 1595–1599.

Hu, E., Liang, P., & Spiegelman, B. M. (1996). AdipoQ is a novel adipose-specific gene dysregulated in obesity. *The Journal of Biological Chemistry, 271*, 10697–10703.

Huang, H., Iida, K. T., Sone, H., Yokoo, T., Yamada, N., & Ajisaka, R. (2006). The effect of exercise training on adiponectin receptor expression in KKAy obese/diabetic mice. *The Journal of Endocrinology*, *189*, 643–653.

Hug, C., Wang, J., Ahmad, N. S., Bogan, J. S., Tsao, T. S., & Lodish, H. F. (2004). T-cadherin is a receptor for hexameric and high-molecular-weight forms of Acrp30/adiponectin. *Proceedings of the National Academy of Sciences of the United States of America*, *101*, 10308–10313.

Inoue, N., Nagao, K., Wang, Y. M., Noguchi, H., Shirouchi, B., & Yanagita, T. (2006). Dietary conjugated linoleic acid lowered tumor necrosis factor-alpha content and altered expression of genes related to lipid metabolism and insulin sensitivity in the skeletal muscle of Zucker rats. *Journal of Agricultural and Food Chemistry*, *54*, 7935–7939.

Inukai, K., Nakashima, Y., Watanabe, M., Takata, N., Sawa, T., Kurihara, S., et al. (2005). Regulation of adiponectin receptor gene expression in diabetic mice. *American Journal of Physiology. Endocrinology and Metabolism*, *288*, E876–E882.

Iwabu, M., Yamauchi, T., Okada-Iwabu, M., Sato, K., Nakagawa, T., Funata, M., et al. (2010). Adiponectin and AdipoR1 regulate PGC-1alpha and mitochondria by Ca(2+) and AMPK/SIRT1. *Nature*, *464*, 1313–1319.

Jortay, J., Senou, M., Delaigle, A., Noel, L., Funahashi, T., Maeda, N., et al. (2010). Local induction of adiponectin reduces lipopolysaccharide-triggered skeletal muscle damage. *Endocrinology*, *151*, 4840–4851.

Kadowaki, T., Yamauchi, T., Kubota, N., Hara, K., Ueki, K., & Tobe, K. (2006). Adiponectin and adiponectin receptors in insulin resistance, diabetes, and the metabolic syndrome. *The Journal of Clinical Investigation*, *116*, 1784–1792.

Kaltenbach, S., Staiger, H., Weisser, M., Haas, C., Stumvoll, M., Machicao, F., et al. (2005). Adiponectin receptor gene expression in human skeletal muscle cells is not regulated by fibrates and thiazolidinediones. *International Journal of Obesity*, *29*, 760–765.

Kelly, K. R., Blaszczak, A., Haus, J. M., Patrick-Melin, A., Fealy, C. E., Solomon, T. P., et al. (2012). A 7-d exercise program increases high-molecular weight adiponectin in obese adults. *Medicine & Science in Sports & Exercise*, *44*, 69–74.

Kishida, K., Funahashi, T., & Shimomura, I. (2012). Molecular mechanisms of diabetes and atherosclerosis: Role of adiponectin. *Endocrine, Metabolic & Immune Disorders Drug Targets*, *12*, 118–131.

Knowler, W. C., Barrett-Connor, E., Fowler, S. E., Hamman, R. F., Lachin, J. M., Walker, E. A., et al. (2002). Reduction in the incidence of type 2 diabetes with lifestyle intervention or metformin. *The New England Journal of Medicine*, *346*, 393–403.

Kosel, D., Heiker, J. T., Juhl, C., Wottawah, C. M., Bluher, M., Morl, K., et al. (2010). Dimerization of adiponectin receptor 1 is inhibited by adiponectin. *Journal of Cell Science*, *123*, 1320–1328.

Krause, M. P., Liu, Y., Vu, V., Chan, L., Xu, A., Riddell, M. C., et al. (2008). Adiponectin is expressed by skeletal muscle fibers and influences muscle phenotype and function. *American Journal of Physiology. Cell Physiology*, *295*, C203–C212.

Kubota, N., Terauchi, Y., Yamauchi, T., Kubota, T., Moroi, M., Matsui, J., et al. (2002). Disruption of adiponectin causes insulin resistance and neointimal formation. *The Journal of Biological Chemistry*, *277*, 25863–25866.

Li, W., Tonelli, J., Kishore, P., Owen, R., Goodman, E., Scherer, P. E., et al. (2007). Insulin-sensitizing effects of thiazolidinediones are not linked to adiponectin receptor expression in human fat or muscle. *American Journal of Physiology. Endocrinology and Metabolism*, *292*, E1301–E1307.

Li, L., Wu, L., Wang, C., Liu, L., & Zhao, Y. (2007). Adiponectin modulates carnitine palmitoyltransferase-1 through AMPK signaling cascade in rat cardiomyocytes. *Regulatory Peptides*, *139*, 72–79.

Liu, Y., Chewchuk, S., Lavigne, C., Brule, S., Pilon, G., Houde, V., et al. (2009). Functional significance of skeletal muscle adiponectin production, changes in animal models of obesity and diabetes, and regulation by rosiglitazone treatment. *American Journal of Physiology. Endocrinology and Metabolism, 297*, E657–E664.

Liu, Y., Michael, M. D., Kash, S., Bensch, W. R., Monia, B. P., Murray, S. F., et al. (2007). Deficiency of adiponectin receptor 2 reduces diet-induced insulin resistance but promotes type 2 diabetes. *Endocrinology, 148*, 683–692.

Liu, B. H., Wang, P. H., Wang, Y. C., Cheng, W. M., Mersmann, H. J., & Ding, S. T. (2008). Fasting regulates the expression of adiponectin receptors in young growing pigs. *Journal of Animal Science, 86*, 3377–3384.

Maeda, K., Okubo, K., Shimomura, I., Funahashi, T., Matsuzawa, Y., & Matsubara, K. (1996). cDNA cloning and expression of a novel adipose specific collagen-like factor, apM1 (AdiPose Most abundant Gene transcript 1). *Biochemical and Biophysical Research Communications, 221*, 286–289.

Maeda, N., Shimomura, I., Kishida, K., Nishizawa, H., Matsuda, M., Nagaretani, H., et al. (2002). Diet-induced insulin resistance in mice lacking adiponectin/ACRP30. *Nature Medicine, 8*, 731–737.

Maeda, N., Takahashi, M., Funahashi, T., Kihara, S., Nishizawa, H., Kishida, K., et al. (2001). PPARgamma ligands increase expression and plasma concentrations of adiponectin, an adipose-derived protein. *Diabetes, 50*, 2094–2099.

Magkos, F., Mohammed, B. S., & Mittendorfer, B. (2010). Enhanced insulin sensitivity after acute exercise is not associated with changes in high-molecular weight adiponectin concentration in plasma. *European Journal of Endocrinology, 162*, 61–66.

Mao, X., Kikani, C. K., Riojas, R. A., Langlais, P., Wang, L., Ramos, F. J., et al. (2006). APPL1 binds to adiponectin receptors and mediates adiponectin signalling and function. *Nature Cell Biology, 8*, 516–523.

Metais, C., Forcheron, F., Abdallah, P., Basset, A., Del Carmine, P., Bricca, G., et al. (2008). Adiponectin receptors: Expression in Zucker diabetic rats and effects of fenofibrate and metformin. *Metabolism, 57*, 946–953.

Mullen, K. L., Tishinsky, J. M., Robinson, L. E., & Dyck, D. J. (2010). Skeletal muscle inflammation is not responsible for the rapid impairment in adiponectin response with high-fat feeding in rats. *American Journal of Physiology. Regulatory, Integrative and Comparative Physiology, 299*, R500–R508.

Nakano, Y., Tobe, T., Choi-Miura, N. H., Mazda, T., & Tomita, M. (1996). Isolation and characterization of GBP28, a novel gelatin-binding protein purified from human plasma. *Journal of Biochemistry, 120*, 803–812.

Nawrocki, A. R., Rajala, M. W., Tomas, E., Pajvani, U. B., Saha, A. K., Trumbauer, M. E., et al. (2006). Mice lacking adiponectin show decreased hepatic insulin sensitivity and reduced responsiveness to peroxisome proliferator-activated receptor gamma agonists. *The Journal of Biological Chemistry, 281*, 2654–2660.

Numao, S., Suzuki, M., Matsuo, T., Nomata, Y., Nakata, Y., & Tanaka, K. (2008). Effects of acute aerobic exercise on high-molecular-weight adiponectin. *Medicine & Science in Sports & Exercise, 40*, 1271–1276.

O'Leary, V. B., Jorett, A. E., Marchetti, C. M., Gonzalez, F., Phillips, S. A., Ciaraldi, T. P., et al. (2007). Enhanced adiponectin multimer ratio and skeletal muscle adiponectin receptor expression following exercise training and diet in older insulin-resistant adults. *American Journal of Physiology. Endocrinology and Metabolism, 293*, E421–E427.

Ordelheide, A. M., Heni, M., Gommer, N., Gasse, L., Haas, C., Guirguis, A., et al. (2011). The myocyte expression of adiponectin receptors and PPARdelta is highly coordinated and reflects lipid metabolism of the human donors. *Experimental Diabetes Research, 2011*, 692536.

Pajvani, U. B., Du, X., Combs, T. P., Berg, A. H., Rajala, M. W., Schulthess, T., et al. (2003). Structure-function studies of the adipocyte-secreted hormone Acrp30/adiponectin. Implications fpr metabolic regulation and bioactivity. *The Journal of Biological Chemistry*, *278*, 9073–9085.

Pajvani, U. B., Hawkins, M., Combs, T. P., Rajala, M. W., Doebber, T., Berger, J. P., et al. (2004). Complex distribution, not absolute amount of adiponectin, correlates with thiazolidinedione-mediated improvement in insulin sensitivity. *The Journal of Biological Chemistry*, *279*, 12152–12162.

Pan, X. R., Li, G. W., Hu, Y. H., Wang, J. X., Yang, W. Y., An, Z. X., et al. (1997). Effects of diet and exercise in preventing NIDDM in people with impaired glucose tolerance. The Da Qing IGT and Diabetes Study. *Diabetes Care*, *20*, 537–544.

Park, H. J., Kang, Y. M., Kim, C. H., & Jung, M. H. (2010). ATF3 negatively regulates adiponectin receptor 1 expression. *Biochemical and Biophysical Research Communications*, *400*, 72–77.

Patel, N. A., Kaneko, S., Apostolatos, H. S., Bae, S. S., Watson, J. E., Davidowitz, K., et al. (2005). Molecular and genetic studies imply Akt-mediated signaling promotes protein kinase CbetaII alternative splicing via phosphorylation of serine/arginine-rich splicing factor SRp40. *The Journal of Biological Chemistry*, *280*, 14302–14309.

Pedersen, B. K., & Febbraio, M. A. (2008). Muscle as an endocrine organ: Focus on muscle-derived interleukin-6. *Physiological Reviews*, *88*, 1379–1406.

Pihlajamaki, J., Lerin, C., Itkonen, P., Boes, T., Floss, T., Schroeder, J., et al. (2011). Expression of the splicing factor gene SFRS10 is reduced in human obesity and contributes to enhanced lipogenesis. *Cell Metabolism*, *14*, 208–218.

Pita, J., Panadero, A., Soriano-Guillen, L., Rodriguez, E., & Rovira, A. (2012). The insulin sensitizing effects of PPAR-gamma agonist are associated to changes in adiponectin index and adiponectin receptors in Zucker fatty rats. *Regulatory Peptides*, *174*, 18–25.

Prior, L. J., Velkoska, E., Watts, R., Cameron-Smith, D., & Morris, M. J. (2008). Undernutrition during suckling in rats elevates plasma adiponectin and its receptor in skeletal muscle regardless of diet composition: A protective effect? *International Journal of Obesity*, *32*, 1585–1594.

Qiao, L., Kinney, B., Yoo, H. S., Lee, B., Schaack, J., & Shao, J. (2012). Adiponectin increases skeletal muscle mitochondrial biogenesis by suppressing mitogen-activated protein kinase phosphatase-1. *Diabetes*, *61*, 1463–1470.

Rodriguez, A., Catalan, V., Becerril, S., Gil, M. J., Mugueta, C., Gomez-Ambrosi, J., et al. (2008). Impaired adiponectin-AMPK signalling in insulin-sensitive tissues of hypertensive rats. *Life Sciences*, *83*, 540–549.

Sanchez-Gurmaches, J., Cruz-Garcia, L., Gutierrez, J., & Navarro, I. (2012). Adiponectin effects and gene expression in rainbow trout: An in vivo and in vitro approach. *The Journal of Experimental Biology*, *215*, 1373–1383.

Satoh, H., Nguyen, M. T., Trujillo, M., Imamura, T., Usui, I., Scherer, P. E., et al. (2005). Adenovirus-mediated adiponectin expression augments skeletal muscle insulin sensitivity in male Wistar rats. *Diabetes*, *54*, 1304–1313.

Scherer, P. E., Williams, S., Fogliano, M., Baldini, G., & Lodish, H. F. (1995). A novel serum protein similar to C1q, produced exclusively in adipocytes. *The Journal of Biological Chemistry*, *270*, 26746–26749.

Shetty, S., Kusminski, C. M., & Scherer, P. E. (2009). Adiponectin in health and disease: Evaluation of adiponectin-targeted drug development strategies. *Trends in Pharmacological Sciences*, *30*, 234–239.

Sigal, R. J., Kenny, G. P., Wasserman, D. H., Castaneda-Sceppa, C., & White, R. D. (2006). Physical activity/exercise and type 2 diabetes: A consensus statement from the American Diabetes Association. *Diabetes Care*, *29*, 1433–1438.

Sixt, S., Beer, S., Bluher, M., Korff, N., Peschel, T., Sonnabend, M., et al. (2010). Long- but not short-term multifactorial intervention with focus on exercise training improves coronary endothelial dysfunction in diabetes mellitus type 2 and coronary artery disease. *European Heart Journal, 31*, 112–119.

Staiger, H., Kaltenbach, S., Staiger, K., Stefan, N., Fritsche, A., Guirguis, A., et al. (2004). Expression of adiponectin receptor mRNA in human skeletal muscle cells is related to in vivo parameters of glucose and lipid metabolism. *Diabetes, 53*, 2195–2201.

Sun, X., He, J., Mao, C., Han, R., Wang, Z., Liu, Y., et al. (2008). Negative regulation of adiponectin receptor 1 promoter by insulin via a repressive nuclear inhibitory protein element. *FEBS Letters, 582*, 3401–3407.

Tan, G. D., Debard, C., Funahashi, T., Humphreys, S. M., Matsuzawa, Y., Frayn, K. N., et al. (2005). Changes in adiponectin receptor expression in muscle and adipose tissue of type 2 diabetic patients during rosiglitazone therapy. *Diabetologia, 48*, 1585–1589.

Tang, Y. T., Hu, T., Arterburn, M., Boyle, B., Bright, J. M., Emtage, P. C., et al. (2005). PAQR proteins: A novel membrane receptor family defined by an ancient 7-transmembrane pass motif. *Journal of Molecular Evolution, 61*, 372–380.

Tomas, E., Tsao, T. S., Saha, A. K., Murrey, H. E., Zhang, C. C., Itani, S. I., et al. (2002). Enhanced muscle fat oxidation and glucose transport by ACRP30 globular domain: Acetyl-CoA carboxylase inhibition and AMP-activated protein kinase activation. *Proceedings of the National Academy of Sciences of the United States of America, 99*, 16309–16313.

Trujillo, M. E., & Scherer, P. E. (2005). Adiponectin–journey from an adipocyte secretory protein to biomarker of the metabolic syndrome. *Journal of Internal Medicine, 257*, 167–175.

Tsao, T. S., Murrey, H. E., Hug, C., Lee, D. H., & Lodish, H. F. (2002). Oligomerization state-dependent activation of NF-kappa B signaling pathway by adipocyte complement-related protein of 30 kDa (Acrp30). *The Journal of Biological Chemistry, 277*, 29359–29362.

Tsuchida, A., Yamauchi, T., Ito, Y., Hada, Y., Maki, T., Takekawa, S., et al. (2004). Insulin/Foxo1 pathway regulates expression levels of adiponectin receptors and adiponectin sensitivity. *The Journal of Biological Chemistry, 279*, 30817–30822.

Tuomilehto, J., Lindstrom, J., Eriksson, J. G., Valle, T. T., Hamalainen, H., Ilanne-Parikka, P., et al. (2001). Prevention of type 2 diabetes mellitus by changes in lifestyle among subjects with impaired glucose tolerance. *The New England Journal of Medicine, 344*, 1343–1350.

Van Berendoncks, A. M., Garnier, A., Beckers, P., Hoymans, V. Y., Possemiers, N., Fortin, D., et al. (2010). Functional adiponectin resistance at the level of the skeletal muscle in mild to moderate chronic heart failure. *Circulation. Heart Failure, 3*, 185–194.

Van Berendoncks, A. M., Garnier, A., Beckers, P., Hoymans, V. Y., Possemiers, N., Fortin, D., et al. (2011). Exercise training reverses adiponectin resistance in skeletal muscle of patients with chronic heart failure. *Heart, 97*, 1403–1409.

Vu, V., Riddell, M. C., & Sweeney, G. (2007). Circulating adiponectin and adiponectin receptor expression in skeletal muscle: Effects of exercise. *Diabetes/Metabolism Research and Reviews, 23*, 600–611.

Waki, H., Yamauchi, T., Kamon, J., Kita, S., Ito, Y., Hada, Y., et al. (2005). Generation of globular fragment of adiponectin by leukocyte elastase secreted by monocytic cell line THP-1. *Endocrinology, 146*, 790–796.

Wang, E. T., Sandberg, R., Luo, S., Khrebtukova, I., Zhang, L., Mayr, C., et al. (2008). Alternative isoform regulation in human tissue transcriptomes. *Nature, 456*, 470–476.

Wang, C., Xin, X., Xiang, R., Ramos, F. J., Liu, M., Lee, H. J., et al. (2009). Yin-Yang regulation of adiponectin signaling by APPL isoforms in muscle cells. *The Journal of Biological Chemistry, 284*, 31608–31615.

Wasserman, D. H. (2009). Four grams of glucose. *American Journal of Physiology. Endocrinology and Metabolism, 296*, E11–E21.

Yamauchi, T., & Kadowaki, T. (2008). Physiological and pathophysiological roles of adiponectin and adiponectin receptors in the integrated regulation of metabolic and cardiovascular diseases. *International Journal of Obesity, 32*(Suppl 7), S13–S18.

Yamauchi, T., Kamon, J., Ito, Y., Tsuchida, A., Yokomizo, T., Kita, S., et al. (2003). Cloning of adiponectin receptors that mediate antidiabetic metabolic effects. *Nature, 423*, 762–769.

Yamauchi, T., Kamon, J., Minokoshi, Y., Ito, Y., Waki, H., Uchida, S., et al. (2002). Adiponectin stimulates glucose utilization and fatty-acid oxidation by activating AMP-activated protein kinase. *Nature Medicine, 8*, 1288–1295.

Yamauchi, T., Kamon, J., Waki, H., Imai, Y., Shimozawa, N., Hioki, K., et al. (2003). Globular adiponectin protected ob/ob mice from diabetes and ApoE-deficient mice from atherosclerosis. *The Journal of Biological Chemistry, 278*, 2461–2468.

Yamauchi, T., Kamon, J., Waki, H., Terauchi, Y., Kubota, N., Hara, K., et al. (2001). The fat-derived hormone adiponectin reverses insulin resistance associated with both lipoatrophy and obesity. *Nature Medicine, 7*, 941–946.

Yamauchi, T., Nio, Y., Maki, T., Kobayashi, M., Takazawa, T., Iwabu, M., et al. (2007). Targeted disruption of AdipoR1 and AdipoR2 causes abrogation of adiponectin binding and metabolic actions. *Nature Medicine, 13*, 332–339.

Yao, H., Ling, H., Wang, H., Zhang, L., Huang, X., & Xia, Z. (2005). Gene expression of adiponectin and adiponectin receptor 1 in type 2 diabetic rats and the relationship with the parameters of glucose and lipid metabolism. *Journal of Huazhong University of Science and Technology. Medical Sciences, 25*, 285–288.

Yoon, M. J., Lee, G. Y., Chung, J. J., Ahn, Y. H., Hong, S. H., & Kim, J. B. (2006). Adiponectin increases fatty acid oxidation in skeletal muscle cells by sequential activation of AMP-activated protein kinase, p38 mitogen-activated protein kinase, and peroxisome proliferator-activated receptor alpha. *Diabetes, 55*, 2562–2570.

Zeng, Q., Isobe, K., Fu, L., Ohkoshi, N., Ohmori, H., Takekoshi, K., et al. (2007). Effects of exercise on adiponectin and adiponectin receptor levels in rats. *Life Sciences, 80*, 454–459.

Zhou, Q., Du, J., Hu, Z., Walsh, K., & Wang, X. H. (2007). Evidence for adipose-muscle cross talk: Opposing regulation of muscle proteolysis by adiponectin and Fatty acids. *Endocrinology, 148*, 5696–5705.

CHAPTER FIVE

Screening for Adiponectin Secretion Regulators

Kyosuke Hino[1], Hidetaka Nagata
Genomic Science Laboratories, Dainippon Sumitomo Pharma Co. Ltd., Konohana-Ku, Osaka, Japan
[1]Corresponding author: e-mail address: kyosuke-hino@ds-pharma.co.jp

Contents

Abstract

Adiponectin is an adipokine secreted from adipocytes and plays important roles in the suppression of metabolic syndromes that can result in type 2 diabetes, obesity, and atherosclerosis. Adiponectin is a promising drug target because a number of studies have shown that upregulation of adiponectin has a number of therapeutic benefits. Extensive efforts have revealed various adiponectin regulators, such as cytokines, transcription factors, and drugs. Cytokines, such as tumor necrosis factor α, IL-6, and IL-18, downregulate adiponectin production. On the other hand, transcription factors such as peroxisome proliferator-activated receptor-γ (PPARγ), CCAAT-enhancer-binding protein α, and forkhead box O1 (FoxO1) upregulate adiponectin expression, although the activating transcription factor 3 and cAMP response element-binding protein downregulate it. Although a number of therapeutic drugs have been reported as adiponectin secretion regulators, most of them act through PPARγ-dependent mechanisms, leaving PPARγ-derived side effects as a concern. Using high-throughput screening, we have identified PPARγ-independent adiponectin secretion regulators as potential drug candidates with a novel mechanism of action.

1. INTRODUCTION

Adiponectin (also called ACRP30, adipoQ, apM1, and GBP28) is an adipokine secreted from adipocytes and plays various roles in the suppression of metabolic disorders that can result in type 2 diabetes, obesity, and

Vitamins and Hormones, Volume 90
ISSN 0083-6729
http://dx.doi.org/10.1016/B978-0-12-398313-8.00005-1

atherosclerosis (Gnacinska, Malgorzewicz, Stojek, Lysiak-Szydlowska, & Sworczak, 2009; Hu, Liang, & Spiegelman, 1996; Maeda et al., 1996; Nakano, Tobe, Choi-Miura, Mazda, & Tomita, 1996; Scherer, Williams, Fogliano, Baldini, & Lodish, 1995; Yu & Ginsberg, 2005). Indeed, patients suffering from obesity, insulin resistance, or type 2 diabetes have been shown to have decreased plasma adiponectin levels (Hotta et al., 2000). On the other hand, administration of adiponectin has been shown to ameliorate insulin resistance and lower glucose level in diabetic mice (Berg, Combs, Du, Brownlee, & Scherer, 2001; Fruebis et al., 2001; Yamauchi et al., 2001). Several studies have revealed that the insulin-sensitizing effects of adiponectin are mediated by increase in fatty acid oxidation through activation of the peroxisome proliferator-activated receptor α (PPARα) as well as through AMP-activated protein kinase (Kahn, Alquier, Carling, & Hardie, 2005; Kersten, Desvergne, & Wahli, 2000; Tomas et al., 2002; Yamauchi et al., 2002, 2003b), both of which involve the adiponectin receptor 1 and 2 (Yamauchi et al., 2003a). Therefore, compounds that enhance adiponectin secretion are considered as potential drugs for diabetes, obesity, atherosclerosis, and other metabolic syndromes.

It is known that inflammatory cytokines, such as tumor necrosis factor α (TNFα), IL-6, and IL-18 suppress adiponectin expression in adipocytes (Liu & Liu, 2010) (Fig. 5.1).

It has been shown that tissue expression levels of TNFα correlate positively with insulin resistance and that chronic exposure of cells, or whole animals, to TNFα induces insulin resistance (Yu & Ginsberg, 2005). TNFα-mediated downregulation of adiponectin is accomplished through protein kinase C (Lim, Kim, & Park, 2008), the c-Jun N-terminal kinase (JNK) (Kim et al., 2005), and IκB kinase-β (IKK-β) (Kamon et al., 2004). TNFα is also known to negatively affect transcription factor which upregulates adiponectin gene expression, such as PPARγ (Zhang et al., 1996), CCAAT-enhancer-binding protein (C/EBP) (Kita et al., 2005; Ron, Brasier, McGehee, & Habener, 1992), sterol regulatory element-binding protein (SREBP) (Sewter et al., 2002), and stimulatory protein 1 (Barth, Langmann, Scholmerich, Schmitz, & Schaffler, 2002). Furthermore, TNFα promotes the expression of IGF-1-binding protein 3, which represses adiponectin transcription and induces insulin resistance (Zappalà & Rechler, 2009).

Serum concentrations of IL-6 are elevated in insulin-resistant states, such as obesity, impaired glucose tolerance, and type 2 diabetes. In addition, IL-6

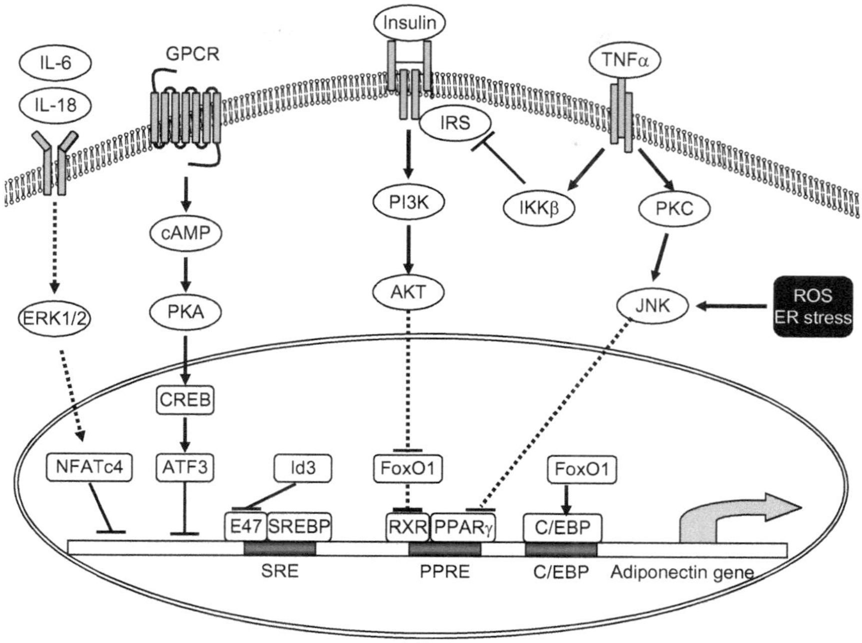

Figure 5.1 Regulation of adiponectin transcription by various signals. Modified Liu and Liu (2010). IRS, insulin receptor substrate; ER, endoplasmic reticulum; GPCR, G protein-coupled receptor; PI3K, phosphoinositide 3-kinase; PKA, protein kinase A; ROS, reactive oxygen species; RXR, retinoid X receptor.

stimulation is known to decrease adiponectin mRNA and its secretion in 3T3-L1 adipocytes. This decrease is reversed by inhibitors of the extracellular-signal-regulated kinase (ERK) 1/2 (Fasshauer et al., 2003).

IL-18 levels inversely correlate with adiponectin plasma levels and are increased in patients with obesity and type 2 diabetes (Liu & Liu, 2010). Stimulation of 3T3-L1 adipocytes with IL-18 reduces adiponectin expression levels (Chandrasekar et al., 2008). IL-18 is also known to induce ERK 1/2-dependent phosphorylation and activation of the nuclear factor of activated T-cells, cytoplasmic 4 (NFATc4), which has been shown to function as a repressor in adiponectin transcription (Kim et al., 2006).

The adiponectin gene promoter contains a large number of transcription factor binding sites. Some of these transcription factors have been shown to regulate adiponectin gene expression (Das, Lin, Widen, Zhang, & Scherer, 2001; Liu & Liu, 2010; Phillips & Kung, 2010) (Fig. 5.1).

PPARγ is a member of the PPAR subfamily of nuclear receptors that also includes PPARα and PPARβ/δ. PPARγ is abundantly expressed in adipose tissue and is well known as positive regulator of adiponectin gene expression (Maeda et al., 2001). Both human (Iwaki et al., 2003) and mouse (Seo et al., 2004) adiponectin promoters contain a PPAR responsive element (PPRE), PPARγ-binding site. Point mutations at this site reduce PPARγ agonist-induced adiponectin promoter transactivation (Iwaki et al., 2003). Another nuclear receptor known as liver receptor homolog-1 (LRH-1) seems to enhance PPARγ activity through LRH-1-binding site (Iwaki et al., 2003).

C/EBPs are a family of transcription factors that interact with the CCAAT box motif and are characterized by a highly conserved basic-leucine zipper (bZIP) domain involved in dimerization and DNA binding (Nerlov, 2007). C/EBPα-binding sites have been identified in the human adiponectin promoter (Park et al., 2004; Segawa et al., 2009), and adiponectin gene transcription is positively regulated by C/EBPα. Overexpression of C/EBPα is known to increase adiponectin mRNA levels, while its suppression decreases adiponectin mRNA levels in cultured adipocytes (Gustafson, Jack, Cushman, & Smith, 2003; Qiao et al., 2005).

SREBPs belong to the basic-helix–loop–helix (bHLH) leucine zipper class of transcription factors that bind to the sterol regulatory element (SRE) DNA sequence TCACNCCAC (Brown & Goldstein, 1997). SRE has been identified in the human adiponectin promoter and is required for human adiponectin promoter activity (Kita et al., 2005). Another bHLH transcription factor E47 potentiates SREBP-1c-mediated adiponectin promoter activation, while the transcription factor inhibitor of differentiation-3 (Id3) dose-dependently inhibits this action via interaction with E47 (Doran et al., 2008).

Forkhead box O1 (FoxO1) is a member of the forkhead box O transcription factor family involved in the regulation of adipocyte differentiation (Nakae et al., 2003). Although FoxO1 seems to positively regulate adiponectin transcription (Nakae et al., 2008; Qiao & Shao, 2006), FoxO1 has also been found to suppress PPARγ gene expression (Fan et al., 2009). These results suggest that regulation of adiponectin production by FoxO1 depends on cell condition and environment.

Activating transcription factor 3 (ATF3) is a member of cAMP response element-binding protein (CREB) family of transcription factors. Binding of

ATF3 to the putative activator protein-1 (AP-1) site located adjacent to the NFAT-binding site represses adiponectin promoter activity (Kim et al., 2006). Similarly, induction of ATF3 by the natural compound thapsigargin reduces adiponectin mRNA and promoter activity. Furthermore, ATF3 binding to the adiponectin promoter is increased in white adipose tissues of ob/ob and db/db mice.

CREB belongs to the bZIP superfamily of transcription factors and binds to certain DNA sequences called cAMP response element. CREB acts as a transcriptional repressor of adiponectin gene expression (Qi et al., 2009) and is known to indirectly inhibit adiponectin transcription through upregulation of the transcriptional repressor ATF3.

NFAT is a family of transcription factors that consists of five members, NFATc1–NFATc4 and NFAT5 (Hogan, Chen, Nardone, & Rao, 2003). Among them, NFATc4 is a negative regulator of adiponectin expression, because removal of the putative NFAT-binding site increases adiponectin promoter activity, whereas overexpression of NFATc4 reduces it (Kim et al., 2006).

In addition to above transcription factors, several other transcription factors, including activating enhancer binding protein-2β (AP-2β) (Ikeda et al., 2006), nuclear factor-Y (Ikeda et al., 2006), and Sp3 (Barth et al., 2002) have been shown to be implicated in the regulation of adiponectin gene expression.

2. PPARγ-DEPENDENT ADIPONECTIN SECRETION REGULATORS

PPARγ agonists, such as pioglitazone and rosiglitazone, are insulin-sensitizing drugs used to treat type 2 diabetes mellitus (Yki-Jarvinen, 2004). They are known to increase adiponectin secretion in cultured adipocytes (Pereira, Leitner, Erickson, & Draznin, 2008), in mice (Yamauchi et al., 2001), and in human (Hirose et al., 2002; Maeda et al., 2001). And that improved adiponectin secretion seems to be one of the major mechanisms of PPARγ agonists (Lihn, Pedersen, & Richelsen, 2005). Compounds that activate PPARγ can upregulate adiponectin secretion (Table 5.1). Pioglitazone and rosiglitazone directly activate PPARγ, and their complex bind to PPRE in the adiponectin promoter (Iwaki et al., 2003). In fact, treatment with PPARγ agonists has been shown to increase circulating

Table 5.1 PPARγ-dependent adiponectin secretion regulators

Category	Compound	References
PPARγ agonists	Pioglitazone and rosiglitazone	Yamauchi et al. (2001)
Sulfonylureas	Glibenclamide, gliclazide, and glimepiride	Fukuen et al. (2005)
Angiostensin receptor blockers	Irbesartan and telmisartan	Clasen et al. (2005)
5-HT2 antagonist	Sarpogrelate	Uchidakitajima et al. (2008)
JNK1 inhibitor	SP600125	Kim et al. (2005)
Small molecule compounds	18F9	Anandharajan et al. (2009)
Flavonoid pigment from plants	Anthocyanins	Tsuda et al. (2004)
Flavonoid from apple tree leaves	Phloretin	Hassan et al. (2007)
Ingredient of *Magnolia officinalis*	Magnolol and honokiol	Choi et al. (2009, 2011)
Components from propolis	Atepillin C	Ikeda, Yanagisawa, et al. (2011)
Polymethoxy flavonoid from Citrus depressa	Nobiletin	Kunimasa et al. (2009)
Synthetic phenolamides	*N*-(2-Phenylethyl) ferulamide	Yamazaki et al. (2008)
Flavonoid from tomato peel	Naringenin chalcone	Horiba et al. (2010)
Extract of a fungus	Norlichexanthone	Ikeda, Kurotobi, et al. (2011)

adiponectin levels by approximate 2- to 3-fold in patients (Tonelli et al., 2004; Yang et al., 2002).

Treatment with sulfonylurea glimepiride has also been shown to increase circulating adiponectin levels in patients, although not to the same extent as PPARγ agonists (Araki et al., 2009; Tsunekawa et al., 2003). The mechanism underlying this increase is based on the fact that glimepiride and other sulfonylureas exhibit weak PPARγ agonistic activity and therefore enhance adiponectin mRNA expression and secretion in 3T3-L1 adipocytes (Fukuen et al., 2005).

Angiotensin type-1 receptor blockers (ARBs), such as telmisartan and irbesartan, have been shown to have partial PPARγ agonistic activity and are thus known to increase adiponectin secretion in 3T3-L1 adipocytes (Clasen et al., 2005). Furthermore, ARBs have been shown to increase circulating adiponectin levels in clinical trials (Negro, Formoso, & Hassan, 2006; Yamada et al., 2008).

It has been shown that mRNA expression of 5-HT2A receptors is upregulated in hypertrophic 3T3-L1 adipocytes and in db/db mice with decreased expression of adiponectin (Uchidakitajima et al., 2008). Sarpogrelate, a 5-HT2A receptor antagonist clinically used for treatment of cutaneous ulcer and ischemic changes resulting from arteriosclerosis, is known to enhance adiponectin expression in 3T3-L1 adipocytes. This enhancement is suppressed by the PPARγ antagonist GW9662. This finding indicates that sarpogrelate mechanism of action is dependent on PPARγ activity. Sarpogrelate is also known to increase circulating adiponectin levels in diabetic patients with arteriosclerosis obliterans (Yamakawa et al., 2003).

TNFα-induced decrease in adiponectin mRNA levels can partially be reversed by treatment with the JNK inhibitor SP600125 (Kim et al., 2005). JNK is activated by oxidative stress and ER stress (Koh et al., 2007; Morgan & Liu, 2010), so JNK might play an important role in adiponectin downregulation. As SP600125 increases PPARγ2 mRNA expression and PPARγ DNA-binding activity, it is considered as an indirect PPARγ activator.

18F9 (4-(3,6-bis (ethoxycarbonyl)-4,5,6,7-tetrahydrothieno (2,3-c) pyridin-2-ylamino)-4-oxobutanoic acid), identified by a morphology-based screening, is known to induce adipogenesis in 3T3-L1 preadipocytes (Choi, Kawazoe, Murakami, Misawa, & Uesugi, 2002). 18F9 is also known to enhance adiponectin mRNA expression in human subcutaneous preadipocytes, to dose-dependently reduce plasma glucose and insulin levels in db/db mice, and to possess PPARγ agonistic activity (Anandharajan et al., 2009).

Various natural compounds capable of regulating adiponectin secretion have been reported (Kawada et al., 2008). Most of them (Table 5.1) regulate adiponectin secretion through PPARγ-dependent mechanisms. These mechanisms can roughly be classified into two types: direct agonistic stimulation of PPARγ and upregulation of PPARγ expression. Direct PPARγ agonists include magnolol, a component of *Magnolia officinalis* used in traditional Chinese and Japanese medicine (Choi et al., 2009); artepillin C, a Brazilian propolis-derived component (Ikeda, Yanagisawa, et al., 2011); synthetic phenolamide *N*-(2-phenylethyl)ferulamide (Yamazaki, Kawano, & Uebayasi, 2008); and naringenin chalcone, a flavonoid from tomato peel (Horiba, Nishimura, Nakai, Abe, & Sato, 2010). PPARγ expression upregulators, on the other hand, include anthocyanins, a group of flavonoid pigments from plants (Tsuda et al., 2004); honokiol, a component of *M. officinalis* (Choi et al., 2011); nobiletin, a citrus polymethoxy flavonoid (Kunimasa et al., 2009); and norlichexanthone, a fungus extract (Ikeda, Kurotobi et al., 2011). Phloretin, a dietary flavonoid from apple tree leaves, shows both activities (Hassan et al., 2007).

Although a number of PPARγ agonists are already marketed, patients taking these drugs could suffer from several side effects, such as weight gain, congestive heart failure, and edema (Guan et al., 2005; Kaul, Bolger, Herrington, Giugliano, & Eckel, 2010). Although it is reported that the use of pioglitazone is associated with bladder cancer (Piccinni, Motola, Marchesini, & Poluzzi, 2011), a recent study has shown that short-term use of pioglitazone does not increase incidence of bladder cancer (Lewis et al., 2011). Therefore, the risk of bladder cancer by pioglitazone remains to be elucidated. In either case, PPARγ-independent adiponectin secretion regulators could become ideal drugs for the management of a number of conditions, including diabetes, obesity, and atherosclerosis. Although a number of adiponectin secretion regulators have been reported (Table 5.1), most of them act through PPARγ activation either directly or indirectly.

Table 5.2 PPARγ-dependent adiponectin secretion regulators

Category	Compound	References
Small molecule compounds	Small molecule compounds from chemical library	Hino et al. (2011)
Triterpene saponins from the medicinal herb	Isoastragaloside I and astragalosides II	Xu et al. (2008)
Biflavonoid from Ginkgo biloba	Isoginkgetin	Liu et al. (2007)

3. PPAR γ-INDEPENDENT ADIPONECTIN SECRETION REGULATORS

A limited number of compounds are known to regulate adiponectin secretion through PPARγ-independent mechanisms (Table 5.2). Isoastragaloside I and astragalosides II, two triterpene saponins from medicinal herbs, selectively increase adiponectin secretion in primary adipocytes without obvious effects on a panel of other adipokines (Xu et al., 2008). These compounds do not induce expression of aP2 and Ero1-Lα, two PPARγ target genes. Another natural compound, isoginkgetin, a biflavonoid from *Ginkgo biloba*, upregulates adiponectin secretion (Liu et al., 2007). Unlike in the case of rosiglitazone, PPARγ activation seems not to be essential for the action of isoginkgetin, which has only a slight effect on adipogenesis.

We have previously reported several PPARγ-independent adiponectin secretion regulators derived from high-throughput screening (HTS) (Hino, Nagata, Shimonishi, & Ido, 2011). We screened our chemical library in 3T3-L1 adipocytes expressing luciferase containing the adiponectin promoter. These hit compounds were assessed by using 3T3-L1 adipocytes expressing luciferase under the control of PPRE, to eliminate PPARγ agonists and indirect PPARγ activators. After confirming hit compounds adiponectin secretion activity, six PPARγ-independent adiponectin secretion regulators were identified. Another feature of our HTS was that insulin-resistant mimic 3T3-L1 adipocytes model induced by TNFα stimulation was used for the screening. Therefore, our hit compounds were effective even in insulin-resistant adipocytes despite no direct inhibition of TNFα. Gene expression patterns regulated by these hit compounds were distinct from those regulated by a PPARγ agonist (K. Hino and H. Nagata, unpublished data), indicating a novel mechanism for regulation of adiponectin secretion. We believe our HTS assay would be useful to screen for other adipokine regulators capable of treating diabetes, atherosclerosis, and other metabolic syndromes.

4. OTHER ADIPONECTIN SECRETION REGULATORS

Although several other adiponectin secretion regulators have been reported, whether their mechanisms of action are PPARγ-dependent or not is still unclear. Some of these regulators are already marketed, while others serve only as pharmacological tools or natural compounds (Table 5.3).

Table 5.3 Other adiponectin secretion regulators

Category	Compound	References
NADPH oxidase inhibitor (antioxidant)	Apocynin	Furukawa et al. (2004)
HMG-CoA reductase inhibitor	Pravastatin	Takagi et al. (2008)
GPR109A agonist	Niacin	Plaisance et al. (2009)
PPARα agonist	Bezafibrate and fenofibrate	Hiuge et al. (2007)
IKK-β inhibitor	IMD-0354	Kamon et al. (2004)
PKD inhibitor	Go6983	Iwamoto and Yokoyama (2011)

Oxidative stress is increased in obese humans and mice (Furukawa et al., 2004) and causes dysregulated production of adiponectin in cultured adipocytes. This dysregulation can be reversed by use of NADPH oxidase inhibitors, such as diphenyleneiodonium, apocynin, and *N*-acetyl-cysteine. Moreover, in obese mice, treatment with apocynin reduced oxidative stress in adipose tissue and attenuated adiponectin downregulation.

In the West of Scotland Coronary Prevention Study (WOSCOPS), treatment of hypercholesterolemic men with pravastatin, an inhibitor of 3-hydroxy-3-methylglutaryl coenzyme A (HMG-CoA) reductase, reduced the likelihood progression to diabetes mellitus by 30% (Freeman et al., 2001). This positive effect was associated with an increase of serum adiponectin level. It has also been reported that treatment with pravastatin increases not only adiponectin production in 3T3-L1 adipocytes but also plasma adiponectin levels in mice (Takagi et al., 2008). Furthermore, analyses of plasma samples from the WOSCOPS indicated a significant increase of plasma adiponectin levels following treatment with pravastatin. These events could be derived from pravastatin-induced attenuation of oxidative stress (Saito et al., 2008).

Niacin stimulates adiponectin secretion through GPR109A receptor in primary rat adipocytes and mice (Plaisance et al., 2009). This effect seems to be caused by inhibition of lipolysis through reduction of cAMP. Adiponectin secretion by niacin stimulation is in accordance with β-adrenoreceptor agonists mediated downregulation of adiponectin, because β-adrenoreceptor agonists increase cAMP and induce lipolysis (Fasshauer, Klein, Neumann, Eszlinger, & Paschke, 2001). Niacin is also known to increase circulating adiponectin levels in human (Linke et al., 2009).

PPARα ligands, such as bezafibrate and fenofibrate, significantly increase adiponectin levels in 3T3-L1 adipocytes and wild-type mice (Hiuge et al., 2007). These effects are diminished in PPARα-deficient adipocytes and mice. Notably, in The Bezafibrate Infarction Prevention study, bezafibrate has been shown to significantly increase serum adiponectin levels in coronary artery disease patients.

Kinase inhibitors have been reported to modulate adiponectin activity. In fact, IMD-0354, an IKK-β inhibitor, has been shown to restore TNFα-mediated downregulation of adiponectin secretion in 3T3-L1 adipocytes (Kamon et al., 2004). Enhancement of insulin-stimulated Akt activity seems to be crucial to regulation of adiponectin secretion in 3T3-L1 adipocytes. In addition, IMD-0354 enhances insulin sensitivity and increase serum adiponectin in KKAy mice. Go6983, a protein kinase D (PKD) inhibitor, upregulates adiponectin gene expression in 3T3-L1 adipocytes and mice (Iwamoto & Yokoyama, 2011). This effect is mediated by inhibition of AP-2β phosphorylation, because AP-2β is activated by PKD, and negatively regulates adiponectin gene expression.

5. CONCLUSIONS

Adiponectin plays important roles in the suppression of a number of conditions, such as type 2 diabetes, obesity, and atherosclerosis. Therefore, adiponectin secretion regulators, especially PPARγ-independent, might become novel drugs for these diseases. Further understanding of adiponectin regulation could lead to the discovery of new drug targets.

REFERENCES

Anandharajan, R., Sayyed, S. G., Doshi, L. S., Dixit, P., Chandak, P. G., Dixit, A. V., et al. (2009). 18F9 (4-(3,6-bis (ethoxycarbonyl)-4,5,6,7-tetrahydrothieno (2,3-c) pyridin-2-ylamino)-4-oxobutanoic acid) enhances insulin-mediated glucose uptake in vitro and exhibits antidiabetic activity in vivo in db/db mice. *Metabolism, 58*, 1503–1516.

Araki, T., Emoto, M., Konishi, T., Ikuno, Y., Lee, E., Teramura, M., et al. (2009). Glimepiride increases high-density lipoprotein cholesterol via increasing adiponectin levels in type 2 diabetes mellitus. *Metabolism, 58*, 143–148.

Barth, N., Langmann, T., Scholmerich, J., Schmitz, G., & Schaffler, A. (2002). Identification of regulatory elements in the human adipose most abundant gene transcript-1 (apM-1) promoter: Role of SP1/SP3 and TNF-alpha as regulatory pathways. *Diabetologia, 45*, 1425–1433.

Berg, A. H., Combs, T. P., Du, X., Brownlee, M., & Scherer, P. E. (2001). The adipocyte-secreted protein Acrp30 enhances hepatic insulin action. *Nature Medicine, 7*, 947–953.

Brown, M. S., & Goldstein, J. L. (1997). The SREBP pathway: Regulation of cholesterol metabolism by proteolysis of a membrane-bound transcription factor. *Cell, 89*, 331–340.

Chandrasekar, B., Patel, D. N., Mummidi, S., Kim, J. W., Clark, R. A., & Valente, A. J. (2008). Interleukin-18 suppresses adiponectin expression in 3T3-L1 adipocytes via a novel signal transduction pathway involving ERK1/2-dependent NFATc4 phosphorylation. *The Journal of Biological Chemistry, 283,* 4200–4209.

Choi, S.-S., Cha, B.-Y., Iida, K., Sato, M., Lee, Y.-S., Teruya, T., et al. (2011). Honokiol enhances adipocyte differentiation by potentiating insulin signaling in 3T3-L1 preadipocytes. *Journal of Natural Medicines, 65,* 424–430.

Choi, S. S., Cha, B. Y., Lee, Y. S., Yonezawa, T., Teruya, T., Nagai, K., et al. (2009). Magnolol enhances adipocyte differentiation and glucose uptake in 3T3-L1 cells. *Life Sciences, 84,* 908–914.

Choi, Y., Kawazoe, Y., Murakami, K., Misawa, H., & Uesugi, M. (2002). Identification of bioactive molecules by adipogenesis profiling of organic compounds. *The Journal of Biological Chemistry, 278,* 7320–7324.

Clasen, R., Schupp, M., Foryst-Ludwig, A., Sprang, C., Clemenz, M., Krikov, M., et al. (2005). PPARgamma-activating angiotensin type-1 receptor blockers induce adiponectin. *Hypertension, 46,* 137–143.

Das, K., Lin, Y., Widen, E., Zhang, Y., & Scherer, P. E. (2001). Chromosomal localization, expression pattern, and promoter analysis of the mouse gene encoding adipocyte-specific secretory protein Acrp30. *Biochemical and Biophysical Research Communications, 280,* 1120–1129.

Doran, A. C., Meller, N., Cutchins, A., Deliri, H., Slayton, R. P., Oldham, S. N., et al. (2008). The helix-loop-helix factors Id3 and E47 are novel regulators of adiponectin. *Circulation Research, 103,* 624–634.

Fan, W., Imamura, T., Sonoda, N., Sears, D. D., Patsouris, D., Kim, J. J., et al. (2009). FOXO1 transrepresses peroxisome proliferator-activated receptor transactivation, coordinating an insulin-induced feed-forward response in adipocytes. *The Journal of Biological Chemistry, 284,* 12188–12197.

Fasshauer, M., Klein, J., Neumann, S., Eszlinger, M., & Paschke, R. (2001). Adiponectin gene expression is inhibited by beta-adrenergic stimulation via protein kinase A in 3T3-L1 adipocytes. *FEBS Letters, 507,* 142–146.

Fasshauer, M., Kralisch, S., Klier, M., Lossner, U., Bluher, M., Klein, J., et al. (2003). Adiponectin gene expression and secretion is inhibited by interleukin-6 in 3T3-L1 adipocytes. *Biochemical and Biophysical Research Communications, 301,* 1045–1050.

Freeman, D. J., Norrie, J., Sattar, N., Neely, R. D., Cobbe, S. M., Ford, I., et al. (2001). Pravastatin and the development of diabetes mellitus: Evidence for a protective treatment effect in the West of Scotland Coronary Prevention Study. *Circulation, 103,* 357–362.

Fruebis, J., Tsao, T. S., Javorschi, S., Ebbets-Reed, D., Erickson, M. R., Yen, F. T., et al. (2001). Proteolytic cleavage product of 30-kDa adipocyte complement-related protein increases fatty acid oxidation in muscle and causes weight loss in mice. *Proceedings of the National Academy of Sciences of the United States of America, 98,* 2005–2010.

Fukuen, S., Iwaki, M., Yasui, A., Makishima, M., Matsuda, M., & Shimomura, I. (2005). Sulfonylurea agents exhibit peroxisome proliferator-activated receptor agonistic activity. *The Journal of Biological Chemistry, 280,* 23653–23659.

Furukawa, S., Fujita, T., Shimabukuro, M., Iwaki, M., Yamada, Y., Nakajima, Y., et al. (2004). Increased oxidative stress in obesity and its impact on metabolic syndrome. *The Journal of Clinical Investigation, 114,* 1752–1761.

Gnacinska, M., Malgorzewicz, S., Stojek, M., Lysiak-Szydlowska, W., & Sworczak, K. (2009). Role of adipokines in complications related to obesity: A review. *Advances in Medical Sciences, 54,* 150–157.

Guan, Y., Hao, C., Cha, D. R., Rao, R., Lu, W., Kohan, D. E., et al. (2005). Thiazolidinediones expand body fluid volume through PPARγ stimulation of ENaC-mediated renal salt absorption. *Nature Medicine, 11,* 861–866.

Gustafson, B., Jack, M. M., Cushman, S. W., & Smith, U. (2003). Adiponectin gene activation by thiazolidinediones requires PPAR gamma 2, but not C/EBP alpha-evidence for differential regulation of the aP2 and adiponectin genes. *Biochemical and Biophysical Research Communications*, *308*, 933–939.

Hassan, M., Yazidi, C., Landrier, J., Lairon, D., Margotat, A., & Amiot, M. (2007). Phloretin enhances adipocyte differentiation and adiponectin expression in 3T3-L1 cells. *Biochemical and Biophysical Research Communications*, *361*, 208–213.

Hino, K., Nagata, H., Shimonishi, M., & Ido, M. (2011). High-throughput screening for small-molecule adiponectin secretion modulators. *Journal of Biomolecular Screening*, *16*, 628–636.

Hirose, H., Kawai, T., Yamamoto, Y., Taniyama, M., Tomita, M., Matsubara, K., et al. (2002). Effects of pioglitazone on metabolic parameters, body fat distribution, and serum adiponectin levels in Japanese male patients with type 2 diabetes. *Metabolism*, *51*, 314–317.

Hiuge, A., Tenenbaum, A., Maeda, N., Benderly, M., Kumada, M., Fisman, E. Z., et al. (2007). Effects of peroxisome proliferator-activated receptor ligands, bezafibrate and fenofibrate, on adiponectin level. *Arteriosclerosis, Thrombosis, and Vascular Biology*, *27*, 635–641.

Hogan, P. G., Chen, L., Nardone, J., & Rao, A. (2003). Transcriptional regulation by calcium, calcineurin, and NFAT. *Genes & Development*, *17*, 2205–2232.

Horiba, T., Nishimura, I., Nakai, Y., Abe, K., & Sato, R. (2010). Naringenin chalcone improves adipocyte functions by enhancing adiponectin production. *Molecular and Cellular Endocrinology*, *323*, 208–214.

Hotta, K., Funahashi, T., Arita, Y., Takahashi, M., Matsuda, M., Okamoto, Y., et al. (2000). Plasma concentrations of a novel, adipose-specific protein, adiponectin, in type 2 diabetic patients. *Arteriosclerosis, Thrombosis, and Vascular Biology*, *20*, 1595–1599.

Hu, E., Liang, P., & Spiegelman, B. M. (1996). AdipoQ is a novel adipose-specific gene dysregulated in obesity. *The Journal of Biological Chemistry*, *271*, 10697–10703.

Ikeda, M., Kurotobi, Y., Namikawa, A., Kuranuki, S., Matsuura, N., Sato, M., et al. (2011). Norlichexanthone isolated from fungus P16 promotes the secretion and expression of adiponectin in cultured ST-13 adipocytes. *Medicinal Chemistry*, *7*, 250–256.

Ikeda, K., Maegawa, H., Ugi, S., Tao, Y., Nishio, Y., Tsukada, S., et al. (2006). Transcription factor activating enhancer-binding protein-2beta. A negative regulator of adiponectin gene expression. *The Journal of Biological Chemistry*, *281*, 31245–31253.

Ikeda, R., Yanagisawa, M., Takahashi, N., Kawada, T., Kumazawa, S., Yamaotsu, N., et al. (2011). Brazilian propolis-derived components inhibit TNF-α-mediated downregulation of adiponectin expression via different mechanisms in 3T3-L1 adipocytes. *Biochimica et Biophysica Acta, General Subjects*, *1810*, 695–703.

Iwaki, M., Matsuda, M., Maeda, N., Funahashi, T., Matsuzawa, Y., Makishima, M., et al. (2003). Induction of adiponectin, a fat-derived antidiabetic and antiatherogenic factor, by nuclear receptors. *Diabetes*, *52*, 1655–1663.

Iwamoto, N., & Yokoyama, S. (2011). Protein kinase D regulates the adiponectin gene expression through phosphorylation of AP-2: A common pathway to the ABCA1 gene regulation. *Atherosclerosis*, *216*, 90–96.

Kahn, B. B., Alquier, T., Carling, D., & Hardie, D. G. (2005). AMP-activated protein kinase: Ancient energy gauge provides clues to modern understanding of metabolism. *Cell Metabolism*, *1*, 15–25.

Kamon, J., Yamauchi, T., Muto, S., Takekawa, S., Ito, Y., Hada, Y., et al. (2004). A novel IKKbeta inhibitor stimulates adiponectin levels and ameliorates obesity-linked insulin resistance. *Biochemical and Biophysical Research Communications*, *323*, 242–248.

Kaul, S., Bolger, A. F., Herrington, D., Giugliano, R. P., & Eckel, R. H. (2010). Thiazolidinedione drugs and cardiovascular risks: A science advisory from the American Heart Association and American College of Cardiology Foundation. *Circulation*, *121*, 1868–1877.

Kawada, T., Goto, T., Hirai, S., Kang, M. S., Uemura, T., Yu, R., et al. (2008). Dietary regulation of nuclear receptors in obesity-related metabolic syndrome. *Asia Pacific Journal of Clinical Nutrition, 17*(Suppl. 1), 126–130.

Kersten, S., Desvergne, B., & Wahli, W. (2000). Roles of PPARs in health and disease. *Nature, 405*, 421–424.

Kim, K., Kim, J., Jeon, J., Yoon, S., Choi, I., & Yang, Y. (2005). c-Jun N-terminal kinase is involved in the suppression of adiponectin expression by TNF-α in 3T3-L1 adipocytes. *Biochemical and Biophysical Research Communications, 327*, 460–467.

Kim, H. B., Kong, M., Kim, T. M., Suh, Y. H., Kim, W. H., Lim, J. H., et al. (2006). NFATc4 and ATF3 negatively regulate adiponectin gene expression in 3T3-L1 adipocytes. *Diabetes, 55*, 1342–1352.

Kita, A., Yamasaki, H., Kuwahara, H., Moriuchi, A., Fukushima, K., Kobayashi, M., et al. (2005). Identification of the promoter region required for human adiponectin gene transcription: Association with CCAAT/enhancer binding protein-β and tumor necrosis factor-α. *Biochemical and Biophysical Research Communications, 331*, 484–490.

Koh, E. H., Park, J. Y., Park, H. S., Jeon, M. J., Ryu, J. W., Kim, M., et al. (2007). Essential role of mitochondrial function in adiponectin synthesis in adipocytes. *Diabetes, 56*, 2973–2981.

Kunimasa, K., Kuranuki, S., Matsuura, N., Iwasaki, N., Ikeda, M., Ito, A., et al. (2009). Identification of nobiletin, a polymethoxyflavonoid, as an enhancer of adiponectin secretion. *Bioorganic & Medicinal Chemistry Letters, 19*, 2062–2064.

Lewis, J. D., Ferrara, A., Peng, T., Hedderson, M., Bilker, W. B., Quesenberry, C. P., Jr., et al. (2011). Risk of bladder cancer among diabetic patients treated with pioglitazone: Interim report of a longitudinal cohort study. *Diabetes Care, 34*, 916–922.

Lihn, A. S., Pedersen, S. B., & Richelsen, B. (2005). Adiponectin: Action, regulation and association to insulin sensitivity. *Obesity Reviews, 6*, 13–21.

Lim, J. Y., Kim, W. H., & Park, S. I. (2008). GO6976 prevents TNF-alpha-induced suppression of adiponectin expression in 3T3-L1 adipocytes: Putative involvement of protein kinase C. *FEBS Letters, 582*, 3473–3478.

Linke, A., Sonnabend, M., Fasshauer, M., Höllriegel, R., Schuler, G., Niebauer, J., et al. (2009). Effects of extended-release niacin on lipid profile and adipocyte biology in patients with impaired glucose tolerance. *Atherosclerosis, 205*, 207–213.

Liu, G., Grifman, M., Macdonald, J., Moller, P., Wong-Staal, F., & Li, Q. X. (2007). Isoginkgetin enhances adiponectin secretion from differentiated adiposarcoma cells via a novel pathway involving AMP-activated protein kinase. *The Journal of Endocrinology, 194*, 569–578.

Liu, M., & Liu, F. (2010). Transcriptional and post-translational regulation of adiponectin. *The Biochemical Journal, 425*, 41–52.

Maeda, K., Okubo, K., Shimomura, I., Funahashi, T., Matsuzawa, Y., & Matsubara, K. (1996). cDNA cloning and expression of a novel adipose specific collagen-like factor, apM1 (AdiPose most abundant gene transcript 1). *Biochemical and Biophysical Research Communications, 221*, 286–289.

Maeda, N., Takahashi, M., Funahashi, T., Kihara, S., Nishizawa, H., Kishida, K., et al. (2001). PPARgamma ligands increase expression and plasma concentrations of adiponectin, an adipose-derived protein. *Diabetes, 50*, 2094–2099.

Morgan, M. J., & Liu, Z. G. (2010). Reactive oxygen species in TNFalpha-induced signaling and cell death. *Molecules and Cells, 30*, 1–12.

Nakae, J., Cao, Y., Oki, M., Orba, Y., Sawa, H., Kiyonari, H., et al. (2008). Forkhead transcription factor FoxO1 in adipose tissue regulates energy storage and expenditure. *Diabetes, 57*, 563–576.

Nakae, J., Kitamura, T., Kitamura, Y., Biggs, W. H., 3rd, Arden, K. C., & Accili, D. (2003). The forkhead transcription factor Foxo1 regulates adipocyte differentiation. *Developmental Cell, 4*, 119–129.

Nakano, Y., Tobe, T., Choi-Miura, N. H., Mazda, T., & Tomita, M. (1996). Isolation and characterization of GBP28, a novel gelatin-binding protein purified from human plasma. *Journal of Biochemistry*, *120*, 803–812.

Negro, R., Formoso, G., & Hassan, H. (2006). The effects of irbesartan and telmisartan on metabolic parameters and blood pressure in obese, insulin resistant, hypertensive patients. *Journal of Endocrinological Investigation*, *29*, 957–961.

Nerlov, C. (2007). The C/EBP family of transcription factors: A paradigm for interaction between gene expression and proliferation control. *Trends in Cell Biology*, *17*, 318–324.

Park, S. K., Oh, S. Y., Lee, M. Y., Yoon, S., Kim, K. S., & Kim, J. W. (2004). CCAAT/enhancer binding protein and nuclear factor-Y regulate adiponectin gene expression in adipose tissue. *Diabetes*, *53*, 2757–2766.

Pereira, R., Leitner, J., Erickson, C., & Draznin, B. (2008). Pioglitazone acutely stimulates adiponectin secretion from mouse and human adipocytes via activation of the phosphatidylinositol 3′-kinase. *Life Sciences*, *83*, 638–643.

Phillips, S. A., & Kung, J. T. (2010). Mechanisms of adiponectin regulation and use as a pharmacological target. *Current Opinion in Pharmacology*, *10*, 676–683.

Piccinni, C., Motola, D., Marchesini, G., & Poluzzi, E. (2011). Assessing the association of pioglitazone use and bladder cancer through drug adverse event reporting. *Diabetes Care*, *34*, 1369–1371.

Plaisance, E. P., Lukasova, M., Offermanns, S., Zhang, Y., Cao, G., & Judd, R. L. (2009). Niacin stimulates adiponectin secretion through the GPR109A receptor. *American Journal of Physiology. Endocrinology and Metabolism*, *296*, E549–E558.

Qi, L., Saberi, M., Zmuda, E., Wang, Y., Altarejos, J., Zhang, X., et al. (2009). Adipocyte CREB promotes insulin resistance in obesity. *Cell Metabolism*, *9*, 277–286.

Qiao, L., Maclean, P. S., Schaack, J., Orlicky, D. J., Darimont, C., Pagliassotti, M., et al. (2005). C/EBPalpha regulates human adiponectin gene transcription through an intronic enhancer. *Diabetes*, *54*, 1744–1754.

Qiao, L., & Shao, J. (2006). SIRT1 regulates adiponectin gene expression through Foxo1-C/enhancer-binding protein alpha transcriptional complex. *The Journal of Biological Chemistry*, *281*, 39915–39924.

Ron, D., Brasier, A. R., McGehee, R. E., Jr., & Habener, J. F. (1992). Tumor necrosis factor-induced reversal of adipocytic phenotype of 3T3-L1 cells is preceded by a loss of nuclear CCAAT/enhancer binding protein (C/EBP). *The Journal of Clinical Investigation*, *89*, 223–233.

Saito, S., Fujiwara, T., Matsunaga, T., Minagawa, K., Fukui, K., Fukuda, I., et al. (2008). Increased adiponectin synthesis in the visceral adipose tissue in men with coronary artery disease treated with pravastatin: A role of the attenuation of oxidative stress. *Atherosclerosis*, *199*, 378–383.

Scherer, P. E., Williams, S., Fogliano, M., Baldini, G., & Lodish, H. F. (1995). A novel serum protein similar to C1q, produced exclusively in adipocytes. *The Journal of Biological Chemistry*, *270*, 26746–26749.

Segawa, K., Matsuda, M., Fukuhara, A., Morita, K., Okuno, Y., Komuro, R., et al. (2009). Identification of a novel distal enhancer in human adiponectin gene. *The Journal of Endocrinology*, *200*, 107–116.

Seo, J. B., Moon, H. M., Noh, M. J., Lee, Y. S., Jeong, H. W., Yoo, E. J., et al. (2004). Adipocyte determination- and differentiation-dependent factor 1/sterol regulatory element-binding protein 1c regulates mouse adiponectin expression. *The Journal of Biological Chemistry*, *279*, 22108–22117.

Sewter, C., Berger, D., Considine, R. V., Medina, G., Rochford, J., Ciaraldi, T., et al. (2002). Human obesity and type 2 diabetes are associated with alterations in SREBP1 isoform expression that are reproduced ex vivo by tumor necrosis factor-alpha. *Diabetes*, *51*, 1035–1041.

Takagi, T., Matsuda, M., Abe, M., Kobayashi, H., Fukuhara, A., Komuro, R., et al. (2008). Effect of pravastatin on the development of diabetes and adiponectin production. *Atherosclerosis*, *196*, 114–121.

Tomas, E., Tsao, T. S., Saha, A. K., Murrey, H. E., Zhang Cc, C., Itani, S. I., et al. (2002). Enhanced muscle fat oxidation and glucose transport by ACRP30 globular domain: Acetyl-CoA carboxylase inhibition and AMP-activated protein kinase activation. *Proceedings of the National Academy of Sciences of the United States of America*, *99*, 16309–16313.

Tonelli, J., Li, W., Kishore, P., Pajvani, U. B., Kwon, E., Weaver, C., et al. (2004). Mechanisms of early insulin-sensitizing effects of thiazolidinediones in type 2 diabetes. *Diabetes*, *53*, 1621–1629.

Tsuda, T., Ueno, Y., Aoki, H., Koda, T., Horio, F., Takahashi, N., et al. (2004). Anthocyanin enhances adipocytokine secretion and adipocyte-specific gene expression in isolated rat adipocytes. *Biochemical and Biophysical Research Communications*, *316*, 149–157.

Tsunekawa, T., Hayashi, T., Suzuki, Y., Matsui-Hirai, H., Kano, H., Fukatsu, A., et al. (2003). Plasma adiponectin plays an important role in improving insulin resistance with glimepiride in elderly type 2 diabetic subjects. *Diabetes Care*, *26*, 285–289.

Uchidakitajima, S., Yamauchi, T., Takashina, Y., Okadaiwabu, M., Iwabu, M., Ueki, K., et al. (2008). 5-Hydroxytryptamine 2A receptor signaling cascade modulates adiponectin and plasminogen activator inhibitor 1 expression in adipose tissue. *FEBS Letters*, *582*, 3037–3044.

Xu, A., Wang, H., Hoo, R. L. C., Sweeney, G., Vanhoutte, P. M., Wang, Y., et al. (2008). Selective elevation of adiponectin production by the natural compounds derived from a medicinal herb alleviates insulin resistance and glucose intolerance in obese mice. *Endocrinology*, *150*, 625–633.

Yamada, S., Ano, N., Toda, K., Kitaoka, A., Shiono, K., Inoue, G., et al. (2008). Telmisartan but not candesartan affects adiponectin expression in vivo and in vitro. *Hypertension Research*, *31*, 601–606.

Yamakawa, J., Takahashi, T., Itoh, T., Kusaka, K., Kawaura, K., Wang, X. Q., et al. (2003). A novel serotonin blocker, sarpogrelate, increases circulating adiponectin levels in diabetic patients with arteriosclerosis obliterans. *Diabetes Care*, *26*, 2477–2478.

Yamauchi, T., Kamon, J., Ito, Y., Tsuchida, A., Yokomizo, T., Kita, S., et al. (2003). Cloning of adiponectin receptors that mediate antidiabetic metabolic effects. *Nature*, *423*, 762–769.

Yamauchi, T., Kamon, J., Minokoshi, Y., Ito, Y., Waki, H., Uchida, S., et al. (2002). Adiponectin stimulates glucose utilization and fatty-acid oxidation by activating AMP-activated protein kinase. *Nature Medicine*, *8*, 1288–1295.

Yamauchi, T., Kamon, J., Waki, H., Imai, Y., Shimozawa, N., Hioki, K., et al. (2003). Globular adiponectin protected ob/ob mice from diabetes and ApoE-deficient mice from atherosclerosis. *The Journal of Biological Chemistry*, *278*, 2461–2468.

Yamauchi, T., Kamon, J., Waki, H., Terauchi, Y., Kubota, N., Hara, K., et al. (2001). The fat-derived hormone adiponectin reverses insulin resistance associated with both lipoatrophy and obesity. *Nature Medicine*, *7*, 941–946.

Yamazaki, Y., Kawano, Y., & Uebayasi, M. (2008). Induction of adiponectin by natural and synthetic phenolamides in mouse and human preadipocytes and its enhancement by docosahexaenoic acid. *Life Sciences*, *82*, 290–300.

Yang, W. S., Jeng, C. Y., Wu, T. J., Tanaka, S., Funahashi, T., Matsuzawa, Y., et al. (2002). Synthetic peroxisome proliferator-activated receptor-gamma agonist, rosiglitazone, increases plasma levels of adiponectin in type 2 diabetic patients. *Diabetes Care*, *25*, 376–380.

Yki-Jarvinen, H. (2004). Thiazolidinediones. *The New England Journal of Medicine*, *351*, 1106–1118.

Yu, Y. H., & Ginsberg, H. N. (2005). Adipocyte signaling and lipid homeostasis: Sequelae of insulin-resistant adipose tissue. *Circulation Research, 96*, 1042–1052.
Zappalà, G., & Rechler, M. M. (2009). IGFBP-3, hypoxia and TNF-α inhibit adiponectin transcription. *Biochemical and Biophysical Research Communications, 382*, 785–789.
Zhang, B., Berger, J., Hu, E., Szalkowski, D., White-Carrington, S., Spiegelman, B. M., et al. (1996). Negative regulation of peroxisome proliferator-activated receptor-gamma gene expression contributes to the antiadipogenic effects of tumor necrosis factor-alpha. *Molecular Endocrinology, 10*, 1457–1466.

CHAPTER SIX

Adiponectin and PPARγ: Cooperative and Interdependent Actions of Two Key Regulators of Metabolism

Olga Astapova, Todd Leff[1]
Department of Pathology, The Cardiovascular Research Institute, Wayne State University School of Medicine, Detroit, Michigan, USA
[1]Corresponding author: e-mail address: tleff@med.wayne.edu

Contents

Abstract

The recent advances in the understanding of adiponectin and other adipokines have highlighted the role of adipose tissue as an active endocrine organ. One of the central regulators of adipocyte biology is peroxisome proliferator-activated receptor gamma (PPARγ), a transcription factor that induces the adipogenic gene expression program during development, promotes adipose remodeling, and regulates the functions of adipocytes in lipid storage, adipokine secretion, and energy homeostasis. Activation of PPARγ results in increased insulin sensitivity in skeletal muscle and liver and improves the secretory profile of adipose tissue, favoring release of insulin-sensitizing adipokines, such as adiponectin, and reducing inflammatory cytokines. Increased adiponectin production is likely a significant mediator of the systemic effects of PPARγ activation. This chapter will review the interplay between PPARγ and adiponectin in regulating metabolism, presenting evidence that PPARγ regulates adiponectin gene expression, processing, and secretion and that the two proteins have overlapping effects on downstream metabolic pathways.

Vitamins and Hormones, Volume 90
ISSN 0083-6729
http://dx.doi.org/10.1016/B978-0-12-398313-8.00006-3

1. PPARγ ACTIVITY CORRELATES WITH ADIPONECTIN HORMONE LEVELS

The idea that there is a physiologically important link between peroxisome proliferator-activated receptor gamma (PPARγ) and adiponectin is based mainly on the observations that under many circumstances changes in PPARγ activity correlate with altered adiponectin hormone levels. This correlative linkage has been observed in patients, animal models, and cell culture and is apparent whether PPARγ activity has been altered by pharmacological or genetic means. In addition, on a more mechanistic level, molecular studies have revealed specific pathways by which PPARγ regulates adiponectin gene expression and hormone production. In this section, we summarize the evidence that links PPARγ activity to adiponectin hormone levels: describing first the pharmacological evidence and then the genetic studies in humans and mice. This will be followed by a review of the molecular mechanisms by which the functional linkage between PPARγ and adiponectin is achieved.

1.1. Pharmacological evidence

In considering the role of PPARγ in adiponectin biology, a key observation is that circulating adiponectin levels rise in diabetic patients after they have been treated with thiazolidinedione (TZD) drugs. TZD compounds (troglitazone, rosiglitazone, and pioglitazone) have been widely prescribed for the treatment of type 2 diabetes since the late 1990s. They act to improve insulin sensitivity and other metabolic pathologies associated with diabetes. Although the exact mechanism by which TZDs exert their antidiabetic activity is not fully understood, these drugs are selective high affinity ligands for the nuclear receptor PPARγ, and it is well accepted that the majority of their metabolic effects are due to the activation of PPARγ.

Although the molecular biology of PPARγ will be described in more detail below, it should be noted here that as a member of the nuclear receptor superfamily, PPARγ functions primarily as a ligand-activated transcription factor. The binding of either a synthetic or a natural ligand to the receptor stimulates its transcriptional activity. This leads to increased expression of the set of genes regulated by that particular nuclear receptor, which in the case of PPARγ are termed PPARγ target genes. One way to frame the question being considered here is whether the adiponectin gene is a PPARγ target gene.

The mechanism by which PPARγ activators ameliorate the metabolic abnormalities associated with diabetes is not fully understood and likely involves multiple pathways in many different tissues. However, a likely contributor is the PPARγ-mediated stimulation of adiponectin hormone production and subsequent increase in the circulating levels of the hormone. This is an attractive idea, especially given our understanding (as described in other chapters in this book), of how adiponectin acts on multiple target tissues to increase fatty acid oxidation, reduce intracellular lipid levels, and improve insulin sensitivity—all of which are known outcomes of TZD treatment.

A very large number of clinical studies have noted that TZD treatment of diabetic patients causes an increase in circulating levels of adiponectin, and it is now well accepted that this is part of the complex profile of metabolic and endocrine changes induced by PPARγ activators. In this section, we describe a small subset of these studies that are representative of the whole body of this literature.

The positive effect of TZD treatment on adiponectin levels was first reported in a series of studies in 2002 and 2003 with the TZDs troglitazone, pioglitazone, and rosiglitazone (Combs et al., 2002; Hirose et al., 2002; Phillips et al., 2003; Yu et al., 2002). In general, these studies showed that TZD treatment stimulated circulating levels of adiponectin, in some cases, up to threefold. Probably, the most comprehensive analysis of the effect of PPARγ activation on adiponectin levels was a meta-analysis of 19 clinical studies in 2008 (Riera-Guardia & Rothenbacher, 2008). These authors concluded that TZD treatment of diabetic patients resulted in a moderate but highly significant increase of serum adiponectin levels. By comparing TZD treatment (pioglitazone) to a non-TZD (e.g., non-PPARγ) treatment (the sulfonylurea glimepiride), Sam et al. found that the beneficial lipoprotein changes associated with pioglitazone treatment were due at least in part to increased expression of adiponectin in the pioglitazone treated subjects (Sam et al., 2012).

Although a smaller number of human studies have directly examined the effect of TZD treatment on adiponectin gene expression (by measuring mRNA levels), the results are consistent with the possibility that the adiponectin gene is a transcriptional target of PPARγ. In a study of insulin-resistant patients, Koenen et al. (2009) showed that pioglitazone treatment increased adiponectin mRNA levels in abdominal subcutaneous adipose tissue explants. Similarly, in the special case of patients with an acquired lipodystrophy, rosiglitazone treatment induced a significant increase in adiponectin mRNA levels in subcutaneous adipose tissue (Sutinen et al., 2004).

While the human studies described above are invaluable in determining the therapeutic relevance of PPARγ-mediated adiponectin expression, they do not provide much opportunity to explore the molecular mechanisms by which PPARγ might regulate adiponectin production. However, a large number of studies in rodent models of diabetes that showed a clear stimulatory effect of TZDs on adiponectin hormone levels have provided an opportunity to examine in more detail the effect of PPARγ activation on adiponectin gene expression. For example, a recent study comparing the PPARγ activators pioglitazone and the newly synthesized TZD rivoglitazone showed that both compounds increased adiponectin gene expression in adipose tissue (Hiuge-Shimizu et al., 2011). Natural compounds thought to be PPARγ ligands have also been shown to stimulate adiponectin gene transcription in treated mice. Fish oil diets in mice were shown to stimulate adiponectin gene expression in a PPARγ-dependent, PPARα-independent fashion (Neschen et al., 2006). These studies are typical of a large body of data clearly demonstrating that pharmacological activation of PPARγ in rodent models leads to a stimulation of adiponectin gene transcription in PPARγ containing tissues. Also in support of direct role for PPARγ in adiponectin production are the numerous *in vitro* studies demonstrating that activation of PPARγ leads to an increase in adiponectin mRNA abundance, protein expression, and secretion in cultured adipocytes, mouse models, and humans (e.g., see Landrier et al., 2009; Maeda et al., 2001).

Although adiponectin has been considered an exclusively adipose-secreted hormone, several studies have recently shown that adiponectin is expressed and modulated by PPARγ in cardiac and skeletal muscle cells. A protective role of PPARγ in cardiomyopathy has been well established, as have the benefits of PPARγ activation in skeletal muscle for insulin sensitivity. Several studies present evidence that adiponectin signaling in these types of cells may in part mediate these effects of PPARγ. In cultured ventricular cardiomyocytes, adiponectin and its receptors are expressed and induced by TZDs (Ding et al., 2007). Activation of PPARγ in cultured cardiomyocytes induces adiponectin secretion from these cells and protects them from alpha-adrenergic-induced hypertrophy via autocrine adiponectin signaling (Amin, Mathews, Alli, et al., 2010). Downstream of adiponectin receptor activation, this effect involves activation of AMPK, an important factor in the regulation of heart metabolism, particularly glucose uptake and fatty acid oxidation (Ding et al., 2007). Similarly, PPARγ activation in cultured C2C12 myocytes and primary myotubes as well as constitutive elevation of PPARγ levels *in vivo* in skeletal muscle results in increased adiponectin expression and secretion from the muscle cells. This muscle action of PPARγ results in protective effects on insulin sensitivity in the context of a high-fat diet (Amin, Mathews, Camp, et al., 2010).

The studies presented above strongly support the idea that PPARγ activation leads to an increase in adiponectin levels in circulation. Given the antidiabetic effects of adiponectin (see additional chapters in this book), these data suggest that at least part of the antidiabetic effect of TZD treatment is due to PPARγ-mediated induction of adiponectin production. Given that PPARγ is a transcription factor, the most likely explanation for this correlation is that PPARγ directly stimulates transcription of the adiponectin gene or that it regulates expression of genes involved in adiponectin processing, secretion, or degradation. Additional evidence that this is indeed the case comes from genetic studies described below.

1.2. Genetic evidence

Although unlikely, it is possible that the TZD-mediated stimulation of adiponectin production in intact organisms described above is due to indirect, PPARγ-independent effects of TZDs. An alternative and potentially more direct approach to exploring the specific role of PPARγ in adiponectin production is to examine the effect of PPARγ mutations (both naturally occurring and artificially generated) on adiponectin production. The following section summarizes human and mouse genetic studies examining the association of PPARγ with adiponectin levels or production.

Although numerous genetic studies have demonstrated associations between both the adiponectin and PPARγ genes and diabetes (Jeninga, Gurnell, & Kalkhoven, 2009; for reviews, see Breitfeld, Stumvoll, & Kovacs, 2012), few of these studies directly addressed the possible genetic relationship between PPARγ and adiponectin. The most common genetic variant in the human PPARγ gene is an alanine substitution for proline at position 12 in the PPARγ2 isoform which occurs with an allele frequency of roughly 15% (Altshuler et al., 2000). Although the effect of this variant (termed Pro12Ala) on human physiology appears to vary between different populations, the majority of the studies, especially those in lean subjects, show that the Pro12Ala variant is associated with improved insulin sensitivity and reduced adiposity (Deeb et al., 1998; e.g., see Altshuler et al., 2000). Evidence for a genetic interaction between the PPARγ and adiponectin genes in these populations would suggest a functional relationship between the two genes. Although a genetic interaction between the Pro12Ala variant and genetic variants in the adiponectin gene has been reported (Yang et al., 2003), this was not observed in other studies (Beckers et al., 2009). More direct evidence of a role for PPARγ in determining adiponectin levels was provided by Takata et al. (2004), who reported that the Pro12Ala allele was associated with altered adiponectin levels in Japanese men.

Although the human genetic studies have not identified a clear functional interaction between PPARγ and adiponectin, a highly related study in mice showed a clear effect of the Pro12Ala mutation on adiponectin expression and function. Heikkinen et al. employed a knock-in strategy in which the endogenous mouse PPARγ gene was replaced with the Pro12Ala variant of the gene (Heikkinen et al., 2009). With regard to the issue at hand, the relevant finding from this study was that the mice containing the knocked-in PPARγ Pro12Ala variant showed an increase in expression of not only the adiponectin gene (in adipose tissue from high-fat fed mice) but also the adiponectin receptor adipoR2 gene in both adipose and skeletal muscle. Physiologically, these mice recapitulated the findings from the set of human studies showing that the Pro12Ala variant is associated with a more beneficial, or antidiabetic, metabolic profile (Deeb et al., 1998; Heikkinen et al., 2009). These findings strongly support the idea that PPARγ is a key regulator not only of adiponectin production but also of adiponectin signaling and that this functional relationship plays a role in maintaining metabolic homeostasis.

Another experimental approach that is valuable in assessing the functional relationship between PPARγ and adiponectin is to genetically alter the level of PPARγ activity in specific tissues either by overexpression of exogenous PPARγ in transgenic mice or by ablating the gene in specific tissues. Although these studies were not specifically designed to address the PPARγ/adiponectin relationship, in many cases, adiponectin levels were measured and, in general, support the notion that PPARγ regulates the expression and production of adiponectin.

Ablation of PPARγ from adipose tissue, using a standard tissue-specific gene ablation approach, resulted in dramatically reduced levels of adiponectin gene expression in adipocytes as well as reduced circulating levels of the hormone (Jones et al., 2005). In a similar study, globally reduced PPARγ expression (caused by alteration of the 3′-UTR of the PPARγ gene) resulted in decreased adiponectin gene expression from perigonadal fat tissue, but not from other adipose tissue beds (Tsai et al., 2009). In contrast to these findings, however, adipose-specific overexpression of a constitutive-active PPARγ did not cause an increase in adiponectin gene expression in adipocytes, even though the animals showed improved insulin sensitivity and improved glucose tolerance (Sugii et al., 2009).

Genetically engineered mice with altered PPARγ levels in nonadipose tissues are also informative. Transgenic overexpression of a constitutive-active PPARγ in skeletal muscle and heart leads to a significant stimulation of myocyte adiponectin gene expression in both tissues in insulin-resistant mice (Amin, Mathews, Alli, et al., 2010; Amin, Mathews, Camp, et al., 2010). Consistent with these findings, ablation of PPARγ specifically from skeletal muscle resulted in a significant reduction in circulating levels of adiponectin, although adiponectin gene expression levels were not determined in this study (Hevener et al., 2003). The idea that PPARγ-dependent expression of adiponectin in tissues other than adipose contributes to the circulating levels of the hormone is further supported by the observation that animals with a macrophage-specific PPARγ knockout have significantly reduced levels of circulating adiponectin (Odegaard et al., 2007) although in this case the effects could be an indirect effect on adiponectin production from adipocytes, which was not measured.

The findings from human and rodent studies described above strongly suggest that PPARγ regulates the circulating levels of adiponectin. This could be accomplished by regulating the expression of the adiponectin gene (presumably at the level of transcription), or by regulating some downstream step in the production of the hormone, such as intracellular processing of the adiponectin hormone or its secretion from the cell. The following section will explore these possibilities in greater detail.

2. PPARγ REGULATES ADIPONECTIN GENE TRANSCRIPTION

If PPARγ regulates the level of adiponectin, as the genetic and pharmacological data described above suggest, then either the adiponectin gene itself or genes encoding proteins involved in the processing and secretion of the hormone are likely to be direct transcriptional targets of PPARγ. The following section is a review of the evidence that this is indeed the case. This will be preceded by a brief overview of the basic molecular biology of PPARγ.

PPARγ is a member of the nuclear receptor superfamily of ligand-activated transcription factors. Typical of all the nuclear receptors, PPARγ regulates gene expression by binding to specific DNA sequences, termed PPAR response elements or PPREs, located near the transcriptional start site of genes regulated by the transcription factor. Once activated by ligand binding, DNA bound PPARγ modulates transcriptional activity by recruitment of transcription coregulators that act to alter local chromatin structure and to promote the assembly of a competent RNA polymerase complex. In contrast to other nuclear receptors, such as the estrogen receptor, PPARγ appears to be bound to the PPRE in both the activated and inactivated forms. When activated by endogenous or synthetic ligands, such as the antidiabetic TZD drugs, PPARγ releases transcriptional corepressors, such as "silencing mediator for retinoid or thyroid-hormone receptors" and nuclear receptor corepressor 1, and recruits coactivators including the Mediator complex, steroid receptor coactivator 1 (SRC1), and transcriptional mediators/intermediary factor 2 (TIF2). These events result in the ligand-dependent transcriptional activation of nearby genes and comprise the general mechanism by which PPARγ ligands stimulate the transcription of PPARγ target genes. For more detail on the molecular mechanisms of transcriptional activation by nuclear receptors, see any of the numerous excellent reviews that are available (Chen & Roeder, 2011; Lonard & O'malley, 2007). The following section outlines the evidence that adiponectin is a PPARγ target gene and that this is responsible for the association of PPARγ and adiponectin hormone levels described above.

The human adiponectin promoter contains several functionally characterized regulatory sequences, including a PPRE in the noncoding strand at 273–285 base pairs upstream of the transcription start site (Fig. 6.1; Iwaki et al., 2003). This PPRE sequence 5′-GGGGCA A AAGTCA-3′ is in line

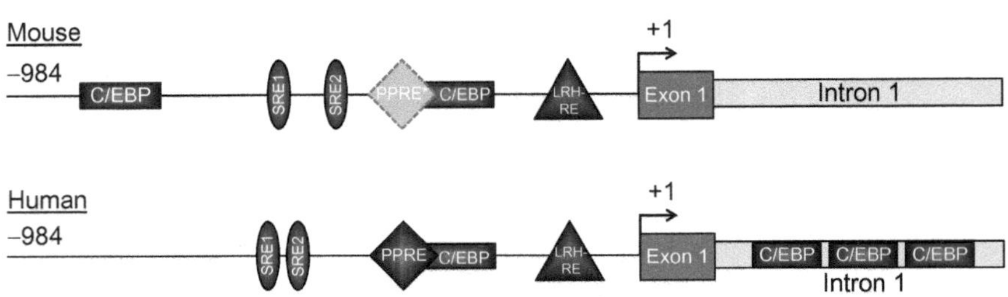

Figure 6.1 Known regulatory binding sites in the human and mouse adiponectin genes. The promoter and start of the transcript (+1) is shown. C/EBP, CCAAT-enhancer-binding protein response element; SRE, sterol regulatory element; LRH-RE, liver receptor homolog response element. The asterisk indicates a putative PPRE in the mouse promoter aligned with the human PPRE. The mouse PPRE has not been shown to be transcriptionally functional.

with the consensus binding site for PPARγ, which is a direct repeat with one spacing nucleotide (AGGTCA A AGGTCA). Unlike many well-characterized PPARγ target genes, the PPRE from the human adiponectin promoter has not been reported to be conserved in other genomes, and in fact, a *bona fide* PPRE regulating adiponectin expression has not been identified in any other species. This is in spite of the fact that at least in mouse cells, pharmacological activation of PPARγ rapidly stimulates adiponectin gene transcription (Maeda et al., 2001). It is possible that in the mouse and other species, PPARγ stimulates adiponectin via an atypical (e.g., unrecognized) PPRE sequence or that PPARγ mediates adiponectin gene transcription indirectly. In any case, the human adiponectin gene clearly contains a functional PPRE that, as described below, is capable of mediating PPARγ-dependent adiponectin gene expression.

The functionality of the human PPAR sequence was verified by demonstrating specific binding of the PPARγ/RXRα heterodimer to the PPRE *in vitro* using a gel shift assay, and robust PPRE-dependent transcriptional induction in luciferase reporter assays of transiently transfected HEK293 cells (Iwaki et al., 2003). Multiple additional studies using the luciferase reporter gene system have confirmed the importance of the PPRE for PPARγ activity on the human adiponectin promoter. In a fully transfected system, activation of PPARγ by its endogenous ligand 15-deoxy-prostaglandin J2 induced luciferase transcription from the intact, but not PPRE-mutated full-length adiponectin promoter, and this induction was abolished by a PPARγ antagonist (Landrier et al., 2009). In mouse embryonic fibroblasts stably expressing PPARγ2, transcriptional activity of the PPARγ–RXRα heterodimer on the adiponectin PPRE was confirmed using a reporter construct driven by three tandem copies of the adiponectin PPRE (Zappalà & Rechler, 2009).

An additional line of evidence consistent with a direct role for PPARγ in the regulation of adiponectin gene expression can be found in studies related to the interaction between PPARγ and inflammatory cytokine signaling. Plasma adiponectin levels are lower in the clinical context of inflammation (Fantuzzi, 2008), even after adjustment for obesity (Engeli et al., 2003). In stable PPARγ expressing fibroblasts, insulin-like growth factor-binding protein 3 (IGFBP-3) reduced the transcriptional activity of PPARγ by binding to and sequestering RXRα (Zappalà & Rechler, 2009). IGFBP-3 is induced at the transcriptional level by inflammatory cytokines, specifically TNFα, and by hypoxia in obesity-related disorders, and may mediate the negative effects of these factors on adiponectin expression.

Multiple lines of evidence point to a role of PPARγ on the adiponectin promoter in inflammation-associated reduction of circulating adiponectin. In cultured adipocytes, treatment with inflammatory cytokine TNFα suppresses adiponectin production and secretion through a mechanism that likely involves activation of the PKC-JNK signaling cascade (Kim et al., 2005) and reduction of PPARγ activity on the PPRE (Lim, Kim, & Park, 2008). In fact, the binding of PPARγ to the adiponectin PPRE is reduced in the presence of TNFα in a gel shift assay (Kita et al., 2005), and treatment of cultured adipocytes with TNFα inhibits adiponectin mRNA expression and secretion (Maeda et al., 2001). JNK signaling is known to play an important role in diet-induced insulin resistance—animals with reduced JNK activity are protected from the effects of high-fat diet on insulin sensitivity (Zhang et al., 2011). Direct modulation of insulin signaling molecules by JNK is a known factor, but regulation of PPARγ activity on the adiponectin promoter may be another mechanism behind this well-documented effect.

PPARα is another member of the PPAR family, which is expressed in the liver, skeletal muscle, and adipose tissue and modulates fatty acid oxidation. Activation of PPARα by fenofibrate treatment in patients with metabolic syndrome elevates plasma adiponectin levels. Evidence from cultured adipocytes with a mutated adiponectin PPRE, or PPARα knockout, reveals that this is likely mediated by transcriptional activity of PPARα on the PPRE in the adiponectin promoter (Hiuge et al., 2007).

In addition to the PPRE, several other regulatory elements have been identified in the adiponectin promoter (Fig. 6.1), some of which interact with the PPRE as part of a transcriptional synergy between PPARγ and other transcription factors. These include transcription factors of well-established importance in the adipose gene expression program: sterol regulatory element-binding protein 1c (SREBP1c) and CCAAT-enhancer-binding protein α (C/EBPα). Insulin-stimulated ADD1/SREBP1c transcriptional activation complex binds to the sterol regulatory elements (SRE1 and SRE2, conserved in mouse and human genomes) and induces gene transcription in both luciferase reporter and *in vivo* contexts (Seo et al., 2004). A C/EBP-binding site overlaps the PPRE in both mouse and human adiponectin gene promoters (Seo et al., 2004) and likely acts in concert with PPARγ. A functional collaboration between C/EBPα and PPARγ has been documented on a global genome scale (Lefterova et al., 2008) and such an interaction could play a role in PPARγ-mediated adiponectin gene regulation. In addition, the human adiponectin gene contains an enhancer within the first intron that is composed of three tandem C/EBP-binding sites which also contribute to modulation of transcription via C/EBPα, although this site is not conserved in the mouse (Qiao et al., 2005). Liver receptor homolog-1, a less widely recognized adipose transcription factor, binds to a cognate site (LRH-RE) downstream of the PPRE (−229 . . . −237), which is transcriptionally functional only when the PPRE is also intact (Iwaki et al., 2003), raising the possibility that the PPARγ/RXRα heterodimer acts in concert with LRH-1 on the adiponectin promoter.

Finally, competition for limited amounts of transcriptional coregulators could be another factor in PPARγ regulation of the adiponectin promoter. Transcription coactivators SRC1 and TIF2 are recruited to the adiponectin promoter by PPARγ in competition with estrogen receptor β (ERβ), another nuclear receptor abundant in adipose tissue. Deficiency of ERβ is protective against insulin resistance. This may be due in part to increased availability of coactivators for PPARγ-mediated recruitment to the adiponectin promoter, resulting in increased adiponectin transcription. Along with other PPARγ target genes, increased adiponectin expression could contribute to improved adipogenesis and insulin sensitivity (Foryst-Ludwig et al., 2008).

Together, the molecular and cellular biology findings presented in this section are fully consistent with the physiological observations that PPARγ activation leads to increased adiponectin production. Furthermore, they suggest that at least part of the effect of PPARγ on adiponectin production is due to direct activation of adiponectin gene transcription. As will be seen in the following section, it is likely that there are additional stimulatory effects of PPARγ on components of the adiponectin secretory pathway that also contribute to the overall positive effect of PPARγ activation on adiponectin hormone levels.

3. PPARγ REGULATES ADIPONECTIN HORMONE SECRETION

Following transcription and translation, the nascent adiponectin peptide undergoes a series of posttranslational modifications before it becomes biologically active, including hydroxylation, glycosylation, oligomerization, secretion, and proteolysis (Wang, Lam, Yau, & Xu, 2008). The primary role of PPARγ in the regulation of this complex process is a point of some debate. While there is evidence of direct transcriptional activation of the adiponectin gene by PPARγ, some groups suggest that this action of PPARγ is negligible, while its primary physiological effect is through modulation of the complex adiponectin secretion machinery (Pajvani et al., 2004; Wang et al., 2007, 2008). Supporting this model, treatment of cultured adipocytes with TZDs increases the amount of secreted adiponectin, and specifically its high-molecular weight (HMW) form, to a much greater extent than its mRNA levels (Pajvani et al., 2004; Qiang, Wang, & Farmer, 2007). Furthermore, not all forms of secreted adiponectin are equally biologically active as mediators of PPARγ activation in the diabetic model. The HMW form of adiponectin, rather than the total circulating amount, correlates strongly with insulin sensitization in women with gestational diabetes following treatment with troglitazone and lowers blood glucose in adiponectin knockout mice (Pajvani et al., 2004). Several studies have revealed specific points in the adiponectin secretory process at which PPARγ may play a role in tipping the balance toward more secretion of the HMW form of the protein.

Within the endoplasmic reticulum, Ero1-Lα and ERp44 play major roles in the posttranslational processing of adiponectin and determine the oligomer composition of the secreted product. ERp44 is a chaperone which retains low-molecular weight and trimer adiponectin within the cell by covalent binding to a thiol group (referred to as thiol-mediated retention; Wolf, 2008). PPARγ has been shown to directly repress the transcription of ERp44 by binding to a PPRE in its promoter in pig kidney and adipose cultured cells (Long et al., 2010), resulting in greater release of adiponectin from the cell. Ero1-Lα is another ER chaperone that facilitates the formation of disulfide bonds to complete the higher-order structure of adiponectin and also competes with adiponectin for thiol binding to ERp44, thereby freeing adiponectin oligomers for secretion (Wang et al., 2008). Adipose Ero1-Lα expression is increased by TZD treatment, coupled with increased secretion of HMW adiponectin, both in mice (Wang et al., 2007) and in cultured adipocytes (Qiang et al., 2007).

GST-kappa/disulfide-bond A oxidoreductase-like protein (DsbA-L) is another ER chaperone that was recently identified in a yeast-two-hybrid screen as an adiponectin-binding protein (Liu et al., 2008). DsbA-L may play a similar role in adiponectin secretion as ERp44, promoting its multimerization and secretion (Wang & Scherer, 2008). DsbA-L expression is induced concomitantly with secreted adiponectin following TZD treatment of 3T3-L1 adipocytes (Liu et al., 2008). While there is no evidence of a direct transcriptional induction of Ero1-Lα or DsbA-L by PPARγ, it is clear that activation of PPARγ affects the ER chaperone-maintained balance of various secreted oligomers of adiponectin.

4. PPARγ AND ADIPONECTIN REGULATE SIMILAR PHYSIOLOGICAL PATHWAYS

In addition to the direct effects of PPARγ on adiponectin hormone production described above, there are other less direct, but not necessarily less important, sites of interaction between these two important metabolic regulators. In the most general sense, both proteins are responsible for inducing similar physiological effects, which can generally be described as improving various aspects of insulin sensitivity in multiple tissues. In part, this is due to the obvious fact that one of the specific effects of PPARγ activation is an increase in adiponectin hormone levels, and as described elsewhere in the book, adiponectin has antidiabetic activity. However, this is due at least partly to the fact that PPARγ also regulates specific components of the adiponectin signaling pathway, including the adiponectin receptors themselves, specific constituents of the intracellular adiponectin signaling pathway, and metabolic enzymes that ultimately mediate the physiological actions of adiponectin. Below, we summarize the evidence for the involvement of PPARγ in three important components of the adiponectin signaling pathway.

Two major adiponectin-sensitive tissues in the body are the liver and skeletal muscle, where both adiponectin receptors adipoR1 and adipoR2 are expressed (Tontonoz & Spiegelman, 2008). PPARγ activation stimulates the expression of adiponectin receptors in various tissues (Ding et al., 2007; Kudoh, Satoh, Hirai, & Watanabe, 2011; Pita, Panadero, Soriano-Guillén, Rodríguez, & Rovira, 2012), thereby effectively increasing adiponectin sensitivity in conjunction with increased circulating adiponectin levels. Intracellular adiponectin signaling is mediated by AMPK, which phosphorylates acyl-intracellularly CoA acetyl-CoA carboxylase (ACC), endothelial nitric oxide synthase (eNOS), and mitogen-activated protein kinase p38 (Heiker, Kosel, & Beck-Sickinger, 2010). While there is no evidence that PPARγ regulates the expression of any of the AMPK subunits or p38 (Schild et al., 2006), some studies suggest that ACC and eNOS may be indirect transcriptional targets of PPARγ.

PPARγ and ACC are often studied together in relation to lipogenesis in obesity and overnutrition (e.g., high-fat diet). Interventions that favor fat storage often induce PPARγ gene expression, along with commonly studied lipogenic enzymes, including ACC, and result in lipid accumulation, typically in the liver. For instance, mice with genetic hypertriglyceridemia have increased PPARγ and ACC expression in the liver, along with other lipogenic genes (Zhang et al., 2006). *In vivo* knockdown of PPARγ2 leads to decreased levels of ACC in mouse models of obesity (Yamazaki, Shiraishi, Kishimoto, Miura, & Ezaki, 2011; Zhang et al., 2006), supporting a regulatory role of PPARγ in ACC expression. Likewise, two copies of the PPARγ gene are required for the high-fat diet-induced increase in ACC and other lipogenic genes in the kidney (Kume et al., 2007). In one study, stable overexpression of PPARγ2 in cultured AML-12 hepatocytes induced ACC protein expression (Kume et al., 2007), suggesting that PPARγ may regulate ACC at the level of gene expression in the liver and possibly other adiponectin-sensitive tissues. In the context of muscle fatty acid uptake and oxidation, however, PPARγ is primarily studied as a regulator of the adiponectin—AMPK—ACC phosphorylation cascade.

The majority of the evidence indicating a role for PPARγ in regulating ACC suggests an indirect mechanism of regulation, involving coinduction or corepression of PPARγ and ACC mRNAs in adipose tissue (Kim & Kim, 2011; Larrieta et al., 2010; Nishiumi, Yabushita, Furuyashiki, Fukuda, & Ashida, 2008) and liver (Djaouti et al., 2010; Nishiumi et al., 2008; Piguet, Stroka, Zimmermann, & Dufour, 2010) in response to nutritional or drug interventions, as well as in cultured adipocytes during differentiation (Lengi & Corl, 2010; Rodríguez et al., 2009), where treatment with TZDs further induces ACC mRNA. One way in which PPARγ activity could indirectly stimulate ACC gene expression is through potentiating the insulin signaling cascade, which induces SREBP, a transcription factor known to be active on the ACC promoter (Zhao et al., 2010).

Similarly, there is limited evidence to support transcriptional regulation of the eNOS gene, either directly or via a mediating factor, by PPARγ. In mice and rats, treatment with partial PPARγ agonist telmisartan results in both PPARγ-dependent gene induction and phosphorylation of eNOS, the latter presumably by PPARγ-regulated adiponectin signaling (Kobayashi et al., 2008; Yuen et al., 2011). As with ACC, some studies report coordinated changes in the expression of PPARγ and eNOS in cultured adipocytes (Shen et al., 2011), diabetic rats (Vasilijević et al., 2010), and brown adipose tissue (Petrović et al., 2010), supporting the gene regulation model. However, more probing experiments reveal that TZDs increase eNOS phosphorylation but fail to induce eNOS gene expression in various animals and cell culture models (Huang et al., 2008; Petrović et al., 2010; Polikandriotis, Mazzella, Rupnow, & Hart, 2005;

Xiao et al., 2010). And even though high doses of troglitazone (Chen, Hu, & Zhang, 2007) and rosiglitazone (Chu et al., 2006) have each been shown to induce eNOS expression, other PPARγ-specific ligands, pioglitazone and 15-deoxy-prostaglandin J2, failed to do so (Goya et al., 2006), suggesting that the former agonists may exert this effect independently of PPARγ.

Thus, in spite of some uncertainty as to exact mechanisms, the overall conclusion from these studies is that there is an overlap between the downstream actions of both PPARγ and adiponectin. In some cases, the two pathways induce the activity and/or amounts of the same enzymes (ACC and eNOS), and in other cases, PPARγ stimulates the amount of specific components of the adiponectin signaling pathway (adipoR1 and adipoR2).

5. CONCLUSION

Given the strong similarity in the general physiological roles of the PPARγ and adiponectin, it might not be surprising that there are numerous functional links between the two proteins. On the one hand, it may be accurate to think of adiponectin and its signaling pathway as simply downstream targets of PPARγ and that PPARγ is the "master regulator" of this process. On the other hand, it may be more correct to consider the two proteins as different components of the same complex physiological network that has evolved to regulate whole-body energy balance.

REFERENCES

Altshuler, D., Hirschhorn, J. N., Klannemark, M., Lindgren, C. M., Vohl, M. C., Nemesh, J., et al. (2000). The common PPARgamma Pro12Ala polymorphism is associated with decreased risk of type 2 diabetes. *Nature Genetics*, *26*, 76–80.

Amin, R. H., Mathews, S. T., Alli, A., & Leff, T. (2010). Endogenously produced adiponectin protects cardiomyocytes from hypertrophy by a PPARgamma-dependent autocrine mechanism. *American Journal of Physiology. Heart and Circulatory Physiology*, *299*, H690–H698.

Amin, R. H., Mathews, S. T., Camp, H. S., Ding, L., & Leff, T. (2010). Selective activation of PPARgamma in skeletal muscle induces endogenous production of adiponectin and protects mice from diet-induced insulin resistance. *American Journal of Physiology. Endocrinology and Metabolism*, *298*, E28–E37.

Beckers, S., Peeters, A. V., de Freitas, F., Mertens, I. L., Verhulst, S. L., Haentjens, D., et al. (2009). Association study and mutation analysis of adiponectin shows association of variants in APM1 with complex obesity in women. *Annals of Human Genetics*, *73*, 492–501.

Breitfeld, J., Stumvoll, M., & Kovacs, P. (2012). Genetics of adiponectin. *Biochimie*, Epub ahead of print. 2012 Mar 15.

Chen, S., Hu, Z., & Zhang, K. (2007). Effect of Troglitazone on expression of adhesion molecules and eNOS in human saphenous vein graft. *Journal of Huazhong University of Science and Technology*, *27*, 657–659.

Chen, W., & Roeder, R. G. (2011). Mediator-dependent nuclear receptor function. *Seminars in Cell & Developmental Biology, 22*, 749–758.

Chu, K., Lee, S.-T., Koo, J.-S., Jung, K.-H., Kim, E.-H., Sinn, D.-I., et al. (2006). Peroxisome proliferator-activated receptor-gamma-agonist, rosiglitazone, promotes angiogenesis after focal cerebral ischemia. *Brain Research, 1093*, 208–218.

Combs, T. P., Wagner, J. A., Berger, J., Doebber, T., Wang, W.-J., Zhang, B. B., et al. (2002). Induction of adipocyte complement-related protein of 30 kilodaltons by PPARgamma agonists: A potential mechanism of insulin sensitization. *Endocrinology, 143*, 998–1007.

Deeb, S. S., Fajas, L., Nemoto, M., Pihlajamäki, J., Mykkänen, L., Kuusisto, J., et al. (1998). A Pro12Ala substitution in PPARgamma2 associated with decreased receptor activity, lower body mass index and improved insulin sensitivity. *Nature Genetics, 20*, 284–287.

Ding, G., Qin, Q., He, N., Francis-David, S., Hou, J., Liu, J., et al. (2007). Adiponectin and its receptors are expressed in adult ventricular cardiomyocytes and upregulated by activation of peroxisome proliferator-activated receptor gamma. *Journal of Molecular and Cellular Cardiology, 43*, 73–84.

Djaouti, L., Jourdan, T., Demizieux, L., Chevrot, M., Gresti, J., Vergès, B., et al. (2010). Different effects of pioglitazone and rosiglitazone on lipid metabolism in mouse cultured liver explants. *Diabetes/Metabolism Research and Reviews, 26*, 297–305.

Engeli, S., Feldpausch, M., Gorzelniak, K., Hartwig, F., Heintze, U., Janke, J., et al. (2003). Association between adiponectin and mediators of inflammation in obese women. *Diabetes, 52*, 942–947.

Fantuzzi, G. (2008). Adiponectin and inflammation: Consensus and controversy. *The Journal of Allergy and Clinical Immunology, 121*, 326–330.

Foryst-Ludwig, A., Clemenz, M., Hohmann, S., Hartge, M., Sprang, C., Frost, N., et al. (2008). Metabolic actions of estrogen receptor beta (ERbeta) are mediated by a negative cross-talk with PPARgamma. *PLoS Genetics, 4*, e1000108.

Goya, K., Sumitani, S., Otsuki, M., Xu, X., Yamamoto, H., Kurebayashi, S., et al. (2006). The thiazolidinedione drug troglitazone up-regulates nitric oxide synthase expression in vascular endothelial cells. *Journal of Diabetes and Its Complications, 20*, 336–342.

Heiker, J. T., Kosel, D., & Beck-Sickinger, A. G. (2010). Molecular mechanisms of signal transduction via adiponectin and adiponectin receptors. *Biological Chemistry, 391*, 1005–1018.

Heikkinen, S., Argmann, C., Feige, J. N., Koutnikova, H., Champy, M.-F., Dali-Youcef, N., et al. (2009). The Pro12Ala PPARgamma2 variant determines metabolism at the gene-environment interface. *Cell Metabolism, 9*, 88–98.

Hevener, A. L., He, W., Barak, Y., Le, J., Bandyopadhyay, G., Olson, P., et al. (2003). Muscle-specific Pparg deletion causes insulin resistance. *Nature Medicine, 9*, 1491–1497.

Hirose, H., Kawai, T., Yamamoto, Y., Taniyama, M., Tomita, M., Matsubara, K., et al. (2002). Effects of pioglitazone on metabolic parameters, body fat distribution, and serum adiponectin levels in Japanese male patients with type 2 diabetes. *Metabolism, 51*, 314–317.

Hiuge, A., Tenenbaum, A., Maeda, N., Benderly, M., Kumada, M., Fisman, E. Z., et al. (2007). Effects of peroxisome proliferator-activated receptor ligands, bezafibrate and fenofibrate, on adiponectin level. *Arteriosclerosis, Thrombosis, and Vascular Biology, 27*, 635–641.

Hiuge-Shimizu, A., Maeda, N., Hirata, A., Nakatsuji, H., Nakamura, K., Okuno, A., et al. (2011). Dynamic changes of adiponectin and S100A8 levels by the selective peroxisome proliferator-activated receptor-gamma agonist rivoglitazone. *Arteriosclerosis, Thrombosis, and Vascular Biology, 31*, 792–799.

Huang, P.-H., Sata, M., Nishimatsu, H., Sumi, M., Hirata, Y., & Nagai, R. (2008). Pioglitazone ameliorates endothelial dysfunction and restores ischemia-induced angiogenesis in diabetic mice. *Biomedicine and Pharmacotherapy*, *62*, 46–52.

Iwaki, M., Matsuda, M., Maeda, N., Funahashi, T., Matsuzawa, Y., Makishima, M., et al. (2003). Induction of adiponectin, a fat-derived antidiabetic and antiatherogenic factor, by nuclear receptors. *Diabetes*, *52*, 1655–1663.

Jeninga, E., Gurnell, M., & Kalkhoven, E. (2009). Functional implications of genetic variation in human PPARgamma. *Trends in Endocrinology and Metabolism*, *20*(8), 380–387.

Jones, J., Barrick, C., Kim, K., Lindner, J., Blondeau, B., Fujimoto, Y., et al. (2005). Deletion of PPARgamma in adipose tissues of mice protects against high fat diet-induced obesity and insulin resistance. *Proceedings of the National Academy of Sciences of the United States of America*, *102*, 6207–6212.

Kim, M.-J., & Kim, H. K. (2011). Effect of garlic on high fat induced obesity. *Acta Biologica Hungarica*, *62*, 244–254.

Kim, K.-Y., Kim, J. K., Jeon, J. H., Yoon, S. R., Choi, I., & Yang, Y. (2005). c-Jun N-terminal kinase is involved in the suppression of adiponectin expression by TNF-alpha in 3T3-L1 adipocytes. *Biochemical and Biophysical Research Communications*, *327*, 460–467.

Kita, A., Yamasaki, H., Kuwahara, H., Moriuchi, A., Fukushima, K., Kobayashi, M., et al. (2005). Identification of the promoter region required for human adiponectin gene transcription: Association with CCAAT/enhancer binding protein-beta and tumor necrosis factor-alpha. *Biochemical and Biophysical Research Communications*, *331*, 484–490.

Kobayashi, N., Ohno, T., Yoshida, K., Fukushima, H., Mamada, Y., Nomura, M., et al. (2008). Cardioprotective mechanism of telmisartan via PPAR-gamma-eNOS pathway in dahl salt-sensitive hypertensive rats. *American Journal of Hypertension*, *21*, 576–581.

Koenen, T. B., Tack, C. J., Kroese, J. M., Hermus, A. R., Sweep, F. C. G., van der Laak, J., et al. (2009). Pioglitazone treatment enlarges subcutaneous adipocytes in insulin-resistant patients. *The Journal of Clinical Endocrinology and Metabolism*, *94*, 4453–4457.

Kudoh, A., Satoh, H., Hirai, H., & Watanabe, T. (2011). Pioglitazone upregulates adiponectin receptor 2 in 3T3-L1 adipocytes. *Life Sciences*, *88*, 1055–1062.

Kume, S., Uzu, T., Araki, S.-I., Sugimoto, T., Isshiki, K., Chin-Kanasaki, M., et al. (2007). Role of altered renal lipid metabolism in the development of renal injury induced by a high-fat diet. *Journal of the American Society of Nephrology*, *18*, 2715–2723.

Landrier, J.-F., Gouranton, E., El Yazidi, C., Malezet, C., Balaguer, P., Borel, P., et al. (2009). Adiponectin expression is induced by vitamin E via a peroxisome proliferator-activated receptor gamma-dependent mechanism. *Endocrinology*, *150*, 5318–5325.

Larrieta, E., Velasco, F., Vital, P., López-Aceves, T., Lazo-de-la-Vega-Monroy, M. L., Rojas, A., et al. (2010). Pharmacological concentrations of biotin reduce serum triglycerides and the expression of lipogenic genes. *European Journal of Pharmacology*, *644*, 263–268.

Lefterova, M., Zhang, Y., Steger, D., Schupp, M., Schug, J., Cristancho, A., et al. (2008). PPARgamma and C/EBP factors orchestrate adipocyte biology via adjacent binding on a genome-wide scale. *Genes & Development*, *22*, 2941–2952.

Lengi, A. J., & Corl, B. A. (2010). Factors influencing the differentiation of bovine preadipocytes in vitro. *Journal of Animal Science*, *88*, 1999–2008.

Lim, J.-Y., Kim, W. H., & Park, S. I. (2008). GO6976 prevents TNF-alpha-induced suppression of adiponectin expression in 3T3-L1 adipocytes: Putative involvement of protein kinase C. *FEBS Letters*, *582*, 3473–3478.

Liu, M., Zhou, L., Xu, A., Lam, K. S. L., Wetzel, M. D., Xiang, R., et al. (2008). A disulfide-bond A oxidoreductase-like protein (DsbA-L) regulates adiponectin multimerization. *Proceedings of the National Academy of Sciences of the United States of America*, *105*, 18302–18307.

Lonard, D. M., & O'malley, B. W. (2007). Nuclear receptor coregulators: Judges, juries, and executioners of cellular regulation. *Molecular Cell*, *27*, 691–700.
Long, Q., Lei, T., Feng, B., Yin, C., Jin, D., Wu, Y., et al. (2010). Peroxisome proliferator-activated receptor-gamma increases adiponectin secretion via transcriptional repression of endoplasmic reticulum chaperone protein ERp44. *Endocrinology*, *151*, 3195–3203.
Maeda, N., Takahashi, M., Funahashi, T., Kihara, S., Nishizawa, H., Kishida, K., et al. (2001). PPARgamma ligands increase expression and plasma concentrations of adiponectin, an adipose-derived protein. *Diabetes*, *50*, 2094–2099.
Neschen, S., Morino, K., Rossbacher, J. C., Pongratz, R. L., Cline, G. W., Sono, S., et al. (2006). Fish oil regulates adiponectin secretion by a peroxisome proliferator-activated receptor-gamma-dependent mechanism in mice. *Diabetes*, *55*, 924–928.
Nishiumi, S., Yabushita, Y., Furuyashiki, T., Fukuda, I., & Ashida, H. (2008). Involvement of SREBPs in 2,3,7,8-tetrachlorodibenzo-p-dioxin-induced disruption of lipid metabolism in male guinea pig. *Toxicology and Applied Pharmacology*, *229*, 281–289.
Odegaard, J. I., Ricardo-Gonzalez, R. R., Goforth, M. H., Morel, C. R., Subramanian, V., Mukundan, L., et al. (2007). Macrophage-specific PPARgamma controls alternative activation and improves insulin resistance. *Nature*, *447*, 1116–1120.
Pajvani, U. B., Hawkins, M., Combs, T. P., Rajala, M. W., Doebber, T., Berger, J. P., et al. (2004). Complex distribution, not absolute amount of adiponectin, correlates with thiazolidinedione-mediated improvement in insulin sensitivity. *The Journal of Biological Chemistry*, *279*, 12152–12162.
Petrović, V., Buzadzić, B., Korać, A., Vasilijević, A., Janković, A., & Korać, B. (2010). NO modulates the molecular basis of rat interscapular brown adipose tissue thermogenesis. *Comparative Biochemistry and Physiology, Toxicology & Pharmacology*, *152*, 147–159.
Phillips, S. A., Ciaraldi, T. P., Kong, A. P. S., Bandukwala, R., Aroda, V., Carter, L., et al. (2003). Modulation of circulating and adipose tissue adiponectin levels by antidiabetic therapy. *Diabetes*, *52*, 667–674.
Piguet, A.-C., Stroka, D., Zimmermann, A., & Dufour, J.-F. (2010). Hypoxia aggravates non-alcoholic steatohepatitis in mice lacking hepatocellular PTEN. *Clinical Science*, *118*, 401–410.
Pita, J., Panadero, A., Soriano-Guillén, L., Rodríguez, E., & Rovira, A. (2012). The insulin sensitizing effects of PPAR-γ agonist are associated to changes in adiponectin index and adiponectin receptors in Zucker fatty rats. *Regulatory Peptides*, *174*, 18–25.
Polikandriotis, J. A., Mazzella, L. J., Rupnow, H. L., & Hart, C. M. (2005). Peroxisome proliferator-activated receptor gamma ligands stimulate endothelial nitric oxide production through distinct peroxisome proliferator-activated receptor gamma-dependent mechanisms. *Arteriosclerosis, Thrombosis, and Vascular Biology*, *25*, 1810–1816.
Qiang, L., Wang, H., & Farmer, S. (2007). Adiponectin secretion is regulated by SIRT1 and the endoplasmic reticulum oxidoreductase Ero1-L alpha. *Molecular and Cellular Biology*, *27*, 4698–4707.
Qiao, L., MacLean, P., Schaack, J., Orlicky, D., Darimont, C., Pagliassotti, M., et al. (2005). C/EBPalpha regulates human adiponectin gene transcription through an intronic enhancer. *Diabetes*, *54*, 1744–1754.
Riera-Guardia, N., & Rothenbacher, D. (2008). The effect of thiazolidinediones on adiponectin serum level: A meta-analysis. *Diabetes, Obesity & Metabolism*, *10*, 367–375.
Rodríguez, A., Gómez-Ambrosi, J., Catalán, V., Gil, M. J., Becerril, S., Sáinz, N., et al. (2009). Acylated and desacyl ghrelin stimulate lipid accumulation in human visceral adipocytes. *International Journal of Obesity and Related Metabolic Disorders*, *33*, 541–552.
Sam, S., Haffner, S., Davidson, M. H., D'Agostino, R., Perez, A., & Mazzone, T. (2012). Pioglitazone-mediated changes in lipoprotein particle composition are predicted by changes in adiponectin level in type 2 diabetes. *The Journal of Clinical Endocrinology and Metabolism*, *97*, E110–E114.

Schild, R. L., Sonnenberg-Hirche, C. M., Schaiff, W. T., Bildirici, I., Nelson, D. M., & Sadovsky, Y. (2006). The kinase p38 regulates peroxisome proliferator activated receptor-gamma in human trophoblasts. *Placenta, 27*, 191–199.

Seo, J. B., Moon, H. M., Noh, M. J., Lee, Y. S., Jeong, H. W., Yoo, E. J., et al. (2004). Adipocyte determination- and differentiation-dependent factor 1/sterol regulatory element-binding protein 1c regulates mouse adiponectin expression. *The Journal of Biological Chemistry, 279*, 22108–22117.

Shen, W., Hao, J., Feng, Z., Tian, C., Chen, W., Packer, L., et al. (2011). Lipoamide or lipoic acid stimulates mitochondrial biogenesis in 3T3-L1 adipocytes via the endothelial NO synthase-cGMP-protein kinase G signalling pathway. *British Journal of Pharmacology, 162*, 1213–1224.

Sugii, S., Olson, P., Sears, D. D., Saberi, M., Atkins, A. R., Barish, G. D., et al. (2009). PPARgamma activation in adipocytes is sufficient for systemic insulin sensitization. *Proceedings of the National Academy of Sciences of the United States of America, 106*, 22504–22509.

Sutinen, J., Kannisto, K., Korsheninnikova, E., Fisher, R. M., Ehrenborg, E., Nyman, T., et al. (2004). Effects of rosiglitazone on gene expression in subcutaneous adipose tissue in highly active antiretroviral therapy-associated lipodystrophy. *American Journal of Physiology. Endocrinology and Metabolism, 286*, E941–E949.

Takata, N., Awata, T., Inukai, K., Watanabe, M., Ohkubo, T., Kurihara, S., et al. (2004). Pro12Ala substitution in peroxisome proliferator-activated receptor gamma 2 is associated with low adiponectin concentrations in young Japanese men. *Metabolism, 53*, 1548–1551.

Tontonoz, P., & Spiegelman, B. (2008). Fat and beyond: The diverse biology of PPARgamma. *Annual Review of Biochemistry, 77*, 289–312.

Tsai, Y.-S., Tsai, P.-J., Jiang, M.-J., Chou, T.-Y., Pendse, A., Kim, H.-S., et al. (2009). Decreased PPAR gamma expression compromises perigonadal-specific fat deposition and insulin sensitivity. *Molecular Endocrinology, 23*, 1787–1798.

Vasilijević, A., Vojcić, L., Dinulović, I., Buzadzić, B., Korać, A., Petrović, V., et al. (2010). Expression pattern of thermogenesis-related factors in interscapular brown adipose tissue of alloxan-treated rats: Beneficial effect of L-arginine. *Nitric Oxide, 23*, 42–50.

Wang, Y., Lam, K. S. L., Yau, M.-H., & Xu, A. (2008). Post-translational modifications of adiponectin: Mechanisms and functional implications. *The Biochemical Journal, 409*, 623–633.

Wang, Z. V., & Scherer, P. E. (2008). DsbA-L is a versatile player in adiponectin secretion. *Proceedings of the National Academy of Sciences of the United States of America, 105*, 18077–18078.

Wang, Z., Schraw, T., Kim, J., Khan, T., Rajala, M., Follenzi, A., et al. (2007). Secretion of the adipocyte-specific secretory protein adiponectin critically depends on thiol-mediated protein retention. *Molecular and Cellular Biology, 27*, 3716–3731.

Wolf, G. (2008). New insights into thiol-mediated regulation of adiponectin secretion. *Nutrition Reviews, 66*, 642–645.

Xiao, X., Su, G., Brown, S., Chen, L., Ren, J., & Zhao, P. (2010). Peroxisome proliferator-activated receptors gamma and alpha agonists stimulate cardiac glucose uptake via activation of AMP-activated protein kinase. *The Journal of Nutritional Biochemistry, 21* (7), 621–626.

Yamazaki, T., Shiraishi, S., Kishimoto, K., Miura, S., & Ezaki, O. (2011). An increase in liver PPARγ2 is an initial event to induce fatty liver in response to a diet high in butter: PPARγ2 knockdown improves fatty liver induced by high-saturated fat. *The Journal of Nutritional Biochemistry, 22*, 543–553.

Yang, W.-S., Hsiung, C. A., Ho, L.-T., Chen, Y.-T., He, C.-T., Curb, J. D., et al. (2003). Genetic epistasis of adiponectin and PPARgamma2 genotypes in modulation of insulin sensitivity: A family-based association study. *Diabetologia, 46*, 977–983.

Yu, J., Javorschi, S., Hevener, A., Kruszynska, Y., Norman, R., Sinha, M., et al. (2002). The effect of thiazolidinediones on plasma adiponectin levels in normal, obese, and type 2 diabetic subjects. *Diabetes*, *51*, 2968–2974.

Yuen, C. Y., Wong, W. T., Tian, X. Y., Wong, S. L., Lau, C. W., Yu, J., et al. (2011). Telmisartan inhibits vasoconstriction via PPARγ-dependent expression and activation of endothelial nitric oxide synthase. *Cardiovascular Research*, *90*, 122–129.

Zappalà, G., & Rechler, M. M. (2009). IGFBP-3, hypoxia and TNF-alpha inhibit adiponectin transcription. *Biochemical and Biophysical Research Communications*, *382*, 785–789.

Zhang, Y.-L., Hernandez-Ono, A., Siri, P., Weisberg, S., Conlon, D., Graham, M. J., et al. (2006). Aberrant hepatic expression of PPARgamma2 stimulates hepatic lipogenesis in a mouse model of obesity, insulin resistance, dyslipidemia, and hepatic steatosis. *The Journal of Biological Chemistry*, *281*, 37603–37615.

Zhang, X., Xu, A., Chung, S. K., Cresser, J. H. B., Sweeney, G., Wong, R. L. C., et al. (2011). Selective inactivation of c-Jun NH2-terminal kinase in adipose tissue protects against diet-induced obesity and improves insulin sensitivity in both liver and skeletal muscle in mice. *Diabetes*, *60*, 486–495.

Zhao, L. F., Iwasaki, Y., Zhe, W., Nishiyama, M., Taguchi, T., Tsugita, M., et al. (2010). Hormonal regulation of acetyl-CoA carboxylase isoenzyme gene transcription. *Endocrine Journal*, *57*, 317–324.

CHAPTER SEVEN

Glucocorticoid Effects on Adiponectin Expression

Siddharth Sukumaran*, Debra C. DuBois*,†, William J. Jusko*,‡, Richard R. Almon*,†,‡,1

*Department of Pharmaceutical Sciences, State University of New York at Buffalo, Buffalo, New York, USA
†Department of Biological Sciences, State University of New York at Buffalo, Buffalo, New York, USA
‡New York State Center of Excellence in Bioinformatics and Life Sciences, State University of New York at Buffalo, Buffalo, New York, USA
[1]Corresponding author: e-mail address: almon@buffalo.edu

Contents

Abstract

Maintenance of energy metabolism and glucose homeostasis is achieved by the regulatory effects of many hormones and their interactions. Glucocorticoids produced from adrenal cortex and adiponectin produced by adipose tissue play important roles in the production, distribution, storage, and utilization of energy substrates. Glucocorticoids are involved in the activation of a number of catabolic processes by affecting the expression of a plethora of genes, while adiponectin acts primarily as an insulin sensitizer. Both are regulated by a number of physiological and pharmacological factors. Although the effects of glucocorticoids on adiponectin expression have been extensively studied in different *in vitro*, animal and clinical study settings, no consensus has been reached. This report reviews the primary literature concerning the effects of glucocorticoids on adiponectin expression and identifies potential reasons for the contradictory results between different studies. In addition, methods to gain better insights pertaining to the regulation of adiponectin expression are discussed.

Vitamins and Hormones, Volume 90
ISSN 0083-6729
http://dx.doi.org/10.1016/B978-0-12-398313-8.00007-5

1. INTRODUCTION

Maintenance of energy metabolism and glucose homeostasis is crucial for proper functioning of an organism. These processes are regulated by a number of complex mechanisms occurring in different tissues and cell types. An extensive network of organs including brain, liver, muscle, adipose tissue, pancreas, and adrenal gland coordinate to maintain this homeostasis and any imbalance in this coordination can lead to metabolic disorders including diabetes, obesity, and metabolic syndrome (Besser & Jeffcoate, 1976; Tirone & Brunicardi, 2001; Yamada, Oka, & Katagiri, 2008). Hormones play an important role in communication between organs and are one of the central components controlling various metabolic processes. Many hormones including insulin, glucagon, glucocorticoids (GCs), leptin, and adiponectin control pathways and gene expression changes related to energy metabolism and its regulation (Eyster, 2011; Heppner et al., 2010; Rose, Vegiopoulos, & Herzig, 2010; Yamada et al., 2008). These different hormones are produced from different organs (insulin and glucagon from pancreatic islets, GCs from adrenal cortex, and leptin and adiponectin from adipose tissue) and affect multiple tissues that express the receptors for these hormones (Kershaw & Flier, 2004; Lukens, 1959; Rose et al., 2010). In addition to regulating overlapping processes in energy metabolism, these hormones can affect the production of each other, thereby causing feedback regulation of homeostasis. This report summarizes previous *in vitro*, preclinical and clinical studies pertaining to the effects of GCs on adiponectin expression and explores their implications.

2. GLUCOCORTICOIDS AND THEIR EFFECTS ON GENE EXPRESSION

GCs are catabolic steroid hormones produced from the adrenal cortex which play an important role in the production, distribution, storage, and utilization of substrates used for energy metabolism (Vegiopoulos & Herzig, 2007). In addition to their effects on energy metabolism, GCs play an important role in regulating inflammatory and immune response processes (Barnes, 1998). GCs suppress these processes, and because of their anti-inflammatory properties, synthetic GCs (such as dexamethasone, prednisolone, methylprednisolone (MP)) are extensively used for the treatment of inflammatory disorders including asthma, rheumatoid arthritis, and lupus erythematosus (Barnes, 1998; Swartz & Dluhy, 1978). However, because of their effects on systemic energy balance, chronic use results in severe side effects including steroid-induced diabetes, dyslipidemia, muscle atrophy, and metabolic syndrome (Rhen & Cidlowski, 2005; Swartz & Dluhy, 1978).

The hypothalamic–pituitary–adrenal axis plays a critical role in regulating the production of GC (Fig. 7.1; Tsigos & Chrousos, 2002). This includes the sequential and feedback regulation by a series of hormones produced by different organs. Corticotropin-releasing hormone (CRH) produced from the hypothalamus positively regulates adrenocorticotropic hormone (ACTH) production from the pituitary which in turn influences GC production in the adrenal cortex. A negative feedback loop in this axis is caused by the negative regulation in the production of CRH and ACTH by GC (Tsigos & Chrousos, 2002).

The molecular mechanism by which GCs regulate gene expression is given in Fig. 7.2 (Schaaf & Cidlowski, 2002; Sukumaran, Jusko, DuBois, & Almon, 2011). Since GCs are highly lipophilic, they can passively diffuse through cell membranes. In the absence of the hormone, the cytosolic receptor for GC is present as an inactive state complexed with heat shock proteins. Binding of GC to its cytosolic receptor results in

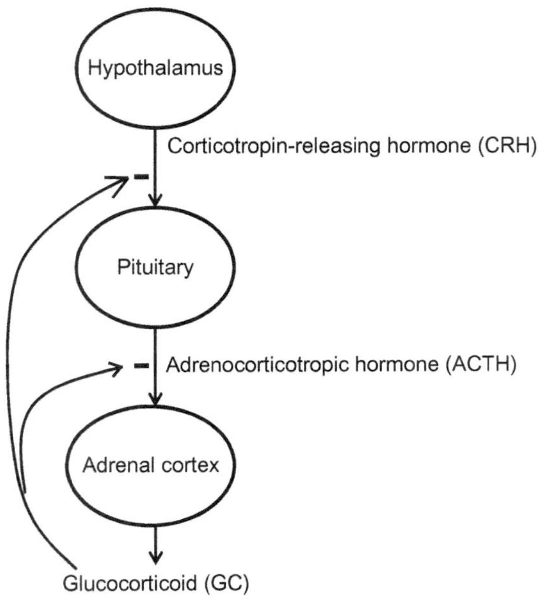

Figure 7.1 Regulation of glucocorticoids production by hypothalamic–pituitary–adrenal axis represents inhibition in the production.

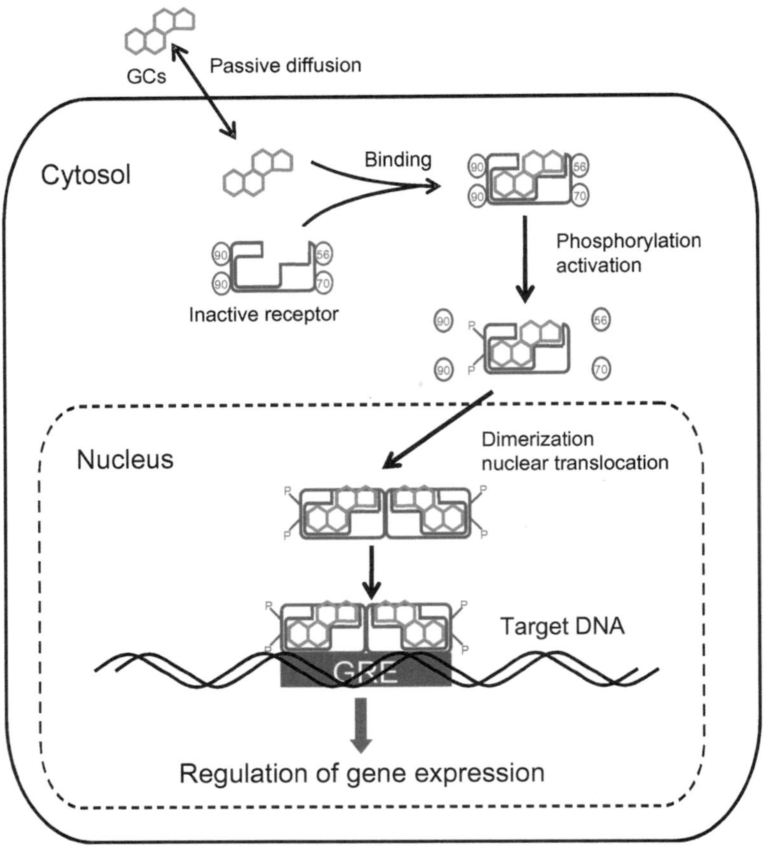

Figure 7.2 Molecular mechanism for the regulation of gene expression by glucocorticoids. (See Color Insert.)

conformational change, phosphorylation, dimerization, and activation of the receptor (Schaaf & Cidlowski, 2002). The activated GC–receptor complex rapidly translocates into the nucleus and regulates gene expression. Multiple mechanisms including direct binding, tethering, and composite binding exist by which GC can positively (transactivation) or negatively (transrepression) regulate gene expression (Oakley & Cidlowski, 2011; Schaaf & Cidlowski, 2002). On the one hand, in the case of direct binding, the activated hormone–receptor complex binds to specific DNA regions referred to as GC response elements, thereby modulating transcription rates (Schaaf & Cidlowski, 2002). On the other hand, GC–receptor complex in the nucleus can interact with other transcription factors, activators, and repressors, thereby modulating the rate of transcription of their regulated genes. For example, activated GC receptors interact with transcription factors like NfκB, AP-1, and STAT and affect the mRNA expression of many genes that are regulated by these transcription factors (Barnes, 1998; Schaaf & Cidlowski, 2002). Microarray and gene expression studies show that GC directly or indirectly regulate a vast array of genes and pathways in many tissues, suggesting its influence on many key biological processes (Almon, DuBois, & Jusko, 2007; Almon, DuBois, Pearson, Stephan, & Jusko, 2003; Almon, DuBois, Piel, & Jusko, 2004; Almon, Lai, DuBois, & Jusko, 2005; Almon et al., 2007).

3. EFFECTS OF GC AND ADIPONECTIN ON ENERGY METABOLISM

One of the most important functions of GC is to maintain proper plasma glucose concentrations to feed neuronal cells under conditions of decreased glucose availability from the diet (Heppner et al., 2010; Vegiopoulos & Herzig, 2007). To accomplish this effect, GCs affect a plethora of processes ranging from production to utilization and disposal of carbohydrates, lipids, and other energy substrates involving different organs and cell types (Vegiopoulos & Herzig, 2007). GC concentrations in plasma increase during starvation and under stress. During normal conditions, GCs show robust circadian oscillations with peak concentrations occurring at the early active period (early morning for humans and early dark period for nocturnal animals), thus preparing the body for the upcoming period of activity (Sukumaran, Almon, DuBois, & Jusko, 2010). GCs increase glucose production by primarily upregulating gluconeogenesis in liver (and sometimes in kidney) (Gunn, Hanson, Meyuhas, Reshef, & Ballard, 1975; Iynedjian, Ballard, & Hanson, 1975; Vegiopoulos & Herzig, 2007). On the utilization side, GCs cause insulin resistance in many tissues including skeletal muscle and adipose tissue, thereby reducing insulin-directed disposition of glucose (Andrews & Walker, 1999). GCs stimulate protein breakdown in skeletal muscle to provide the carbon backbone from amino acids to fuel gluconeogenesis in liver (Hasselgren, 1999). In addition, GCs stimulate lipolysis in white adipose tissue producing free fatty acids for many tissues and glycerol for gluconeogenesis. In addition, GCs also inhibit lipid synthesis (Vegiopoulos & Herzig, 2007). All these processes are regulated through changes in gene expression (which eventually causes change in protein expression) and signaling cascades caused by GC (Vegiopoulos & Herzig, 2007).

Adiponectin, the adipokine produced primarily from white adipose tissue, is a very important hormone involved in metabolic processes (Kadowaki & Yamauchi, 2005). The effects of adiponectin are primarily mediated through their membrane-bound G protein-coupled receptors, AdipoR1 and AdipoR2 are expressed in multiple tissues and organs and regulate MAP kinase, AMP kinase, and PPARγ signaling cascades (Dridi & Taouis, 2009; Kadowaki & Yamauchi, 2005). The primary function of adiponectin is to act as an endogenous insulin sensitizer, and it has a major influence on free fatty acids metabolism. Studies show that adiponectin increases glucose utilization in skeletal muscle, decreases gluconeogenesis in liver, and increases fatty acid oxidation in both liver and skeletal muscle (Dridi & Taouis, 2009). The plasma concentrations of adiponectin are inversely proportional to the fat mass present and hence adiponectin concentrations are decreased in obese patients which serves as one cause for their insulin resistance (Swarbrick & Havel, 2008).

Thus, GC and adiponectin play opposite roles in regulating energy metabolism with GC primarily involved in activating catabolic processes and insulin resistance, but adiponectin acting primarily as an insulin sensitizer (Kadowaki & Yamauchi, 2005; Vegiopoulos & Herzig, 2007). However, with respect to immune and inflammatory processes, both GC and adiponectin are anti-inflammatory in their effects (Barnes, 1998; Ouchi & Walsh, 2007).

4. *IN VITRO* AND ANIMAL STUDIES ON THE EFFECTS OF GC ON ADIPONECTIN EXPRESSION

Initial studies on the regulation of adiponectin expression by GC and other hormones were done in the early 2000s, mainly in *in vitro* systems using human or murine adipocytes. Such studies produced conflicting results. For example, studies by Halleux et al. showed that in human visceral adipose tissue explants dexamethasone (dex) treatment downregulated adiponectin mRNA and protein expression, while insulin had the opposite effect of upregulating the expression in a dose-dependent manner (Halleux et al., 2001). Similarly, dex treatment decreased adiponectin mRNA expression in 3T3 murine adipocytes (Fasshauer, Klein, Neumann, Eszlinger, & Paschke, 2002). In contrast, in a study by Oliveira et al., dex treatment did not cause any change in adiponectin mRNA expression in 3T3-L1 adipocytes (de Oliveira et al., 2011b). Similar results were observed in studies on canine adipocytes, where dex treatment did not cause any changes in adiponectin expression (Ryan et al., 2010). Likewise, in a recent study using human bone marrow adipocytes, dex treatment did not have any effect on either adiponectin gene expression or secretion (Hozumi et al., 2010). The results from these *in vitro* studies are summarized in Table 7.1. Data from animal studies also yield inconsistent results on the effects of GC on adiponectin expression (Table 7.2). A study by Combs et al. shows that Swiss Webster mice treated with prednisolone from a slow release pellet for 28 days showed a significant increase in plasma adiponectin concentrations (Combs et al., 2003). Similarly, in neonatal Sprague–Dawley rats, dex treatment resulted in increased plasma adiponectin concentrations (Raff & Bruder, 2006). However, other studies give contrasting results on the effects of GC on adiponectin expression. 11β-Hydroxysteroid dehydrogenases (11βHSD) are enzymes involved in tissue-specific activation or deactivation of GC (Tomlinson et al., 2004). 11βHSD1 is involved in the activation of GCs and 11βHSD2 is involved in the inactivation of GC. Inactivation of 11βHSD1 results in decreased intracellular GC concentrations. Morton et al. showed that tissue-specific knockout of 11βHSD1 in adipose tissue in mice results in increased adiponectin gene expression (Morton et al., 2004). Similarly, Kershaw et al. showed that transgenic tissue-specific overexpression of 11βHSD2 in adipose tissue in high-fat fed mice results in a significant increase in adiponectin gene expression, although no significant changes in expression were observed in normal chow fed animals (Kershaw et al., 2005). In addition, a study by Shi et al. showed that chronic GC treatment results in decreased adiponectin mRNA expression and serum adiponectin concentrations (Shi et al., 2010). In this study, 7-week-old Sprague–Dawley rats were given either 5 or 15 mg/kg/day of hydrocortisone via intraperitoneal injection for 20 days with normal chow diet or high-fat diet. Low-dose treatment in normal chow diet fed animals (5 mg/kg) showed only modest decreases in mRNA expression and serum concentrations which were not statistically significant, but high-dose treatment in normal chow diet fed and both dosing regimens in high-fat fed groups showed significant decreases in both mRNA expression and serum concentrations of adiponectin.

4.1. Effects of adrenalectomy on adiponectin expression in murine models

Adrenalectomy is the surgical removal of adrenal gland which produces GC. In addition to GC, the adrenal gland also produces mineralocorticoids, epinephrine, and norepinephrine (Randall, 2004). Makimura et al. first explored the effects of adrenalectomy on adiponectin expression in wild

Table 7.1 Summary of *in vitro* studies on the effects of GC on adiponectin expression

References	Type of GC	Concentration and duration of exposure	Culture system	Effects on adiponectin mRNA /protein expression
Downregulation				
Fasshauer et al. (2003)	Dexamethasone	100 nM for 16 h	3T3-L1 preadipocytes	Decrease in protein expression/secretion
Fasshauer et al. (2002)	Dexamethasone	100 nM for 16 h	3T3-L1 preadipocytes	Decrease in mRNA expression
Halleux et al. (2001)	Dexamethasone	50 nM for 20 h	Isolated adipocytes from human visceral adipose tissue	Decrease in mRNA expression
Degawa-Yamauchi et al. (2005)	Dexamethasone	100 nM for 48 h	Human subcutaneous adipocytes	Decrease in protein secretion
No regulation				
Hozumi et al. (2010)	Dexamethasone	1 μM for 24 h	Human bone marrow adipocytes	No change in protein secretion
Sakamoto et al. (2011)	Dexamethasone	1 μM for 24 h	Human bone marrow adipocytes	No change in mRNA expression
Ryan et al. (2010)	Dexamethasone	2 and 20 nM for 2 and 24 h	Differentiated canine adipocytes	No change in mRNA expression
de Oliveira, Iwanaga-Carvalho, et al. (2011b)	Dexamethasone	1 μM for 24 h	3T3-L1 adipocytes	No change in mRNA expression

Table 7.2 Summary of animal studies on the effects of GC on adiponectin expression

References	Type of GC	Dosing regimen	Model system	Effects on adiponectin mRNA/ protein expression
Negative regulation				
Shi et al. (2010)	Hydrocortisone	Peritoneal injection of 5 or 15 mg/kg/day for 20 days	SPF Sprague–Dawley rats	Decrease in mRNA expression and plasma protein concentrations
Morton et al. (2004)	Predicted decrease in intracellular GC in adipose tissue	NA	11βHSD1 knockout in adipose tissue and matched wild type controls	11βHSD1 KO increases adiponectin mRNA expression
Kershaw et al. (2005)	Predicted decrease in intracellular GC in adipose tissue	NA	11βHSD2 overexpression in adipose tissue with high-fat feeding	11βHSD2 overexpression in adipose tissue of high-fat fed animals results in increase in adiponectin mRNA expression
Positive regulation				
Combs et al. (2003)	Prednisolone	2.1 mg/kg/day for 28 days	Swiss Webster mice	Increase in plasma protein concentrations
Raff and Bruder (2006)	Dexamethasone	Subcutaneous tapering dosing regimen (from 0.5 mg/kg to 0.05 mg/kg)	Neonatal Sprague–Dawley rats	Increase in plasma protein concentrations

Continued

Table 7.2 Summary of animal studies on the effects of GC on adiponectin expression—cont'd

References	Type of GC	Dosing regimen	Model system	Effects on adiponectin mRNA/ protein expression
Effects of adrenalectomy				
Makimura, Mizuno, Bergen, and Mobbs (2002)	Removal of endogenous GC	NA	ob/ob and matched wild type mice	Increase in serum protein and mRNA expression in ob/ob mice but decrease in serum protein and no change in mRNA expression in wild type
de Oliveira et al. (2011a)	Removal of endogenous GC and supplementation with dexamethasone	Subcutaneous dosing of 2 mg/kg, twice daily for 3 days	Male Wistar rats	Decrease in serum adiponectin and mRNA expression in epididymal fat with dex dosing
de Oliveira, Iwanaga-Carvalho, et al. (2011b)	Removal of endogenous GC and supplementation with dexamethasone	Subcutaneous dosing of 2 mg/kg, twice daily for 3 days	Male Wistar rats	Decrease in serum adiponectin with dex dosing and decrease in mRNA expression in epididymal fat with adrenalectomy

type and obese ob/ob mice (Makimura et al., 2002). Adrenalectomy did not have any significant effect on adiponectin mRNA expression in epididymal fat pads from wild type animals but causes a significant increase in the mRNA expression in obese ob/ob rats. However, serum adiponectin concentrations were decreased in wild-type animals after adrenalectomy, while the concentrations are increased in ob/ob mice after adrenalectomy in accordance with its mRNA expression. Recent studies by Oliveira et al. in male Wistar rats show some interesting results on the effect of adrenalectomy and exogenous GC supplementation on the regulation of adiponectin expression (de Oliveira, de Mattos, et al., 2011a,b, de Oliveira, Iwanaga-Carvalho, et al., 2011). Adrenalectomy did not have any significant effect on the serum adiponectin concentrations, but subcutaneous injection of 2 mg/kg of dex, twice a day for 3 days to these adrenalectomized rats, caused a significant decrease in adiponectin. However, the effects of adrenalectomy and dex on adiponectin mRNA expression in adipose tissue are highly dependent on the type of fat depot studied. For example, adrenalectomy has no significant effect on adiponectin mRNA expression in retroperitoneal and epididymal adipose tissue but significantly upregulates the expression in subcutaneous adipose tissue. In addition, dex supplementation does not affect adiponectin mRNA expression in retroperitoneal and subcutaneous adipose tissue but significantly downregulates the expression in epididymal fat. These results suggest that regulation of adiponectin expression is not only tissue specific but also depot specific in adipose tissue.

5. CLINICAL STUDIES ON THE EFFECTS OF GC ON ADIPONECTIN EXPRESSION

Because of the clinical importance of adiponectin and GC in conditions like insulin resistance, diabetes, and metabolic syndrome, many studies on the effects of GC on adiponectin were done in healthy human volunteers and patients (Table 7.3). Similar to *in vitro* and animal studies, results from clinical studies are also somewhat contradictory, although many of the studies actually suggest that GCs increase adiponectin expression. Jang et al. studied the effects of type 2 diabetes and GC treatment on plasma adiponectin concentrations (Jang et al., 2008). Plasma adiponectin was lower in diabetes patients compared to healthy volunteers, and the concentrations were significantly increased after 4 mg daily oral dosing of dexamethasone for 4 days in both healthy volunteers and diabetics. Similarly, in a double-blinded, randomized crossover study by Rieth et al., prednisolone dosing caused a

Table 7.3 Summary of clinical studies on the effects of GC on adiponectin expression

References	Type of GC	Dosing regimen	Model system	Effects on adiponectin mRNA/protein expression
Negative regulation				
Fallo et al. (2004)	Hydrocortisone for normal volunteers and increased cortisol in Cushing's syndrome patients	25 mg I.V bolus and blood taken from 0–180 min	Healthy volunteers and Cushing's syndrome patients	Decrease in plasma adiponectin in the time series study with concentrations significantly low at 30 and 60 min after drug dosing. Plasma concentrations low in Cushing's syndrome patients
Krsek et al. (2004)	Increased cortisol in Cushing's syndrome patients	NA	Female patients with Cushing's syndrome	Lower adiponectin concentrations in patients but difference is not significant
Positive regulation				
Uchida et al. (2006)	Methylprednisolone	500 mg/day dosing for 3 days	Patients with IgA nephropathy	Increase in serum adiponectin concentrations
Hjelmesaeth et al. (2006)	Prednisolone	15 ± 7 mg	Patients with renal transplantation	Dosing associated with increased serum adiponectin concentrations
Jang, Inder, Obeyesekere, and Alford (2008)	Dexamethasone	4 mg daily oral dosing for 4 days	Healthy volunteers and diabetes patients	Increase in plasma adiponectin concentrations
Rieth et al. (2009)	Prednisolone	60 mg/day for 1 week	Healthy volunteers with regular sport practice	Increase in blood adiponectin concentrations

Cimmino et al. (2010)	Prednisone	25 mg/day for 1 month	Polymyalgia rheumatic patients	Increase in plasma adiponectin concentrations
Kreiner and Galbo (2010)	Prednisolone	20 mg/day for 14 days	Normal and polymyalgia rheumatic patients	Increase in plasma adiponectin concentrations
No regulation				
Libe et al. (2005)	Increased cortisol in Cushing's syndrome patients	NA	Female patients with Cushing's syndrome and normal volunteers	No difference in plasma adiponectin concentrations
Lewandowski, Szosland, and Lewinski (2006)	Dexamethasone	Oral dosing of 0.5 mg every 6 h for 1 or 2 days	Normal female volunteers	No change in serum adiponectin concentrations
Darmon et al. (2006)	Hydrocortisone	3 h infusion of 1.5 $\mu g \cdot kg^{-1}$ min^{-1}	Normal and obese subjects	No change in plasma adiponectin concentrations

significant increase in the blood adiponectin concentrations compared to placebo (Rieth et al., 2009). In another study by Uchida et al., 500 mg/day MP dosing for 3 days resulted in a significant increase in serum adiponectin concentrations in patients with IgA nephropathy (Uchida et al., 2006). Correlation studies by Hjelmesaeth et al. in patients with renal transplantation suggest that GC treatment is associated with increased serum adiponectin concentrations (Hjelmesaeth et al., 2006). Kreiner et al. showed that plasma adiponectin concentrations were lower in polymyalgia rheumatic patients, and 20 mg/day of prednisolone dosing for 14 days significantly increases plasma concentrations of adiponectin in both polymyalgia rheumatic and control patients (Kreiner & Galbo, 2010). Similar observations were reported in the study done by Cimmino et al. where volunteers were dosed with 25 mg/day prednisone for a period of 1 month (Cimmino et al., 2010).

In contrast to the above studies, Fallo et al. showed that GC dosing decreased plasma adiponectin concentrations in healthy volunteers (Fallo et al., 2004). This was a time series study where a single dose of 25 mg hydrocortisone was given intravenously with blood samples taken from −15 to 180 min after dosing. Hydrocortisone caused a slight but significant decrease in plasma adiponectin at 30 and 60 min after dosing, and the levels eventually returned to baseline control values. This study also showed that plasma adiponectin concentrations were lower in patients with Cushing's syndrome who have elevated endogenous GC concentrations. Another study looking at adipokine concentrations in Cushing's syndrome patients shows that adiponectin concentrations are lower in these patients compared to normal subjects although the differences were not statistically significant (Krsek et al., 2004). A study by Libe et al. reported that the plasma adiponectin concentrations were not different between Cushing's syndrome and healthy volunteers (Libe et al., 2005). In another study by Darmon et al., 3-h infusion of 1.5 $\mu g \cdot kg^{-1}$ min^{-1} hydrocortisone in normal and obese subjects did not cause any significant change in plasma adiponectin concentrations (Darmon et al., 2006). Similarly, a dose of 0.5 mg of dexamethasone every 6 h for 48 h in human subjects did not cause any significant change in serum adiponectin concentrations (Lewandowski et al., 2006).

6. OTHER HORMONES, FACTORS, AND CONDITIONS THAT REGULATE ADIPONECTIN EXPRESSION AND THEIR RELATIONSHIP TO GC

One of the main reasons for the contradictory results for the effects of GC on adiponectin expression is due to the fact that adiponectin is regulated by many different hormones, transcription factors, and disease conditions

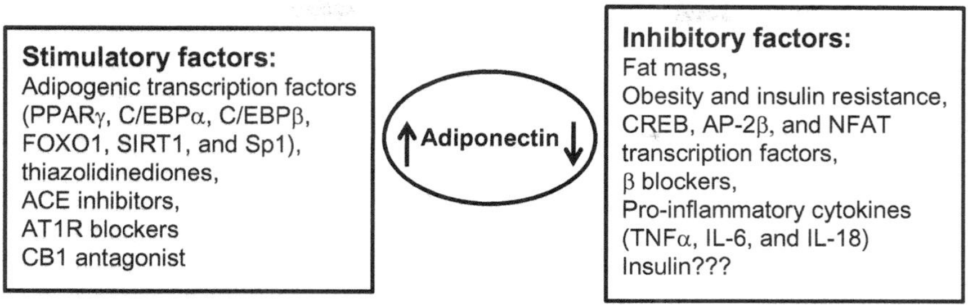

Figure 7.3 Factors regulating adiponectin expression.

(Fig. 7.3). GC can regulate some of these factors, and hence based on the specific condition, the effects of GC on adiponectin expression can differ. One of the main factors that regulate adiponectin expression is the amount of adipose tissue present (Swarbrick & Havel, 2008). Adiponectin expression is negatively correlated to the fat mass, and increases in fat mass result in decreased expression of adiponectin. This is one of the primary reasons for reduced concentrations of plasma adiponectin in obese patients. Similarly, adiponectin concentrations are found to be decreased in insulin-resistant conditions and diabetes (Spranger et al., 2003; Ziemke & Mantzoros, 2010). Many of the adipogenic transcription factors including PPARγ, C/EBPα, C/EBPβ, FOXO1, SIRT1, and Sp1 are found to enhance adiponectin gene expression (Liu & Liu, 2010). In contrast, CREB, AP-2β, and NFAT are found to negatively regulate its expression (Liu & Liu, 2010). PPARγ agonists like thiazolidinediones which are extensively used as antidiabetic drugs are found to stimulate adiponectin gene expression in both *in vitro* and *in vivo* studies (Iwaki et al., 2003; Yu et al., 2002). Similarly, ACE inhibitors, AT1R blockers, and CB1 antagonist are also found to stimulate adiponectin expression, while β blockers are found to have inhibitory effects (Swarbrick & Havel, 2008). Plasma adiponectin concentrations are also sexually dimorphic with higher concentrations observed in females compared to males (Swarbrick & Havel, 2008).

Inflammatory states are another important factor found to regulate adiponectin expression (Swarbrick & Havel, 2008). Increased macrophage infiltration in adipose tissue because of increased fat mass and obesity is implicated with many clinical conditions including diabetes, insulin resistance, and metabolic syndrome (Wang, Mariman, Renes, & Keijer, 2008). Various proinflammatory cytokines including TNFα, IL-6, and IL-18 produced from these infiltrated macrophages are found to adversely affect the expression of adiponectin from adipose tissue (Fasshauer & Paschke, 2003; Liu & Liu, 2010). TNFα can inhibit adiponectin expression by suppressing the expression of adipogenic transcription factors including PPARγ, C/EBP, and SREBP, while IL-6 and IL-18 effects are mediated through MAPK signaling cascades (Liu & Liu, 2010). In spite of the fact that inflammatory factors reduce adiponectin expression, plasma adiponectin concentrations are found to be higher in inflammatory and autoimmune diseases including osteoarthritis, rheumatoid arthritis, systemic lupus erythematosus, type 1 diabetes, and inflammatory bowel disease, except for polymyalgia rheumatic where the concentrations are found to be lower in patients compared to normal subjects (Kreiner & Galbo, 2010; Otero et al., 2006; Stofkova, 2009).

Several studies have been performed on the effect of insulin on adiponectin expression, but like GC, the role of insulin on adiponectin remains controversial and different studies suggests different results ranging from positive regulation to negative regulation. Studies by Fasshauer et al. and Xu et al. in 3T3-L1 adipocytes show that insulin treatment decreases adiponectin gene expression (Fasshauer et al., 2002; Xu et al., 2004). However, the study by Halleux et al. showed that insulin treatment increased adiponectin gene expression in a concentration-dependent manner in human visceral adipose tissue (Halleux et al., 2001). Hyperinsulinemic-euglycemic clamp studies in humans suggest that the increase in insulin concentrations decreases circulating adiponectin concentrations (Brame, Considine, Yamauchi, Baron, & Mather, 2005; Mohlig et al., 2002). In addition, tissue-specific knockout of insulin receptor in adipose tissue resulted in an increase in plasma adiponectin, suggesting that insulin has inhibitory effects on adiponectin expression (Bluher et al., 2002).

7. POTENTIAL REASONS FOR CONTRADICTORY RESULTS ON THE EFFECTS OF GC ON ADIPONECTIN EXPRESSION AND METHODS THAT CAN BE USEFUL IN GAINING BETTER INSIGHT ABOUT THIS REGULATION

There are multiple potential reasons why different studies on the effects of GC on adiponectin expression show contradictory results. Some of the reasons are as follows:

1. A plethora of mechanisms (different transcription factors, signaling cascades, and hormones) regulate adiponectin expression (Fasshauer & Paschke, 2003; Liu & Liu, 2010; Swarbrick & Havel, 2008). GCs are known to regulate, directly or indirectly, the expression of many different genes, signaling cascades, and transcription factors, some of

which are involved in regulating adiponectin expression. For example, GCs increase plasma glucose concentrations which subsequently increase plasma insulin which can regulate adiponectin expression (Jin & Jusko, 2009). Similarly, GCs influence adipogenesis and adipocyte differentiation and are also known to affect adipogenic transcription factors including C/EBP and PPARγ, all of which can influence adiponectin expression (MacDougald & Mandrup, 2002; Ringold, Chapman, Knight, & Torti, 1986). In addition, GCs are potent anti-inflammatory agents and cause reduction in proinflammatory cytokines, which again can indirectly regulate adiponectin expression (Barnes, 1998; Rhen & Cidlowski, 2005). Furthermore, the levels of these regulatory factors, their response to GC, and their effects on adiponectin can vary for different conditions, and hence the final outcome of the study of GC on adiponectin expression will be conditionally dependent. This is one of the main reasons why it is difficult to predict the outcome of the effects *in vivo* just based on *in vitro* data, to extrapolate data from animal studies to humans, and to predict the effects in disease conditions from data based on normal subjects and vice versa.

2. Only one-third of adipose tissue is made up of mature adipocytes, and the remaining tissue is made up of preadipocytes, mesenchymal stem cells, immune cells (predominantly macrophages), and vasculature (Wang et al., 2008). In addition, adipose tissue can be broadly classified into white and brown adipose tissue (Bjorndal, Burri, Staalesen, Skorve, & Berge, 2011; Wang et al., 2008). White adipose tissue can be divided into subcutaneous and intraabdominal adipose tissue, and intraabdominal fat is further divided into mesenteric, epididymal, and retroperitoneal fat based on their location (Bjorndal et al., 2011). All these variations can cause difference in response to external signals and hence affect the outcome of the regulation in expression. A good example was described in an earlier section, where the effects of adrenalectomy and dex treatment on adiponectin expression were highly dependent on the type of fat depot studied (de Oliveira, de Mattos, et al., 2011a).
3. One major problem in almost all of these reports is that these are single time point studies. Especially, with *in vivo* experiments, because of the complex matrix of factors involved in regulating adiponectin expression and since these factors have their own time course response profiles after GC dosing, adiponectin expression can show complex time course response patterns after drug dosing (with upregulation at some time points and down or no regulation other time points depending on the collective

effects of the factors affected by the drug dosing). Hence the time at which the measurement was made with respect to the dosing time could make a difference.

4. Other factors that could possibly cause inconsistency in the results between different studies could be the dose amount, differences between the synthetic GC used for the study, routes of administration, strain, and species differences, which again could produce a difference in response based on the changes in the factors that can regulate adiponectin expression.

To address some of these concerns and to gain better insights to understand the regulation of adiponectin expression by GC, a rich time course study in rats was performed in our laboratory (Sukumaran et al., 2011). In this study, 54 male Wistar rats were given 50 mg/kg MP as an intramuscular injection and the animals were sacrificed at 18 different time points (from 0.25 to 96 h). Six animals treated with saline were used as controls. After sacrifice, different organs and plasma were harvested and stored at −80 °C. A wide range of measurements including plasma MP, glucose, free fatty acids and insulin concentrations, and gene expression of GC receptor in white adipose tissue were made along with adiponectin gene expression and plasma protein concentrations. All measurements were made at 18 time points after MP administration (three animal replicates per time point) to understand the dynamic changes in their concentrations and to characterize the direct and indirect effects of GC dosing on adiponectin expression. Because of the complexities involved in the time-dependent regulation of all these factors, a mathematical model reflecting the mechanism of GC-mediated gene expression was developed to describe these dynamics in order to better characterize the effects of GC on adiponectin expression. After intramuscular dosing, plasma concentrations of MP decline biexponentially and are not detected in plasma 8 h after dosing. Plasma glucose and free fatty acid concentrations were increased after drug dosing, which resulted in an increase in plasma insulin concentrations. The model showed that the plasma insulin concentrations peak at 16 h after MP dosing, which eventually went back to its baseline. The GCs are known to cause downregulation of their own receptor expression, which was incorporated in the model to obtain a more quantitative understanding of the drug effects. MP dosing shows an interesting effect on adiponectin gene expression and plasma adiponectin concentrations as shown in Fig. 7.4. Adiponectin gene expression showed a very quick increase with the peak occurring at around 2 h after drug dosing followed by a delayed downregulation with the concentrations falling below

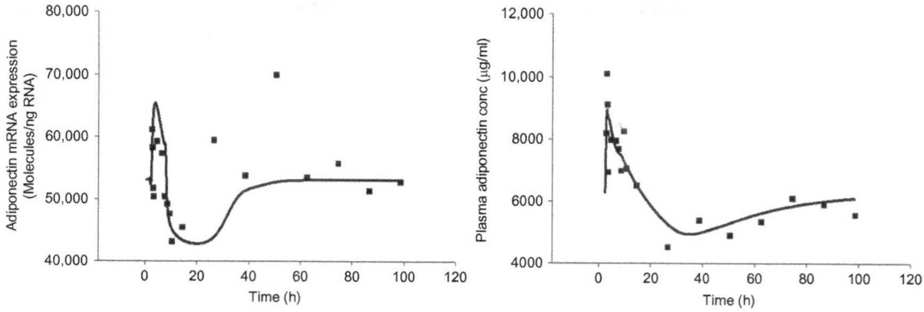

Figure 7.4 Effects of methylprednisolone dosing on adiponectin mRNA expression in white adipose tissue (left) and plasma adiponectin concentrations (right) in Wistar rats. Solid squares represent the measured data and solid curve represents the prediction from the mathematical model (Sukumaran et al., 2011).

the controls, reaching its nadir at 15 h. Subsequently, the expression goes back to the baseline at around 36 h and stays at the baseline values after that. The plasma adiponectin concentrations showed a similar profile after GC dosing, with peak occurring almost immediately after drug dosing, the nadir occurring at 24 h with reestablishment of the baseline only after 84 h. These results suggest that GC acting through its receptor in white adipose tissue directly upregulates adiponectin gene expression, while the increase in plasma insulin after GC dosing results in the downregulation of adiponectin expression, which is supported by the data and the mathematical model. This study shows the importance of doing time series experiments because if the measurements of adiponectin mRNA expression and plasma concentrations were made only at the early time points after drug dosing, then the conclusion will be that GC upregulates the expression, while if the measurement were made only after several hours, then the conclusion will be that GC downregulates the expression, and if the measurements were made only at the later time points, then the conclusion would have been no regulation. Similar approaches can be taken to incorporate other physiological factors to get a quantitative assessment of the regulation of adiponectin gene expression by GC, which will give a better insight about the mechanisms involved in these processes.

8. CONCLUSION

Both GC and adiponectin are important hormones involved in the regulation of energy metabolism and are implicated in many clinical conditions including metabolic syndrome, diabetes, and obesity. Although extensively studied, no consensus exists on the exact effects of GC on adiponectin expression. Complex regulation of adiponectin expression by many physiological factors, conditions, tissue type, and depot-specific differences can be the cause for the discrepancy in the observation between different studies performed using different model systems, and hence caution should be exerted in extrapolating data from one system to another. One way to achieve better insights concerning the regulation of adiponectin expression is to perform rich time series experiments after GC dosing and measure all relevant physiological factors to develop a quantitative model that will differentiate the effects of different factors affected by GC on the expression of adiponectin.

ACKNOWLEGMENT

This work was supported by grant GM24211 from the National Institutes of Health.

REFERENCES

Almon, R. R., DuBois, D. C., & Jusko, W. J. (2007). A microarray analysis of the temporal response of liver to methylprednisolone: A comparative analysis of two dosing regimens. *Endocrinology, 148*, 2209–2225.

Almon, R. R., DuBois, D. C., Pearson, K. E., Stephan, D. A., & Jusko, W. J. (2003). Gene arrays and temporal patterns of drug response: Corticosteroid effects on rat liver. *Functional & Integrative Genomics, 3*, 171–179.

Almon, R. R., DuBois, D. C., Piel, W. H., & Jusko, W. J. (2004). The genomic response of skeletal muscle to methylprednisolone using microarrays: Tailoring data mining to the structure of the pharmacogenomic time series. *Pharmacogenomics, 5*, 525–552.

Almon, R. R., DuBois, D. C., Yao, Z., Hoffman, E. P., Ghimbovschi, S., & Jusko, W. J. (2007). Microarray analysis of the temporal response of skeletal muscle to methylprednisolone: Comparative analysis of two dosing regimens. *Physiological Genomics, 30*, 282–299.

Almon, R. R., Lai, W., DuBois, D. C., & Jusko, W. J. (2005). Corticosteroid-regulated genes in rat kidney: Mining time series array data. *American Journal of Physiology. Endocrinology and Metabolism, 289*, E870–E882.

Andrews, R. C., & Walker, B. R. (1999). Glucocorticoids and insulin resistance: Old hormones, new targets. *Clinical Science (London, England), 96*, 513–523.

Barnes, P. J. (1998). Anti-inflammatory actions of glucocorticoids: Molecular mechanisms. *Clinical Science (London, England), 94*, 557–572.

Besser, G. M., & Jeffcoate, W. J. (1976). Endocrine and metabolic diseases. Adrenal diseases. *British Medical Journal, 1*, 448–451.

Bjorndal, B., Burri, L., Staalesen, V., Skorve, J., & Berge, R. K. (2011). Different adipose depots: Their role in the development of metabolic syndrome and mitochondrial response to hypolipidemic agents. *Journal of Obesity, 2011*, 490650.

Bluher, M., Michael, M. D., Peroni, O. D., Ueki, K., Carter, N., Kahn, B. B., et al. (2002). Adipose tissue selective insulin receptor knockout protects against obesity and obesity-related glucose intolerance. *Developmental Cell, 3*, 25–38.

Brame, L. A., Considine, R. V., Yamauchi, M., Baron, A. D., & Mather, K. J. (2005). Insulin and endothelin in the acute regulation of adiponectin in vivo in humans. *Obesity Research, 13*, 582–588.

Cimmino, M. A., Andraghetti, G., Briatore, L., Salani, B., Parodi, M., Cutolo, M., et al. (2010). Changes in adiponectin and leptin concentrations during glucocorticoid treatment: A pilot study in patients with polymyalgia rheumatica. *Annals of the New York Academy of Sciences, 1193*, 160–163.

Combs, T. P., Berg, A. H., Rajala, M. W., Klebanov, S., Iyengar, P., Jimenez-Chillaron, J. C., et al. (2003). Sexual differentiation, pregnancy, calorie restriction, and aging affect the adipocyte-specific secretory protein adiponectin. *Diabetes, 52*, 268–276.

Darmon, P., Dadoun, F., Boullu-Ciocca, S., Grino, M., Alessi, M. C., & Dutour, A. (2006). Insulin resistance induced by hydrocortisone is increased in patients with abdominal obesity. *American Journal of Physiology. Endocrinology and Metabolism, 291*, E995–E1002.

de Oliveira, C., de Mattos, A. B., Biz, C., Oyama, L. M., Ribeiro, E. B., & do Nascimento, C. M. (2011). High-fat diet and glucocorticoid treatment cause hyperglycemia associated with adiponectin receptor alterations. *Lipids in Health and Disease, 10*, 11.

de Oliveira, C., Iwanaga-Carvalho, C., Mota, J. F., Oyama, L. M., Ribeiro, E. B., & Oller do Nascimento, C. M. (2011). Effects of adrenal hormones on the expression of adiponectin and adiponectin receptors in adipose tissue, muscle and liver. *Steroids, 76*, 1260–1267.

Degawa-Yamauchi, M., Moss, K. A., Bovenkerk, J. E., Shankar, S. S., Morrison, C. L., Lelliott, C. J., et al. (2005). Regulation of adiponectin expression in human adipocytes: Effects of adiposity, glucocorticoids, and tumor necrosis factor alpha. *Obesity Research, 13*, 662–669.

Dridi, S., & Taouis, M. (2009). Adiponectin and energy homeostasis: Consensus and controversy. *The Journal of Nutritional Biochemistry, 20*, 831–839.

Eyster, K. M. (2011). Dysfunctional hormonal regulation of metabolism in obesity. *South Dakota Medicine*, Spec No: 18–21.

Fallo, F., Scarda, A., Sonino, N., Paoletta, A., Boscaro, M., Pagano, C., et al. (2004). Effect of glucocorticoids on adiponectin: A study in healthy subjects and in Cushing's syndrome. *European Journal of Endocrinology, 150*, 339–344.

Fasshauer, M., Klein, J., Neumann, S., Eszlinger, M., & Paschke, R. (2002). Hormonal regulation of adiponectin gene expression in 3T3-L1 adipocytes. *Biochemical and Biophysical Research Communications, 290*, 1084–1089.

Fasshauer, M., Kralisch, S., Klier, M., Lossner, U., Bluher, M., Klein, J., et al. (2003). Adiponectin gene expression and secretion is inhibited by interleukin-6 in 3T3-L1 adipocytes. *Biochemical and Biophysical Research Communications, 301*, 1045–1050.

Fasshauer, M., & Paschke, R. (2003). Regulation of adipocytokines and insulin resistance. *Diabetologia, 46*, 1594–1603.

Gunn, J. M., Hanson, R. W., Meyuhas, O., Reshef, L., & Ballard, F. J. (1975). Glucocorticoids and the regulation of phosphoenolpyruvate carboxykinase (guanosine triphosphate) in the rat. *The Biochemical Journal, 150*, 195–203.

Halleux, C. M., Takahashi, M., Delporte, M. L., Detry, R., Funahashi, T., Matsuzawa, Y., et al. (2001). Secretion of adiponectin and regulation of apM1 gene expression in human visceral adipose tissue. *Biochemical and Biophysical Research Communications, 288*, 1102–1107.

Hasselgren, P. O. (1999). Glucocorticoids and muscle catabolism. *Current Opinion in Clinical Nutrition and Metabolic Care, 2*, 201–205.

Heppner, K. M., Habegger, K. M., Day, J., Pfluger, P. T., Perez-Tilve, D., Ward, B., et al. (2010). Glucagon regulation of energy metabolism. *Physiology & Behavior, 100*, 545–548.

Hjelmesaeth, J., Flyvbjerg, A., Jenssen, T., Frystyk, J., Ueland, T., Hagen, M., et al. (2006). Hypoadiponectinemia is associated with insulin resistance and glucose intolerance after renal transplantation: Impact of immunosuppressive and antihypertensive drug therapy. *Clinical Journal of the American Society of Nephrology, 1*, 575–582.

Hozumi, A., Osaki, M., Sakamoto, K., Goto, H., Fukushima, T., Baba, H., et al. (2010). Dexamethasone-induced plasminogen activator inhibitor-1 expression in human primary bone marrow adipocytes. *Biomedical Research, 31*, 281–286.

Iwaki, M., Matsuda, M., Maeda, N., Funahashi, T., Matsuzawa, Y., Makishima, M., et al. (2003). Induction of adiponectin, a fat-derived antidiabetic and antiatherogenic factor, by nuclear receptors. *Diabetes, 52*, 1655–1663.

Iynedjian, P. B., Ballard, F. J., & Hanson, R. W. (1975). The regulation of phosphoenolpyruvate carboxykinase (GTP) synthesis in rat kidney cortex. The role of acid-base balance and glucocorticoids. *The Journal of Biological Chemistry, 250*, 5596–5603.

Jang, C., Inder, W. J., Obeyesekere, V. R., & Alford, F. P. (2008). Adiponectin, skeletal muscle adiponectin receptor expression and insulin resistance following dexamethasone. *Clinical Endocrinology, 69*, 745–750.

Jin, J. Y., & Jusko, W. J. (2009). Pharmacodynamics of glucose regulation by methylprednisolone. II. Normal rats. *Biopharmaceutics & Drug Disposition, 30*, 35–48.

Kadowaki, T., & Yamauchi, T. (2005). Adiponectin and adiponectin receptors. *Endocrine Reviews, 26*, 439–451.

Kershaw, E. E., & Flier, J. S. (2004). Adipose tissue as an endocrine organ. *The Journal of Clinical Endocrinology and Metabolism, 89*, 2548–2556.

Kershaw, E. E., Morton, N. M., Dhillon, H., Ramage, L., Seckl, J. R., & Flier, J. S. (2005). Adipocyte-specific glucocorticoid inactivation protects against diet-induced obesity. *Diabetes, 54*, 1023–1031.

Kreiner, F., & Galbo, H. (2010). Insulin sensitivity and related cytokines, chemokines, and adipokines in polymyalgia rheumatica. *Scandinavian Journal of Rheumatology, 39*, 402–408.

Krsek, M., Silha, J. V., Jezkova, J., Hana, V., Marek, J., Weiss, V., et al. (2004). Adipokine levels in Cushing's syndrome; elevated resistin levels in female patients with Cushing's syndrome. *Clinical Endocrinology, 60*, 350–357.

Lewandowski, K. C., Szosland, K., & Lewinski, A. (2006). Short-term dexamethasone administration does not alter serum adiponectin or resistin concentrations in overweight and obese subjects despite an increase in insulin resistance. *Clinical Endocrinology, 65*, 551–552.

Libe, R., Morpurgo, P. S., Cappiello, V., Maffini, A., Bondioni, S., Locatelli, M., et al. (2005). Ghrelin and adiponectin in patients with Cushing's disease before and after successful transsphenoidal surgery. *Clinical Endocrinology, 62*, 30–36.

Liu, M., & Liu, F. (2010). Transcriptional and post-translational regulation of adiponectin. *The Biochemical Journal, 425*, 41–52.

Lukens, F. D. (1959). The pancreas: Insulin and glucagon. *Annual Review of Physiology, 21*, 445–474.

MacDougald, O. A., & Mandrup, S. (2002). Adipogenesis: Forces that tip the scales. *Trends in Endocrinology and Metabolism, 13*, 5–11.

Makimura, H., Mizuno, T. M., Bergen, H., & Mobbs, C. V. (2002). Adiponectin is stimulated by adrenalectomy in ob/ob mice and is highly correlated with resistin mRNA. *American Journal of Physiology. Endocrinology and Metabolism, 283*, E1266–E1271.

Mohlig, M., Wegewitz, U., Osterhoff, M., Isken, F., Ristow, M., Pfeiffer, A. F., et al. (2002). Insulin decreases human adiponectin plasma levels. *Hormone and Metabolic Research, 34*, 655–658.

Morton, N. M., Paterson, J. M., Masuzaki, H., Holmes, M. C., Staels, B., Fievet, C., et al. (2004). Novel adipose tissue-mediated resistance to diet-induced visceral obesity in 11 beta-hydroxysteroid dehydrogenase type 1-deficient mice. *Diabetes, 53*, 931–938.

Oakley, R. H., & Cidlowski, J. A. (2011). Cellular processing of the glucocorticoid receptor gene and protein: New mechanisms for generating tissue-specific actions of glucocorticoids. *The Journal of Biological Chemistry, 286*, 3177–3184.

Otero, M., Lago, R., Gomez, R., Lago, F., Dieguez, C., Gomez-Reino, J. J., et al. (2006). Changes in plasma levels of fat-derived hormones adiponectin, leptin, resistin and visfatin in patients with rheumatoid arthritis. *Annals of the Rheumatic Diseases, 65*, 1198–1201.

Ouchi, N., & Walsh, K. (2007). Adiponectin as an anti-inflammatory factor. *Clinica Chimica Acta, 380*, 24–30.

Raff, H., & Bruder, E. D. (2006). Adiponectin and resistin in the neonatal rat: Effects of dexamethasone and hypoxia. *Endocrine, 29*, 341–344.

Randall, D. C. (2004). Discovering the role of the adrenal gland in the control of body function. *American Journal of Physiology. Regulatory, Integrative and Comparative Physiology, 287*, R1007–R1008.

Rhen, T., & Cidlowski, J. A. (2005). Antiinflammatory action of glucocorticoids—New mechanisms for old drugs. *The New England Journal of Medicine, 353*, 1711–1723.

Rieth, N., Jollin, L., Le Panse, B., Lecoq, A. M., Arlettaz, A., De Ceaurriz, J., et al. (2009). Effects of short-term corticoid ingestion on food intake and adipokines in healthy recreationally trained men. *European Journal of Applied Physiology, 105*, 309–313.

Ringold, G. M., Chapman, A. B., Knight, D. M., & Torti, F. M. (1986). Hormonal control of adipogenesis. *Annals of the New York Academy of Sciences, 478*, 109–119.

Rose, A. J., Vegiopoulos, A., & Herzig, S. (2010). Role of glucocorticoids and the glucocorticoid receptor in metabolism: Insights from genetic manipulations. *The Journal of Steroid Biochemistry and Molecular Biology, 122*, 10–20.

Ryan, V. H., German, A. J., Wood, I. S., Hunter, L., Morris, P., & Trayhurn, P. (2010). Adipokine expression and secretion by canine adipocytes: Stimulation of inflammatory adipokine production by LPS and TNFalpha. *Pflügers Archiv, 460*, 603–616.

Sakamoto, K., Osaki, M., Hozumi, A., Goto, H., Fukushima, T., Baba, H., et al. (2011). Simvastatin suppresses dexamethasone-induced secretion of plasminogen activator inhibitor-1 in human bone marrow adipocytes. *BMC Musculoskeletal Disorders, 12*, 82.

Schaaf, M. J., & Cidlowski, J. A. (2002). Molecular mechanisms of glucocorticoid action and resistance. *The Journal of Steroid Biochemistry and Molecular Biology, 83*, 37–48.

Shi, J. H., Du, W. H., Liu, X. Y., Fan, Y. P., Hu, X. L., Zhou, H. Y., et al. (2010). Glucocorticoids decrease serum adiponectin level and WAT adiponectin mRNA expression in rats. *Steroids, 75*, 853–858.

Spranger, J., Kroke, A., Mohlig, M., Bergmann, M. M., Ristow, M., Boeing, H., et al. (2003). Adiponectin and protection against type 2 diabetes mellitus. *Lancet, 361*, 226–228.

Stofkova, A. (2009). Leptin and adiponectin: From energy and metabolic dysbalance to inflammation and autoimmunity. *Endocrine Regulations, 43*, 157–168.

Sukumaran, S., Almon, R. R., DuBois, D. C., & Jusko, W. J. (2010). Circadian rhythms in gene expression: Relationship to physiology, disease, drug disposition and drug action. *Advanced Drug Delivery Reviews, 62*, 904–917.

Sukumaran, S., Jusko, W. J., DuBois, D. C., & Almon, R. R. (2011). Mechanistic modeling of the effects of glucocorticoids and circadian rhythms on adipokine expression. *The Journal of Pharmacology and Experimental Therapeutics, 337*, 734–746.

Swarbrick, M. M., & Havel, P. J. (2008). Physiological, pharmacological, and nutritional regulation of circulating adiponectin concentrations in humans. *Metabolic Syndrome and Related Disorders, 6*, 87–102.

Swartz, S. L., & Dluhy, R. G. (1978). Corticosteroids: Clinical pharmacology and therapeutic use. *Drugs, 16*, 238–255.

Tirone, T. A., & Brunicardi, F. C. (2001). Overview of glucose regulation. *World Journal of Surgery, 25*, 461–467.

Tomlinson, J. W., Walker, E. A., Bujalska, I. J., Draper, N., Lavery, G. G., Cooper, M. S., et al. (2004). 11beta-hydroxysteroid dehydrogenase type 1: A tissue-specific regulator of glucocorticoid response. *Endocrine Reviews, 25*, 831–866.

Tsigos, C., & Chrousos, G. P. (2002). Hypothalamic-pituitary-adrenal axis, neuroendocrine factors and stress. *Journal of Psychosomatic Research*, *53*, 865–871.
Uchida, H. A., Nakamura, Y., Kaihara, M., Norii, H., Hanayama, Y., Sugiyama, H., et al. (2006). Steroid pulse therapy impaired endothelial function while increasing plasma high molecule adiponectin concentration in patients with IgA nephropathy. *Nephrology, Dialysis, Transplantation*, *21*, 3475–3480.
Vegiopoulos, A., & Herzig, S. (2007). Glucocorticoids, metabolism and metabolic diseases. *Molecular and Cellular Endocrinology*, *275*, 43–61.
Wang, P., Mariman, E., Renes, J., & Keijer, J. (2008). The secretory function of adipocytes in the physiology of white adipose tissue. *Journal of Cellular Physiology*, *216*, 3–13.
Xu, A., Wong, L. C., Wang, Y., Xu, J. Y., Cooper, G. J., & Lam, K. S. (2004). Chronic treatment with growth hormone stimulates adiponectin gene expression in 3T3-L1 adipocytes. *FEBS Letters*, *572*, 129–134.
Yamada, T., Oka, Y., & Katagiri, H. (2008). Inter-organ metabolic communication involved in energy homeostasis: Potential therapeutic targets for obesity and metabolic syndrome. *Pharmacology & Therapeutics*, *117*, 188–198.
Yu, J. G., Javorschi, S., Hevener, A. L., Kruszynska, Y. T., Norman, R. A., Sinha, M., et al. (2002). The effect of thiazolidinediones on plasma adiponectin levels in normal, obese, and type 2 diabetic subjects. *Diabetes*, *51*, 2968–2974.
Ziemke, F., & Mantzoros, C. S. (2010). Adiponectin in insulin resistance: Lessons from translational research. *The American Journal of Clinical Nutrition*, *91*, 258S–261S.

CHAPTER EIGHT

Adiponectin and Reproduction

Esther Dos Santos[1], René Pecquery, Philippe de Mazancourt, Marie-Noëlle Dieudonné

Université de Versailles-St-Quentin, Service de Biochimie et Biologie Moléculaire, UPRES-EA 2493, UFR Paris-Ile de France Ouest, PRES Universud Paris, Centre Hospitalier de Poissy-Saint Germain, Poissy Cedex, France

[1]Corresponding author: e-mail address: esther.dos-santos@uvsq.fr

Contents

Abstract

Adiponectin is an adipocyte-derived cytokine that acts as a major regulator of insulin sensitivity. Adiponectin deficiency can result in severe diabetes and metabolic disorders in humans. Since its discovery, our understanding of adiponectin's biological functions has expanded from insulin sensitization properties to new effects on inflammation, immunology, and human reproduction. Indeed, both obesity and excessive leanness are associated with reproductive dysfunction. The objective of this chapter is to review such biological actions and the potential roles of adiponectin on human reproduction. There is accumulative evidence for direct effects of this adipokine on the late stages of folliculogenesis and on the development of a functional placenta. In addition, clinical and genomic studies associate hypoadiponectinemia with pregnancy-related disorders, including polycystic ovarian syndrome.

ABBREVIATIONS

BMI body mass index
FSH follicle-stimulating hormone
HCG human chorionic gonadotropin
IL interleukin
IUGR intrauterine growth restriction
LH luteinizing hormone
PCOS polycystic ovarian syndrome
TNFα tumor necrosis factor α

Vitamins and Hormones, Volume 90
ISSN 0083-6729
http://dx.doi.org/10.1016/B978-0-12-398313-8.00008-7

1. INTRODUCTION

Adipose tissue has long been considered a depot for triglycerides during times of metabolic affluence and/or for mobilization during periods of nutrient deprivation. Over the past two decades, adipose tissue has emerged as an endocrine gland that modulates events, including appetite, lipid uptake, and metabolism, and even the growth of adipose precursor cells. Both the excess of fat mass in obesity and its paucity in anorexia are known to result in dysfunctions of reproduction. In women, obesity correlates with menstrual disorders and infertility, gestational failure, and obstetric complications. In contrast, excessive leanness is associated with amenorrhea, implantation miscarriage, and poor fetal growth. Adipose tissue secretes hormones known as adipokines, reflecting their origin and their effects as cytokines on target tissues. Among the best examples of an adipokine that plays a pivotal role in reproductive process in mammals is leptin. Indeed, leptin-null mice are obese and infertile (Zhang et al., 1994), and administration of recombinant leptin to these animals corrects obesity and restores full fertility (Chehab, Lim, & Lu, 1996). Furthermore, leptin stimulates gonadotropin secretion at the hypothalamic and pituitary levels to directly modulate ovarian steroidogenesis, the embryonic implantation process, and placental functions (Cervero et al., 2006; Guerre-Millo, 2008; Henson & Castracane, 2006). Adiponectin is another adipokine that is known for its important role in the regulation of energy homeostasis. More precisely, adiponectin is described as a molecule that has antidiabetic, antiatherogenic, anti-inflammatory, and angiogenic properties. Recently, it has been proposed to be involved in the control of reproductive functions. Adiponectin may be an important signal indicating the adequacy of nutritional status for reproductive function (Campos, Palin, Bordignon, & Murphy, 2008; Mitchell, Armstrong, Robker, & Norman, 2005). The present review summarizes the structure, expression regulations, transduction signaling, and the implication of adiponectin in the reproductive processes.

2. ADIPONECTIN AND ADIPONECTIN SIGNALING

Adiponectin was independently identified by four research groups using different methods two decades ago and was thus assigned a series of names, including apM1 (adipose most abundant gene transcript 1), Acrp30 (adipocyte complement-related protein of 30 kDa), AdipoQ, and GBP28 (gelatin-binding protein of 28 kDa; Maeda et al., 1996; Nakano, Tobe, Choi-Miura, Mazda, & Tomita, 1996; Scherer, Williams, Fogliano, Baldini, & Lodish, 1995). The *adiponectin* gene consists of three exons and two introns, spanning a 17-kb region, and is located on chromosome 3q27 in a region recently mapped as a susceptibility locus for type II diabetes and adiposity (Hu, Liang, & Spiegelman, 1996; Scherer et al., 1995). Human adiponectin contains 244 amino acid residues and consists of a 20-residue signal sequence, an N-terminal region, a collagen-like region, and a C-terminal globular domain. This polypeptide of 30 kDa is assembled into an array of complexes composed of multimers of 30 kDa polypeptide. Adiponectin subunits assemble into trimers, called low molecular weight (LMW) complexes, hexamers or middle molecular weight (MMW), or a more elaborate high molecular weight (HMW) complex composed of nine hexamers (>300 kDa; Liu & Liu, 2010). Full-length adiponectin (fAd) can be cleaved by leucocyte elastase and can exist in serum as a globular form, thus creating trimers (gAd; Hada et al., 2007). The formation of HMW adiponectin in cells is regulated by posttranslational modifications such as hydroxylation, glycosylation, and disulfide bond formation. Impairments in such modifications lead to a marked reduction in the intracellular levels of adiponectin and its subsequent secretion (Wang, Qiang, & Farmer, 2008). Even if adiponectin protein contains a signal peptide, the molecular mechanisms of adiponectin secretion are not, as of yet, completely understood. The NAD-dependent deacetylase SIRT 1 activity seems to be implicated (Qiang, Wang, & Farmer, 2007). Once secreted, adiponectin circulates at relatively high concentrations in human circulation (5–15 μg/ml), which accounts for 0.01% of total plasma protein. The HMW form is predominant in human circulation (Hada et al., 2007) and corresponds to the active form of adiponectin (Kobayashi et al., 2004). In contrast to leptin, adiponectin levels are inversely correlated with visceral fat accumulation (Arita et al., 1999; Weyer et al., 2001). Additionally, weight reduction in obese individuals is accompanied by an increase in plasma adiponectin concentration (Yang et al., 2001), suggesting that adipose tissue can exert negative feedback on adiponectin production and/or secretion. In obese subjects, adipose tissue is characterized by heightened oxidative stress, chronic inflammation, and macrophage infiltration (Guerre-Millo, 2008). Reactive oxygen species (ROS) and proinflammatory cytokines such as tumor necrosis factor α (TNFα) and interleukin-6 (IL-6) are potent inhibitors of *adiponectin* gene expression in cultured adipocytes (Fasshauer, Klein, Neumann, Eszlinger, & Paschke, 2002; Fasshauer, Paschke, & Stumvoll, 2004). These negative regulations of adiponectin mRNA expression could therefore contribute to lowering adiponectin release observed in adipose tissue of obese patients. Inversely, the expression of adiponectin is stimulated by the thiazolidinediones class of insulin-sensitizing drugs used to treat type II diabetes, suggesting that the *adiponectin* gene is a target of peroxisome proliferator-activated receptor γ (Combs et al., 2002). Furthermore, plasma concentrations of adiponectin show clear gender differences, with females having higher levels (especially of HMW isoforms) than males (Combs et al., 2003; Nishizawa et al., 2002). Another *in vitro* study revealed that testosterone significantly reduced the expression and the secretion of adiponectin in rat adipocytes (Combs et al., 2003; Nishizawa et al., 2002).

The biological actions of adiponectin are carried out through interactions with its specific membrane adiponectin receptor subtypes, AdipoR1 and AdipoR2. Both receptors contain seven transmembrane domains but are structurally and functionally distinct from G-protein-coupled receptors because the intra/extracellular orientations of N-terminus and C-terminus are opposite (Kadowaki & Yamauchi, 2005). There are differences in the binding preference of the receptors to the adiponectin isoforms as well as differences in the tissue distribution of the two AdipoR. Indeed, AdipoR1 is ubiquitously expressed and displays a high affinity for gAd. AdipoR2 is predominantly expressed in liver and has a higher affinity for fAd (Yamauchi et al., 2003). Emerging evidence shows that downstream signaling of the AdipoR involves APPL1, an adaptor protein containing a pleckstrin homology domain, phosphotyrosine-binding domain, and a leucine zipper motif (Deepa & Dong, 2009). APPL1 also functions in the insulin-signaling pathway and is an important mediator of, for example, adiponectin-dependent insulin sensitization in skeletal muscle (Deepa & Dong, 2009). An additional receptor for adiponectin, T-cadherin, has been recently described. T-cadherin binds hexameric and HMW species of adiponectin but not the trimeric or globular forms (Hug et al., 2004). Furthermore, T-cadherin is a glycosylphosphatidylinositol-anchored extracellular protein that may act as a coreceptor as no signaling pathway has yet been identified (Wheelock & Johnson, 2003).

The most common transduction pathway by which adiponectin acts on target tissues is the AMP-activated protein kinase (AMPK), which acts as a key regulator of energy balance (Kahn, Alquier, Carling, & Hardie, 2005).

Notably, AMPK is activated once the cellular ATP/AMP ratio is reduced and then this kinase (i) activates catabolic pathways such as glycolysis, fatty-acid oxidation and (ii) inhibits anabolic pathways like lipogenesis, cholesterol synthesis (Foretz et al., 2006). The binding of adiponectin to its receptors, AdipoR, also activates downstream targets such as the protein kinase A, phosphoinositol 3 kinase, p38/p42/p44 mitogen-activated phospho-kinase, peroxisome proliferator-activated receptor-α, nuclear factor-κB, and JUN-kinase pathway (Kadowaki & Yamauchi, 2005).

3. ADIPONECTIN: MAIN FUNCTIONS IN HUMAN

In vivo and *in vitro* investigations have revealed that adiponectin exerts numerous effects on target tissues with the best studied effects being sensitization to insulin and antiatherosclerotic and anti-inflammatory properties (Tilg & Moschen, 2006; Yamauchi et al., 2001).

The mechanisms underlying the increase in insulin sensitivity in response to adiponectin remains to be fully established and may include a multitude of pathways. For example, in skeletal muscle, adiponectin stimulates glucose transport by increasing GLUT4 translocation (Ceddia et al., 2005); activates insulin signaling; and facilitates fatty-acid transport, fatty-acid oxidation, and energy dissipation, resulting in decreased triglyceride content (Yamauchi et al., 2001). In the liver, adiponectin reduces glucose production by enhancing insulin action (Berg, Combs, Du, Brownlee, & Scherer, 2001; Combs, Berg, Obici, Scherer, & Rossetti, 2001). Furthermore, adiponectin reduces the activity of gluconeogenic enzymes such as phosphoenolpyruvate carboxykinase and glucose-6-phosphatase and increases fatty-acid oxidation (Combs et al., 2001; Yamauchi et al., 2002). Consequently, adiponectin reduces fatty-acid influx and liver triglyceride content (Yamauchi et al., 2001). Finally, adiponectin also suppresses the secretion of TNFα by adipose tissue, which is a factor that is known to contribute to insulin resistance (Hotamisligil, 1999).

In addition to its role in the modulation of glucose and lipid metabolism, adiponectin has been reported to have potent anti-inflammatory properties. Adiponectin is able to (i) induce the production of anti-inflammatory cytokines such as IL-10 and (ii) inhibit the production of proinflammatory cytokines like IFNγ, IL-6, and TNFα via macrophages (Wolf, Wolf, Rumpold, Enrich, & Tilg, 2004; Wulster-Radcliffe, Ajuwon, Wang, Christian, & Spurlock, 2004). Moreover, adiponectin is able to stabilize atherosclerotic plaques and seems to prevent their rupture by blocking TNFα effects on immune cell recruitment, vascular smooth muscle cell proliferation, and lipid-engorged foam cell formation (Ouchi et al., 2001). Adiponectin is also described as a proangiogenic factor. Indeed, in mice, adiponectin deficiency impairs revascularization. Moreover, in humans, hypoadiponectinemia is associated with impaired endothelial dysfunction related to vasorelaxation (Ouchi et al., 2003). Notably, adiponectin's protective effects on endothelial function are mediated through an increase in the expression of endothelial nitric oxide synthase (Chen, Montagnani, Funahashi, Shimomura, & Quon, 2003). Finally, a study on mice bearing null mutations for the *adiponectin* gene clearly indicates that the ablation of this cytokine provokes metabolic syndrome, which is associated with increased cardiovascular risk, hyperinsulinemia, insulin resistance, atherogenic dislipidemia, and elevated blood pressure (Maeda et al., 2002). Reversal of metabolic syndrome symptoms in these mice can be achieved by ectopic expression of adiponectin (Kadowaki et al., 2006). Similarly, AdipoR1/R2 double-KO mice also develop insulin resistance (Yamauchi et al., 2007). In humans, genetic studies described an association between variants of the *adiponectin* gene (T45G and G276T) and the development of metabolic syndrome (Takahashi et al., 2000; Yang et al., 2003). These different observations led to the conclusion that adiponectin deficiency, due to either fat accretion or genetic variation, is a principal factor in the prevalence of metabolic syndrome

Adiponectin is also known to exert a protective action against cancer development. Circulating adiponectin levels are inversely correlated with an increased risk of gastric, prostate, and hormonal cancers like mammary, endometrial, and prostate carcinoma (Bub, Miyazaki, & Iwamoto, 2006; Cong et al., 2007; Ishikawa et al., 2007; Mantzoros, Petridou, Dessypris, et al., 2004). In breast cancer, an inverse correlation was found between plasma adiponectin levels and the histological grade of the tumor (Miyoshi et al., 2003). Different *in vitro* studies, including ours, also reported that adiponectin reduces cell proliferation and promotes apoptosis of human mammary cancer cells (Dieudonne et al., 2006; Dos Santos et al., 2008; Kelesidis, Kelesidis, & Mantzoros, 2006). These antiproliferative effects of adiponectin could be explained by (i) direct regulation of the cell cycle by the cytokine and/or (ii) a sequestration of growth factors like EGF as already observed in human aortic smooth muscle cells (Wang et al., 2005).

4. ADIPONECTIN AND STEROIDOGENESIS

Multiple studies in the literature revealed that adiponectin is able to affect the reproductive system through central effects on the hypothalamus and/or peripheral actions on the ovary (Campos et al., 2008; Mitchell et al.,

2005). The adiponectin system (adiponectin, AdipoR1, and AdipoR2) was recently described in rat and human hypothalami (Kos et al., 2007), in pituitary glands (Rodriguez-Pacheco et al., 2007), and in ovaries of various species including pigs (Lord, Ledoux, Murphy, Beaudry, & Palin, 2005), chicken (Chabrolle, Tosca, Crochet, Tesseraud, & Dupont, 2007), rats (Chabrolle, Tosca, & Dupont, 2007), and humans (Chabrolle et al., 2009). AdipoR were found to be abundantly expressed in the hypothalamus, and their expression levels were comparable to those in the liver (Yamauchi et al., 2003). Interestingly, unlike in the serum, only trimers and hexamers, and not HMW multimers, were found in the central nervous system (Kubota et al., 2007).

First, it was demonstrated in mouse gonadotropic cells and in rat pituitary cells that adiponectin inhibits basal and GnRH-stimulated luteinizing hormone (LH) secretion (Lu, Tang, Olefsky, Mellon, & Webster, 2008; Rodriguez-Pacheco et al., 2007). Second, in rat primary pituitary cells, recombinant adiponectin regulates GnRH receptor expression (Rodriguez-Pacheco et al., 2007). In mouse adrenal cells, adiponectin affects the adrenocortical function as the secretion of corticosterone and aldosterone were inhibited by adiponectin (Li et al., 2009). Additionally, *in vitro* studies have shown that physiologically relevant levels of recombinant adiponectin induce ovarian gene expression and steroidogenesis in mammalian ovaries. Indeed, the expression of a cluster of proteins associated with the process of ovulation including cyclooxygenase-2, prostaglandin E2, and vascular endothelial growth factor are induced in response to adiponectin (Ledoux et al., 2006). In rat ovaries, the synthesis of the steroid hormones, progesterone and estrogen, is regulated by adiponectin (Chabrolle, Tosca, & Dupont, 2007). In this study, responses to adiponectin were additive to gonadotropins, and Ledoux et al. (2006) further demonstrated an interaction with insulin. Similarly, in rats, adiponectin acts synergically with insulin-like growth factor-1 to induce steroidogenic gene expression and steroidogenesis (Chabrolle, Tosca, & Dupont, 2007). Physiological levels of recombinant adiponectin (5 or 10 μg/ml) are able to increase progesterone and/or estrogen secretions in response to insulin-like growth factor I in human-cultured granulosa cells (Chabrolle et al., 2009). Recent data clearly demonstrate that adiponectin alters bovine granulosa cell steroidogenesis, oocyte maturation, and embryo development (Maillard et al., 2010). Thus, while it is important to note that adiponectin can directly induce ovarian gene expression, the interactions between adiponectin and insulin or insulin-like growth factor-1 are consistent with the role of adiponectin

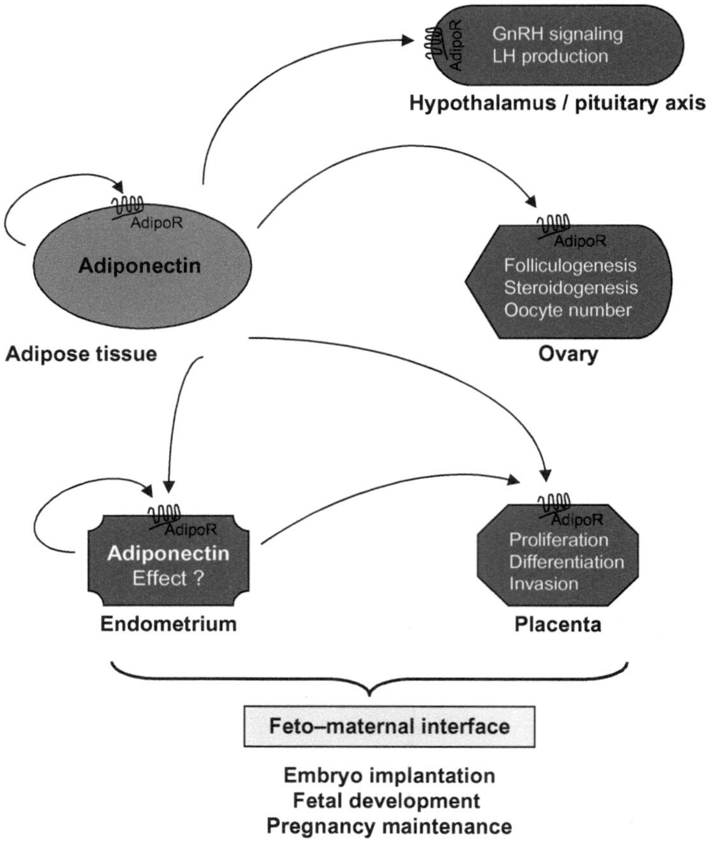

Figure 8.1 The potential interactions between adiponectin and female reproduction.

in other tissues, that is, the sensibilization to insulin. In support of a role for adiponectin in human folliculogenesis and ovulation, circulating adiponectin positively correlates with the number of oocytes retrieved in women treated with follicle-stimulating hormone (FSH) to induce superovulation for the purpose of *in vitro* fertilization (Liu et al., 2006). Finally, recombinant human adiponectin decreases LH, insulin-induced progesterone, and androstenedione secretions by bovine theca cells from large follicles (Lagaly, Aad, Grado-Ahuir, Hulsey, & Spicer, 2008; Fig. 8.1).

5. ADIPONECTIN AND PLACENTAL FUNCTIONS

Cumulative evidence suggests that adiponectin could regulate reproductive and placental processes (Campos et al., 2008; Mitchell et al., 2005). Adiponectin receptors (AdipoR1 and AdipoR2) are present in human oocytes, placenta, embryo, and endometrium (Benaitreau et al., 2009; Caminos et al., 2005; Liu et al., 2006; McDonald & Wolfe, 2009; Takemura et al., 2006). These results suggest that the different tissues and cell types that cooperatively participate in the establishment of a pregnancy are targets for adiponectin.

It was demonstrated that the overexpression of adiponectin impairs female fertility in mice (Combs et al., 2004), whereas the absence of adiponectin has no effect (Campos et al., 2008; Mitchell et al., 2005). Otherwise, studies of null mutation of AdipoR1 and R2 indicated that both offspring and their crosses (knockout of both receptors) are capable of successfully breeding, but extensive breeding trials have not been conducted (Kadowaki & Yamauchi, 2005). Given these contradictions, this *in vivo* approach was performed in rodents and completed using *ex vivo* and *in vitro* experiments.

Although earlier reports suggested that adiponectin was produced and secreted by the human placenta (Caminos et al., 2005; Chen et al., 2006), more recent studies, including our published data, showed that adiponectin mRNA is not expressed in the placenta (Benaitreau, Dos Santos, Leneveu, Alfaidy, et al., 2010; Corbetta et al., 2005; Pinar et al., 2008). In contrast, we and others observed that adiponectin is expressed (mRNA and protein levels) in human endometrium (Benaitreau, Dos Santos, Leneveu, De Mazancourt, et al., 2010; Takemura et al., 2006). Second, adiponectin at the feto–maternal interface originates from maternal blood. Adiponectin is thus found in this interface very early when remodeling of spiral arteries operates. Finally, adiponectin presence at the feto–maternal barrier is relayed by the production and secretion from fetal tissues (Corbetta et al., 2005). These different observations may explain the fact that, in cord blood, adiponectinemia increases with gestational age, reaching a more than 20-fold higher level at term compared with 24 weeks of gestation (Corbetta et al., 2005; Kajantie, Hytinantti, Hovi, & Andersson, 2004). A recent study revealed that maternal adiponectin plasma concentration may play a role in placental growth as placenta thickness was positively associated with maternal adiponectin levels (Bomba-Opon et al., 2010). Furthermore, some studies indicated that adiponectin levels are related to the occurrence of fetal macrosomia. The authors speculate that the positive association of birth weight and adiponectin may reflect the direct effect of adiponectin on the sensitivity of tissues to insulin and components of the IGF system (Kotani et al., 2004; Mantzoros, Petridou, Alexe, et al., 2004). Surprisingly, in primary human trophoblast cells from term placenta, fAd seems to attenuate insulin-stimulated amino acid transport, which may have important implications for fetal growth (Jones, Jansson, & Powell, 2010). These recent data disagree with the well-known insulin-sensitizing actions of adiponectin in liver and muscle reported in the literature. A possible explanation of such discrepancies is the fact that a novel isoform of the receptor adaptor protein, APPL2, was identified and shown to suppress adiponectin signaling by sequestering APPL1 (Wang et al., 2009). Thus, it can be speculated that differences between placenta and muscle in the expression and/or function of both APPL1 and APPL2 result in distinct responses to adiponectin.

Presently, only some reports demonstrated a direct role of adiponectin during the preimplantation period and during the first steps of placental development. Indeed, recent results provide evidence that adiponectin may play a regulatory role on oocyte proliferation, as recombinant adiponectin stimulates meiotic maturation and early *in vitro* embryo development of porcine oocytes (Chappaz et al., 2008). In addition, circulating adiponectin positively correlates with the number of oocytes retrieved in women treated with FSH for the purpose of *in vitro* fertilization (Liu et al., 2006). Furthermore, in human first-trimester placenta, our laboratory clearly demonstrated that adiponectin not only attenuates trophoblastic growth but also enhances invasion capacities of cytotrophoblasts by inducing the activity of metalloproteinases (Benaitreau et al., 2009; Benaitreau, Dos Santos, Leneveu, Alfaidy, et al., 2010; Benaitreau, Dos Santos, Leneveu, De Mazancourt, et al., 2010). Furthermore, it was demonstrated that adiponectin promoted the syncytialization capacities of human first-trimester cytotrophoblasts as reflected by increase in human chorionic gonadotropin (HCG) secretion and the upregulation of two trophoblast differentiation markers, leptin and syncytin 2 (Benaitreau, Dos Santos, Leneveu, De Mazancourt, et al., 2010). Inversely, in human-term placenta, a study revealed that adiponectin treatment of primary cytotrophoblasts attenuated the endocrine function of human trophoblastic cells, as a significant drop in the expression of HCG, placental lactogen, and some steroidogenic enzymes like p450-aromatase was observed (McDonald & Wolfe, 2009). These observations revealed that adiponectin seems to exert gestational age-dependent effects in human placentas, as already described for another placental growth factor, EGF (Maruo, Matsuo, Murata, & Mochizuki, 1992). Finally, in human placental explants collected at term, adiponectin has been described to stimulate the production of proinflammatory cytokines IL-1β, IL-6, IL-8, TNFα, and the prostaglandin PGE_2 (Lappas, Permezel, & Rice, 2005; McDonald & Wolfe,

2009). The authors suggested that adiponectin may play a role in the processes of human labor and delivery, as the release of prostaglandins and proinflammatory cytokines produced within the intrauterine environment participate in the regulation of myometrial contractility, cervical ripening, and rupture of membranes (Rice, 2001). Numerous studies have identified the anti-inflammatory role of adiponectin in a variety of cell types, such as macrophages (Wolf et al., 2004; Wulster-Radcliffe et al., 2004). Interestingly, the proinflammatory versus anti-inflammatory effects of adiponectin have been suggested to be mediated through different forms of the hormone. HMW adiponectin has been reported to be pro- or anti-inflammatory, while the cleaved form, the globular adiponectin, has been implicated in mediating only proinflammatory processes (Haugen & Drevon, 2007; Tomizawa, Hattori, Kasai, & Nakano, 2008).

In summary, these *in vitro* studies reveal that adiponectin seems to participate in the development of a functional placenta characterized by secretory and endometrial-anchoring capacities and will then be considered a favorable factor of the human implantation process (Fig. 8.1).

6. ADIPONECTIN AND PREGNANCY-ASSOCIATED PATHOLOGIES

In maternal plasma, adiponectin concentrations remain constant during the course of normal pregnancy (Cortelazzi et al., 2007; Nien et al., 2007a). Inversely, changes in adiponectin levels have been described in some pathologies of the female reproductive system such as gestational diabetes mellitus, preeclampsia, polycystic ovarian syndrome (PCOS), intrauterine growth-restricted fetuses (IUGR), and endometriosis (Cortelazzi et al., 2007; Lu et al., 2006; Thyfault et al., 2005).

Gestational diabetes mellitus is the most frequent metabolic disorder in pregnancy, affecting 1–10% of all pregnancies (Beischer, Wein, Sheedy, & Dargaville, 1996). Lower circulating adiponectin concentrations have been reported in patients with gestational diabetes compared to patients with normal pregnancy, independent of prepregnancy body mass index (BMI) and insulin sensitivity (Choi et al., 2008; Cortelazzi et al., 2007; Horosz, Bomba-Opon, Szymanska, & Wielgos, 2011; Nien et al., 2007b; Worda et al., 2004). Fetuses of diabetic mothers exhibit significantly lower adiponectin levels compared to normal fetuses of the same gestational age, independent of birth weight (Cortelazzi et al., 2007; Lindsay et al., 2003). Lain, Daftary, Ness, and Roberts (2008) clearly demonstrated that women with first-trimester adiponectin concentrations below the 25th percentile are 10 times more likely to be diagnosed with gestational diabetes mellitus compared to women with higher adiponectin levels.

In women with PCOS, hypoadiponectinemia is also observed (Panidis et al., 2004; Sir-Petermann et al., 2007; Toulis et al., 2009). PCOS is a common, heterogeneous endocrine disorder in women that is characterized by a clustering of hyperandrogenism, chronic anovulation, and polycystic ovaries, affecting 5–10% of women at a reproductive age (Franks, 1995). In such women, therapy with insulin-sensitizing drugs such as metformin and troglitazone improves ovulatory function and fertility (De Leo, la Marca, & Petraglia, 2003). Moreover, genetic investigations revealed that variants of the *adiponectin* gene (single T45G nucleotide polymorphism in exon 2) are more prevalent in women with PCOS (Heinonen et al., 2005).

In preeclampsia, data concerning plasma adiponectin levels are presently controversial. Indeed, some studies revealed reduced adiponectin concentrations in women with preeclampsia (Cortelazzi et al., 2007; Ouyang & Chen, 2007; Suwaki et al., 2006). In addition, two studies report that hypoadiponectinemia and preeclampsia correlate only when obesity and gestational diabetes are present as complicating factors (Suwaki et al., 2006; Thyfault et al., 2005). Others demonstrated no significant differences in adiponectin levels between preeclamptic and control pregnant women (Dalamaga, Srinivas, Elovitz, Chamberland, & Mantzoros, 2011; O'Sullivan, Kriketos, Martin, & Brown, 2006). Finally, a paradoxical and significant increase of circulating adiponectin was described (Fasshauer et al., 2008; Haugen et al., 2006; Kajantie, Kaaja, Ylikorkala, Andersson, & Laivuori, 2005; Naruse et al., 2005; Nien et al., 2007a; Ramsay, Jamieson, Greer, & Sattar, 2003). The reasons for these conflicting results are presently unclear. However, differences in preeclampsia definition, the ethnic background of the patients, BMI, and smoking might contribute to these discrepancies. More pertinent to this discussion, there is genomic evidence that preeclampsia is associated with a polymorphism of *adiponectin* gene (single G276T nucleotide polymorphism in intron 2; Saarela, Hiltunen, Helisalmi, Heinonen, & Laakso, 2006). So, additional studies are needed to identify a real association with circulating hypoadiponectinemia during the first trimester with preeclampsia.

Given the significance of glucose and insulin in fetal growth and the fundamental role of adiponectin in insulin metabolism, it is reasonable to assume that adiponectin may play a regulatory role in IUGR (Briana & Malamitsi-Puchner, 2009). IUGR fetuses were effectively demonstrated to make metabolic adaptations that benefit them in the short time. For example, they are able to shift their adiponectin pattern toward the HMW isoform, which specifically correlates with insulin sensitivity, thus sensitizing their body to insulin and preparing for neonatal catch-up growth (Ibanez et al., 2008). In contrast, lower adiponectin concentrations have been previously demonstrated in IUGR fetuses, and this downregulation has been proposed to represent a predisposing factor for later development of insulin resistance and metabolic syndrome (Kamoda, Saitoh, Saito, Sugiura, & Matsui, 2004; Takaya, Yamato, Higashino, & Kaneko, 2007). On the other hand, normal or higher adiponectin concentrations in IUGR insulin-resistant children have been recently reported (Evagelidou et al., 2007). A possible explanation for these contradictory results may rely on the fact that all the above studies have not consistently characterized IUGR. Alternatively, discrepancies could be, to a large extent, explained by differences in specific methodological aspects. Taken together, these data imply that adiponectin deficiency may be a plausible and attractive explanation for the metabolic abnormalities observed in IUGR children.

Moreover, concentrations of adiponectin in peritoneal fluid are decreased in women suffering from endometriosis (Takemura et al., 2005). The authors even suggested that adiponectin would be an antiendometriotic factor, possibly due to its anti-inflammatory, antiangiogenic, and antifibrotic properties.

Finally, some studies suggest the potential of maternal adiponectin to serve as a marker for fertility. High circulating adiponectin levels seemed to be effectively associated with *in vitro* fertilization success (Bersinger & Wunder, 2010; Bersinger, Birkhauser, & Wunder, 2006).

Little is known concerning the role of adiponectin in human endometrium. It was demonstrated that adiponectin and its receptors are expressed in human uteri (Takemura et al., 2006). Another study revealed high AdipoR expression during the luteal period, which corresponds to the embryo implantation period (Takemura et al., 2006). Recent experiments performed in our laboratory indicate an implication for the adiponectin system in the unexplained recurrent implantation failure. Indeed, while adiponectin was expressed to the same extent in fertile and infertile women, the expression of both receptors, AdipoR1 and AdipoR2, was reduced in endometrium of infertile women compared to the control group (Dos Santos et al., 2012). These results strongly suggested that adiponectin plays a role in human uterine receptivity.

In summary, implantation failure and pregnancy loss resulting from diabetes, PCOS, or endometriosis may result in aberrant adiponectin secretion (Table 8.1).

Table 8.1 Adiponectin levels in different pregnancy-associated pathologies

Complicated pregnancy	Adiponectin levels	Authors
Gestational diabetes mellitus	Serum ↘	Lindsay et al. (2003) Worda et al. (2004) Cortelazzi et al. (2007) Lain et al. (2008) Choi et al. (2008) Horosz et al. (2011)
Preeclampsia	Serum ↘	Suwaki et al. (2006) Ouyang & Chen (2007) Cortelazzi et al. (2007)
	↔	O'Sullivan et al. (2006) Dalamaga et al. (2011)
	↗	Ramsay et al. (2003) Kajantie et al. (2005) Naruse et al. (2005) Haugen et al. (2006) Nien et al. (2007b) Fasshauer et al. (2008)
PCOS	Serum ↘	De Leo et al. (2003) Panidis et al. (2004) Heinonen et al. (2005) Sir-Petermann et al. (2007) Toulis et al. (2009)
Intrauterine growth restriction	Infant serum ↘	Kamoda et al. (2004) Takaya et al. (2007)
	↗	Evagelidou et al. (2007)
Endometriosis	Peritoneal fluid ↘	Takemura et al. (2005)

7. CONCLUSIONS AND PERSPECTIVES

Since adiponectin and its initial association with insulin sensibility was discovered, it is now evident that this "fat hormone" also plays important roles in reproductive biology. So, this chapter summarizes the findings that associate adiponectin and the physiology of pregnancy. Adiponectin is now known to play a wide range of important roles, which extend from maternal physiology to embryo implantation and from paracrine effects in the placenta to regulation of conceptus development and fetal growth. However, leptin, another adipokine, has dominated the literature on female fertility and reproductive complications (Gonzalez et al., 2000; Henson & Castracane, 2006). Many recent studies emphasize the importance of considering not only the separate effects of adiponectin or leptin but also the ratio between these two adipokines. Indeed, growing evidence indicates that the adiponectin–leptin ratio can serve as a clinically useful marker for detecting some pathologies. For example, the adiponectin–leptin ratio seems to be a good candidate for predicting future insulin resistance, atherosclerosis, or metabolic syndrome, and eventually cardiovascular disease, in high-risk subjects (Kotani & Sakane, 2011; Zaletel, Barlovic, & Prezelj, 2010). Other studies underline that lower adiponectin–leptin ratios were found to be significantly associated with an increased risk of hormonal cancer development, such as endometrial, prostate, and breast cancers (Ashizawa et al., 2010; Grossmann et al., 2010; Mistry, Digby, Desai, & Randeva, 2008). In breast cancer, the adiponectin–leptin ratio is even correlated to tumor size and tumor aggressiveness (Chen et al., 2006). These observations suggest that the balance between adiponectin and leptin and adiponectin and other cytokines (IL-18, TNFα, leukemia inhibitor factor, etc.) must be considered in the complex endo/para/autocrine regulation of human reproduction. Thus, further investigations are clearly needed to precisely identify the specific role of adiponectin in such cytokine networks.

REFERENCES

Arita, Y., Kihara, S., Ouchi, N., Takahashi, M., Maeda, K., Miyagawa, J., et al. (1999). Paradoxical decrease of an adipose-specific protein, adiponectin, in obesity. *Biochemical and Biophysical Research Communications*, *257*, 79–83.

Ashizawa, N., Yahata, T., Quan, J., Adachi, S., Yoshihara, K., & Tanaka, K. (2010). Serum leptin-adiponectin ratio and endometrial cancer risk in postmenopausal female subjects. *Gynecologic Oncology*, *119*, 65–69.

Beischer, N. A., Wein, P., Sheedy, M. T., & Dargaville, R. M. (1996). Effect of follow-up of women with gestational diabetes on the ratio of IDDM to NIDDM in pregnancy. *Diabetes Care*, *19*, 653–655.

Benaitreau, D., Dieudonne, M. N., Dos Santos, E., Leneveu, M. C., Mazancourt, P., & Pecquery, R. (2009). Antiproliferative effects of adiponectin on human trophoblastic cell lines JEG-3 and BeWo. *Biology of Reproduction*, *80*, 1107–1114.

Benaitreau, D., Dos Santos, E., Leneveu, M. C., Alfaidy, N., Feige, J. J., de Mazancourt, P., et al. (2010). Effects of adiponectin on human trophoblast invasion. *The Journal of Endocrinology*, *207*, 45–53.

Benaitreau, D., Dos Santos, E., Leneveu, M. C., De Mazancourt, P., Pecquery, R., & Dieudonne, M. N. (2010). Adiponectin promotes syncytialisation of BeWo cell line and primary trophoblast cells. *Reproductive Biology and Endocrinology*, *8*, 128.

Berg, A. H., Combs, T. P., Du, X., Brownlee, M., & Scherer, P. E. (2001). The adipocyte-secreted protein Acrp30 enhances hepatic insulin action. *Nature Medicine*, *7*, 947–953.

Bersinger, N. A., Birkhauser, M. H., & Wunder, D. M. (2006). Adiponectin as a marker of success in intracytoplasmic sperm injection/embryo transfer cycles. *Gynecological Endocrinology*, *22*, 479–483.

Bersinger, N. A., & Wunder, D. M. (2010). Adiponectin isoform distribution in serum and in follicular fluid of women undergoing treatment by ICSI. *Acta Obstetricia et Gynecologica Scandinavica*, *89*, 782–788.

Bomba-Opon, D. A., Wielgos, M., Horosz, E., Bartkowiak, R., Kosinski, P., & Bochenska, K. (2010). Placental first trimester's measurements in relation to maternal plasma adiponectin, leptin and insulin concentrations. *Neuro Endocrinology Letters*, *31*, 573–576.

Briana, D. D., & Malamitsi-Puchner, A. (2009). Intrauterine growth restriction and adult disease: The role of adipocytokines. *European Journal of Endocrinology*, *160*, 337–347.

Bub, J. D., Miyazaki, T., & Iwamoto, Y. (2006). Adiponectin as a growth inhibitor in prostate cancer cells. *Biochemical and Biophysical Research Communications*, *340*, 1158–1166.

Caminos, J. E., Nogueiras, R., Gallego, R., Bravo, S., Tovar, S., Garcia-Caballero, T., et al. (2005). Expression and regulation of adiponectin and receptor in human and rat placenta. *The Journal of Clinical Endocrinology and Metabolism*, *90*, 4276–4286.

Campos, D. B., Palin, M. F., Bordignon, V., & Murphy, B. D. (2008). The 'beneficial' adipokines in reproduction and fertility. *International Journal of Obesity*, *32*, 223–231.

Ceddia, R. B., Somwar, R., Maida, A., Fang, X., Bikopoulos, G., & Sweeney, G. (2005). Globular adiponectin increases GLUT4 translocation and glucose uptake but reduces glycogen synthesis in rat skeletal muscle cells. *Diabetologia*, *48*, 132–139.

Cervero, A., Dominguez, F., Horcajadas, J. A., Quinonero, A., Pellicer, A., & Simon, C. (2006). The role of the leptin in reproduction. *Current Opinion in Obstetrics & Gynecology*, *18*, 297–303.

Chabrolle, C., Tosca, L., Crochet, S., Tesseraud, S., & Dupont, J. (2007). Expression of adiponectin and its receptors (AdipoR1 and AdipoR2) in chicken ovary: Potential role in ovarian steroidogenesis. *Domestic Animal Endocrinology*, *33*, 480–487.

Chabrolle, C., Tosca, L., & Dupont, J. (2007). Regulation of adiponectin and its receptors in rat ovary by human chorionic gonadotrophin treatment and potential involvement of adiponectin in granulosa cell steroidogenesis. *Reproduction*, *133*, 719–731.

Chabrolle, C., Tosca, L., Rame, C., Lecomte, P., Royere, D., & Dupont, J. (2009). Adiponectin increases insulin-like growth factor I-induced progesterone and estradiol secretion in human granulosa cells. *Fertility and Sterility*, *92*, 1988–1996.

Chappaz, E., Albornoz, M. S., Campos, D., Che, L., Palin, M. F., Murphy, B. D., et al. (2008). Adiponectin enhances in vitro development of swine embryos. *Domestic Animal Endocrinology*, *35*, 198–207.

Chehab, F. F., Lim, M. E., & Lu, R. (1996). Correction of the sterility defect in homozygous obese female mice by treatment with the human recombinant leptin. *Nature Genetics*, *12*, 318–320.

Chen, H., Montagnani, M., Funahashi, T., Shimomura, I., & Quon, M. J. (2003). Adiponectin stimulates production of nitric oxide in vascular endothelial cells. *The Journal of Biological Chemistry*, *278*, 45021–45026.

Chen, J., Tan, B., Karteris, E., Zervou, S., Digby, J., Hillhouse, E. W., et al. (2006). Secretion of adiponectin by human placenta: Differential modulation of adiponectin and its receptors by cytokines. *Diabetologia*, *49*, 1292–1302.

Choi, S. H., Kwak, S. H., Youn, B. S., Lim, S., Park, Y. J., Lee, H., et al. (2008). High plasma retinol binding protein-4 and low plasma adiponectin concentrations are associated with severity of glucose intolerance in women with previous gestational diabetes mellitus. *The Journal of Clinical Endocrinology and Metabolism*, *93*, 3142–3148.

Combs, T. P., Berg, A. H., Obici, S., Scherer, P. E., & Rossetti, L. (2001). Endogenous glucose production is inhibited by the adipose-derived protein Acrp30. *The Journal of Clinical Investigation*, *108*, 1875–1881.

Combs, T. P., Berg, A. H., Rajala, M. W., Klebanov, S., Iyengar, P., Jimenez-Chillaron, J. C., et al. (2003). Sexual differentiation, pregnancy, calorie restriction, and aging affect the adipocyte-specific secretory protein adiponectin. *Diabetes*, *52*, 268–276.

Combs, T. P., Pajvani, U. B., Berg, A. H., Lin, Y., Jelicks, L. A., Laplante, M., et al. (2004). A transgenic mouse with a deletion in the collagenous domain of adiponectin displays elevated circulating adiponectin and improved insulin sensitivity. *Endocrinology*, *145*, 367–383.

Combs, T. P., Wagner, J. A., Berger, J., Doebber, T., Wang, W. J., Zhang, B. B., et al. (2002). Induction of adipocyte complement-related protein of 30 kilodaltons by PPARgamma agonists: A potential mechanism of insulin sensitization. *Endocrinology*, *143*, 998–1007.

Cong, L., Gasser, J., Zhao, J., Yang, B., Li, F., & Zhao, A. Z. (2007). Human adiponectin inhibits cell growth and induces apoptosis in human endometrial carcinoma cells, HEC-1-A and RL95 2. *Endocrine-Related Cancer*, *14*, 713–720.

Corbetta, S., Bulfamante, G., Cortelazzi, D., Barresi, V., Cetin, I., Mantovani, G., et al. (2005). Adiponectin expression in human fetal tissues during mid- and late gestation. *The Journal of Clinical Endocrinology and Metabolism*, *90*, 2397–2402.

Cortelazzi, D., Corbetta, S., Ronzoni, S., Pelle, F., Marconi, A., Cozzi, V., et al. (2007). Maternal and foetal resistin and adiponectin concentrations in normal and complicated pregnancies. *Clinical Endocrinology*, *66*, 447–453.

Dalamaga, M., Srinivas, S. K., Elovitz, M. A., Chamberland, J., & Mantzoros, C. S. (2011). Serum adiponectin and leptin in relation to risk for preeclampsia: Results from a large case-control study. *Metabolism*, *60*, 1539–1544.

De Leo, V., la Marca, A., & Petraglia, F. (2003). Insulin-lowering agents in the management of polycystic ovary syndrome. *Endocrine Reviews*, *24*, 633–667.

Deepa, S. S., & Dong, L. Q. (2009). APPL1: Role in adiponectin signaling and beyond. *American Journal of Physiology. Endocrinology and Metabolism*, *296*, E22–E36.

Dieudonne, M. N., Bussiere, M., Dos Santos, E., Leneveu, M. C., Giudicelli, Y., & Pecquery, R. (2006). Adiponectin mediates antiproliferative and apoptotic responses in human MCF7 breast cancer cells. *Biochemical and Biophysical Research Communications*, *345*, 271–279.

Dos Santos, E., Benaitreau, D., Dieudonne, M. N., Leneveu, M. C., Serazin, V., Giudicelli, Y., et al. (2008). Adiponectin mediates an antiproliferative response in human MDA-MB 231 breast cancer cells. *Oncology Reports*, *20*, 971–977.

Dos Santos, E., Serazin, V., Morvan, C., Torre, A., Wainer, B., De Mazancourt, P., & Dieudonné, M. N. (2012). Adiponectin and leptin systems in human endometrium during window of implantation. *Fertility and Sterility*, *97*, 771–778.

Evagelidou, E. N., Giapros, V. I., Challa, A. S., Kiortsis, D. N., Tsatsoulis, A. A., & Andronikou, S. K. (2007). Serum adiponectin levels, insulin resistance, and lipid profile in children born small for gestational age are affected by the severity of growth retardation at birth. *European Journal of Endocrinology, 156*, 271–277.

Fasshauer, M., Klein, J., Neumann, S., Eszlinger, M., & Paschke, R. (2002). Hormonal regulation of adiponectin gene expression in 3T3-L1 adipocytes. *Biochemical and Biophysical Research Communications, 290*, 1084–1089.

Fasshauer, M., Paschke, R., & Stumvoll, M. (2004). Adiponectin, obesity, and cardiovascular disease. *Biochimie, 86*, 779–784.

Fasshauer, M., Waldeyer, T., Seeger, J., Schrey, S., Ebert, T., Kratzsch, J., et al. (2008). Circulating high-molecular-weight adiponectin is upregulated in preeclampsia and is related to insulin sensitivity and renal function. *European Journal of Endocrinology, 158*, 197–201.

Foretz, M., Taleux, N., Guigas, B., Horman, S., Beauloye, C., Andreelli, F., et al. (2006). Regulation of energy metabolism by AMPK: A novel therapeutic approach for the treatment of metabolic and cardiovascular diseases. *Med. Sci. (Paris), 22*, 381–388.

Franks, S. (1995). Polycystic ovary syndrome. *The New England Journal of Medicine, 333*, 853–861.

Gonzalez, R. R., Simon, C., Caballero-Campo, P., Norman, R., Chardonnens, D., Devoto, L., et al. (2000). Leptin and reproduction. *Human Reproduction Update, 6*, 290–300.

Grossmann, M. E., Ray, A., Nkhata, K. J., Malakhov, D. A., Rogozina, O. P., Dogan, S., et al. (2010). Obesity and breast cancer: Status of leptin and adiponectin in pathological processes. *Cancer Metastasis Reviews, 29*, 641–653.

Guerre-Millo, M. (2008). Adiponectin: An update. *Diabetes & Metabolism, 34*, 12–18.

Hada, Y., Yamauchi, T., Waki, H., Tsuchida, A., Hara, K., Yago, H., et al. (2007). Selective purification and characterization of adiponectin multimer species from human plasma. *Biochemical and Biophysical Research Communications, 356*, 487–493.

Haugen, F., & Drevon, C. A. (2007). Activation of nuclear factor-kappaB by high molecular weight and globular adiponectin. *Endocrinology, 148*, 5478–5486.

Haugen, F., Ranheim, T., Harsem, N. K., Lips, E., Staff, A. C., & Drevon, C. A. (2006). Increased plasma levels of adipokines in preeclampsia: Relationship to placenta and adipose tissue gene expression. *American Journal of Physiology. Endocrinology and Metabolism, 290*, E326–E333.

Heinonen, S., Korhonen, S., Helisalmi, S., Koivunen, R., Tapanainen, J., Hippelainen, M., et al. (2005). Associations between two single nucleotide polymorphisms in the adiponectin gene and polycystic ovary syndrome. *Gynecological Endocrinology, 21*, 165–169.

Henson, M. C., & Castracane, V. D. (2006). Leptin in pregnancy: An update. *Biology of Reproduction, 74*, 218–229.

Horosz, E., Bomba-Opon, D. A., Szymanska, M., & Wielgos, M. (2011). Third trimester plasma adiponectin and leptin in gestational diabetes and normal pregnancies. *Diabetes Research and Clinical Practice, 93*, 350–356.

Hotamisligil, G. S. (1999). The role of TNFalpha and TNF receptors in obesity and insulin resistance. *Journal of Internal Medicine, 245*, 621–625.

Hu, E., Liang, P., & Spiegelman, B. M. (1996). AdipoQ is a novel adipose-specific gene dysregulated in obesity. *The Journal of Biological Chemistry, 271*, 10697–10703.

Hug, C., Wang, J., Ahmad, N. S., Bogan, J. S., Tsao, T. S., & Lodish, H. F. (2004). T-cadherin is a receptor for hexameric and high-molecular-weight forms of Acrp30/adiponectin. *Proc. Natl. Acad. Sci. USA, 101*, 10308–10313.

Ibanez, L., Sebastiani, G., Lopez-Bermejo, A., Diaz, M., Gomez-Roig, M. D., & de Zegher, F. (2008). Gender specificity of body adiposity and circulating adiponectin, visfatin, insulin, and insulin growth factor-I at term birth: Relation to prenatal growth. *The Journal of Clinical Endocrinology and Metabolism, 93*, 2774–2778.

Ishikawa, M., Kitayama, J., Yamauchi, T., Kadowaki, T., Maki, T., Miyato, H., et al. (2007). Adiponectin inhibits the growth and peritoneal metastasis of gastric cancer through its specific membrane receptors AdipoR1 and AdipoR2. *Cancer Science, 98*, 1120–1127.

Jones, H. N., Jansson, T., & Powell, T. L. (2010). Full-length adiponectin attenuates insulin signaling and inhibits insulin-stimulated amino Acid transport in human primary trophoblast cells. *Diabetes, 59*, 1161–1170.

Kadowaki, T., & Yamauchi, T. (2005). Adiponectin and adiponectin receptors. *Endocrine Reviews, 26*, 439–451.

Kadowaki, T., Yamauchi, T., Kubota, N., Hara, K., Ueki, K., & Tobe, K. (2006). Adiponectin and adiponectin receptors in insulin resistance, diabetes, and the metabolic syndrome. *The Journal of Clinical Investigation, 116*, 1784–1792.

Kahn, B. B., Alquier, T., Carling, D., & Hardie, D. G. (2005). AMP-activated protein kinase: Ancient energy gauge provides clues to modern understanding of metabolism. *Cell Metabolism, 1*, 15–25.

Kajantie, E., Hytinantti, T., Hovi, P., & Andersson, S. (2004). Cord plasma adiponectin: A 20-fold rise between 24 weeks gestation and term. *The Journal of Clinical Endocrinology and Metabolism, 89*, 4031–4036.

Kajantie, E., Kaaja, R., Ylikorkala, O., Andersson, S., & Laivuori, H. (2005). Adiponectin concentrations in maternal serum: Elevated in preeclampsia but unrelated to insulin sensitivity. *Journal of the Society for Gynecologic Investigation, 12*, 433–439.

Kamoda, T., Saitoh, H., Saito, M., Sugiura, M., & Matsui, A. (2004). Serum adiponectin concentrations in newborn infants in early postnatal life. *Pediatric Research, 56*, 690–693.

Kelesidis, I., Kelesidis, T., & Mantzoros, C. S. (2006). Adiponectin and cancer: A systematic review. *British Journal of Cancer, 94*, 1221–1225.

Kobayashi, H., Ouchi, N., Kihara, S., Walsh, K., Kumada, M., Abe, Y., et al. (2004). Selective suppression of endothelial cell apoptosis by the high molecular weight form of adiponectin. *Circulation Research, 94*, e27–e31.

Kos, K., Harte, A. L., da Silva, N. F., Tonchev, A., Chaldakov, G., James, S., et al. (2007). Adiponectin and resistin in human cerebrospinal fluid and expression of adiponectin receptors in the human hypothalamus. *The Journal of Clinical Endocrinology and Metabolism, 92*, 1129–1136.

Kotani, K., & Sakane, N. (2011). Leptin:adiponectin ratio and metabolic syndrome in the general Japanese population. *The Korean Journal of Laboratory Medicine, 31*, 162–166.

Kotani, Y., Yokota, I., Kitamura, S., Matsuda, J., Naito, E., & Kuroda, Y. (2004). Plasma adiponectin levels in newborns are higher than those in adults and positively correlated with birth weight. *Clinical Endocrinology, 61*, 418–423.

Kubota, N., Yano, W., Kubota, T., Yamauchi, T., Itoh, S., Kumagai, H., et al. (2007). Adiponectin stimulates AMP-activated protein kinase in the hypothalamus and increases food intake. *Cell Metabolism, 6*, 55–68.

Lagaly, D. V., Aad, P. Y., Grado-Ahuir, J. A., Hulsey, L. B., & Spicer, L. J. (2008). Role of adiponectin in regulating ovarian theca and granulosa cell function. *Molecular and Cellular Endocrinology, 284*, 38–45.

Lain, K. Y., Daftary, A. R., Ness, R. B., & Roberts, J. M. (2008). First trimester adipocytokine concentrations and risk of developing gestational diabetes later in pregnancy. *Clinical Endocrinology, 69*, 407–411.

Lappas, M., Permezel, M., & Rice, G. E. (2005). Leptin and adiponectin stimulate the release of proinflammatory cytokines and prostaglandins from human placenta and maternal adipose tissue via nuclear factor-kappaB, peroxisomal proliferator-activated receptor-gamma and extracellularly regulated kinase 1/2. *Endocrinology, 146*, 3334–3342.

Ledoux, S., Campos, D. B., Lopes, F. L., Dobias-Goff, M., Palin, M. F., & Murphy, B. D. (2006). Adiponectin induces periovulatory changes in ovarian follicular cells. *Endocrinology*, *147*, 5178–5186.

Li, P., Sun, F., Cao, H. M., Ma, Q. Y., Pan, C. M., Ma, J. H., et al. (2009). Expression of adiponectin receptors in mouse adrenal glands and the adrenocortical Y-1 cell line: Adiponectin regulates steroidogenesis. *Biochemical and Biophysical Research Communications*, *390*, 1208–1213.

Lindsay, R. S., Walker, J. D., Havel, P. J., Hamilton, B. A., Calder, A. A., & Johnstone, F. D. (2003). Adiponectin is present in cord blood but is unrelated to birth weight. *Diabetes Care*, *26*, 2244–2249.

Liu, M., & Liu, F. (2010). Transcriptional and post-translational regulation of adiponectin. *The Biochemical Journal*, *425*, 41–52.

Liu, Y. H., Tsai, E. M., Chen, Y. L., Chen, H. S., Chen, Y. C., Wu, L. C., et al. (2006). Serum adiponectin levels increase after human chorionic gonadotropin treatment during in vitro fertilization. *Gynecologic and Obstetric Investigation*, *62*, 61–65.

Lord, E., Ledoux, S., Murphy, B. D., Beaudry, D., & Palin, M. F. (2005). Expression of adiponectin and its receptors in swine. *Journal of Animal Science*, *83*, 565–578.

Lu, M., Tang, Q., Olefsky, J. M., Mellon, P. L., & Webster, N. J. (2008). Adiponectin activates adenosine monophosphate-activated protein kinase and decreases luteinizing hormone secretion in LbetaT2 gonadotropes. *Molecular Endocrinology*, *22*, 760–771.

Lu, D., Yang, X., Wu, Y., Wang, H., Huang, H., & Dong, M. (2006). Serum adiponectin, leptin and soluble leptin receptor in pre-eclampsia. *International Journal of Gynaecology and Obstetrics*, *95*, 121–126.

Maeda, K., Okubo, K., Shimomura, I., Funahashi, T., Matsuzawa, Y., & Matsubara, K. (1996). cDNA cloning and expression of a novel adipose specific collagen-like factor, apM1 (AdiPose most abundant gene transcript 1). *Biochemical and Biophysical Research Communications*, *221*, 286–289.

Maeda, N., Shimomura, I., Kishida, K., Nishizawa, H., Matsuda, M., Nagaretani, H., et al. (2002). Diet-induced insulin resistance in mice lacking adiponectin/ACRP30. *Nature Medicine*, *8*, 731–737.

Maillard, V., Uzbekova, S., Guignot, F., Perreau, C., Rame, C., Coyral-Castel, S., et al. (2010). Effect of adiponectin on bovine granulosa cell steroidogenesis, oocyte maturation and embryo development. *Reproductive Biology and Endocrinology*, *8*, 23.

Mantzoros, C., Petridou, E., Alexe, D. M., Skalkidou, A., Dessypris, N., Papathoma, E., et al. (2004). Serum adiponectin concentrations in relation to maternal and perinatal characteristics in newborns. *European Journal of Endocrinology*, *151*, 741–746.

Mantzoros, C., Petridou, E., Dessypris, N., Chavelas, C., Dalamaga, M., Alexe, D. M., et al. (2004). Adiponectin and breast cancer risk. *The Journal of Clinical Endocrinology and Metabolism*, *89*, 1102–1107.

Maruo, T., Matsuo, H., Murata, K., & Mochizuki, M. (1992). Gestational age-dependent dual action of epidermal growth factor on human placenta early in gestation. *The Journal of Clinical Endocrinology and Metabolism*, *75*, 1362–1367.

McDonald, E. A., & Wolfe, M. W. (2009). Adiponectin attenuation of endocrine function within human term trophoblast cells. *Endocrinology*, *150*, 4358–4365.

Mistry, T., Digby, J. E., Desai, K. M., & Randeva, H. S. (2008). Leptin and adiponectin interact in the regulation of prostate cancer cell growth via modulation of p53 and bcl-2 expression. *BJU International*, *101*, 1317–1322.

Mitchell, M., Armstrong, D. T., Robker, R. L., & Norman, R. J. (2005). Adipokines: Implications for female fertility and obesity. *Reproduction*, *130*, 583–597.

Miyoshi, Y., Funahashi, T., Kihara, S., Taguchi, T., Tamaki, Y., Matsuzawa, Y., et al. (2003). Association of serum adiponectin levels with breast cancer risk. *Clinical Cancer Research*, *9*, 5699–5704.

Nakano, Y., Tobe, T., Choi-Miura, N. H., Mazda, T., & Tomita, M. (1996). Isolation and characterization of GBP28, a novel gelatin-binding protein purified from human plasma. *Journal of Biochemistry*, *120*, 803–812.

Naruse, K., Yamasaki, M., Umekage, H., Sado, T., Sakamoto, Y., & Morikawa, H. (2005). Peripheral blood concentrations of adiponectin, an adipocyte-specific plasma protein, in normal pregnancy and preeclampsia. *Journal of Reproductive Immunology*, *65*, 65–75.

Nien, J. K., Mazaki-Tovi, S., Romero, R., Erez, O., Kusanovic, J. P., Gotsch, F., et al. (2007a). Plasma adiponectin concentrations in non-pregnant, normal and overweight pregnant women. *Journal of Perinatal Medicine*, *35*, 522–531.

Nien, J. K., Mazaki-Tovi, S., Romero, R., Erez, O., Kusanovic, J. P., Gotsch, F., et al. (2007b). Adiponectin in severe preeclampsia. *Journal of Perinatal Medicine*, *35*, 503–512.

Nishizawa, H., Shimomura, I., Kishida, K., Maeda, N., Kuriyama, H., Nagaretani, H., et al. (2002). Androgens decrease plasma adiponectin, an insulin-sensitizing adipocyte-derived protein. *Diabetes*, *51*, 2734–2741.

O'Sullivan, A. J., Kriketos, A. D., Martin, A., & Brown, M. A. (2006). Serum adiponectin levels in normal and hypertensive pregnancy. *Hypertension in Pregnancy*, *25*, 193–203.

Ouchi, N., Kihara, S., Arita, Y., Nishida, M., Matsuyama, A., Okamoto, Y., et al. (2001). Adipocyte-derived plasma protein, adiponectin, suppresses lipid accumulation and class A scavenger receptor expression in human monocyte-derived macrophages. *Circulation*, *103*, 1057–1063.

Ouchi, N., Ohishi, M., Kihara, S., Funahashi, T., Nakamura, T., Nagaretani, H., et al. (2003). Association of hypoadiponectinemia with impaired vasoreactivity. *Hypertension*, *42*, 231–234.

Ouyang, Y., & Chen, H. (2007). Reduced plasma adiponectin and elevated leptin in preeclampsia. *International Journal of Gynaecology and Obstetrics*, *98*, 110–114.

Panidis, D., Kourtis, A., Kukuvitis, A., Farmakiotis, D., Xita, N., Georgiou, I., et al. (2004). Association of the T45G polymorphism in exon 2 of the adiponectin gene with polycystic ovary syndrome: Role of Delta4-androstenedione. *Human Reproduction*, *19*, 1728–1733.

Pinar, H., Basu, S., Hotmire, K., Laffineuse, L., Presley, L., Carpenter, M., et al. (2008). High molecular mass multimer complexes and vascular expression contribute to high adiponectin in the fetus. *The Journal of Clinical Endocrinology and Metabolism*, *93*, 2885–2890.

Qiang, L., Wang, H., & Farmer, S. R. (2007). Adiponectin secretion is regulated by SIRT1 and the endoplasmic reticulum oxidoreductase Ero1-L alpha. *Molecular and Cellular Biology*, *27*, 4698–4707.

Ramsay, J. E., Jamieson, N., Greer, I. A., & Sattar, N. (2003). Paradoxical elevation in adiponectin concentrations in women with preeclampsia. *Hypertension*, *42*, 891–894.

Rice, G. E. (2001). Cytokines and the initiation of parturition. *Frontiers of Hormone Research*, *27*, 113–146.

Rodriguez-Pacheco, F., Martinez-Fuentes, A. J., Tovar, S., Pinilla, L., Tena-Sempere, M., Dieguez, C., et al. (2007). Regulation of pituitary cell function by adiponectin. *Endocrinology*, *148*, 401–410.

Saarela, T., Hiltunen, M., Helisalmi, S., Heinonen, S., & Laakso, M. (2006). Adiponectin gene haplotype is associated with preeclampsia. *Genetic Testing*, *10*, 35–39.

Scherer, P. E., Williams, S., Fogliano, M., Baldini, G., & Lodish, H. F. (1995). A novel serum protein similar to C1q, produced exclusively in adipocytes. *The Journal of Biological Chemistry*, *270*, 26746–26749.

Sir-Petermann, T., Echiburu, B., Maliqueo, M. M., Crisosto, N., Sanchez, F., Hitschfeld, C., et al. (2007). Serum adiponectin and lipid concentrations in pregnant women with polycystic ovary syndrome. *Human Reproduction*, *22*, 1830–1836.

Suwaki, N., Masuyama, H., Nakatsukasa, H., Masumoto, A., Sumida, Y., Takamoto, N., et al. (2006). Hypoadiponectinemia and circulating angiogenic factors in overweight patients complicated with pre-eclampsia. *American Journal of Obstetrics and Gynecology, 195*, 1687–1692.

Takahashi, M., Arita, Y., Yamagata, K., Matsukawa, Y., Okutomi, K., Horie, M., et al. (2000). Genomic structure and mutations in adipose-specific gene, adiponectin. *International Journal of Obesity and Related Metabolic Disorders, 24*, 861–868.

Takaya, J., Yamato, F., Higashino, H., & Kaneko, K. (2007). Intracellular magnesium and adipokines in umbilical cord plasma and infant birth size. *Pediatric Research, 62*, 700–703.

Takemura, Y., Osuga, Y., Harada, M., Hirata, T., Koga, K., Yoshino, O., et al. (2005). Concentration of adiponectin in peritoneal fluid is decreased in women with endometriosis. *American Journal of Reproductive Immunology, 54*, 217–221.

Takemura, Y., Osuga, Y., Yamauchi, T., Kobayashi, M., Harada, M., Hirata, T., et al. (2006). Expression of adiponectin receptors and its possible implication in the human endometrium. *Endocrinology, 147*, 3203–3210.

Thyfault, J. P., Hedberg, E. M., Anchan, R. M., Thorne, O. P., Isler, C. M., Newton, E. R., et al. (2005). Gestational diabetes is associated with depressed adiponectin levels. *Journal of the Society for Gynecologic Investigation, 12*, 41–45.

Tilg, H., & Moschen, A. R. (2006). Adipocytokines: Mediators linking adipose tissue, inflammation and immunity. *Nature Reviews. Immunology, 6*, 772–783.

Tomizawa, A., Hattori, Y., Kasai, K., & Nakano, Y. (2008). Adiponectin induces NF-kappaB activation that leads to suppression of cytokine-induced NF-kappaB activation in vascular endothelial cells: Globular adiponectin vs. high molecular weight adiponectin. *Diabetes & Vascular Disease Research, 5*, 123–127.

Toulis, K. A., Goulis, D. G., Farmakiotis, D., Georgopoulos, N. A., Katsikis, I., Tarlatzis, B. C., et al. (2009). Adiponectin levels in women with polycystic ovary syndrome: A systematic review and a meta-analysis. *Human Reproduction Update, 15*, 297–307.

Wang, Y., Lam, K. S., Xu, J. Y., Lu, G., Xu, L. Y., Cooper, G. J., et al. (2005). Adiponectin inhibits cell proliferation by interacting with several growth factors in an oligomerization-dependent manner. *The Journal of Biological Chemistry, 280*, 18341–18347.

Wang, H., Qiang, L., & Farmer, S. R. (2008). Identification of a domain within peroxisome proliferator-activated receptor gamma regulating expression of a group of genes containing fibroblast growth factor 21 that are selectively repressed by SIRT1 in adipocytes. *Molecular and Cellular Biology, 28*, 188–200.

Wang, C., Xin, X., Xiang, R., Ramos, F. J., Liu, M., Lee, H. J., et al. (2009). Yin-Yang regulation of adiponectin signaling by APPL isoforms in muscle cells. *The Journal of Biological Chemistry, 284*, 31608–31615.

Weyer, C., Funahashi, T., Tanaka, S., Hotta, K., Matsuzawa, Y., Pratley, R. E., et al. (2001). Hypoadiponectinemia in obesity and type 2 diabetes: Close association with insulin resistance and hyperinsulinemia. *The Journal of Clinical Endocrinology and Metabolism, 86*, 1930–1935.

Wheelock, M. J., & Johnson, K. R. (2003). Cadherins as modulators of cellular phenotype. *Annual Review of Cell and Developmental Biology, 19*, 207–235.

Wolf, A. M., Wolf, D., Rumpold, H., Enrich, B., & Tilg, H. (2004). Adiponectin induces the anti-inflammatory cytokines IL-10 and IL-1RA in human leukocytes. *Biochemical and Biophysical Research Communications, 323*, 630–635.

Worda, C., Leipold, H., Gruber, C., Kautzky-Willer, A., Knofler, M., & Bancher-Todesca, D. (2004). Decreased plasma adiponectin concentrations in women with gestational diabetes mellitus. *American Journal of Obstetrics and Gynecology*, *191*, 2120–2124.

Wulster-Radcliffe, M. C., Ajuwon, K. M., Wang, J., Christian, J. A., & Spurlock, M. E. (2004). Adiponectin differentially regulates cytokines in porcine macrophages. *Biochemical and Biophysical Research Communications*, *316*, 924–929.

Yamauchi, T., Kamon, J., Ito, Y., Tsuchida, A., Yokomizo, T., Kita, S., et al. (2003). Cloning of adiponectin receptors that mediate antidiabetic metabolic effects. *Nature*, *423*, 762–769.

Yamauchi, T., Kamon, J., Minokoshi, Y., Ito, Y., Waki, H., Uchida, S., et al. (2002). Adiponectin stimulates glucose utilization and fatty-acid oxidation by activating AMP-activated protein kinase. *Nature Medicine*, *8*, 1288–1295.

Yamauchi, T., Kamon, J., Waki, H., Terauchi, Y., Kubota, N., Hara, K., et al. (2001). The fat-derived hormone adiponectin reverses insulin resistance associated with both lipoatrophy and obesity. *Nature Medicine*, *7*, 941–946.

Yamauchi, T., Nio, Y., Maki, T., Kobayashi, M., Takazawa, T., Iwabu, M., et al. (2007). Targeted disruption of AdipoR1 and AdipoR2 causes abrogation of adiponectin binding and metabolic actions. *Nature Medicine*, *13*, 332–339.

Yang, W. S., Hsiung, C. A., Ho, L. T., Chen, Y. T., He, C. T., Curb, J. D., et al. (2003). Genetic epistasis of adiponectin and PPARgamma2 genotypes in modulation of insulin sensitivity: A family-based association study. *Diabetologia*, *46*, 977–983.

Yang, W. S., Lee, W. J., Funahashi, T., Tanaka, S., Matsuzawa, Y., Chao, C. L., et al. (2001). Weight reduction increases plasma levels of an adipose-derived anti-inflammatory protein, adiponectin. *The Journal of Clinical Endocrinology and Metabolism*, *86*, 3815–3819.

Zaletel, J., Barlovic, D. P., & Prezelj, J. (2010). Adiponectin-leptin ratio: A useful estimate of insulin resistance in patients with Type 2 diabetes. *Journal of Endocrinological Investigation*, *33*, 514–518.

Zhang, Y., Proenca, R., Maffei, M., Barone, M., Leopold, L., & Friedman, J. M. (1994). Positional cloning of the mouse obese gene and its human homologue. *Nature*, *372*, 425–432.

CHAPTER NINE

Adiponectin and Its Receptors in Preimplantation Embryo Development

Štefan Čikoš[1]
Institute of Animal Physiology, Slovak Academy of Sciences, Šoltésovej 4, Košice, Slovak Republic
[1]Corresponding author: e-mail address: cikos@saske.sk

Contents

Abstract

Adiponectin can play an important role in regulating the female reproductive function and embryo development and can affect the embryo at very early stages of pregnancy—during the preimplantation period. Disturbances in the maternal adiponectin system are associated with several diseases, including diabetes type 2, obesity, and some female reproductive disorders. Adiponectin receptors are expressed in oocytes and preimplantation embryos and can be activated by adiponectin produced by maternal adipose tissue or organs of the female reproductive tract. Adiponectin can affect proliferation and survival of cells in preimplantation embryos, and these effects are isoform dependent. Experimental results suggest involvement of various protein kinases, including mitogen-activated protein kinases, in the regulation of these processes by adiponectin. Actions of adiponectin on lipid and glucose metabolism can increase the energy supply to the embryo, and final targets of adiponectin signaling are metabolic enzymes, glucose transporters, and fatty acid transporters. The involvement of several signaling molecules, such as AMPK/PRKA, PI3K, or AKT/PKB, in the regulation of

Vitamins and Hormones, Volume 90
ISSN 0083-6729
http://dx.doi.org/10.1016/B978-0-12-398313-8.00009-9

metabolic processes by adiponectin has been demonstrated in preimplantation embryos. In summary, adiponectin produced in an endocrine/paracrine/autocrine manner can significantly influence preimplantation embryo development, uterine receptivity, and embryo implantation.

1. INTRODUCTION

The preimplantation period of development (i.e., the period before implantation of an embryo into the uterus) is one of the most sensitive phases in mammalian ontogeny, and disturbances at this developmental stage can result in poor pregnancy outcomes in humans as well as in animals (Ford & Youngs, 1993; Humblot, 2001; Nepomnaschy et al., 2006). Thus, knowledge of factors that can influence preimplantation embryo development is important for understanding both human and animal reproduction processes.

Adiponectin belongs to the group of proteins called "adipokines" which are secreted from adipose tissue into the bloodstream and influence many physiological processes including reproduction. Most information about the roles of adipokines in reproduction refers to leptin, but much less is known about the role of adiponectin and other adipokines (Campos, Palin, Bordignon, & Murphy, 2008; Mitchell, Armstrong, Robker, & Norman, 2005; Zavalza-Gómez, Anaya-Prado, Rincón-Sánchez, & Mora-Martínez, 2008). Recent data suggest that adiponectin can play an important role in maintaining normal reproductive functions and in the communication between mother and embryo. The influence of adiponectin on ovarian, endometrial, and placental functions as well as on early embryo development has been reported in several mammalian species (Archanco, Gómez-Ambrosi, Tena-Sempere, Frühbeck, & Burrell, 2007; Caminos et al., 2005; Chabrolle, Tosca, & Dupont, 2007; Chappaz et al., 2008; Čikoš et al., 2010; Fischer, Santos, Thieme, Ramin, & Fischer, 2010; Kim et al., 2011; Ledoux et al., 2006; Takemura et al., 2006).

In comparison to other adipokines, adiponectin has some unique features. Plasma concentrations of adiponectin are much higher than those of other adipokines and show clear gender differences with females having higher levels (especially of high molecular adiponectin isoforms) than males (Combs et al., 2003; Nishizawa et al., 2002). In contrast to other adipokines, the plasma level of adiponectin is inversely correlated with body fat amount in adults, that is, it is decreased in obesity and in diseases associated with obesity, such as type 2 diabetes and cardiovascular diseases. Adiponectin can form several isoforms. Full-length adiponectin can be found in circulation as trimer, hexamer, or large multimer, and it also can be cleaved to produce truncated globular adiponectin, forming trimers (Kadowaki & Yamauchi, 2005; Oh, Ciaraldi, & Henry, 2007; Trujillo & Scherer, 2005; Wang, Lam, Yau, & Xu, 2008). Several studies have shown different effects of adiponectin in various cell types, and the data clearly demonstrate that effects of adiponectin are cell-type and isoform dependent (Bub, Miyazaki, & Iwamoto, 2006; Hada et al., 2007; Haugen & Drevon, 2007; Kobayashi et al., 2004; Neumeier et al., 2006; Song, Chan, & Rovin, 2009; Tsao, Murrey, Hug, Lee, & Lodish, 2002, 2003, Tsao et al., 2003). Thus, investigation of the effects of particular adiponectin isoforms in cells of preimplantation embryos is necessary to reveal the role of adiponectin signaling in the regulation of early embryo development.

2. ADIPONECTIN AND FERTILITY

Adiponectin knockout mice are fertile (Kubota et al., 2002; Ma et al., 2002; Maeda et al., 2002), which suggests that adiponectin is not absolutely essential for the reproductive process. However, adiponectin paralogs (C1q/TNFα-related proteins) identified in mouse and rabbit (Schmidt et al., 2008; Wong, Wang, Hug, Tsao, & Lodish, 2004) may compensate for adiponectin in null mice. Moreover, several studies indicate the importance of adiponectin for female reproductive health. Changes in adiponectin levels have been shown to be associated with some pathologies of the female reproductive system such as polycystic ovarian syndrome (Ardawi, Path, & Rouzi, 2005), endometriosis (Takemura et al., 2005), endometrial cancer (Dal Maso et al., 2004), preeclampsia (Mazaki-Tovi et al., 2009), and breast cancer (Vona-Davis & Rose, 2007). Obesity associated with decreased adiponectin levels is considered as a risk factor for these reproductive disorders, although the decreased serum adiponectin level is not always connected with the increased body mass index in these diseases. Other obesity-related disorders, such as type 2 diabetes and cardiovascular disease, are also connected with decreased maternal adiponectin levels (Kadowaki et al., 2006; Oh et al., 2007; Whitehead, Richards, Hickman, Macdonald, & Prins, 2006) and can result in reduced adiponectin availability for the developing embryo.

Obesity can negatively affect fertility by contributing to anovulation, reduced conception rate, reduced response to fertility treatment, and impaired pregnancy outcomes. Obese women have a high prevalence of infertility compared with the lean population, and there are data suggesting that some negative influences of obesity on reproductive health can be mediated by adipokines, including adiponectin (Campos et al., 2008; Fedorcsák et al., 2004; Mitchell et al., 2005; Norman et al., 2004; Van der Steeg et al., 2008). Some studies even suggest the potential of maternal adiponectin to serve as a marker for fertility (Bersinger & Wunder, 2010; Bersinger, Birkhäuser, & Wunder, 2006). In addition, recent data indicate that adiponectin can influence the embryo at the very early stages of development (Chappaz et al., 2008; Čikoš et al., 2010; Fischer et al., 2010; Kim et al., 2011).

3. OVERVIEW OF PREIMPLANTATION EMBRYO DEVELOPMENT

The fusion of spermatozoon with matured oocyte results in the formation of a one-cell preimplantation embryo called the zygote. Maturation of the oocyte takes place in the ovarian follicle which develops from the stage of small primordial follicle (containing a small oocyte) to larger primary and secondary follicles (containing a fully grown but still immature oocyte), and finally, it reaches the stage of large tertiary follicle (containing a cavity filled with follicular fluid). In the tertiary follicle, the oocyte completes meiosis (to the metaphase II stage) and then is released into the oviduct (in a process called ovulation). The ovulated oocyte can then be fertilized by sperm to form a zygote. The zygote undergoes a series of cleavage divisions that subdivide the embryo into smaller and smaller compartments. The first cell divisions are under control of maternal mRNAs and proteins synthesized during the oocyte development. The embryonic genome begins to be activated (transcribed) between the 2-cell and 16-cell stage depending on the species (e.g., at the 2-cell stage in the mouse and at the 4–8-cell stage in humans). A few cell divisions after the embryonic genome activation (three in the mouse, four in humans) cells maximize their intercellular contacts and compaction occurs. Several further rounds of mitotic cell divisions then lead to the formation of a ball of cells called the morula. At the late morula stage (around the 32- to 64-cell stage, depending upon the species), a fluid-filled cavity (blastocoel) appears and the preimplantation embryo reaches the blastocyst stage. At this stage, the first events of cellular differentiation can be observed: the embryo consists of two cell lineages, the inner cell mass (ICM, embryoblast) and the trophoectoderm (TE, trophoblast). The pluripotent ICM cells are destined to constitute the embryo proper, while the more differentiated polarized TE cells form later extraembryonal tissues. At the final stage of preimplantation development, the blastocyst escapes (hatches) from its protective glycoprotein coat (called the zona pellucida), and subsequently implants in the uterus (for review, see Heikinheimo & Gibbons, 1998; Kidder, 1992; Nothias, Majumder, Kaneko, & DePamphilis, 1995; Fleming et al., 2004; Duranthon, Watson, & Lonergan, 2008; Hardy & Spanos, 2002; Carson et al., 2000). In some species (such as rabbits, cattle, and pigs), blastocysts undergo dramatic increases in cell number and size after hatching, and gastrulation (the next step in embryo development which follows after the blastocyst formation) begins before the embryo implantation into the uterus (Blomberg, Hashizume, & Viebahn, 2008; Maddox-Hyttel et al., 2003; Oestrup et al., 2009; Vejlsted, Du, Vajta, & Maddox-Hyttel, 2006; Viebahn, Mayer, & Hrabé de Angelis, 1995).

The wall of the uterus is composed of a mucosal layer, the endometrium (which lines the uterine cavity) and a fibromuscular layer, the myometrium. The endometrium consists of a surface epithelium (where the embryo attaches at the beginning of the implantation process) and an underlying connective tissue stroma (containing cells mainly of fibroblastic type). The mucosa is invaginated into many tubular uterine glands formed by epithelial cells (Jiménez-Ayala & Jiménez-Ayala Portillo, 2008). The implanting embryo induces changes in the endometrial stromal cells (accumulation of glycogen and lipid droplets, etc.), and this process, necessary for embryo implantation and survival during early pregnancy, is called decidualization (Abrahamsohn & Zorn, 1993; Kierszenbaum, 2001; Kim, Jaffe, & Fazleabas, 1999).

4. EXPRESSION OF ADIPONECTIN AND ADIPONECTIN RECEPTORS IN OOCYTES AND PREIMPLANTATION EMBRYOS

4.1. Rodent models

The most comprehensive information on the expression of adiponectin receptors during the preimplantation developmental period is available for the mouse at present. Transcripts for AdipoR1 and AdipoR2 have been detected in ovulated oocytes and in preimplantation embryos at all tested developmental stages (from the 2-cell stage to the blastocyst stage, Table 9.1), indicating the presence of these mRNAs in cells of embryos throughout the preimplantation developmental period (Čikoš et al., 2010; Kim et al., 2011; Schmidt et al., 2008). However, variant results were obtained in two experiments quantitating AdipoR1 and AdipoR2 mRNAs (using real-time RT-PCR) during the mouse preimplantation period. Čikoš et al. (2010) demonstrated that mRNAs for AdipoR1 and AdipoR2 were synthesized in the mouse oocyte, then the amount of both mRNAs dropped (at the 4-cell stage), but it increased above the oocyte level from the morula stage, reaching the highest level in morulas and blastocysts. In contrast, Kim et al. (2011) found a decrease in AdipoR1 and AdipoR2 mRNA levels between the 8-cell and blastocyst developmental stages, and the mRNA levels of both receptors were even lower in blastocysts than in 2-cell embryos. The discrepancy between these results could reflect strain-specific differences in expression profiles of the receptors (ICR mice used in the Čikoš experiment vs. C57BL6 mice used in the Kim experiment), or it could occur for technical reasons. Čikoš et al. (2010) normalized AdipoR1 and AdipoR2 mRNA quantity with external control RNA expressing the receptor mRNA quantity per embryo/oocyte, and used three different methods for transformation of real-time RT-PCR fluorescence data (the relative standard curve method, the LinRegPCR-Ct method, and the SCF method)—all the methods showed similar results indicating their reliability. Kim et al. (2011) normalized AdipoR1 and AdipoR2 mRNA quantity with GAPDH reference gene expressing the receptor mRNA quantity per GAPDH mRNA quantity, and used the "delta-delta cycle threshold" method to calculate the relative quantity of receptor mRNAs. There is no generally accepted standard for normalization of transcript quantity in real-time PCR experiments comparing gene expression between different stages of preimplantation embryos. Since mouse preimplantation embryos (from ovulated oocyte/zygote to blastocyst stage) are similar in their size and biological mass, comparison of the relative amounts of transcripts and proteins per embryo seems be most suitable for determining gene expression pattern. Normalization with reference (housekeeping) genes is probably less convenient because of questionable expression stability of candidate reference genes during all of the preimplantation period. Advantages and disadvantages of various normalization strategies and techniques for transformation of real-time PCR fluorescence data to the target gene quantity have been widely discussed (Bower, Moser, Hill, &

Table 9.1 Expression of adiponectin and its receptors in oocytes and preimplantation embryos

Gene	Species	Protein, mRNA: developmental stage[a]	References
AdipoR1	Mouse	mRNA: oocyte, 2c, 4c, 8c, Mor, Blast	Čikoš et al. (2010), Schmidt et al. (2008), Kim et al. (2011)
		Protein: oocyte, 1c, 4c, 8c, Blast	Čikoš et al. (2010), Kim et al. (2011)
	Rat	mRNA: oocyte	Chabrolle et al. (2007)
		Protein: oocyte	Chabrolle et al., 2007
	Rabbit	mRNA and protein: Blast	Schmidt et al. (2008), Fischer et al. (2010)
	Pig	mRNA: oocyte, 4c, Blast	Whitworth et al. (2005)
		mRNA and protein: oocyte, 2c, Blast	Chappaz et al. (2008)
	Cow	mRNA and protein: oocyte	Tabandeh et al. (2010), Maillard et al. (2010)
		Protein: Blast	Maillard et al. (2010)
AdipoR2	Mouse	mRNA: oocyte, 2c, 4c, 8c, Mor, Blast	Čikoš et al. (2010), Schmidt et al. (2008), Kim et al. (2011)
		Protein: oocyte, 1c, 4c, 8c, Blast	Čikoš et al. (2010), Kim et al. (2011)
	Rat	mRNA: oocyte	Chabrolle et al. (2007)
		Protein: oocyte	Chabrolle et al. (2007)
	Rabbit	mRNA and protein: Blast	Schmidt et al. (2008), Fischer et al. (2010)
	Pig	mRNA: oocyte, 4c, Blast	Whitworth et al. (2005)
		mRNA and protein: oocyte, 2c, Blast	Chappaz et al. (2008)
	Cow	mRNA and protein: oocyte Protein: Blast	Tabandeh et al. (2010), Maillard et al. (2010) Maillard et al. (2010)

Continued

Table 9.1 Expression of adiponectin and its receptors in oocytes and preimplantation embryos—cont'd

Gene	Species	Protein, mRNA: developmental stage[a]	References
Adiponectin	Mouse	mRNA: 2c, 8c, Blast	Čikoš et al. (2010), Kim et al. (2011)
		Protein: 1c, 4c, 8c, Blast	Kim et al. (2011)
	Rat	mRNA: oocyte	Chabrolle et al. (2007)
		Protein: oocyte	Chabrolle et al. (2007)
	Rabbit	mRNA and protein: Blast	Schmidt et al. (2008)
	Cow	mRNA and protein: oocyte	Tabandeh et al. (2010), Maillard et al. (2010)

[a] 1c, 1-cell embryo; 2c, 2-cell embryo; 4c, 4-cell embryo; 8c, 8-cell embryo; Mor, morula; Blast, blastocyst.

Lehnert, 2007; Čikoš & Koppel, 2009; Huggett, Dheda, Bustin, & Zumla, 2005; Jeong et al., 2005; Mamo, Gal, Bodo, & Dinnyes, 2007; Thellin et al., 1999) and are beyond the scope of the present study.

Immunofluorescent staining using specific antibodies against AdipoR1 and AdipoR2 revealed the presence of these proteins in ovulated oocytes and in preimplantation embryos at all tested developmental stages (from the 1-cell stage to the blastocyst stage, Table 9.1), and the signal intensity was higher for AdipoR2 than for AdipoR1 at most stages (Čikoš et al., 2010; Kim et al., 2011). Confocal microscopy showed strong fluorescent signal mainly at the cell periphery suggesting that receptors are localized in the cytoplasm membrane. This localization is required for the extracellular ligand (adiponectin) binding and receptor functioning. The finding of surface expression of AdipoR1 and AdipoR2 in trophoblast stem (TS) cells (derived from blastocysts) supports the opinion that these receptors can be localized in the cytoplasm membrane of mouse blastocyst cells (Kim et al., 2011).

Studies examining expression of adiponectin in mouse preimplantation embryos have produced controversial results. Schmidt et al. (2008) did not find adiponectin mRNA in mouse blastocysts (the expression at the protein level was not examined at this stage), but they found adiponectin protein after embryo implantation (on day 5 post coitum). Čikoš et al. (2010) detected adiponectin mRNA in blastocysts but not at earlier developmental stages (from ovulated oocytes to morulas); the presence of adiponectin protein in mouse blastocysts was examined by immunofluorescence staining and showed no specific signal. However, Kim et al. (2011) found adiponectin mRNA in 2-cell and 8-cell mouse embryos but not in blastocysts. Interestingly, these authors detected adiponectin protein from the 1-cell to blastocyst stage, despite the fact that they did not find adiponectin mRNA at the blastocyst stage itself. The adiponectin signal demonstrated in Kim's study was localized at the cell periphery which suggests that the detected protein could represent adiponectin bound to its receptors (AdipoR1 and AdipoR2). Thus, adiponectin protein detected in preimplantation embryos could originate from an extraembryonal source such as oviduct, uterus, or maternal adipose tissue.

In summary, expression of adiponectin receptors AdipoR1 and AdipoR2 in mouse preimplantation embryos is well documented, although more thorough characterization of their expression pattern during the preimplantation period is needed. However, expression of adiponectin in cells of mouse preimplantation embryos must be verified by further studies.

In rats, the expression of adiponectin and its receptors has not been examined in preimplantation embryos till now, but adiponectin, AdipoR1 and AdipoR2 transcripts and proteins were detected in preovulatory as well as ovulated oocytes (Chabrolle et al., 2007).

4.2. Other mammalian species

In rabbits, gastrulation begins before the embryo implantation in the uterus, and several gastrulation stages can be distinguished (Viebahn et al., 1995). Examination of AdipoR1 and AdipoR2 transcripts in rabbit blastocysts at different gastrulation stages showed a time-dependent increase in their expression. Transcripts of both receptors were expressed at higher levels in the trophoblast, especially in the early stages; the expression of the receptors in the embryoblast rose significantly shortly prior to implantation. Expression of AdipoR1 and AdipoR2 in both cell lineages (embryoblast and trophoblast) of rabbit blastocysts was documented also at the protein level (Fischer et al., 2010; Schmidt et al., 2008). In addition, adiponectin mRNA was detected in rabbit blastocyst and immunohistochemistry showed the presence of adiponectin protein in both cell lineages, with stronger signal in the embryoblast than in the trophoblast (Schmidt et al., 2008).

In pigs, expression of AdipoR1 and AdipoR2 has been examined during preimplantation development *in vivo* and *in vitro*. Transcripts for both receptors were detected in preovulatory oocytes, 4-cell embryos, and blastocysts (Whitworth et al., 2005). In addition, Chappaz et al. (2008) demonstrated the presence of AdipoR1 and AdipoR2 protein in preovulatory oocytes (with stronger immunofluorescent signal for AdipoR2 compared to AdipoR1) and in parthenogenetically activated 2-cell embryos and blastocysts. The immunofluorescent signal for both receptors was more intense in blastocysts than in 2-cell embryos which suggests rising expression of AdipoR1 and AdipoR2 receptor protein in the course of pig preimplantation development.

In cows, transripts for adiponectin, AdipoR1 and AdipoR2 were detected in preovulatory oocytes isolated from small and large ovarian follicles as well as in *in vitro* matured oocytes (Maillard et al., 2010; Tabandeh, Hosseini, Saeb, Kafi, & Saeb, 2010). Preovulatory oocytes also expressed adiponectin and both receptors at the protein level. However, *in vitro* derived bovine blastocysts expressed clearly only AdipoR1 protein, while AdipoR2 and adiponectin were weakly present and undetectable, respectively (Maillard et al., 2010).

Expression of adiponectin and its receptors documented so far in mammalian oocytes and preimplantation embryos is summarized in Table 9.1.

5. POTENTIAL ROLES OF ADIPONECTIN IN OOCYTE AND EARLY EMBRYO DEVELOPMENT

5.1. Adiponectin in oocyte maturation

The oocyte develops in ovarian follicles and expression of adiponectin receptors, and/or adiponectin has been demonstrated in ovarian follicular cells (granulosa and cumulus cells) of several mammalian species (Chabrolle et al., 2007, 2009; Lord, Ledoux, Murphy, Beaudry, & Palin, 2005; Maillard et al., 2010; Tabandeh et al., 2010). Moreover, adiponectin protein has been detected in porcine, bovine, and human follicular fluid (Bersinger & Wunder, 2010; Ledoux et al., 2006; Maillard et al., 2010).

The influence of adiponectin on oocyte maturation has been investigated in pigs and cows. Significant decrease in frequency of meiotic immature oocytes was found in porcine cumulus-oocyte complexes isolated from large ovarian follicles and then matured *in vitro* in the presence of full-length porcine recombinant adiponectin. Examination of potential intracellular pathways showed that adiponectin enhanced maturation of porcine oocytes through the p38 mitogen-activated protein kinase (p38 MAPK).However, adiponectin had no stimulatory effect in oocytes derived from small follicles, and this result was found in pigs as well as in cows (Chappaz et al., 2008; Maillard et al., 2010). These results suggest that oocytes at different developmental stages have differing ability to respond to adiponectin. Since adiponectin receptors were detected in porcine and bovine oocytes/cumulus cells from both small and large follicles (Chappaz et al., 2008; Maillard et al., 2010), the absence of some components of adiponectin intracellular signaling in immature oocytes/cumulus cells could be the reason for the adiponectin insensitivity in small follicles. However, some interspecies differences in the adiponectin effects on oocyte maturation cannot be excluded either. The results of the two experiments (Chappaz et al., 2008; Maillard et al., 2010) suggest the possibility that adiponectin could be effective in pig but not in cow. Moreover, some differences in the expression of components of the adiponectin system in oocytes were found in mice (no expression of adiponectin, Čikoš et al., 2010), and rats and cows (expression of adiponectin, Chabrolle et al., 2007; Maillard et al., 2010; Tabandeh et al., 2010).

Besides oocyte maturation, adiponectin has been shown to influence several ovarian functions. For instance, porcine recombinant adiponectin stimulated expression of several genes (including genes involved in steroidogenesis) in porcine granulosa cells, and the effects of adiponectin were mediated by ERK1/2 mitogen-activated protein kinase cascade (Ledoux et al., 2006). The stimulatory effect of human recombinant adiponectin on insulin-like growth factor-I- induced production of progesterone and estradiol was found in human and rat but not in cow granulosa cells. Activation of several protein kinases by adiponectin, including p38 MAPK, protein kinase B (AKT/PKB), and adenosine monophosphate-activated protein kinase (AMPK/PRKA), was demonstrated in these cells (Chabrolle et al., 2007, 2009; Maillard et al., 2010). These results indicate that adiponectin can play an important role in modulation of ovarian function, influencing periovulatory remodeling of the ovarian follicle, and steroid synthesis/secretion.

5.2. Adiponectin in embryo preimplantation development and implantation

As described above, preimplantation embryos express adiponectin receptors, which is a prerequisite for activation of intracellular signaling cascades and induction of a cell response by adiponectin stimulus. Most of the preimplantation embryo development takes place in the oviduct, and the source of adiponectin in the oviduct fluid can be maternal blood circulation (where adiponectin is present in relatively high concentrations), oviduct epithelium or follicular fluid (Archanco et al., 2007; Bersinger & Wunder, 2010; Buhi, Alvarez, & Kouba, 2000; Guérin & Ménézo, 2011; Ledoux et al., 2006; Leese, 1988; Maillard et al., 2010; Rodgers & Irving-Rodgers, 2010). The distribution of adiponectin isoforms in the oviduct fluid is unknown, and there are results showing differences in the distribution of adiponectin isoforms between the serum and the follicular fluid (Bersinger & Wunder, 2010).

The role of adiponectin signaling in regulation of preimplantation development has been examined in experiments where adiponectin receptors were activated by addition of adiponectin protein to the embryos cultured *in vitro*. It is necessary to note that adiponectin protein used in *in vitro* experiments can originate from various sources (usually recombinant proteins prepared in mammalian or bacterial cells), and can represent various isoforms/posttranslational modification states of the protein. Many publications do not specify the isoform of adiponectin applied, so comparison of adiponectin effects by analyzing published data is rather problematic. Full-length recombinant adiponectin generated in mammalian cells is expected to be correctly glycosylated and hydroxylated and be able to form all three isoforms.

Table 9.2 Effects of adiponectin on oocyte maturation and preimplantation embryo development *in vitro*

Adiponectin protein	Species	Effect (↑ stimulatory, ↓inhibitory)	References
Full-length (10 μg/ml, [a]M)	Mouse	↑ preimplantation development	Čikoš et al. (2010)
Globular (10 μg/ml, [a]Ec)	Mouse	↔ preimplantation development	Čikoš et al. (2010)
Mutated trimeric full-length (10 μg/ml, [a]M)	Mouse	↓ preimplantation development	Čikoš et al. (2010)
[b]Full-length (10 μg/ml, [a]M)	Cow	↔ preimplantation development	Maillard et al. (2010)
[b]Full-length (10 μg/ml, [a]M)	Cow	↔ oocyte maturation	Maillard et al. (2010)
Full-length (30 μg/ml, [a]Ec)	Pig	↑ preimplantation development	Chappaz et al. (2008)
Full-length (30 μg/ml, [a]Ec)	Pig	↑ oocyte maturation	Chappaz et al. (2008)
[b]Full-length (1 μg/ml, [a]M)	Rabbit	[c]↑AMPK, [c]↓PCK2, [c]↑SLC2A4, [c]↑Glucupt blastocyst	Fischer et al. (2010)

The applied adiponectin concentration and the cell type used for production of recombinant adiponectin are indicated in brackets.
[a]M, mammalian cells; Ec, *Escherichia coli*.
[b]Probably full-length adiponectin, but not clearly stated in the paper.
[c]↑AMPK, increase of AMPK phosphorylation; ↓PCK2, decrease of PCK2 mRNA level; ↑SLC2A4, stimulation of SLC2A4 translocation; ↑Glucupt, increase of glucose uptake.

However, bacterially produced full-length recombinant adiponectin (lacking posttranslational modifications) can form only hexamers and trimers (Tsao et al., 2003; Wang et al., 2008). Effects of adiponectin on oocyte maturation and preimplantation embryo development found in *in vitro* experiments are summarized in Table 9.2.

5.2.1 Cell proliferation and survival

The capability of adiponectin to influence cell proliferation and survival has been reported in various cell types, and the resulting cell response can be cell type dependent. The involvement of several protein kinases, including mitogen-activated protein kinases (MAPKs) and AKT/PKB, have been found in the regulation of proliferation and apoptosis, and some differences in the effectiveness of particular adiponectin isoforms have been documented. For instance, the stimulatory effect of adiponectin on cell proliferation has been shown in osteoblasts and hemopoietic stem cells (Berner et al., 2004; DiMascio et al., 2007; Luo et al., 2005), whereas a growth-inhibiting effect of adiponectin has been demonstrated in several cancer cell lines (breast, prostate, human choriocarcinoma; Benaitreau et al., 2009; Bub et al., 2006; Dos Santos et al., 2008; Grossmann, Nkhata, Mizuno, Ray, & Cleary, 2008). Similarly, stimulatory or inhibitory effects of adiponectin on cell survival and/or apoptosis have been found in various cell types (Cong et al., 2007; Grossmann et al., 2008; Körner et al., 2007; Shibata et al., 2005; Sun & Chen, 2010; Wijesekara et al., 2010).

As in other cell types, different adiponectin isoforms can have different effects on development of preimplantation embryos. Čikoš et al. (2010) compared effects of full-length adiponectin with the effects of two other protein isoforms—truncated globular adiponectin and mutated full-length adiponectin (with cysteine 39 replaced by alanine) which can only form trimer, but not hexamer or high molecular weight forms. The full-length adiponectin administered to mouse preimplantation embryos *in vitro* significantly changed the embryo distribution, increasing the proportion of embryos with high cell numbers and decreasing the proportion of embryos with lower cell numbers. In contrast, administration of trimeric adiponectin decreased the proportion of embryos with higher cell numbers and increased the proportion of embryos with lower cell numbers. Analysis of cell numbers in embryos that reached the blastocyst stage confirmed the opposite effects of the two adiponectin isoforms: incubation of embryos with full-length adiponectin led to significant increase in embryo cell number, whereas incubation of embryos with trimeric adiponectin led to significant reduction of the cell number in blastocysts. Moreover, the administration of trimeric adiponectin significantly increased the proportion of dead cells in the blastocysts. However, no significant effect of globular adiponectin on mouse preimplantation embryos was found.

The positive effect of full-length adiponectin on preimplantation embryo development has also been demonstrated in pigs. Administration of full-length porcine recombinant adiponectin stimulated *in vitro* development of parthenogenetically activated porcine embryos to the blastocyst stage, and examination of porcine oocytes suggested that p38 MAPK could be involved in the adiponectin signaling (Chappaz et al., 2008). However, Maillard et al. (2010) found no effect of human recombinant adiponectin on bovine preimplantation embryo development *in vitro*, which suggest some possible species-specific effects of adiponectin. Another explanation for this result could be the difference in amino acid sequences between human and bovine adiponectin which could result in lower effectiveness of human adiponectin in bovine cells.

The different effects of the three adiponectin isoforms found in the experiment of Čikoš et al. (2010) could result from different intracellular signaling cascades induced by particular adiponectin isoforms, or they also could relate to differences in the abundance of AdipoR1 and AdipoR2 receptor in the embryo. It has been demonstrated that AdipoR1 is a high-affinity receptor for globular adiponectin and a low-affinity receptor for full-length adiponectin, whereas AdipoR2 is an intermediate-affinity receptor for full-length and globular adiponectin (Yamauchi et al., 2003). The lower immunofluorescent signal intensity for AdipoR1 than for AdipoR2 found at most tested developmental stages (Čikoš et al. 2010; Kim et al., 2011) suggests lower abundance of the AdipoR1 receptor, which could explain at least in part the lower effectiveness of the globular adiponectin found in mouse preimplantation embryos. The opposite effects of the two full-length adiponectin isoforms suggest that trimeric adiponectin and the adiponectin capable of forming higher multimers can trigger different signaling pathways in preimplantation embryo cells. Moreover, the modulation of actions of growth factors by adiponectin, which can be isoform-specific (Wang et al., 2005), could contribute to adiponectin effects *in vivo*.

5.2.2 Lipid and glucose metabolism

Experiments examining signaling pathways triggered by adiponectin in cells of preimplantation embryos demonstrated expression and/or activity of several molecules which are involved in adiponectin signaling in other cell types. Peroxisome proliferator-activated receptors (PPARs, nuclear receptors/transcription factors involved in regulation of lipid and carbohydrate metabolism) function as downstream as well as upstream components of the adiponectin signaling cascade (Amin, Mathews, Alli, & Leff, 2010; Kadowaki et al., 2006; Tsuchida et al., 2005; Varga, Czimmerer, & Nagy, 2011; Yamauchi et al., 2003; Yoon et al., 2006), and transcripts for PPARα and PPARδ have been detected in mouse 2-cell embryos, 8-cell embryos, and blastocysts (Kim et al., 2011). Adenosine monophosphate-activated protein kinase (AMPK/PRKA, a key regulator of energy homeostasis in the cell) has been shown to mediate effects of adiponectin on carbohydrate and lipid metabolism in different tissues (Kahn, Alquier, Carling, & Hardie, 2005; Palanivel et al., 2007; Yamauchi et al., 2002, 2003), and mouse blastocysts were found to express the activated (Thr172-phosphorylated) form of this kinase (Kim et al., 2011). The ability of adiponectin to activate AMPK/PRKA in cells of early embryos was demonstrated in rabbit blastocysts, where adiponectin treatment significantly increased AMPK/PRKA phosphorylation at Thr172 (Fischer et al., 2010). These results indicate that adiponectin receptors expressed in blastocysts can respond to adiponectin stimulus by activation of signaling pathways through AMPK/PRKA. Among various downstream targets of AMPK/PRKA those connected with carbohydrate and lipid metabolism have been studied most extensively. In general, activation of AMPK/PRKA switches on catabolic pathways that generate ATP (i.e., fatty acid uptake and oxidation, glucose uptake and glycolysis) and switches off ATP-consuming anabolic pathways (i.e., fatty acid synthesis, gluconeogenesis). These effects are achieved by rapid phosphorylation of metabolic enzymes and via long-term influence on gene expression (Kahn et al., 2005; Steinberg & Kemp, 2009)

Examination of components of lipid metabolism in preimplantation embryos has shown that transcripts for fatty acid transport protein-1 (FATP1/SLC27A1) and acyl CoA oxidase (ACOX1), the first enzyme of the fatty-acid beta oxidation pathway in peroxisomes, are expressed in mouse 2-cell embryos, 8-cell embryos, and blastocysts (Kim et al., 2011). Upregulation of FATP-1 and ACOX mRNAs by adiponectin has been demonstrated in muscle cells (Maeda et al., 2002; Yamauchi et al., 2001). In addition, increased fatty acid oxidation induced by sequential activation of AMPK/PRKA, p38 MAPK, and PPARα (connected with increased expression ACOX and other genes) has been found in skeletal muscle cells (Yoon et al., 2006). Based on these observations, it may be hypothesized that adiponectin stimulates fatty-acid beta oxidation in preimplantation embryos.

The influence of adiponectin on glucose metabolism, via changes in gene expression and glucose transport, has been demonstrated in rabbit blastocysts. Adiponectin added to culture medium decreased mRNA levels of phosphoenolpyruvate carboxykinase (PKC2, a key enzyme of gluconeogenesis) in embryoblast as well as in trophoblast cells, and the PCK2 decrease was blocked by a specific AMPK/PRKA inhibitor. These results indicate that adiponectin can inhibit gluconeogenesis in cells of rabbit blastocysts by activation of a signaling pathway which involves AMPK/PRKA and results in decrease of PKC2 gene transcription or/and mRNA stability

(Fischer et al., 2010). Preimplantation embryos of various mammalian species express several facilitative glucose transporters (members of solute carrier family 2A, SLC2A/GLUT; Augustin et al., 2001; Dan-Goor, Sasson, Davarashvili, & Almagor, 1997; Fischer et al., 2010; Kim et al., 2011; Korgun et al., 2001; Navarrete Santos, Tonack, Kirstein, Kietz, & Fischer, 2004; Tonack, Fischer, & Navarrete Santos, 2004), and experiments using rabbit blastocysts have shown that adiponectin can stimulate translocation of glucose transporter SLC2A4/GLUT4 from cytoplasm to the cell membrane. In addition, the treatment of rabbit blastocysts with adiponectin significantly stimulated glucose uptake, which corresponds with the glucose transporter translocation, and is also consistent with the inhibitory effect of adiponectin on gluconeogenesis (Fischer et al., 2010). Interestingly, insulin stimulus neither induced SLC2A4/GLUT4 translocation nor increased glucose uptake in rabbit blastocysts (Navarrete Santos, Tonack, et al., 2004), which contrasts with the mouse, where insulin is able to stimulate blastocyst glucose uptake via IGF-I receptor-mediated translocation of SLC2A8/GLUT8 glucose transporter (Carayannopoulos et al., 2000; Chi, Schlein, & Moley, 2000; Pantaleon & Kaye, 1996; Pinto, Carayannopoulos, Hoehn, Dowd, & Moley, 2002). These results suggest possible interspecies differences in regulation of preimplantation embryo glucose uptake, which seems to be insulin and IGFI dependent in the mouse and insulin independent/ adiponectin stimulated in the rabbit. Based on these observations, it has been proposed that adiponectin could replace classical insulin effects in rabbit preimplantation embryos (Fischer et al., 2010).

Signaling pathways leading from activated adiponectin receptors to membrane translocation of glucose transporters are not known in preimplantation embryos, and there are data indicating that besides AMPK, AKT/PKB can participate in this process. AKT/PKB is a well-known mediator of insulin metabolic effects. Occupation of the insulin receptor by its ligand (insulin) triggers a signaling cascade which involves activation of phosphatidyl inositol 3-kinase (PI3K) and AKT/PKB, and results in several effects including SLC2A4/GLUT4 translocation and increased glucose uptake (Watson & Pessin, 2001; Welsh et al., 2005). The importance of the PI3K–AKT/PKB pathway in regulation of glucose utilization and preimplantation embryo development has been clearly demonstrated in the mouse (Riley et al., 2005; Riley, Carayannopoulos, Wyman, Chi, & Moley, 2006). Moreover, the possibility of AKT/PKB activation by AMPK/PRKA in the process of glucose uptake regulation has been demonstrated in insulin-resistant TS cells derived from mouse blastocysts (Louden, Chi, & Moley, 2008). The data from several cell types indicate that AKT/PKB can be activated by adiponectin too (Coope et al., 2008; Grossmann et al., 2008; Maruyama et al., 2011; Wijesekara et al., 2010). Moreover, the activation of both AKT/PKB and AMPK/PRKA by adiponectin has been found in endothelial and prostate cancer cells (Barb, Neuwirth, Mantzoros, & Balk, 2007; Chen, Montagnani, Funahashi, Shimomura, & Quon, 2003; Xiao et al., 2011), and results obtained in endothelial cells indicate that AMPK/PRKA can be upstream of AKT/PKB in the adiponectin signaling cascade (Ouchi et al., 2004). In accordance with these observations, adiponectin treatment led to increased phosphorylation of both AMPK/PRKA and AKT1/PKBα in rabbit blastocysts (Fischer et al., 2010).

These results provide first insights into our understanding of adiponectin signaling (especially in the process of regulation of lipid and glucose metabolism) in preimplantation embryos. Examination of adiponectin signaling in various mammalian cell types has revealed that APPL1 (adaptor protein containing pleckstrin homology domain-1) is the first protein interacting with adiponectin receptors, and this protein has been shown to activate a variety of downstream targets, including AMPK/PRKA, Rab5 (a small GTPase which can induce GLUT4 membrane translocation), PI3K, and AKT/PKB (Deepa & Dong, 2009; Mao et al., 2006). Considering the knowledge of APPL1 function and the available information on adiponectin signaling in preimplantation embryos, we can suppose that several signaling cascades may be involved in adiponectin regulation of fatty acid oxidation and glucose uptake in the early embryo. Adiponectin may stimulate fatty-acid beta oxidation in preimplantation embryos via AMPK/PRKA and PPARs, upregulating enzymes of fatty-acid oxidation. Regulation of glucose uptake (by membrane translocation of glucose transporters) can involve AdipoR1/2-AMPK/PRKA-AKT/PGB-GLUTs signaling cascade, but AMPK/PRKA-independent AKT/PKB signaling cannot be excluded either. Moreover, the direct interaction of APPL1 with Rab5 (after adiponectin receptor activation) could eliminate the necessity for other mentioned signaling molecules in the process of glucose uptake regulation by adiponectin. Finally, it is necessary to note that some interspecies differences in mechanisms of adiponectin action in preimplantation embryos can be expected. Differences in regulation of the PI3K–AKT/PKB pathway and glucose uptake by insulin have been demonstrated in mouse and rabbit blastocysts, for instance (Carayannopoulos et al., 2000; Navarrete Santos, Tonack, et al., 2004; Pantaleon & Kaye, 1996; Riley et al., 2005).

5.2.3 Embryo implantation

Adiponectin can also play an important role in the implantation process which finishes the preimplantation developmental period. Transcripts and proteins for adiponectin, AdipoR1 and AdipoR2 have been detected in uterine tissue of pregnant mice and rabbits during the periimplantation period (Kim et al., 2011; Schmidt et al., 2008). In the time before embryo implantation, adiponectin protein was detected mainly in the epithelial cells of mouse uterine glands. In rabbits, adiponectin protein was detected in glandular and luminal epithelial cells as well as in stromal cells of the endometrium in this period. After the start of rabbit embryo implantation, adiponectin protein showed strong expression in glands close to the implantation site, and the protein was localized in the implanting blastocyst as well (Schmidt et al., 2008). After the start of mouse embryo implantation, adiponectin, AdipoR1 and AdipoR2 were mainly detected in decidual cells of the implantation sites and in epithelial cells of the interimplantation sites. Expression of all three genes was significantly higher at the implantation sites than at the interimplantation sites (Kim et al., 2011).

In humans, transcripts and proteins for both adiponectin receptors (AdipoR1 and AdipoR2) and adiponectin mRNA have been detected in endometrial tissue, endometrial stromal, and epithelial cells. Comparison of endometrial AdipoR1 and AdipoR2 mRNA expression throughout the menstrual cycle revealed increased levels of both transcripts in the mid-luteal phase, which corresponds with the period of embryo implantation (Takemura et al., 2006).

Examination of the distribution of adiponectin and its receptors within the endometrium of pregnant mice and rabbits indicates that adiponectin signaling is higher at the implantation sites (localized mainly in decidual cells) than at the interimplantation sites (localized mainly in epithelial cells). These results suggest that adiponectin and its receptors are important for the implantation process, but they are also basally expressed in epithelial cells (Kim et al., 2011; Schmidt et al., 2008). The stimulatory effect of implantation and decidualization on the expression of adiponectin, AdipoR1 and AdipoR2 in the uterine tissue was confirmed with the use of mouse model of delayed implantation, artificial decidualization, and *in vitro* decidualization system. The results showed that mainly decidual cells are responsible for the adiponectin, AdipoR1 and AdipoR2 increase in the mouse uterine tissue, and that decidual cells are able to secrete adiponectin protein (Kim et al., 2011). The secreted adiponectin can then influence the implanting embryo as well as endometrial functions in a paracrine/autocrine manner. Activation of AMPK/PRKA and inhibition of proinflammatory cytokine secretion found in cultured human endometrial cells after adiponectin treatment suggests that adiponectin may regulate the energy supply and inflammatory response of endometrial cells (Takemura et al., 2006). Moreover, adiponectin could contribute to the process of decidualization by increasing glucose intake (via GLUTs membrane translocation) in endometrial stromal cells. Thus, adiponectin may play an important role in both preimplantation embryo development and endometrial changes which render the uterus favorable for embryo implantation. Adiponectin and its receptors are further expressed in embryonic as well as placental tissues, which indicate the continuous role of adiponectin in prenatal development. The function of adiponectin in the regulation of glucose and triglycerides transport to the embryo as well as in the regulation of proliferation and invasion of trophoblast cells has been proposed (Caminos et al., 2005; Chen et al., 2006; Corbetta et al., 2005; Struwe, Berzl, Schild, & Dötsch, 2009; Zhou et al., 2005).

6. CONCLUSIONS

Experimental data indicate that adiponectin can play an important role in regulating the female reproductive function and embryo development. Adiponectin can affect embryonal development at very early stages of pregnancy—during the preimplantation period. This developmental period is one of the most sensitive phases in mammalian ontogeny and can be decisive for the pregnancy outcome. Disturbances in the maternal adiponectin system are associated with several diseases, including diabetes type 2, obesity, and some female reproductive disorders, and can negatively influence early embryo development.

Adiponectin can play multiple roles in preimplantation embryo development and implantation in an autocrine/paracrine/endocrine manner. Available data indicate that the actions of adiponectin on lipid and glucose metabolism can result in the increase of energy supply to the embryo. This is significant because sufficient glucose intake can be important not only for energy production but also for cell survival of the early embryo. Glucose deprivation can trigger an apoptotic cascade in the cells of mouse preimplantation embryos, and may manifest later in pregnancy as a malformation or miscarriage (Chi, Pingsterhaus, Carayannopoulos, & Moley, 2000; Riley et al., 2006; Riley & Moley, 2006). Experimental results have demonstrated that adiponectin effects on proliferation and survival of cells of preimplantation embryos are isoform dependent. The molecular mechanisms of these processes are unknown at present, however, and can only be anticipated based on the information available from other cell types. Further studies are necessary to complete our understanding of adiponectin signaling in cells of preimplantation embryos and the role of adiponectin in early embryo development.

ACKNOWLEDGMENT

This work was supported by the Slovak Research and Development Agency under contract APVV-0815-11.

REFERENCES

Abrahamsohn, P. A., & Zorn, T. M. (1993). Implantation and decidualization in rodents. *The Journal of Experimental Zoology*, *266*, 603–628.

Amin, R. H., Mathews, S. T., Alli, A., & Leff, T. (2010). Endogenously produced adiponectin protects cardiomyocytes from hypertrophy by a PPARgamma-dependent autocrine mechanism. *American Journal of Physiology. Heart and Circulatory Physiology*, *299*, 690–698.

Archanco, M., Gómez-Ambrosi, J., Tena-Sempere, M., Frühbeck, G., & Burrell, M. A. (2007). Expression of leptin and adiponectin in the rat oviduct. *The Journal of Histochemistry and Cytochemistry*, *55*, 1027–1037.

Ardawi, M. S. M., Path, F. R. C., & Rouzi, A. A. (2005). Plasma adiponectin and insulin resistance in women with polycystic ovary syndrome. *Fertility and Sterility*, *83*, 1708–1715.

Augustin, R., Pocar, P., Navarrete-Santos, A., Wrenzycki, C., Gandolfi, F., Niemann, H., et al. (2001). Glucose transporter expression is developmentally regulated in in vitro derived bovine preimplantation embryos. *Molecular Reproduction and Development*, *60*, 370–376.

Barb, D., Neuwirth, A., Mantzoros, C. S., & Balk, S. P. (2007). Adiponectin signals in prostate cancer cells through Akt to activate the mammalian target of rapamycin pathway. *Endocrine-Related Cancer*, *14*, 995–1005.

Benaitreau, D., Dieudonné, M. N., Santos, E. D., Leneveu, M. Ch., de Mazancourt, P., & Pecquery, R. (2009). Antiproliferative effects of adiponectin on human trophoblastic cell lines JEG-3 and BeWo. *Biology of Reproduction*, *80*, 1107–1114.

Berner, H. S., Lyngstadaas, S. P., Spahr, A., Monjo, M., Thommesen, L., Drevon, Ch. A., et al. (2004). Adiponectin and its receptors are expressed in bone-forming cells. *Bone*, *35*, 842–849.

Bersinger, N. A., Birkhäuser, M. H., & Wunder, D. M. (2006). Adiponectin as a marker of success in intracytoplasmic sperm injection/embryo transfer cycles. *Gynecological Endocrinology*, *22*, 479–483.

Bersinger, N. A., & Wunder, D. M. (2010). Adiponectin isoform distribution in serum and follicular fluid of women undergoing treatment by ICSI. *Acta Obstetricia et Gynecologica Scandinavica, 89*, 782–788.

Blomberg, L., Hashizume, K., & Viebahn, C. (2008). Blastocyst elongation, trophoblastic differentiation, and embryonic pattern formation. *Reproduction, 135*, 181–195.

Bower, N. I., Moser, R. J., Hill, J. R., & Lehnert, S. A. (2007). Universal reference method for real-time PCR gene expression analysis of preimplantation embryos. *BioTechniques, 42*, 199–206.

Bub, J. D., Miyazaki, T., & Iwamoto, Y. (2006). Adiponectin as a growth inhibitor in prostate cancer cells. *Biochemical and Biophysical Research Communications, 340*, 1158–1166.

Buhi, W. C., Alvarez, I. M., & Kouba, A. J. (2000). Secreted proteins of the oviduct. *Cells, Tissues, Organs, 166*, 165–179.

Caminos, J. E., Nogueiras, R., Gallego, R., Bravo, S., Tovar, S., Garcia-Caballero, T., et al. (2005). Expression and regulation of adiponectin and receptor in human and rat placenta. *The Journal of Clinical Endocrinology and Metabolism, 90*, 4276–4286.

Campos, D. B., Palin, M. F., Bordignon, V., & Murphy, B. D. (2008). The "beneficial" adipokines in reproduction and fertility. *International Journal of Obesity, 32*, 223–231.

Carson, D. D., Bagchi, I., Dey, S. K., Enders, A. C., Fazleabas, A. T., Lessey, B. A., et al. (2000). Embryo implantation. *Developmental Biology, 223*, 217–237.

Carayannopoulos, M. O., Chi, M. M., Cui, Y., Pingsterhaus, J. M., McKnight, R. A., Mueckler, M., et al. (2000). GLUT8 is a glucose transporter responsible for insulin-stimulated glucose uptake in the blastocyst. *Proceedings of the National Academy of Sciences of the United States of America, 97*, 7313–7318.

Chabrolle, Ch., Tosca, L., & Dupont, J. (2007). Regulation of adiponectin and its receptors in rat ovary by human chorionic gonadotrophin treatment and potential involment of adiponectin in granulosa cell steroidogenesis. *Reproduction, 133*, 719–731.

Chabrolle, Ch., Tosca, L., Ramé, C., Lecomte, P., Royere, D., & Dupont, J. (2009). Adiponectin increases insulin-like growth factor I-induced progesterone and estradiol in human granulosa cells. *Fertility and Sterility, 92*, 1988–1996.

Chappaz, E., Albornoz, M. S., Campos, D., Che, L., Palin, M. F. D., Murphy, B., et al. (2008). Adiponectin enhances in vitro development of swine embryos. *Domestic Animal Endocrinology, 35*, 198–207.

Chen, H., Montagnani, M., Funahashi, T., Shimomura, I., & Quon, M. J. (2003). Adiponectin stimulates production of nitric oxide in vascular endothelial cells. *The Journal of Biological Chemistry, 278*, 45021–45026.

Chen, J., Tan, B., Karteris, E., Zervou, S., Digby, J., Hillhouse, E. W., et al. (2006). Secretion of adiponectin by human placenta: Differential modulation of adiponectin and its receptors by cytokines. *Diabetologia, 49*, 1292–1302.

Chi, M. M., Schlein, A. L., & Moley, K. H. (2000). High insulin-like growth factor 1 (IGF-1) and insulin concentrations trigger apoptosis in the mouse blastocyst via down-regulation of the IGF-1 receptor. *Endocrinology, 141*, 4784–4792.

Chi, M. M., Pingsterhaus, J., Carayannopoulos, M., & Moley, K. H. (2000). Decreased glucose transporter expression triggers BAX-dependent apoptosis in the murine blastocyst. *The Journal of Biological Chemistry, 275*, 40252–40257.

Čikoš, Š., Burkuš, J., Bukovská, A., Fabian, D., Rehák, P., & Koppel, J. (2010). Expression of adiponectin receptors and effects of adiponectin isoforms in mouse preimplantation embryos. *Human Reproduction, 25*, 2247–2255.

Čikoš, Š., & Koppel, J. (2009). Transformation of real-time PCR fluorescence data to target gene quantity. *Analytical Biochemistry, 384*, 1–10.

Combs, T. P., Berg, A. H., Rajala, M. W., Klebanov, S., Iyengar, P., Jimenez-Chillaron, J. C., et al. (2003). Sexual differentiation, pregnancy, calorie restriction, and aging affect the adipocyte-specific secretory protein adiponectin. *Diabetes, 52*, 268–274.

Cong, L., Gasser, J., Zhao, J., Yang, B., Li, F., & Zhao, A. Z. (2007). Human adiponectin inhibits cell growth and induces apoptosis in human endometrial carcinoma cells, HEC-1-A and RL95-2. *Endocrine-Related Cancer, 14*, 713–720.
Coope, A., Milanski, M., Araújo, E. P., Tambascia, M., Saad, M. J., Geloneze, B., et al. (2008). AdipoR1 mediates the anorexigenic and insulin/leptin-like actions of adiponectin in the hypothalamus. *FEBS Letters, 582*, 1471–1476.
Corbetta, S., Bulfamante, G., Cortelazzi, D., Barresi, V., Cetin, I., Mantovani, G., et al. (2005). Adiponectin expression in human fetal tissues during mid- and late gestation. *The Journal of Clinical Endocrinology and Metabolism, 90*, 2397–2402.
Dan-Goor, M., Sasson, S., Davarashvili, A., & Almagor, M. (1997). Expression of glucose transporter and glucose uptake in human oocytes and preimplantation embryos. *Human Reproduction, 12*, 2508–2510.
Dal Maso, L., Augustin, L. S. A., Karalis, A., Talamini, R., Franceschi, S., Trichopoulos, D., et al. (2004). Circulating adiponectin and endometrial cancer risk. *The Journal of Clinical Endocrinology and Metabolism, 89*, 1160–1163.
Deepa, S. S., & Dong, L. Q. (2009). APPL1: Role in adiponectin signaling and beyond. *American Journal of Physiology. Endocrinology and Metabolism, 296*, E22–E36.
DiMascio, L., Voermans, C., Uqoezwa, M., Duncan, A., Lu, D., Wu, J., et al. (2007). Identification of adiponectin as a novel hemopoietic stem cell growth factor. *Journal of Immunology, 178*, 3511–3520.
Dos Santos, E., Benaitreau, D., Dieudonne, M. N., Leneveu, M. Ch., Serazin, V., Giudicelli, Y., et al. (2008). Adiponectin mediates an antiproliferative response in human MDA-MB 231 breast cancer cells. *Oncology Reports, 20*, 971–977.
Duranthon, V., Watson, A. J., & Lonergan, P. (2008). Preimplantation embryo programming: Transcription, epigenetics, and culture environment. *Reproduction, 135*, 141–150.
Fedorcsák, P., Dale, P. O., Storeng, R., Ertzeid, G., Bjercke, S., Oldereid, N., et al. (2004). Impact of overweight and underweight on assisted reproduction treatment. *Human Reproduction, 19*, 2523–2528.
Fischer, S., Santos, A. N., Thieme, R., Ramin, N., & Fischer, B. (2010). Adiponectin stimulates glucose uptake in rabbit blastocysts. *Biology of Reproduction, 83*, 859–865.
Fleming, T. P., Wilkins, A., Mears, A., Miller, D. J., Thomas, F., Ghassemifar, M. R., et al. (2004). Society for Reproductive Biology Founders' Lecture 2003. The making of an embryo: Short-term goals and long-term implications. *Reproduction, Fertility, and Development, 16*, 325–337.
Ford, S. P., & Youngs, C. R. (1993). Early embryonic development in prolific Meishan pigs. *Journal of Reproduction and Fertility. Supplement, 48*, 271–278.
Grossmann, M. E., Nkhata, K. J., Mizuno, N. K., Ray, A., & Cleary, M. P. (2008). Effects of adiponectin on breast cancer cell growth and signaling. *British Journal of Cancer, 98*, 370–379.
Guérin, P., & Ménézo, Y. (2011). Review: Role of tubal environment in preimplantation embryogenesis: Application to co-culture assays. *Zygote, 19*, 47–54.
Hada, Y., Yamauchi, T., Waki, H., Tsuchida, A., Hara, K., Yago, H., et al. (2007). Selective purification and characterization of adiponectine multimer species from human plasma. *Biochemical and Biophysical Research Communications, 356*, 487–493.
Hardy, K., & Spanos, S. (2002). Growth factor expression and function in the human and mouse preimplantation embryo. *The Journal of Endocrinology, 172*, 221–236.
Haugen, F., & Drevon, Ch. (2007). Activation of nuclear factor-κB by high molecular weight and globular adiponectin. *Endocrinology, 148*, 5478–5486.
Heikinheimo, O., & Gibbons, W. E. (1998). The molecular mechanisms of oocyte maturation and early embryonic development are unveiling new insights into reproductive medicine. *Molecular Human Reproduction, 4*, 745–756.
Huggett, J., Dheda, K., Bustin, S., & Zumla, A. (2005). Real-time RT-PCR normalisation; strategies and considerations. *Genes and Immunity, 6*, 279–284.

Humblot, P. (2001). Use of pregnancy specific proteins and progesterone assays to monitor pregnancy and determine the timing, frequencies and sources of embryonic mortality in ruminants. *Theriogenology, 56*, 1417–1433.

Jeong, Y. J., Choi, H. W., Shin, H. S., Cui, X. S., Kim, N. H., Gerton, G. L., et al. (2005). Optimization of real time RT-PCR methods for the analysis of gene expression in mouse eggs and preimplantation embryos. *Molecular Reproduction and Development, 71*, 284–289.

Jiménez–Ayala, M., & Jiménez–Ayala Portillo, B. (2008). Cytology of the normal endometrium – cycling and postmenopausal. In: M. Jiménez–Ayala, B. Jiménez–Ayala Portillo, E. Iglesias & M. R. Vallejo (Eds.), *"Endometrial adenocarcinoma: prevention and early diagnosis, Monogr. Clin. Cytol."*, Vol. 17, (pp. 32–39). Basel: Karger AG.

Kadowaki, T., & Yamauchi, T. (2005). Adiponectin and adiponectin receptors. *Endocrine Reviews, 26*, 439–451.

Kadowaki, T., Yamauchi, T., Kubota, N., Hara, K., Ueki, K., & Tobe, K. (2006). Adiponectin and adiponectin receptors in insulin resistance, diabetes, and the metabolic syndrome. *The Journal of Clinical Investigation, 116*, 1784–1792.

Kahn, B. B., Alquier, T., Carling, D., & Hardie, D. G. (2005). AMP-activated protein kinase: Ancient energy gauge provides clues to modern understanding of metabolism. *Cell Metabolism, 1*, 15–25.

Kidder, G. M. (1992). The genetic program for preimplantation development. *Developmental Genetics, 13*, 319–325.

Kierszenbaum, A. L. (2001). Decidualization and implantation: embryo-uterine bioinformatics at work. *Molecular Reproduction and Development, 59*, 123–125.

Kim, J. J., Jaffe, R. C., & Fazleabas, A. T. (1999). Blastocyst invasion and the stromal response in primates. *Human Reproduction, 2*, 45–55.

Kim, S. T., Marquard, K., Stephens, S., Louden, E., Allsworth, J., & Moley, K. H. (2011). Adiponectin and adiponectin receptors in the mouse preimplantation embryo and uterus. *Human Reproduction, 26*, 82–95.

Kobayashi, H., Ouchi, N., Kihara, S., Walsh, K., Kumada, M., Abe, Y., et al. (2004). Selective suppression of endothelial cell apoptosis by the high molecular weight form oj adiponectin. *Circulation Research, 94*, 27–31.

Korgun, E. T., Demir, R., Hammer, A., Dohr, G., Desoye, G., Skofitsch, G., et al. (2001). Glucose transporter expression in rat embryo and uterus during decidualization, implantation, and early postimplantation. *Biology of Reproduction, 65*, 1364–1370.

Körner, A., Pazaitou-Panayiotou, K., Kelesidis, T., Kelesidis, I., Williams, C. J., Kaprara, A., et al. (2007). Total and high-molecular-weight adiponectin in breast cancer: in vitro and in vivo studies. *The Journal of Clinical Endocrinology and Metabolism, 92*, 1041–1048.

Kubota, N., Terauchi, Y., Yamauchi, T., Kubota, T., Moroi, M., Matsui, J., et al. (2002). Disruption of adiponectin causes insulin resistance and neointimal formation. *The Journal of Biological Chemistry, 277*, 25863–25866.

Ledoux, S., Campos, D. B., Lopes, F. L., Dobias-Goff, M., Palin, M. F., & Murphy, B. D. (2006). Adiponectin induces periovulatory changes in ovarian follicular cells. *Endocrinology, 147*, 5178–5186.

Leese, H. J. (1988). The formation and function of oviduct fluid. *Journal of Reproduction and Fertility, 82*, 843–856.

Lord, E., Ledoux, S., Murphy, B. D., Beaudry, D., & Palin, M. F. (2005). Expression of adiponectin and its receptors in swine. *Journal of Animal Science, 83*, 565–578.

Louden, E., Chi, M. M., & Moley, K. H. (2008). Crosstalk between the AMP-activated kinase and insulin signaling pathways rescues murine blastocyst cells from insulin resistance. *Reproduction, 136*, 335–344.

Luo, X. H., Guo, L. J., Yuan, L. Q., Xie, H., De Zhou, H., Wu, X. P., et al. (2005). Adiponectin stimulates human osteoblasts proliferation and differentiation via the MAPK signaling pathway. *Experimental Cell Research, 309*, 99–109.

Ma, K., Cabrero, A., Saha, P. K., Kojima, H., Li, L., Chang, B. H. J., et al. (2002). Increased β-oxidation but no insulin resistance or glucose intolerance in mice lacking adiponectin. *The Journal of Biological Chemistry*, *277*, 34658–34661.

Maddox-Hyttel, P., Alexopoulos, N. I., Vajta, G., Lewis, I., Rogers, P., Cann, L., et al. (2003). Immunohistochemical and ultrastructural characterization of the initial post-hatching development of bovine embryos. *Reproduction*, *125*, 607–623.

Maeda, N., Shimomura, I., Kishida, K., Nishizawa, H., Matsuda, M., Nagaretani, H., et al. (2002). Diet-induced insulin resistance in mice lacking adiponectin/ACRP30. *Nature Medicine*, *8*, 731–737.

Maillard, V., Uzbekova, S., Guignot, F., Perreau, Ch., Ramé, Ch., Coyral-Castel, S., et al. (2010). Effect of adiponectin on bovine granulose cell steroidogenesis, oocyte maturation and embryo development. *Reproductive Biology and Endocrinology*, *8*, 23.

Mamo, S., Gal, A. B., Bodo, S., & Dinnyes, A. (2007). Quantitative evaluation and selection of reference genes in mouse oocytes and embryos cultured in vivo and in vitro. *BMC Developmental Biology*, *7*, 14.

Mao, X., Kikani, C. K., Riojas, R. A., Langlais, P., Wang, L., Ramos, F. J., et al. (2006). APPL1 binds to adiponectin receptors and mediates adiponectin signalling and function. *Nature Cell Biology*, *8*, 516–523.

Maruyama, S., Shibata, R., Ohashi, K., Ohashi, T., Daida, H., Walsh, K., et al. (2011). Adiponectin ameliorates doxorubicin-induced cardiotoxicity through Akt protein-dependent mechanism. *The Journal of Biological Chemistry*, *286*, 32790–32800.

Mazaki-Tovi, S., Romero, R., Vaisbuch, E., Kusanovic, J. P., Erez, O., Gotsch, F., et al. (2009). Maternal serum adiponectin multimers in preeclampsia. *Journal of Perinatal Medicine*, *37*, 349–363.

Mitchell, M., Armstrong, D. T., Robker, R. L., & Norman, R. J. (2005). Adipokines: Implications for female fertility and obesity. *Reproduction*, *130*, 583–597.

Navarrete Santos, A., Tonack, S., Kirstein, M., Kietz, S., & Fischer, B. (2004). Two insulin-responsive glucose transporter isoforms and the insulin receptor are developmentally expressed in rabbit preimplantation embryos. *Reproduction*, *128*, 503–516.

Navarrete Santos, A., Tonack, S., Kirstein, M., Pantaleon, M., Kaye, P., & Fischer, B. (2004). Insulin acts via mitogen-activated protein kinase phosphorylation in rabbit blastocysts. *Reproduction*, *128*, 517–526.

Nepomnaschy, P. A., Welch, K. B., McConnell, D. S., Low, B. S., Strassmann, I., & England, B. G. (2006). Cortisol levels and very early pregnancy loss in humans. *Proc. Nat. Acad. Sci. U.S.A.*, *103*, 3938–3942.

Neumeier, M., Weigert, J., Schäffler, A., Wehrwein, G., Müller-Ladner, U., Schölmerich, J., et al. (2006). Different effects of adiponectin isoforms in human monocytic cells. *Journal of Leukocyte Biology*, *79*, 803–808.

Nishizawa, H., Shimomura, I., Kishida, K., Maeda, N., Kuriyama, H., Nagaretani, H., et al. (2002). Androgens decrease plasma adiponectin, an insulin-senzitizing adipocyte-derived protein. *Diabetes*, *51*, 2734–2741.

Norman, R. J., Noakes, M., Wu, R., Davies, M. J., Moran, L., & Wang, J. X. (2004). Improving reproductive performance in overweight/obese women with effective weight management. *Human Reproduction Update*, *10*, 267–280.

Nothias, J. Y., Majumder, S., Kaneko, K. J., & DePamphilis, M. L. (1995). Regulation of gene expression at the beginning of mammalian development. *The Journal of Biological Chemistry*, *270*, 22077–22080.

Oestrup, O., Hall, V., Petkov, S. G., Wolf, X. A., Hyldig, S., & Hyttel, P. (2009). From zygote to implantation: Morphological and molecular dynamics during embryo development in the pig. *Reproduction in Domestic Animals*, *3*, 39–49.

Oh, D. K., Ciaraldi, T., & Henry, R. R. (2007). Adiponectin in health and disease. *Diabetes, Obesity & Metabolism*, *9*, 282–289.

Ouchi, N., Kobayashi, H., Kihara, S., Kumada, M., Sato, K., Inoue, T., et al. (2004). Adiponectin stimulates angiogenesis by promoting cross-talk between AMP-activated protein kinase and Akt signaling in endothelial cells. *The Journal of Biological Chemistry*, *279*, 1304–1309.

Palanivel, R., Fang, X., Park, M., Eguchi, M., Pallan, S., De Girolamo, S., et al. (2007). Globular and full-length forms of adiponectin mediate specific changes in glucose and fatty acid uptake and metabolism in cardiomyocytes. *Cardiovascular Research*, *75*, 148–1157.

Pantaleon, M., & Kaye, P. L. (1996). IGF-I and insulin regulate glucose transport in mouse blastocysts via IGF-I receptor. *Molecular Reproduction and Development*, *44*, 71–776.

Pinto, A. B., Carayannopoulos, M. O., Hoehn, A., Dowd, L., & Moley, K. H. (2002). Glucose transporter 8 expression and translocation are critical for murine blastocyst survival. *Biology of Reproduction*, *66*, 1729–1733.

Riley, J. K., Carayannopoulos, M. O., Wyman, A. H., Chi, M., Ratajczak, C. K., & Moley, K. H. (2005). The PI3K/Akt pathway is present and functional in the preimplantation mouse embryo. *Developmental Biology*, *284*, 377–386.

Riley, J. K., Carayannopoulos, M. O., Wyman, A. H., Chi, M., & Moley, K. H. (2006). Phosphatidylinositol 3-kinase activity is critical for glucose metabolism and embryo survival in murine blastocysts. *The Journal of Biological Chemistry*, *281*, 6010–6019.

Riley, J. K., & Moley, K. H. (2006). Glucose utilization and the PI3-K pathway: Mechanisms for cell survival in preimplantation embryos. *Reproduction*, *131*, 823–835.

Rodgers, R. J., & Irving-Rodgers, H. F. (2010). Formation of the ovarian follicular antrum and follicular fluid. *Biology of Reproduction*, *82*, 1021–1029.

Schmidt, T., Fischer, S., Tsikolia, N., Santos, A. N., Rohrbach, S., Ramin, N., et al. (2008). Expression of adipokines in preimplantation rabbit and mice embryos. *Histochemistry and Cell Biology*, *129*, 817–825.

Shibata, R., Sato, K., Pimentel, D. R., Takemura, Y., Kihara, S., Ohashi, K., et al. (2005). Adiponectin protects against myocardial ischemia-reperfusion injury through AMPK- and COX-2–dependent mechanisms. *Nature Medicine*, *11*, 1096–1103.

Song, H., Chan, J., & Rovin, B. H. (2009). Induction of chemokine expression by adiponectin *in vitro* is isoform dependent. *Translational Research*, *154*, 18–26.

Steinberg, G. R., & Kemp, B. E. (2009). AMPK in Health and Disease. *Physiological Reviews*, *89*, 1025–1078.

Struwe, E., Berzl, G. M., Schild, R. L., & Dötsch, J. (2009). Gene expression of placental hormones regulating energy balance in small for gestational age neonates. *European Journal of Obstetrics, Gynecology, and Reproductive Biology*, *142*, 38–42.

Sun, Y., & Chen, X. (2010). Effect of adiponectin on apoptosis: Proapoptosis or anti-apoptosis? *BioFactors*, *36*, 179–186.

Tabandeh, M. R., Hosseini, A., Saeb, M., Kafi, M., & Saeb, S. (2010). Changes in the gene expression of adiponectin and adiponectin receptors (AdipoR1 and AdipoR2) in ovarian follicular cells of dairy cow at different stages of development. *Theriogenology*, *73*, 659–669.

Takemura, Y., Osuga, Y., Harada, M., Hirata, T., Koga, K., Morimoto, Ch., et al. (2005). Serum adiponectin concentrations are decreased in women with andometriosis. *Human Reproduction*, *20*, 3510–3513.

Takemura, Y., Osuga, Y., Yamauchi, T., Kobayashi, M., Harada, M., Hirata, T., et al. (2006). Expression of adiponectin receptors and its possible implication in the human endometrium. *Endocrinolgy*, *147*, 3203–3210.

Thellin, O., Zorzi, W., Lakaye, B., De Borman, B., Coumans, B., Hennen, G., et al. (1999). Housekeeping genes as internal standards: Use and limits. *Journal of Biotechnology*, *75*, 291–295.

Tonack, S., Fischer, B., & Navarrete Santos, A. (2004). Expression of the insulin-responsive glucose transporter isoform 4 in blastocysts of C57/BL6 mice. *Anat. Embryol. (Berl)*, *208*, 225–230.

Trujillo, M. E., & Scherer, P. E. (2005). Adiponectin—journey from an adipocyte secretory protein to biomarker of the metabolic syndrome. *Journal of Internal Medicine, 257*, 167–175.

Tsao, T. S., Murrey, H. E., Hug, Ch., Lee, D. H., & Lodish, H. F. (2002). Oligomerization state-dependent activation of NF-$_{\kappa}$B signaling pathway by adipocyte complement-related protein of 30kDa (Acrp30). *The Journal of Biological Chemistry, 277*, 29359–29362.

Tsao, T. S., Tomas, E., Murrey, H. E., Hug, Ch., Lee, D. H., Ruderman, N. B., et al. (2003). Role of disulfide bonds in Acrp30/Adiponectin structure and signaling specificity. *The Journal of Biological Chemistry, 278*, 50810–50817.

Tsuchida, A., Yamauchi, T., Takekawa, S., Hada, Y., Ito, Y., Maki, T., et al. (2005). Peroxisome proliferator-activated receptor (PPAR)alpha activation increases adiponectin receptors and reduces obesity-related inflammation in adipose tissue: Comparison of activation of PPARalpha, PPARgamma, and their combination. *Diabetes, 54*, 3358–3370.

Van der Steeg, J. W., Steures, P., Eijkemans, M. J. C., Habbema, J. D. F., Hompes, P. G. A., Burggraaff, J. M., et al. (2008). Obesity affects spontaneous pregnancy chances in subfertile, ovulatory women. *Human Reproduction, 23*, 324–328.

Varga, T., Czimmerer, Z., & Nagy, L. (2011). PPARs are a unique set of fatty acid regulated transcription factors controlling both lipid metabolism and inflammation. *Biochimica et Biophysica Acta, 1812*, 1007–1022.

Vejlsted, M., Du, Y., Vajta, G., & Maddox-Hyttel, P. (2006). Post-hatching development of the porcine and bovine embryo–defining criteria for expected development in vivo and in vitro. *Theriogenology, 65*, 153–165.

Viebahn, C., Mayer, B., & Hrabé de Angelis, M. (1995). Signs of the principle body axes prior to primitive streak formation in the rabbit embryo. *Anat. Embryol. (Berl), 192*, 159–169.

Vona-Davis, L., & Rose, D. P. (2007). Adipokines as endocrine, paracrine, and autocrine factors in breast cancer risk and progression. *Endocrine-Related Cancer, 14*, 189–206.

Wang, Y., Lam, K. S. L., Xu, J. Y., Lu, G., Xu, L. Y., Cooper, G. J. S., et al. (2005). Adiponectin inhibits cell proliferation by interacting with several growth factors in an oligomerization-dependent manner. *The Journal of Biological Chemistry, 280*, 18341–18347.

Wang, Y., Lam, K. S. L., Yau, M. H., & Xu, A. (2008). Post-translational modifications of adiponectin: Mechanisms and functional implications. *The Biochemical Journal, 409*, 623–633.

Watson, R. T., & Pessin, J. E. (2001). Intracellular organization of insulin signaling and GLUT4 translocation. *Recent Progress in Hormone Research, 56*, 175–193.

Welsh, G. I., Hers, I., Berwick, D. C., Dell, G., Wherlock, M., Birkin, R., et al. (2005). Role of protein kinase B in insulin-regulated glucose uptake. *Biochemical Society Transactions, 33*, 346–349.

Whitehead, J. P., Richards, A. A., Hickman, I. J., Macdonald, G. A., & Prins, J. B. (2006). Adiponectin–a key adipokine in the metabolic syndrome. *Diabetes, Obesity & Metabolism, 8*, 264–280.

Whitworth, K. M., Agca, C., Kim, J. G., Patel, R. V., Springer, G. K., Bivens, N. J., et al. (2005). Transcriptional profiling of pig embryogenesis by using a 15-K member unigene set specific for pig reproductive tissues and embryos. *Biology of Reproduction, 72*, 1437–1451.

Wijesekara, N., Krishnamurthy, M., Bhattacharjee, A., Suhail, A., Sweeney, G., & Wheeler, M. B. (2010). Adiponectin-induced ERK and Akt phosphorylation protects against pancreatic beta cell apoptosis and increases insulin gene expression and secretion. *The Journal of Biological Chemistry, 285*, 33623–33631.

Wong, G. W., Wang, J., Hug, C., Tsao, T. S., & Lodish, H. F. (2004). A family of Acrp30/adiponectin structural and functional paralogs. *Proceedings of the National Academy of Sciences of the United States of America, 101*, 10302–10307.

Xiao, X., Dong, Y., Zhong, J., Cao, R., Zhao, X., Wen, G., et al. (2011). Adiponectin protects endothelial cells from the damages induced by the intermittent high level of glucose. *Endocrine, 40,* 386–393.

Yamauchi, T., Kamon, J., Waki, H., Terauchi, Y., Kubota, N., Hara, K., et al. (2001). The fat-derived hormone adiponectin reverses insulin resistance associated with both lipoatrophy and obesity. *Nature Medicine, 7,* 941–946.

Yamauchi, T., Kamon, J., Minokoshi, Y., Ito, Y., Waki, H., Uchida, S., et al. (2002). Adiponectin stimulates glucose utilization and fatty-acid oxidation by activating AMP-activated protein kinase. *Nature Medicine, 8,* 1288–1295.

Yamauchi, T., Kamon, J., Ito, Y., Tsuchida, A., Yokomizo, T., Kita, S., et al. (2003). Cloning of adiponectin receptors that mediate antidiabetic metabolic effects. *Nature, 423,* 762–768.

Yoon, M. J., Lee, G. Y., Chung, J. J., Ahn, Y. H., Hong, S. H., & Kim, J. B. (2006). Adiponectin increases fatty acid oxidation in skeletal muscle cells by sequential activation of AMP-activated protein kinase, p38 mitogen-activated protein kinase, and peroxisome proliferator-activated receptor alpha. *Diabetes, 55,* 2562–2570.

Zavalza-Gómez, A. B., Anaya-Prado, R., Rincón-Sánchez, A. R., & Mora-Martínez, J. M. (2008). Adipokines and insulin resistance during pregnancy. *Diabetes Research and Clinical Practice, 80,* 8–15.

Zhou, Y., Sun, X., Jin, L., Stringfield, T., Lin, L., & Chen, Y. (2005). Expression profiles of adiponectin receptors in mouse embryos. *Gene Expression Patterns, 5,* 711–715.

CHAPTER TEN

Adiponectin and the Control of Female Reproductive Functions

Marie-France Palin*[,1], V. Vilceu Bordignon[†], Bruce D. Murphy[‡]
*Agriculture and Agri-Food Canada, Sherbrooke, Quebec, Canada
[†]Department of Animal Science, McGill University, Ste-Anne-de-Bellevue, Quebec, Canada
[‡]Université de Montréal, St-Hyacinthe, Quebec, Canada
[1]Corresponding author: e-mail address: mariefrance.palin@agr.gc.ca

Contents

Abstract

Adiponectin is the most abundant protein secreted by the white adipose tissue. It circulates at high levels in the bloodstream and its serum concentration is inversely correlated with body fat mass. The wide distribution of adiponectin receptors (AdipoR1, AdipoR2, and T-cadherin) in peripheral tissues and organs allows adiponectin to exert pleiotropic effects on whole-body metabolism. Besides its well-known antidiabetic, antiatherogenic, and anti-inflammatory properties, accumulating evidence suggests a direct role for adiponectin in reproductive tissues. The mammalian ovary and the ovarian follicle express AdipoR1 and AdipoR2, and treating pig granulosa cells with adiponectin induces changes characteristic of the periovulatory period. Moreover, additive effects are observed between adiponectin and insulin in induction of granulosa cell gene expression, thus suggesting that adiponectin actions on the ovary may be mediated through its insulin-sensitizing effects. Adiponectin receptors are also detected

Vitamins and Hormones, Volume 90
ISSN 0083-6729
http://dx.doi.org/10.1016/B978-0-12-398313-8.00010-5

in the uterus. In women, higher *AdipoR1* and *AdipoR2* gene expression was observed during the mid-secretory phase of the menstrual cycle, suggesting that adiponectin is implicated in the endometrial changes in preparation for embryo implantation. Adiponectin receptors are found in oocytes and early developing pig, rabbit, and mice embryos, and it has been demonstrated that adiponectin can increase the success of porcine embryo development to the blastocyst stage *in vitro*. Moreover, adiponectin concentration is two to three times greater in human fetal circulation and in umbilical cord blood, compared to adult plasma. This further indicates a role for adiponectin in fetal growth. It has been further suggested that adipose-derived and locally produced adiponectin may act as a key neuromodulator of reproductive functions. For example, the inhibition of LH and GnRH release from rat pituitary and hypothalamic cells following treatment with adiponectin provides evidence that adiponectin may also act on the release of gonadotropins.

Adipose tissue is now recognized as an important factor in the complex equation by which the nutritional status regulates female reproductive functions. For example, underweight women have delayed puberty and higher risk of premature delivery, whereas overweight and obese women have early puberty and are prone to develop polycystic ovary syndrome (PCOS), gestational diabetes mellitus (GDM), and preeclampsia. Because hypoadiponectinemia is often associated with the abovementioned disorders, it has been suggested that this adipokine might play a role in the development of these pregnancy-related complications. Moreover, as these reproductive disorders often come with metabolic complications such as insulin and glucose resistance, the insulin-sensitizing effects of adiponectin may explain the observed association of this adipokine with PCOS, GDM, and preeclampsia. This review summarizes current knowledge on the role of adiponectin in female reproductive tissues and highlights mechanisms where information is available. We also discuss about the known and potential roles of adiponectin in the development of reproductive disorders.

1. INTRODUCTION

It is now well accepted that metabolic dysregulations, such as obesity and leanness, often lead to reproductive dysfunctions. For example, overweight or obese women are at higher risk of developing menstrual disorders, polycystic ovaries, and obstetric complications, whereas underweight women have delayed menarche, irregular menstrual cycle, and present an increased likelihood of miscarriage (Brewer & Balen, 2010; Budak et al., 2006; The ESHRE Capri Workshop Group, 2006; Van Binsbergen, Voelingh Bennink, Odink, Haspels, & Koppeschaar, 1990). The adipose tissue and its secreted factors have emerged as key players in the complex interplay by which the female nutritional status regulates reproductive functions (Campos, Palin, Bordignon, & Murphy, 2008). Indeed, the adipose tissue is now considered as a major endocrine organ secreting numerous factors including the adipokines (Trayhurn & Wood, 2004). Among these adipokines, adiponectin has gained more attention because of its insulin-sensitizing, antiatherosclerotic, and anti-inflammatory properties (Brochu-Gaudreau et al., 2010). Accumulating evidence suggests that adiponectin also have key roles in regulating female reproductive functions.

Adiponectin, also known as Acrp30, apM1, GBP28, and AdipoQ, is a 30-kDa protein that is mainly secreted by the white adipose tissue (Arita et al., 1999; Hu, Liang, & Spiegelman, 1996; Maeda et al., 1996; Nakano, Tobe, Choi-Miura, Mazda, & Tomita, 1996; Scherer, Williams, Fogliano, Baldini, & Lodish, 1995). Unlike most of the other adipokines, adiponectin circulating levels correlate negatively with the body mass index (BMI) and with abdominal fat accretion (Cnop et al., 2003; Lara-Castro, Luo, Wallace, Klein, & Garvey, 2006; Matsubara, Maruoka, & Katayose, 2002). Systemic adiponectin levels are reduced in obesity and type 2 diabetes (Weyer et al., 2001), whereas weight reduction leads to increase in circulating adiponectin levels (Yang et al., 2001). Adiponectin belongs to the C1q/TNF-α (tumor necrosis factor alpha) superfamily (Wong, Wang, Hug, Tsao, & Lodish, 2004). It is composed of an N-terminal signal peptide, a collagenous domain, and a C-terminal globular C1q-like domain (Scherer et al., 1995). In the plasma, adiponectin can be found in different oligomeric forms, the hexamers (medium molecular weight, MMW) and high-molecular weight (HMW) forms being the major ones (Nakano et al., 1996). These oligomers present different activities with the HMW form being more active in liver (Trujillo & Scherer, 2005) and endothelial (Ouchi et al., 2004) cells. Moreover, circulating levels of HMW adiponectin are more closely related with its insulin-sensitizing effects (Lara-Castro et al., 2006).

Three adiponectin receptors have been identified based on their capability to bind adiponectin. Adiponectin receptors 1 and 2 are seven-transmembrane receptors, members of the progestin AdipoQ receptors family (Tang et al., 2005). Although AdipoR1 and AdipoR2 share 67% protein identity, gene knockout studies demonstrated that they can have opposing effects on energy metabolism, with *AdipoR1*$^{-/-}$ mice being obese and glucose intolerant and *AdipoR2*$^{-/-}$ mice being lean and resistant to high-fat-induced obesity (Bjursell et al., 2007). Differences were also reported with respect to adiponectin binding affinity. Indeed, AdipoR1 was identified as a high-affinity receptor for globular adiponectin and low-affinity receptor for the full-length adiponectin, whereas AdipoR2 shows an intermediate affinity for full-length and globular adiponectin (Yamauchi et al., 2003). Moreover, these two receptors are regulated differentially by insulin, which can repress *AdipoR1* gene expression but has no effect on *AdipoR2* in muscle cells (Sun et al., 2008). A third adiponectin receptor, T-cadherin (also known as cadherin 13 (CDH13)), has been identified (Hug et al., 2004). This receptor is a glycosylphosphatidylinositol-anchored protein that lacks the transmembrane and intracellular domains (Hug et al., 2004). T-cadherin can bind the MMW and HMW forms of adiponectin but not the trimeric or globular species (Hug et al., 2004).

The binding of adiponectin to AdipoR1 and AdipoR2 provokes the activation of different intracellular pathways. In skeletal muscle and liver, the insulin-sensitizing effects of adiponectin are mediated through the activation of the adenosine monophosphate (AMP)-activated protein kinase (AMPK), p38 mitogen-activated kinase (MAPK), and the peroxisome proliferators-activated receptor-α (Deepa & Dong, 2009). The adaptor protein-containing pleckstrin homology domain, phosphotyrosine-binding domain, and leucine zipper motif (APPL1) protein acts as a mediator of adiponectin intracellular signaling and was identified as the "missing link" between adiponectin and its downstream targets (Deepa & Dong, 2009). The involvement of APPL1 in adiponectin signaling cascade is supported by the observation that its overexpression in skeletal muscle cells increases AMPK and p38 MAPK phosphorylation, whereas its suppression prevents the adiponectin-mediated activation of AMPK and p38 MAPK (Mao et al., 2006). Adiponectin proliferative effects are also mediated through the APPL1 protein, which can activate the extracellular-signal-regulated kinase 1/2 (Erk1/2)-MAPK signaling pathway (Lee, Klein, El-Shewy, Luttrell, & Luttrell, 2008). Finally, in endothelial cells, the antiproliferative and cytoprotective effects of adiponectin seem to be mediated through an APPL1-dependent signaling pathway involving the phosphatidylinositol 3-kinase (PI3K) and the v-akt murine thymoma viral oncogene homolog (Akt) (Chandrasekar et al., 2008).

The identification of AdipoR1 and AdipoR2 receptors in different reproductive tissues (Lord, Ledoux, Murphy, Beaudry, & Palin, 2005) and the observed sexual dimorphism in circulating adiponectin levels (Pajvani et al., 2003; Peake, Kriketos, Campbell, Shen, & Charlesworth, 2005) were the first indications that adiponectin might be involved in reproductive functions. The recent findings that adiponectin is locally produced in different reproductive tissues further suggested that this adipokine may play key roles in reproduction. The following sections will address current knowledge on adiponectin actions in female reproductive functions and its potential involvement in the development of reproductive disorders such as polycystic ovary syndrome (PCOS), gestational diabetes mellitus (GDM), intrauterine growth restriction (IUGR), and preeclampsia.

2. ADIPONECTIN IN OVARIAN FUNCTIONS

2.1. Role of adiponectin in ovulation and luteinization processes

The first evidence that adiponectin could exert a biological effect on the ovary was provided by the detection of AdipoR1 and AdipoR2 receptors in the whole pig ovary (Lord et al., 2005). Since then, the presence of adiponectin receptor transcripts and proteins has been reported in human (Chabrolle et al., 2009), rodent (Chabrolle, Tosca, & Dupont, 2007), chicken (Chabrolle, Tosca, Crochet, Tesseraud, & Dupont, 2007; Maddineni, Metzger, Ocon, Hendricks, & Ramachandran, 2005; Ramachandran, Ocon-Grove, & Metzger, 2007), and bovine (Maillard et al., 2010; Tabandeh, Hosseini, Saeb, Kafi, & Saeb, 2010) ovaries. At the cellular level, AdipoR1 and AdipoR2 mRNA and/or proteins were detected in both granulosa and theca cells (Chabrolle, Tosca, & Dupont, et al., 2007; Chabrolle, Tosca, Crochet, Tesseraud, & Dupont, et al., 2007; Chabrolle et al., 2009; Maillard et al., 2010; Tabandeh et al., 2010) as well as in the oocyte (Chabrolle, Tosca, & Dupont, et al., 2007; Maillard et al., 2010), cumulus cells (Tabandeh et al., 2010), and in the human granulosa tumor cell line KGN (Pierre et al., 2009). Ovarian cells are also able to produce adiponectin, as demonstrated by the presence of adiponectin mRNA and protein in theca (Chabrolle, Tosca, & Dupont, et al., 2007; Chabrolle, Tosca, Crochet, Tesseraud, & Dupont, 2007; Chabrolle et al., 2009; Maillard et al., 2010; Tabandeh et al., 2010) and cumulus (Maillard et al., 2010) cells and in the oocyte (Chabrolle, Tosca, & Dupont, 2007; Maillard et al., 2010). Moreover, adiponectin from local and/or from circulation accumulates in follicular fluid (Chabrolle et al., 2009; Gutman et al., 2009; Ledoux et al., 2006; Maillard et al., 2010). Species differences exist with respect to the quantity of adiponectin mRNA and/or protein in granulosa cells, as this occurs in rat, cattle, chickens, and sheep (Chabrolle, Tosca, & Dupont, 2007; Chabrolle, Tosca, Crochet, Tesseraud, & Dupont, 2007; Maillard et al., 2010; Ortega, Rey, Velazquez, & Padmanabhan, 2010; Tabandeh et al., 2010). However, adiponectin is undetectable both in human granulosa cells and in the KGN cell line (Chabrolle, Tosca, & Dupont, 2007; Chabrolle et al., 2009). Adiponectin expression levels vary according to the ovarian cell types, with a much higher mRNA and/or protein expression found in theca than in granulosa and cumulus cells in the rat, chicken, and cow (Chabrolle, Tosca, & Dupont, 2007, Chabrolle, Tosca, Crochet, Tesseraud, & Dupont, 2007; Tabandeh et al., 2010). However, the opposite was reported in sheep where the highest adiponectin expression was found in granulosa cells (Ortega et al., 2010). The presence of adiponectin transcripts and proteins in ovarian cell compartments suggest an autocrine and/or paracrine effect of locally produced adiponectin in the ovary.

Few studies have investigated the expression of adiponectin and its receptors in follicular cells throughout the different developmental stages. In dairy cows, adiponectin mRNA abundance measured in granulosa and cumulus cells and in oocytes was higher in large than in small follicles. In contrast, lower adiponectin mRNA levels were found in theca cells of large follicles compared with the small follicles (Tabandeh et al., 2010). In the same study, *AdipoR1* and *AdipoR2* mRNA levels were higher in large than in small follicles in the granulosa, theca, and cumulus cells and in oocytes. It is, therefore, possible that changes in adiponectin and adiponectin receptors mRNA abundance may be under the influence of hormonal or biological factors associated with follicular development. However, further work is needed to confirm these results, as Ledoux et al. (2006) reported no change in porcine adiponectin protein levels in the follicular fluid of small, medium, or large follicles and no changes in *AdipoR1* and *AdipoR2* mRNA levels in small, medium, and large follicles. Moreover, similar levels of adiponectin protein were found in granulosa cells of sheep primordial, primary, small preantral, and large preantral follicles, whereas it was much higher in antral follicles (Ortega et al., 2010).

To study the effects of the gonadotropins luteinizing hormone (LH) and follicle-stimulating hormone (FSH) on the ovarian expression of adiponectin and its receptors, Chabrolle, Tosca, and Dupont, et al. (2007) treated rats with pregnant mare serum gonadotropin (PMSG) for 24 h to induce follicular growth, followed by human chorionic gonadotropin (hCG) to provoke ovulation and luteinization. Interestingly, they observed higher levels of adiponectin protein and of *AdipoR1* mRNA and protein in response to the hCG treatment, whereas *AdipoR2* mRNA and protein levels were unchanged (Chabrolle, Tosca, & Dupont, 2007). Moreover, there was no effect of the PMSG treatment on the expression of adiponectin and its receptors. In humans, women receiving recombinant

LH treatment had higher adiponectin protein in follicular fluid compared with the placebo group (Gutman et al., 2009). In bovine theca cells, the addition of LH increased *AdipoR2* mRNA abundance by 17–27% (Lagaly, Aad, Grado-Ahuir, Hulsey, & Spicer, 2008). These results show that ovarian adiponectin, AdipoR1, and AdipoR2 can be modulated by LH, thus suggesting a possible role for adiponectin in LH-regulated processes including follicle maturation and ovulation.

Adiponectin, AdipoR1, and AdipoR2 transcripts and proteins were also detected in the rat (Chabrolle, Tosca, & Dupont, 2007) and bovine (Maillard et al., 2010) corpora lutea (CL). Moreover, Ledoux et al. (2006) showed that recombinant porcine adiponectin increased vascular endothelial growth factor (VEGF) mRNA and protein abundance in granulosa cell cultures and that this activation is abrogated by the addition of U0126, a specific Erk1/2 inhibitor. As inactivation of VEGF completely abolishes luteal vascularization and progesterone production in rats, primates, and cows (Robinson et al., 2009), these results are interpreted to indicate that adiponectin, through the activation of VEGF, plays a key role in luteal growth and development.

To determine whether adiponectin can participate in the periovulatory remodeling of the ovarian follicle, *in vitro* studies were conducted. Using a porcine granulosa cell culture model, Ledoux et al. (2006) demonstrated a direct effect of recombinant porcine adiponectin on ovarian genes known to be involved in prostaglandin synthesis, steroidogenesis, and angiogenesis. Indeed, there was a time-dependent increase in cyclooxygenase 2 (*COX-2*), prostaglandin E synthase (*PGES*), and *VEGF* mRNA abundance with the addition of adiponectin (25 μg/ml) in cultured porcine granulosa cells. This was accompanied by increased secretion of prostaglandin E_2 and VEGF. Moreover, the addition of recombinant adiponectin to granulosa cells increased the abundance of the steroidogenic acute regulatory protein (*StAR*) and the cytochrome P450, family 19, subfamily A, polypeptide 1 (*CYP19A1*) mRNA levels, two genes involved in ovarian steroidogenesis (Ledoux et al., 2006). In contrast, Chabrolle, Tosca, and Dupont (2007) and Chabrolle et al. (2009) were unable to find any effect on cytochrome P450, family 11, subfamily A, polypeptide 1 (CYP11A1), 3-β-hydroxysteroid dehydrogenase, StAR, and CYP19A1 protein levels when human recombinant adiponectin (5 μg/ml) was added to human or rat granulosa cell cultures. Similar results were reported by Lagaly et al. (2008), with no effect of human recombinant adiponectin (3 μg/ml) on *CYP19A1* and *CYP11A1* mRNA levels in bovine granulosa cell cultures. However, in this last study, adiponectin, in combination with insulin-like growth factor I (IGF-I) and LH, provoked decreases in *CYP19A1* and *CYP11A1* mRNA levels in bovine theca cell cultures, thus suggesting that adiponectin effects on ovarian steroidogenesis might primarily be localized in theca cells. The observed discrepancies in steroidogenic gene expression may be explained by species differences or by the amount of recombinant protein that was added in cell cultures: 25 μg/ml for the porcine and 5 or 3 μg/ml for the human recombinant adiponectin.

It was earlier reported that the IGF-I-induced secretion of progesterone and estradiol by human granulosa cells was increased with the addition of recombinant human adiponectin and this was paralleled with increases in CYP19A1 protein levels (Chabrolle et al., 2009). Similar results were reported in chickens (Chabrolle, Tosca, Crochet, Tesseraud, & Dupont, 2007) and rats (Chabrolle, Tosca, & Dupont, 2007), but there was no effect of adiponectin on IGF-I-induced production of progesterone and estradiol in cattle (Maillard et al., 2010). In chicken granulosa cells, recombinant human adiponectin reduced the LH- and FSH-induced secretion of progesterone depending on the stage of the follicular hierarchy, that is, the effect was seen in F3/4 but not in F1 or F2 granulosa cells (F1, F2, F3, and F4 are referring to the first, second, third, and fourth largest preovulatory follicles, respectively) (Ramachandran et al., 2007). In contrast, there was no effect of adiponectin on FSH-induced production of progesterone and estradiol in human (Chabrolle et al., 2009) and rat (Chabrolle, Tosca, & Dupont, 2007) granulosa cells. In human, rat, bovine, and chicken granulosa cells, adiponectin alone had no effect on the secretion of progesterone and estradiol (Chabrolle, Tosca, & Dupont, 2007; Chabrolle, Tosca, Crochet, Tesseraud, & Dupont, 2007; Chabrolle et al., 2009; Maillard et al., 2010). In bovine theca cell cultures, adiponectin decreased the LH + insulin-induced production of progesterone and androstenedione (Lagaly et al., 2008). Collectively, the above-mentioned findings suggest that adiponectin alone has limited effects on sex steroids secretion by ovarian cells, but that it can modulate the steroid production induced by IGF-I, FSH, or LH, thereby influencing the ovulatory processes. Moreover, the observed discrepancies may be explained by species differences and also by the fact that the granulosa cells studied in culture were extracted from preovulatory follicles from different developmental stages.

Given that insulin can interact with gonadotropins in inducing steroidogenesis (Sekar, Garmey, & Veldhuis, 2000) and that adiponectin increases insulin sensitivity in peripheral tissues (Fasshauer & Paschke, 2003), interactions among LH, insulin, and adiponectin were investigated in ovarian cells. Ledoux et al. (2006) observed that treating porcine granulosa cells with combinations of insulin + adiponectin, LH + adiponectin, or insulin + LH + adiponectin increased the *COX-2* mRNA abundance when compared with adiponectin alone. Surprisingly, combining adiponectin and insulin resulted in decreased secretion of progesterone and estradiol in bovine granulosa cells, compared with the insulin treatment alone (Maillard et al., 2010). This was associated with a reduction in insulin-induced phosphorylation of Erk1/2 MAPK when granulosa cells were pretreated with recombinant adiponectin. These results are in agreement with those of Lagaly et al. (2008) who reported lower production of progesterone and androstenedione when adiponectin was added to bovine theca cells treated with LH + insulin. The mechanisms are unclear as the intracellular signaling pathways were not investigated in that study.

Adiponectin actions on ovarian steroidogenesis and gene and protein expression are mediated through multiple intracellular pathways. Previous studies have reported the implication of the MAPK Erk1/2 signaling pathway in FSH-induced progesterone secretion by the granulosa cells (Moore, Otsuka, & Shimasaki, 2001; Tosca et al., 2005). Interestingly, adiponectin was also found to increase rapidly and transiently the phosphorylation of MAPK Erk1/2 in rat, human, and porcine granulosa cells as well as in the human KGN cell line (Chabrolle, Tosca, & Dupont, 2007; Chabrolle et al., 2009; Ledoux et al., 2006; Pierre et al., 2009). Increased phosphorylation of AMPKα was also observed after treating chicken (Chabrolle, Tosca, Crochet, Tesseraud, & Dupont, 2007), porcine (Ledoux et al., 2006), and rat (Chabrolle, Tosca, & Dupont, 2007) granulosa cells and KGN cells with recombinant adiponectin, whereas there was no such effect on human granulosa cells (Chabrolle et al., 2009). Recombinant adiponectin also increased the phosphorylation of p38 MAPK in human and rat granulosa cells and KGN cells (Chabrolle, Tosca, & Dupont, 2007; Chabrolle et al., 2009; Pierre et al., 2009). As AMPK is considered as a sensor of cellular energy status (Dupont, Chabrolle, Ramé, Tosca, & Coyral-Castel, 2008), these results suggest that adiponectin may have a role in regulating the amount of energy needed for follicular growth. Accumulating evidence suggests that adiponectin, AMPK, and PPARs work in concert to coordinate the ovarian cell functions. The PPARγ transcription factor is highly expressed in granulosa cells and weakly expressed in theca cells and CL (Dupont et al., 2008). Its protein expression levels throughout the follicular development are similar to those of adiponectin in sheep, increasing with follicle size and reaching a maximum in antral follicles (Ortega et al., 2010). Of interest, the adiponectin promoter region contains a peroxisome proliferator response element (Iwaki et al., 2003) and circulating adiponectin increases during treatment with pioglitazone, an agonist of PPARγ (Tsuchiya, Akaza, Yoshimoto, & Hirata, 2009). Moreover, PPARγ can be phosphorylated by AMPK (Leff, 2003), and, as noted above, this phosphorylation is first activated by adiponectin in ovarian cells. Interestingly, previous studies reported that PPARs can activate the transcription of genes playing key roles in ovarian functions (Komar, 2005). Among those genes are *COX-2* and *VEGF*, genes that are upregulated in porcine granulosa cell cultures treated with recombinant adiponectin (Ledoux et al., 2006). Finally, adiponectin was found to increase Akt phosphorylation in the rat granulosa cells (Chabrolle, Tosca, & Dupont, 2007) but not in human granulosa cells (Chabrolle et al., 2009).

In overview, the available evidence suggests that adiponectin interacts with and enhances the effects of endocrine factors, including insulin and gonadotropins. Its actions are mediated by multiple intracellular pathways, and signaling appears to vary depending on the species and the developmental status of the ovarian follicle (Fig. 10.1).

2.2. Adiponectin across the menstrual cycle

It is well known that female ovarian function can be affected by the nutritional status (Frish, 1984). For example, underweight women have low estradiol levels, delayed menarche, and irregular menstrual cycle (Budak et al., 2006; Van Binsbergen et al., 1990), whereas overweight and obese subjects present early menarche and are more prone to PCOS (Brewer & Balen, 2010). Moreover, menstrual cyclicity can be regularized and fertility improved in obese women by modest weight losses, as little as 5–10% of total body weight (Crosignani et al., 2003). Of interest, circulating adiponectin levels are higher in females than in males and it has been suggested that ovarian steroids might regulate adiponectin secretion (Bottner et al., 2004; Combs et al., 2003; Gui, Silha, & Murphy, 2004). However, conflicting results have been reported on the effects of exogenous estrogen administration on circulating adiponectin levels. In some studies, there were no effect (Chalvatzas et al., 2009; Sumino et al., 2004), while others reported an increase (Chu, Cosper, Nakhuda, & Lobo, 2006) or a decrease (Im, Lee, Lee, & Lee, 2006) in systemic adiponectin levels. The observation that plasma adiponectin levels remain stable throughout the menstrual cycle (Asimakopoulos et al., 2009; Dafopoulos,

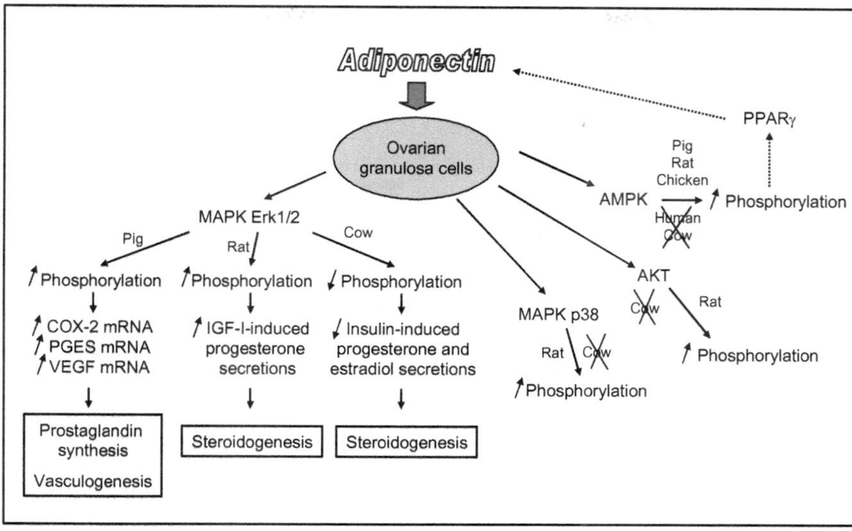

Figure 10.1 In ovarian granulosa cells, adiponectin action is mediated through different intracellular signaling pathways. AMPK, adenosine monophosphate (AMP)-activated protein kinase; AKT, v-akt murine thymoma viral oncogene homolog 1; COX-2, cyclooxygenase 2; Erk1/2, extracellular-signal-regulated kinase; MAPK, mitogen-activated protein kinase; PGES, prostaglandin E synthase; PPARγ, peroxisome proliferator-activated receptor γ; VEGF, vascular endothelial growth factor.

Sourlas, Kallitsaris, Pournaras, & Messinis, 2009; Hall, White, & O'Sullivan, 2009; Kleiblova, Springer, & Haluzik, 2006; Rouen, Lukacs, & Reame, 2010) further suggests that factors other than estrogens might be responsible for the observed sexual dimorphism in adiponectin levels. However, none of these studies have investigated variant adiponectin isoform levels during the menstrual cycle. This may be of importance because sexual dimorphism has been observed in terms of the proportion of the different adiponectin isoforms in circulation with a much higher proportion of the HMW form of adiponectin in females relative to males (Pajvani et al., 2003; Peake et al., 2005). It was recently reported that the HMW isoform of adiponectin is negatively associated with estradiol after adjusting for BMI and age in healthy premenopausal women, whereas there was no association with total adiponectin serum levels (Merki-Feld, Imthurn, Rosselli, & Spanaus, 2011). A negative association between estradiol and total, HMW, and MMW adiponectin serum concentrations was also observed in a study comparing premenopausal, postmenopausal, and pregnant women (Leung et al., 2009). Taken together, the above findings suggest that female ovarian hormones might not have a major influence on the regulation of adiponectin secretion and that other serum factors might account for the observed sexual dimorphism in circulating adiponectin levels. This was demonstrated *in vitro* by decreases in adiponectin mRNA levels in human adipocytes that were treated with 10% of female pooled serum and an even stronger downregulation of adiponectin expression when 10% male pooled serum was added (Horenburg, Fischer-Posovszky, Debatin, & Wabitsch, 2008). In this study, the addition of estradiol or testosterone had no effect on adiponectin mRNA levels and there was no effect on the expression of the HMW, MMW, and low-molecular weight (LMW) adiponectin isoforms (Horenburg et al., 2008), thus suggesting that serum factors other than sex steroids might be involved in adiponectin regulation. Even though ovarian hormones are not considered as the main regulators of adiponectin secretion, they may have effects on adiponectin receptors expression. Indeed, this has been demonstrated by Takemura et al. (2006) who reported higher endometrial *AdipoR1* and *AdipoR2* mRNA levels during the mid-secretory phase of the human menstrual cycle, compared with the other phases. This indicates that ovarian hormones may regulate the expression of adiponectin receptors in the uterus.

2.3. Adiponectin at puberty and in pre- and postmenopausal women

Although genetic factors are important determinants of puberty onset, changes in body composition and hormonal environment are also considered as key signals (Brufani et al., 2009; Kaplowitz, Slora, Wasserman, & Pedlow, 2001). Moreover, the pubertal transition is also associated with a reduction in insulin sensitivity (Goran & Gower, 2001; Moran et al., 1999), which may be explained by increases in adiposity (Brufani et al., 2009; Travers, Jeffers, Bloch, Hill, & Eckel, 1995). Because adiponectin is a known insulin sensitizer and its circulating levels are closely related to body fat distribution, it was hypothesized that this adipokine could play a role in puberty attainment. Circulating total adiponectin levels are similar in prepuberty and at the onset of puberty (Andersen, Frystyk, Wolthers, Heuck, & Flyvbjerg, 2007; Bottner et al., 2004; Martos-Moreno, Barrios, & Argente, 2006; Woo, Dolan, Daniels, Goodman, & Martin, 2005) between healthy boys and girls. At the mid- and postpuberty stages, total adiponectin levels decrease in boys, whereas it remains stable in girls, leading to lower adiponectin levels in boys compared with girls (Andersen et al., 2007; Bottner et al., 2004; Martos-Moreno et al., 2006;

Woo et al., 2005). Significant decreases in adiponectin HMW ratio have been reported in postpuberty compared with prepuberty in boys but it remained unchanged in girls (Andersen et al., 2007). The abovementioned gender differences in total and HMW adiponectin are most likely due to changes in androgen levels as negative correlations were observed between serum testosterone and the total and HMW ratio adiponectin (Andersen et al., 2007; Bottner et al., 2004). A positive correlation was also observed between total adiponectin levels and the sex hormone-binding globulin, a marker of testosterone bioavailability (Martos-Moreno et al., 2006). The pubertal transition corresponds to a period of rapid growth and development, changes in body fat, and decreases in insulin sensitivity. Different studies have reported associations between circulating adiponectin levels and anthropometric and metabolic parameters. For example, Punthakee et al. (2006) observed that adiponectin variations related with age were mainly explained by changes of BMI in girls, whereas it explained half of the age effect in boys, the other half being explained by androgen effects. After adjustment for age and gender, Huang et al. (2004) found negative correlations between plasma adiponectin levels and parameters of total body and central fatness such as fat mass, waist circumference, and waist-to-hip ratio. Other studies have reported similar associations (Bottner et al., 2004), but Schoppen et al. (2010) were unable to find any association between systemic adiponectin levels and body composition data. During the pubertal period, serum concentrations of adiponectin are also negatively correlated with indexes of insulin resistance, including the homeostasis model assessment (HOMA-IR) and fasting plasma insulin (Bottner et al., 2004). However, central body fat in girls and lean mass accretion in boys were identified as the main factors predicting insulin sensitivity in obese children and adolescents, independently of serum adiponectin (Brufani et al., 2009).

Across the menopausal transition, women gain weight and accumulate body fat (Leung et al., 2009) with an increased tendency to accumulate intra-abdominal fat (Colombel & Charbonnel, 1997; Kanaley et al., 2001; Lee et al., 2009; Sieminska et al., 2005). This is often accompanied by dyslipidemia (Schnatz & Schnatz, 2006) and insulin resistance (Kaaja, 2008; Leung et al., 2009), and female sex steroids can prevent or reduce these deleterious changes (Lobo, 2008). Given that low-circulating adiponectin levels are associated with increased body fat (Gavrila et al., 2003; Maury et al., 2007) and increased insulin resistance (Weyer et al., 2001), it was hypothesized that this adipokine may contribute to the metabolic changes occurring during the transition to menopause. Contradictory findings have been reported with respect to adiponectin levels across the menopausal transition. Although some studies found decreases in adiponectin levels in postmenopausal women (Chu, Cosper, Orio, Carmina, & Lobo, 2006; Sumino et al., 2004), others were either unable to detect any changes (Hong et al., 2007; Nishizawa et al., 2002; Sieminska et al., 2005) or even reported higher levels of circulating adiponectin (Lee et al., 2009; Leung et al., 2009). These discrepancies may be explained by women being at different menopausal stages at the time of sampling, as lower adiponectin levels were observed in perimenopausal women compared with those in pre- or postmenopausal stages (Sowers et al., 2008). Adiposity levels may also account for these differences, as lower serum adiponectin levels were earlier reported in obese than in normal weight postmenopausal women (Rouen et al., 2010). Moreover, Soni, Conroy, Mackey, and Kuller (2011) found higher circulating adiponectin levels with increasing weight loss in overweight postmenopausal women undergoing an 18-month lifestyle intervention trial. Of interest was that adiponectin was found to correlate negatively with changes in intra-abdominal fat but not with subcutaneous adipose tissue or total percent body fat in healthy pre- and postmenopausal women (Lee et al., 2009). As central obesity is often linked with insulin resistance and type 2 diabetes, the association between circulating adiponectin levels and parameters of insulin resistance were also investigated in postmenopausal women. One study reported a negative correlation between the HOMA-IR index and plasma adiponectin in postmenopausal women (Im et al., 2006), whereas Leung et al. (2009) reported no significant association between these two parameters in pre- and postmenopausal women. Sex steroid concentration variations in study subjects may also explain some of the observed discrepancies in adiponectin levels because postmenopausal women being on long-term estrogen therapy had lower levels of plasma adiponectin (Im et al., 2006). However, short-term (3–6 months) hormone replacement therapy had no effect on postmenopausal plasma adiponectin levels (Sumino et al., 2004). In animal studies, lower adiponectin levels were reported after administration of estrogens, and this effect was reversed with estrogen receptor antagonist (Combs et al., 2003). Negative associations were found between estradiol and total, HMW and MMW adiponectin serum concentrations across the menopausal period (Leung et al., 2009). This is in accordance with the finding of Sumino et al. (2004) who reported a negative correlation between estradiol and total plasma adiponectin in postmenopausal women. A negative correlation between total adiponectin and estradiol was also observed in postmenopausal women that were on long-term hormonal therapy, but not in postmenopausal women who never had hormone therapy (Im et al., 2006). However, Sieminska et al. (2005) failed to detect any association between estradiol and adiponectin levels in pre- and postmenopausal women. Collectively, the abovementioned studies demonstrate the complex interplay between adiponectin secretion, female sex hormones, adiposity, and insulin resistance across the menopausal transition. Larger studies are needed to determine whether adiponectin is a key player in the development of insulin resistance, increased adiposity, and cardiovascular disease in postmenopausal women.

2.4. Adiponectin and the PCOS

The PCOS is a common endocrine and metabolic disorder that affects 5–10% of women of reproductive age and is responsible for as many as 70% of cases of anovulatory infertility (Brassard, AinMelk, & Baillargeon, 2008; Norman, Dewailly, Legro, & Hicket, 2007). In accordance with the Rotterdam PCOS Consensus Workshop Group (The Rotterdam ESHRE/ASRM-Sponsored PCOS Consensus Workshop Group, 2004), PCOS diagnosis is confirmed when at least two of the following criteria are observed: (1) the presence of polycystic ovaries, (2) clinical and/or biochemical hyperandrogenism, and (3) ovulatory dysfunction (oligomenorrhea or amenorrhea). About 50% of PCOS women are overweight or obese (Norman et al., 2007). The abdominal and visceral fat tissue seems to play a key role in the development of this disorder, as abdominal obesity is observed in both obese and normal weight PCOS patients (Barber, McCarthy, Wass, & Franks, 2006; Douchi et al., 1995; Kirchengast & Huber, 2001). Moreover, a large proportion of PCOS women present a cluster of abnormalities that are often associated with the metabolic syndrome, including insulin resistance and impaired glucose intolerance, dyslipidemia, hypertension, and an increased risk of type 2 diabetes and cardiovascular disease (Diamanti-Kandarakis, Papavassiliou, Kandarakis, & Chrousos, 2007; Moran, Misso, Wild, & Norman, 2010; Tomlinson, Millward, Stenhouse, & Pinkey, 2010). As adiponectin possesses insulin-sensitizing, antidiabetic, and antiatherogenic properties, and because its circulating levels are reduced with obesity and type 2 diabetes mellitus (Weyer et al., 2001), it has been suggested that this adipokine may play a role in the pathogenesis of PCOS. Several studies have reported decreased plasma levels of adiponectin with PCOS (Ardawi & Rouzi, 2005; Aroda et al., 2008; Barber et al., 2008;

Carmina et al., 2005, 2006; Escobar-Morreale et al., 2006; Sepilian & Nagamani, 2005; Wickham, Cheang, Clore, Baillargeon, & Nestler, 2011), while others failed to demonstrate differences between PCOS women and healthy controls (Bik et al., 2007; Gulcelik, Aral, Serter, & Koc, 2008; Haap et al., 2005; Orio et al., 2003; Shroff et al., 2007). To better understand whether hypoadiponectinemia is a key player in PCOS development, Toulis et al. (2009) published a systematic review of manuscripts comparing circulating adiponectin levels in women affected or not by PCOS and performed a meta-analysis using studies involving only women matched for similar BMI. This analysis revealed that, after controlling for BMI, adiponectin levels are lower in women affected by PCOS than in non-PCOS controls and that hypoadiponectinemia is probably related with insulin resistance in PCOS women. Studies reporting increases in circulating adiponectin levels in PCOS women treated with insulin sensitizers further suggest a strong association between adiponectin levels and insulin sensitivity. Indeed, treating PCOS women with metformin (Agarwal et al., 2010; Elkind-Hirsch, Marrioneaux, Bhushan, Vernor, & Bhushan, 2008; Jakubowska et al., 2008; Steiner et al., 2007), rosiglitazone (Majuri et al., 2007; Steiner et al., 2007), or pioglitazone (Glintborg et al., 2008) increase circulating adiponectin levels that also correlate with reductions in insulin secretion (Elkind-Hirsch et al., 2008) and with improvement of insulin action on glucose metabolism and lipid oxidation (Glintborg et al., 2008).

Given that hyperandrogenism is a characteristic feature of PCOS, it was hypothesized that androgens may also influence circulating adiponectin levels in PCOS women. It was earlier reported that systemic adiponectin levels correlate negatively with testosterone (Bik et al., 2007; Escobar-Morreale et al., 2006; Spranger et al., 2004), Δ4-andostenedione (Ardawi & Rouzi, 2005; Panidis et al., 2003), dehydroepiandrosterone, dehydroepiandrosterone sulfate (Ardawi & Rouzi, 2005), and the free androgen index (Sieminska et al., 2004) in PCOS women, whereas other studies found positive correlations between adiponectin and testosterone (Ardawi & Rouzi, 2005; Glintborg et al., 2006; Vrbikova et al., 2005) or no association at all with testosterone (Gulcelik, Aral, Serter, Demir, & Culha, 2006; Panidis et al., 2003; Sieminska et al., 2004; Trolle, Lauszus, Frystyk, & Flyvbjerg, 2010). Of interest, Luque-Ramirez, Alvarez-Blasco, and Escobar-Morreale (2009) observed an increase in circulating adiponectin in obese PCOS women that were treated with an antiandrogenic contraceptive. However, there were no correlations between the changes in adiponectin levels during the antiandrogenic treatment and testosterone or Δ4-andostenedione concentrations (Luque-Ramirez et al., 2009). Interestingly, the addition of recombinant human adiponectin to bovine ovarian theca cells decreased the production of androstenedione as well as the 17α-hydroxylase (CYP17A1) mRNA abundance, the enzyme responsible for androgen synthesis (Lagaly et al., 2008). These authors hypothesized that the low systemic adiponectin levels observed in PCOS may decrease androgen production through a decrease in the inhibitory effects of adiponectin on LH and insulin/IGF-I-stimulated androstenedione production by the ovary (Lagaly et al., 2008). It is currently unclear whether circulating androgens are important modulators of adiponectin serum levels in PCOS or if changes in adiponectin levels precede variations in androgen levels. Further work on the mechanism of action of adiponectin in ovarian cells and of androgens in the adipose tissue of PCOS women is required to address this question.

The close association of circulating HMW adiponectin with insulin sensitivity (Lara-Castro et al., 2006) and the progression of type 2 diabetes mellitus (Nakashima et al., 2006) suggest that levels of HMW adiponectin might be a better predictor of PCOS development and may explain discrepancies observed in previous studies with respect to adiponectin levels in PCOS women. A limited number of studies have investigated plasma levels of HMW adiponectin in women with PCOS. O'Connor et al. (2010) observed a selective reduction of HMW adiponectin in PCOS women compared with BMI- and insulin resistance-matched controls. Similar to these findings, Aroda et al. (2008) reported lower HMW adiponectin serum levels in women with PCOS than in control women matched for age and BMI. Interestingly, circulating HMW adiponectin levels were found to be a strong independent predictor of insulin sensitivity in PCOS women (Wickham et al., 2011). Contrasting studies reported no difference in HMW adiponectin levels between PCOS and control women (Glintborg et al., 2008) and, although Barber et al. (2008) observed lower total and HMW adiponectin in unadjusted PCOS cases compared to control women, these differences disappeared after adjusting for age and fat mass. In the abovementioned studies, different methods were used to measure HMW adiponectin (Wickham et al., 2011) and a number of PCOS women were treated with the insulin sensitizer metformin (Barber et al., 2008), which may explain some of the observed discrepancies. Nevertheless, further work is needed to establish whether HMW adiponectin shows stronger associations relative to total adiponectin with PCOS development and with measures of insulin sensitivity in affected women.

Accumulating evidence supports the role of genetic factors in the etiology of PCOS. Familial clustering of the disease has been reported with first-degree relatives showing an increased risk of developing PCOS (Crosignani & Nicolosi, 2001; Vink, Sadrzadeh, Lambalk, & Boomsma, 2006). Moreover, it was reported that normal-weight prepubertal daughters of PCOS women present early metabolic abnormalities of PCOS, such as hypoadiponectinemia and hyperinsulinemia, when compared with daughters of control women (Sir-Petermann et al., 2007). Upon reaching puberty, daughters of PCOS women display higher levels of triglycerides and are more prone to develop hyperandrogenism. A large number of functional candidate gene polymorphisms have been studied with the aim of identifying association with the risk of developing PCOS (Unluturk, Harmanci, Kocaefe, & Yildiz, 2007). Among those, two adiponectin single-nucleotide polymorphisms (SNPs) (T45G in exon 2 and G276T in intron 2) have received the most attention (Unluturk et al., 2007). Indeed, these two SNPs have been identified as promising candidates, based on their association with obesity, insulin sensitivity, and the increased risk of developing type 2 diabetes (Filippi et al., 2004; Hara et al., 2002; Stumvoll et al., 2002). Some studies have reported a higher prevalence of the G allele for the T45G polymorphism (Haap et al., 2005; Panidis et al., 2004) and a reduced frequency of the T allele for the G276T polymorphism (Heinonen et al., 2005) in PCOS women. Associations between specific genotypes of the T45G and G276T SNPs and the susceptibility to PCOS were also reported (Li et al., 2011; Ranjzad et al., 2011; Zhang et al., 2008). However, conflicting results were reported by other groups with no association between the T45G and G276T polymorphisms and the development of PCOS (Escobar-Morreale et al., 2006; Xita et al., 2005; Yoshihara, Yahata, Kashima, Mikami, & Tanaka, 2009). Nevertheless, when analyses are restricted to the PCOS groups, these two polymorphisms are found to influence some metabolic variables of PCOS such as insulin resistance (Xita et al., 2005; Yoshihara et al., 2009; Zhang et al., 2008), obesity, and circulating adiponectin levels (Xita et al., 2005). Although these results suggest that SNPs in the adiponectin gene may influence the phenotypic expression of PCOS, further work is needed before adiponectin gene polymorphisms can be used as a mean of identifying women at risk of developing PCOS.

3. ADIPONECTIN IN PREGNANCY

3.1. Adiponectin at the maternal–fetal interface

During the peri-implantation period, a complex interplay takes place between the mother and the fetus. This period is associated with an extensive growth and remodeling of the endometrium and placenta along with the secretion of numerous cytokines and growth factors needed to create an optimal environment for uterine receptivity and blastocyst growth. The presence of AdipoR1 and AdipoR2 in the human and pig endometrium (Lord et al., 2005; Takemura et al., 2006) and their elevated expression levels observed during the mid-secretory phase of the menstrual cycle provided the first evidence that adiponectin may play a key role in embryo implantation (Takemura et al., 2006). Since then, AdipoR1 and AdipoR2 were also found to be expressed in rabbit and mouse endometrium and preimplantation embryos (Kim et al., 2011; Schmidt et al., 2008). Moreover, AdipoR1 and AdipoR2 transcripts and proteins have been detected in the human placenta and in JEG-3 and BeWo trophoblastic cell lines, whereas CDH13, the third adiponectin receptor, was undetectable (Benaitreau et al., 2009). In rabbits, adiponectin protein was detected in stromal and epithelial cells of the endometrium as well as in endothelial and smooth muscle cells of the myometrium (Schmidt et al., 2008). Adiponectin mRNA is further present in human endometrial epithelial and stromal cells (Takemura et al., 2006), and adiponectin protein was found in epithelial cells of the uterine glands in mice (Schmidt et al., 2008). Moreover, in the rat oviduct, the strongest adiponectin protein signal was detected in the secretory granules of the isthmus, thus suggesting that adiponectin is secreted in the oviduct fluid (Archanco, Gomez-Ambrosi, Tena-Sempere, Fruhbeck, & Burrell, 2007). These findings further suggest a role for adiponectin in oocytes and spermatozoa final maturation and in fertilization.

Conflicting results were reported with respect to adiponectin expression in the placenta. A local production of adiponectin was observed in human term placenta (Caminos et al., 2005; Chen et al., 2006; Jarvenpaa et al., 2009), but it was absent in placental tissue samples collected between 6 and 12 weeks (Benaitreau et al., 2009) or at mid-gestation (25 weeks) (Corbetta et al., 2005). Although it is tempting to explain these discrepancies by the different gestational ages of study subjects, there are at least two studies where Northern blot and real-time PCR analyses both failed to detect adiponectin transcript in human term placenta (Haugen et al., 2006; McDonald & Wolfe, 2009). Thus, further investigation is needed to identify the factors controlling adiponectin transcription in placental tissue. Interestingly, Chen et al. (2006) reported that placental adiponectin, *AdipoR1* and *AdipoR2* transcript levels are modulated by cytokines such as the TNF-α, interferon gamma, leptin, and interleukin 6 (IL-6). They hypothesized that these proinflammatory cytokines may regulate placental adiponectin secretion in pregnancy-associated insulin-resistant states. Despite the existence of controversial results over the placental adiponectin secretion, the presence of adiponectin receptors suggest that locally produced and/or systemic adiponectin may play key roles in regulating placental function. Indeed, treating cytotrophoblast cells with recombinant globular adiponectin resulted in decreased secretion of chorionic gonadotropin α- and β-subunits and of placental lactogen accompanied by lower *CYP11A1* and aromatase transcript levels (McDonald & Wolfe, 2009). The influence of adiponectin at the maternal–fetal interface was further demonstrated by its antiproliferative (Benaitreau et al., 2009) and prodifferentiation effects on trophoblastic cells (Benaitreau, Dos Santos, Leneveu, de Mazancourt, et al., 2010). It was also shown that adiponectin can have proinvasive effects in human trophoblastic cells through an increased activity of the matrix metalloproteinases MMP2 and MMP9 and reduced expression of expression of *TIMP2*, an inhibitor of metalloproteinases (Benaitreau, Dos Santos, Leneveu, Alfaidy, et al., 2010). This further demonstrates that adiponectin may play a role in the early steps of human conceptus implantation. Using a mouse model of delayed and activated implantation, Kim et al. (2011) observed increased expression of adiponectin, AdipoR1, and AdipoR2 proteins in the luminal epithelia at the termination of delayed implantation, thus suggesting a putative role for adiponectin in uterine receptivity and implantation. The observed increase in secretion of the proinflammatory cytokines IL-6, IL-8, and monocyte chemotactic protein 1, when treating endometrial epithelial and stromal cells with recombinant adiponectin, further suggested that adiponectin may have anti-inflammatory effects in the human endometrium (Takemura et al., 2006).

Few studies have investigated the intracellular signaling pathways activated by adiponectin at the maternal–fetal interface (Fig. 10.2). Takemura et al. (2006) first reported that human recombinant adiponectin was able to induce phosphorylation of AMPK in endometrial epithelial and stromal cells. As AMPK is known for its implication in the regulation of cellular

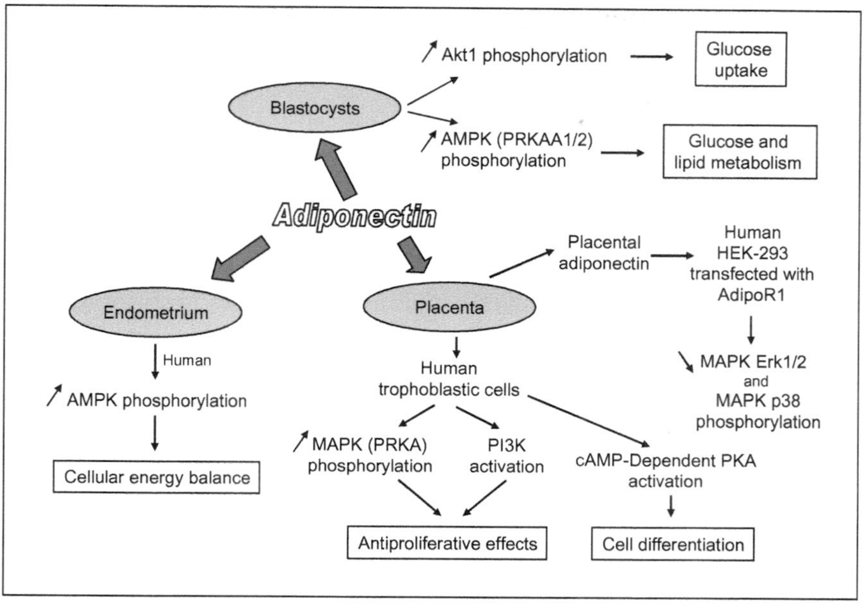

Figure 10.2 Schematic representation of intracellular signaling pathways activated by adiponectin at the maternal–fetal interface. AdipoR1, adiponectin receptor 1; AMPK, adenosine monophosphate (AMP)-activated protein kinase; AKT, v-akt murine thymoma viral oncogene homolog 1; Erk1/2, extracellular-signal-regulated kinase; HEK, human embryonic kidney; MAPK, mitogen-activated protein kinase; PI3K, phosphoinositide-3-kinase; PKA, protein kinase A; PRKAA1/2, protein kinase, AMP-activated.

energy balance, it was suggested that adiponectin might regulate energy supply in the human endometrium (Takemura et al., 2006). In the human choriocarcinoma cell lines JEG-3 and BeWo, which can be used as a first-trimester trophoblast cell model, human recombinant adiponectin activated the phosphorylation of AMPK (Benaitreau et al., 2009). Moreover, it was demonstrated that the observed antiproliferative effects of adiponectin on JEG-3 and BeWo cells were mediated, at least in part, by the AMPK and PI3K pathways (Benaitreau et al., 2009).

Treating BeWo cells with recombinant human adiponectin increased the intracellular cAMP production and, the addition of H89, a specific inhibitor of the protein kinase A (PKA) pathway, suppressed the adiponectin-induced production of hCG (Benaitreau, Dos Santos, Leneveu, de Mazancourt, et al., 2010). As the hCG secretion can be used as a maker of cytotrophoblast differentiation (Shi, Lei, Rao, & Lin, 1993), it was suggested that the positive effects of adiponectin on trophoblastic cell differentiation are mediated through the cAMP-dependent PKA signaling pathway (Benaitreau, Dos Santos, Leneveu, de Mazancourt, et al., 2010). The addition of human placental explants conditioned media to human embryonic kidney 293 cells (HEK-293), previously transfected with the *AdipoR1* gene, decreased the phosphorylation of p38 and Erk1/2 MAPK (Chen et al., 2006). However, there was no change in p38 and Erk1/2 phosphorylation when using adiponectin-free conditioned media, thus confirming that adiponectin is indeed the placental factor responsible for the activation of the MAPK signaling pathway (Chen et al., 2006). This further suggested that locally produced adiponectin may be involved in placental development, as it was previously demonstrated that the p38 and Erk1/2 kinases are associated with the regulation of trophoblastic cell differentiation (Daoud et al., 2005).

3.2. Adiponectin and embryo development

The presence of the adiponectin system, including adiponectin, AdipoR1, and AdipoR2, has been confirmed at both transcripts and protein levels in ovarian follicular cells, oocytes, and embryos of different species, such as pigs (Chappaz et al., 2008; Lord et al., 2005; Whitworth et al., 2005), rats (Chabrolle, Tosca, & Dupont, 2007), mice(Cikos et al., 2010; Kim et al., 2011; Schmidt et al., 2008), rabbits (Schmidt et al., 2008), bovine (Maillard et al., 2010; Tabandeh et al., 2010), and also in humans (Chabrolle et al., 2009). There is also evidence that the expression of adiponectin and its receptors is regulated during the follicular growth in the ovary and during early embryo development (Chabrolle, Tosca, & Dupont, 2007; Cikos et al., 2010; Kim et al., 2011; Schmidt et al., 2008; Tabandeh et al., 2010). As noted above, proteins of the adiponectin system are present in the oviduct (Archanco et al., 2007) and endometrium (Lord et al., 2005; Schmidt et al., 2008; Takemura et al., 2006). Moreover, both mRNA and protein levels of adiponectin, AdipoR1, and AdipoR2 were found to be higher in the endometrium at the sites of embryo implantation compared to the interimplantation sites (Kim et al., 2011). All these observations support the concept that adiponectin may play a local (autocrine and/or paracrine) role in the regulation of follicular and oocyte growth, and during early embryo development and implantation.

Although some progress has been obtained while investigating the mechanistic roles of adiponectin in follicular cells, little information has been produced regarding its potential effects on oocytes and embryos (Brochu-Gaudreau et al., 2010; Michalakis & Segars, 2010). Observations from women undergoing assisted reproductive treatments revealed that the serum level of adiponectin was positively correlated with the average number of oocytes recovered from each patient. Also, serum adiponectin was significantly higher in the patients that conceived after *in vitro* fertilization and embryo transfer compared to those that did not conceive (Liu et al., 2005). In a similar study, it was shown that the circulating levels of adiponectin near the time of oocyte collection were higher in patients that became pregnant after intracytoplasmic sperm injection and embryo transfer compared to the patients where the treatment was not successful (Bersinger, Birkhauser, & Wunder, 2006). These observations suggest that adiponectin may have positive effects on oocyte maturation and embryo development.

To date, there are only few controlled studies that have been conducted to evaluate the effect of adiponectin during oocyte maturation and embryo culture *in vitro*. In the first study, Chappaz and collaborators showed that the presence of recombinant adiponectin (30 μg/ml) enhanced nuclear maturation of porcine oocytes (Chappaz et al., 2008). In the same study, the inclusion of adiponectin during oocyte maturation or embryo culture promoted blastocyst formation of swine embryos produced by parthenogenetic activation. Beneficial effects were also observed in another study where mouse embryos collected at the four-cell stage were cultured *in vitro* for 65 h in the presence of adiponectin at the concentration of 10 μg/ml (Cikos et al., 2010). The presence of full-length adiponectin increased the proportion of embryos with high cell numbers (> 128 cells) and decreased the proportion of those with lower cell numbers (< 65 cells). However, in a bovine study, the addition of 10 μg/ml adiponectin during *in vitro* maturation did not affect oocyte maturation, embryo cleavage, and development to the blastocyst stage (Maillard et al., 2010). Yet, the effect of adiponectin during embryo culture was not evaluated in this study. Comparisons between these studies conducted *in vitro* are difficult due to a number of differences, including the species, the concentration of adiponectin tested, and the different composition of culture medium used for oocyte maturation and embryo culture.

Given the known effects of adiponectin on different biological functions including lipid and glucose metabolism, insulin sensitivity, angiogenesis, and inflammation, adiponectin may influence oocytes and embryos in many ways. The effect of adiponectin on glucose metabolism was assessed in rabbit blastocysts collected at day-6 *post coitum* (Fischer, Santos, Thieme, Ramin, & Fischer, 2010). In this study, the exposure to 1 μg/ml of adiponectin induced a fourfold increase in phosphorylation of the AMP-activated protein kinase alpha 1/2 (PRKAA1/2), which is known to regulate glucose and lipid metabolism. Treatment also decreased the expression of phosphoenolpyruvate carboxykinase 2 (*PCK2*), which is an important regulator of gluconeogenesis. Moreover, adiponectin also increased the Akt1 phosphorylation, improved embryonic glucose uptake, and induced the translocation of the solute carrier family 2 (facilitated glucose transporter), member 4, which affects the insulin-sensitive uptake of glucose. These observations confirm that early developing embryos are responsive to adiponectin and support a functional role of this adipokine in the regulation of embryo metabolism (Fig. 10.2).

Another likely effect of adiponectin is on embryo implantation. In support to this are the observations that adiponectin system is more abundantly expressed in human endometrium during the luteal period, which corresponds to the embryo implantation period (Takemura et al., 2006), and in the sites of embryo implantation in the mouse uterus (Kim et al., 2011). Moreover, adiponectin was shown to modulate the secretion of proinflammatory cytokines by endometrial stromal cells (Takemura et al., 2006), as well as the migration and invasion of human trophoblast cells (Benaitreau, Dos Santos, Leneveu, Alfaidy, et al., 2010). These findings support a local role for the adiponectin system on embryo development and uterine receptivity. Despite these recent findings, additional studies are necessary to further dissect the mechanisms by which adiponectin affects embryo development.

3.3. The role of adiponectin in preeclampsia

In developing countries, preeclampsia is a leading cause of maternal and fetal mortality (WHO, 2005). From a clinical point of view, preeclampsia is characterized by hypertension and proteinuria, conditions that usually develop after 20 weeks of pregnancy (Maynard & Karumanchi, 2011). Several risk factors associated with the development of preeclampsia have been identified and many of those share features of the metabolic syndrome such as obesity, insulin resistance, and dyslipidemia (Sibai, Dekker, & Kupferminc, 2005). The central initiating cause of preeclampsia is believed to be an inadequate placentation, which predisposes to placental hypoperfusion, ischemia, and endothelial cell dysfunction (Masuyama & Hiramatsu, 2010; Tuuli & Odibo, 2010). The presence of adiponectin and its receptors in the placenta (Caminos et al., 2005; Yamauchi et al., 2003) and the close association of circulating adiponectin with manifestations of the metabolic syndrome suggest that this adipokine may be involved in the pathogenesis of preeclampsia.

Recent investigations have reported higher (Fasshauer et al., 2008; Haugen et al., 2006; Kajantie, Kaaja, Ylikorkala, Andersson, & Laivuori, 2005; Lu et al., 2006; Naruse et al., 2005; Nien et al., 2007a; Ramsay, Jamieson, Greer, & Sattar, 2003), lower (Cortelazzi et al., 2007; D'Anna et al., 2006; Herse et al., 2009; Ichida et al., 2007; Mazaki-Tovi, Romero, Vaisbuch, Kusanovic, et al., 2009; Mori, Shinohara, Wakatsuki, Watanabe, & Fujimaki, 2010; Ouyang, Chen, & Chen, 2007), and not different (Dalamaga, Srinivas, Elovitz, Chamberland, & Mantzoros, 2011) systemic adiponectin concentrations in women with preeclampsia. These discrepancies may be explained by differences in plasma adiponectin levels observed between early- and late-onset preeclampsia (D'Anna et al., 2006). Indeed, Masuyama et al. (2010) recently reported higher circulating adiponectin in women with late-onset (after 32 weeks) preeclampsia when compared with healthy controls, whereas there was no difference between early-onset preeclampsia (before 32 weeks) and controls. Gestational age may also explain differences in adiponectin levels because women with preeclampsia had lower circulating adiponectin levels during the first trimester, whereas higher adiponectin concentrations were observed in the third trimester (D'Anna et al., 2006). Interestingly, Ogland et al. (2010) were unable to find significant difference in cord blood adiponectin levels between women with or without preeclampsia but, higher adiponectin levels were observed in women with preeclampsia who delivered during gestational weeks 32–36. Differences in study population characteristics such as the use of mild and/or severe preeclamptic patients may also account for some of the observed discrepancies.

As the risk of developing preeclampsia increases with obesity and insulin resistance (O'Brien, Ray, & Chan, 2003; Sibai et al., 2005), the influence of these two parameters on adiponectin levels was also investigated in preeclampsia. Hendler et al. (2005) earlier reported that among women with severe preeclampsia, overweight women have lower serum adiponectin levels, while normal weight women show higher adiponectin levels. On the other hand, several studies have reported disruption in the expected negative association between adiponectin and BMI and/or insulin resistance in preeclamptic patients. For example, although significant negative correlations were observed between plasma adiponectin and BMI in normal pregnancies, this relationship was lost in women with preeclampsia (Herse et al., 2009; Nien et al., 2007a). Moreover, Ramsay et al. (2003) found inverse correlations between plasma adiponectin and early pregnancy BMI and fasting insulin concentrations in normal pregnancy, while these correlations disappeared in preeclamptic women. An altered regulation of adiponectin circulating levels was also observed by Mazaki-Tovi, Romero, Vaisbuch, Kusanovic, et al. (2009), with preeclamptic patients having similar total, HMW, MMW, and LMW serum adiponectin levels between normal weight and overweight/obese women. Similarly, Haugen et al. (2006) reported that plasma adiponectin did not correlate with BMI or markers of insulin resistance in preeclamptic women. In a study performed by Kajantie et al. (2005), there was no association between circulating adiponectin levels and insulin sensitivity in both normotensive and preeclamptic women. Interestingly, it was suggested that the association between insulin resistance and adiponectin levels can vary between early-onset and late-onset preeclamptic subgroups and also among different gestational age (D'Anna et al., 2006). Indeed, a negative correlation was observed between adiponectin and HOMA-IR values in late-onset preeclamptic patients, while there was no association in the early-onset preeclampsia group. Moreover, this inverse correlation was observed in the first trimester but disappeared in the third trimester (D'Anna et al., 2006).

The reduced circulating adiponectin levels observed in preeclamptic women after delivery, compared with third trimester pregnancy levels, is consistent with a possible role of placental adiponectin in the pathophysiology of preeclampsia (Kajantie et al., 2005). However, the contribution of placental adiponectin to circulating adiponectin levels is likely to be limited, based on the absence of detectable mRNA (Haugen et al., 2006) or low expression levels of adiponectin mRNA (Jarvenpaa et al., 2009) and protein (Ichida et al., 2007) in placental tissues of normal and preeclamptic women. Nevertheless, the observed increase in *AdipoR1* mRNA abundance (Jarvenpaa et al., 2009) and of adiponectin protein (Ichida et al., 2007) in placental tissues from preeclamptic women, when compared with healthy pregnant controls, suggest a possible role for adiponectin in placental angiogenesis. Of interest, placental adiponectin proteins are only detected in the endothelial cells of chorionic vessels (Ichida et al., 2007), but it is not known whether this adiponectin is from locally produced or systemic origin. Ledoux et al. (2006) previously demonstrated the direct effect of recombinant adiponectin on the angiogenic factor VEGF in granulosa cell culture. Interestingly, the VEGF-neutralizing molecule, sFlt-1, is expressed by the placenta and is abnormally high in the plasma of preeclamptic women (Koga et al., 2003; Maynard et al., 2003). Moreover, the exogenous expression of sFlt-1 in healthy pregnant rats provokes symptoms associated with preeclampsia such as endothelial dysfunction, hypertension, and proteinuria (Maynard et al., 2003). Thus, it can be hypothesized that the observed upregulation of adiponectin protein and *AdipoR1* mRNA with preeclampsia may be part of the mechanisms in place to restore endothelial damages. However, further work is needed before concluding on the possible involvement of adiponectin in restoring placental endothelial damages.

3.4. Adiponectin and IUGR

IUGR is the failure of the fetus to reach its expected growth potential due to adverse intrauterine environment. This may lead to early fetal programming of diseases that will be expressed in later life such as insulin resistance, obesity, hypertension, and dyslipidemia (Levy-Marchal & Jacquet, 2004). Reduced fetal growth is also accompanied by an altered development of the adipose tissue, which translates into a reduction of total body fat mass and an increase in central adiposity (Briana & Malamitsi-Puchner, 2009; Petersen, Gotfredsen, & Knudsen, 1988; Yajnik et al., 2002). The majority of small for gestational age (SGA) newborns will experience rapid postnatal catch-up growth and accumulate more abdominal fat when compared with normal birth size newborns (Levy-Marchal & Jacquet, 2004). Moreover, SGA children who experience rapid postnatal catch-up growth are more prone to develop insulin resistance and type 2 diabetes (Ong et al., 2004; Soto et al., 2003). As adiponectin correlates negatively with body fat (Gavrila et al., 2003) and promotes insulin sensitivity (Fasshauer & Paschke, 2003), it was hypothesized that this adipokine may be involved in IUGR.

Maternal adiponectin circulating levels were found to be higher (Wang et al., 2010), lower (Kyriakakou et al., 2008), or to show no difference (Fasshauer et al., 2007; Sawidou, Sotiriadis, Kaihura, Nicolaides, & Sattar, 2008) in IUGR when compared with appropriate for gestational age (AGA) groups. Differences in gestational age may account for some of the observed discrepancies as it was earlier reported that plasma adiponectin concentrations decrease with advancing pregnancy (Nien et al., 2007b). Discrepancies also exist in fetal or neonatal adiponectin levels with decreased (Kamoda, Saitoh, Saito, Sugiura, & Matsui, 2004; Martos-Moreno et al., 2009; Street, Volta, Ziveri, Viani, & Bernasconi, 2009; Takaya, Yamato, Higashino, & Kaneko, 2007) or increased (Wang et al., 2010) adiponectin levels in SGA than in AGA, whereas other studies failed to report differences between SGA and AGA (Kyriakakou et al., 2008; Martinez-Cordero, Amador-Licona, Guizar-Mendoza, Hernandez-Mendez, & Ruelas-Orozco, 2006). All of the above studies assessed total adiponectin levels. However, a shift in adiponectin isoforms toward the HMW adiponectin was recently reported in SGA compared with AGA newborns (Ibanez et al., 2008). Of note, HMW adiponectin is the dominant isoform in newborn umbilical cord serum (Odden & Morkrid, 2007) and it was previously reported that serum HMW adiponectin is a more relevant indicator of insulin sensitivity (Pajvani et al., 2004). Therefore, the higher level of cord blood HMW adiponectin found in SGA individuals (Ibanez et al., 2008) may increase insulin sensitivity, thus preparing SGA newborns for rapid postnatal catch-up growth.

In SGA and AGA newborns, Takaya et al. (2007) observed a positive correlation between cord plasma adiponectin concentrations and the quantitative insulin sensitivity check index (Katz et al., 2000), which is a measure of insulin sensitivity. However, other studies reported no correlation between placental adiponectin concentrations and cord serum or placental insulin levels (Street et al., 2009) and no correlation between newborn serum adiponectin and blood glucose levels (Kamoda et al., 2004). None of these studies have investigated the association between parameters of insulin sensitivity and HMW adiponectin concentrations, which is considered as a better indicator of insulin sensitivity. Moreover, these correlation analyses were performed on whole study population with no distinction between SGA and AGA individuals. Therefore, further work is needed to determine whether adiponectin (total and HMW) concentrations are correlated with insulin sensitivity parameters in SGA newborns.

In SGA and AGA term neonates, positive correlations were reported between circulating adiponectin concentrations and birth weight (Kamoda et al., 2004; Takaya et al., 2007) and placental adiponectin levels correlated positively with birth weight and placenta weight (Street et al., 2009). On the opposite, Wang et al. (2010) found negative correlations between umbilical serum adiponectin and birth weight, placental weight, and BMI. In the last study, negative correlations were also observed between placental adiponectin and birth weight, placental weight, and BMI. In another study analyzing data from preterm and full-term newborns, there was no correlation between cord blood adiponectin levels and birth weight (Martos-Moreno et al., 2009). Differences in the above-mentioned studies may be explained by different gestational ages, as a strong positive correlation was reported between cord blood adiponectin and gestational age, after controlling for birth weight (Martos-Moreno et al., 2009). Moreover, correlation analyses between adiponectin and birth weight were conducted on the whole study populations; thus, it would be interesting to determine whether these associations differ among SGA or AGA newborns.

3.5. Adiponectin and GDM

GDM is a common complication of pregnancy characterized by insulin resistance and decreased insulin secretion by the pancreatic β-cells (Ben-Ziv & Hod, 2008). Short-term adverse outcomes for the mother include preeclampsia and higher risk of cesarean section, whereas long-term adverse effects are an increased risk of developing type 2 diabetes mellitus and cardiovascular diseases (Ben-Ziv & Hod, 2008). An increased risk of perinatal mortality and abnormal fetal growth (macrosomia and IUGR) is also associated with GDM (Ben-Ziv & Hod, 2008). Type 2 diabetes mellitus and GDM share common pathogenic features. As circulating adiponectin levels are decreased in obese and type 2 diabetes mellitus patients (Brochu-Gaudreau et al., 2010), it was suggested that similar alterations in systemic adiponectin may be observed in GDM. Indeed, several studies have reported decreased circulating adiponectin levels in women with GDM compared with normal controls (Cortelazzi et al., 2007; Georgiou et al., 2008; Mazaki-Tovi, Romero, Vaisbuch, Erez, et al., 2009; Retnakaran et al., 2007; Sawidou et al., 2010; Soheilykhah et al., 2009; Tsai et al., 2005; Vitoratos et al., 2008; Worda et al., 2004). Moreover, serum HMW, MMW, and LMW adiponectin isoforms are also lower in women with GDM than in those having a normal pregnancy (Mazaki-Tovi, Romero, Vaisbuch, Erez, et al., 2009; Retnakaran et al., 2007). Low plasma adiponectin levels were also reported by Ranheim et al. (2004) in GDM women having prepregnancy BMI <25, but not in GDM individuals with BMI ≥ 25. In this last study, the adiponectin mRNA levels from abdominal subcutaneous adipose tissue were lower in GDM women than in healthy pregnant controls. Only one study reported similar circulating adiponectin concentrations between GDM and euglycemic pregnant women (Paradisi et al., 2010).

Of interest, an inverse correlation was observed between maternal plasma adiponectin levels and the neonatal birth weight in women who developed GDM (Tsai et al., 2005), and adiponectin concentrations measured in the fetal circulation was lower in newborns from GDM mother than in those from normal mothers (Cortelazzi et al., 2007). Therefore, the modulation of adiponectin levels with GDM may affect the fetal energy metabolism and the associated long-term adverse effects (obesity and diabetes) observed in infants of GDM mothers. In a recent study investigating adiponectin DNA methylation in the human placenta, negative correlations between fetal side CpGs and maternal glucose levels and between maternal side DNA methylation and HOMA-IR were observed (Bouchard et al., 2011). These findings suggest that adiponectin DNA methylation profile can be modulated by the mothers' glycemic levels and that these epigenetic adaptations may be associated with an increased risk of developing obesity and type 2 diabetes in newborns of GDM mothers.

In women with GDM, there was no correlation between circulating adiponectin levels and maternal BMI (Tsai et al., 2005; Vitoratos et al., 2008) and there was no difference in total, HMW, MMW, and LMW serum adiponectin levels between obese/overweight and normal weight GDM women (Mazaki-Tovi, Romero, Vaisbuch, Erez, et al., 2009). These results demonstrate that the expected negative association between adiponectin and obesity may be disrupted in women with GDM. Similar alterations in the regulation of adiponectin circulating levels were also observed in women with preeclampsia (see above). However, further work is needed to confirm these findings as two studies rather reported negative correlations between maternal circulating adiponectin levels and BMI in individuals with GDM (Cseh et al., 2004; Soheilykhah et al., 2009).

As GDM women present insulin resistance (Ben-Ziv & Hod, 2008), the relationship between serum adiponectin and parameters of insulin resistance was investigated. Vitoratos et al. (2008) reported a negative correlation between serum adiponectin levels and the HOMA-IR in pregnant women with no distinction between individuals affected or not by GDM. Only two studies have evaluated the relationship between adiponectin and insulin resistance parameters in GDM patients. Cseh et al. (2004) were the first to report negative correlations between plasma adiponectin and fasting C-peptide concentrations and also with the fasting C-peptide/blood glucose ratio, an indirect measure of insulin resistance. In contrast, Tsai et al. (2005) were unable to find such correlations in GDM women despite the fact that plasma adiponectin correlated negatively with different insulin resistance parameters in normal pregnant women.

Evidence is accumulating to indicate that a genetic component may be associated with the risk of developing GDM (Robitaille & Grant, 2008). Genetic association studies were carried on to examine the possible association between adiponectin polymorphisms and the development of type 2 diabetes. Among the most common polymorphisms, SNP45T > G, identified in exon 2 of the adiponectin gene, was found to be associated with insulin sensitivity (Ruchat et al., 2008) and lower adiponectin levels were found in carriers of the G allele (Petrone et al., 2006). More recently, a higher frequency of the G allele was observed in GDM women (32.7%), compared with women having a normal pregnancy (17.9%) (Low, Tohit, Chong, & Idris, 2011). However, further work is needed before confirming this association as this study used a very low effective (GDM, $n=26$; normal patients, $n=53$).

4. ADIPONECTIN ACTION ON THE CENTRAL NERVOUS SYSTEM AND THE PITUITARY GLAND

The sum of current research findings suggests that reproductive functions are closely related to energy homeostasis and that metabolic dysregulation, such as obesity and anorexia nervosa, often leads to reproductive abnormalities. The presence of AdipoR1 and AdipoR2 in the central nervous system (CNS) and the pituitary gland suggests that adiponectin may be involved in the signaling by adipose tissue to the brain to regulate reproductive functions. The first indications that adiponectin might act directly on the CNS came from studies reporting the presence of *AdipoR1* and *AdipoR2* transcripts in human and mouse whole brain tissues (Neumeier et al., 2007; Yamauchi et al., 2003). Since then, these two receptors have been detected in the human and rat hypothalamus (Guillod-Maximin et al., 2009; Hoyda, Fry, Ahima, & Ferguson, 2007; Kos et al., 2007), in the human, rat, and chicken pituitary gland (Rodriguez-Pacheco et al., 2007), in GT1-7 hypothalamic gonadotropin-releasing hormone (GnRH) neurons (Wen et al., 2008), and also in LβT2 pituitary gonadotroph cells (Lu, Tang, Olefsky, Mellon, & Webster, 2008). Qi et al. (2004) were the first to detect adiponectin in mice cerebrospinal fluid (CSF), though adiponectin concentrations represented only 1–4% of that found in the mice serum. These findings were then confirmed in human (Ebinuma et al., 2007; Kos et al., 2007; Kusminski et al., 2007; Neumeier et al., 2007) and rats (Caja, Torrente, Martinez, Abelenda, & Puerta, 2005), where adiponectin concentration in CSF was found to be approximately 1000-fold lower when compared with serum levels. Moreover, it was earlier reported that the trimeric LMW adiponectin is the predominant form found in human CSF (Ebinuma et al., 2007; Kusminski et al., 2007) and that HMW adiponectin inefficiently crosses the blood-to-brain barrier. Higher CSF adiponectin levels were found in females than in males (Neumeier et al., 2007; Qi et al., 2004), which is in agreement with gender differences observed in serum (Combs et al., 2003). However, further work is needed to confirm sexual dimorphism as Kos et al. (2007) reported similar CSF adiponectin concentrations in men and women.

Although several studies detected adiponectin in CSF, conflicting results have been reported with respect to the possible translocation of adiponectin across the blood–brain barrier. Qi et al. (2004) observed that an increase in CSF adiponectin concentrations after intravenous injection of recombinant adiponectin, thus suggesting a blood-to-CSF transport. On the opposite, Spranger et al. (2006) were unable to detect adiponectin in human CSF, and they demonstrated that radiolabeled globular adiponectin was unable to cross the blood-to-brain barrier in mice. Similar results were reported by Pan, Tu, and Kastin (2006), who reported the absence of blood-to-brain barrier permeation by adiponectin. The presence of *AdipoR1* and *AdipoR2* transcripts in brain endothelial cells and the observed reduction in inflammatory cytokine secretion by adiponectin in those cells further suggested that adiponectin does not need to cross the blood-to-brain barrier to exert its effect on the CNS (Spranger et al., 2006). The expression of *AdipoR1* and *AdipoR2* genes in the rat area postrema (AP), which lacks a blood-to-brain barrier, and the modification of AP neurons membrane properties with the addition of globular adiponectin suggest another possible site of adiponectin action in the CNS (Fry et al., 2006).

Accumulating evidence suggests that adiponectin may regulate reproductive functions through the modulation of gonadotropic hormone secretion. For example, adiponectin can inhibit LH release and GnRH-stimulated LH secretion in rat pituitary cell cultures (Rodriguez-Pacheco et al., 2007). However, these effects were only observed after short-term exposures (4 h) to recombinant adiponectin. Interestingly, the GnRH receptor transcript levels were reduced by 50% in pituitary cells treated with adiponectin, thus suggesting that adiponectin may reduce LH secretions through reduced gonadotroph sensitivity to GnRH (Rodriguez-Pacheco et al., 2007). Similar results were observed in mouse LβT2 gonadotroph cells, where an acute treatment (30 min) with globular and full-length adiponectin inhibited both the basal and GnRH-stimulated secretions of LH, whereas chronic exposure to adiponectin (48 h) had no effect on LH secretion (Lu et al., 2008). It was further demonstrated that adiponectin effect on LβT2 cells LH secretions is mediated through an increased activation of AMPK (Lu et al., 2008). In the rat hypothalamus, intracerebroventricular injections of human globular adiponectin also induced AMPK phosphorylation (Guillod-Maximin et al., 2009). It was further demonstrated that globular adiponectin can inhibit GnRH secretion in GT1-7 hypothalamic GnRH neurons through the activation of the AMPK signaling pathway (Wen et al., 2008). Collectively, these results suggest that hypothalamic and pituitary AMPK may act as a whole-body energy sensor mediating adiponectin action on gonadotropin release and reproductive function.

5. CONCLUSION

The above experimental and clinical investigations clearly demonstrate that adiponectin is an important regulator of female reproduction. Indeed, the presence of adiponectin receptors in numerous peripheral and central reproductive tissues and the capacity of adiponectin to activate multiple intracellular signal transduction pathways (see Figs. 10.1 and 10.2) support a role for adiponectin in female fertility and reproductive success. However, it remains to be determined whether adiponectin is essential for female reproduction and if the observed beneficial effects of adiponectin on reproductive functions are mainly due to its insulin-sensitizing effects. Interestingly, the inactivation of adiponectin in different null-mouse models had no effect on fertility (Maeda et al., 2002; Nawrocki et al., 2006), thus suggesting that adiponectin is facultative in reproduction. On the other hand, specific feedback mechanisms may exist to compensate for the lack of adiponectin. The recent identification of adiponectin paralogs, members of the C1q/tumor necrosis factor-α-related proteins (CTRPs), along with the increased serum levels of CTRP1 and CTRP6 in adiponectin-null mice are the first evidence that compensatory mechanisms may indeed be in place (Wong et al., 2004, 2008). Recent studies have also demonstrated that some of these paralogs share overlapping functions with adiponectin, including blood glucose lowering and the suppression of hepatic gluconeogenesis (Peterson, Wei, & Wong, 2010; Wong et al., 2009). These findings open a new research area to establish whether and how adiponectin and the different CTRPs are working in concert to regulate reproductive functions.

As mentioned above, further work is needed before using circulating adiponectin as a diagnostic tool in indentifying women with increased risk of developing reproductive abnormalities such as PCOS, preeclampsia, IUGR, and GDM. Most studies addressing this question were performed on total adiponectin levels, which may explain some of the observed discrepancies. Indeed, the adiponectin HMW isoform may prove to be a better predictor, and it will also be of interest to investigate the possible association between circulating levels of the different adiponectin paralogs and the development of those disorders. Other adipokines, such as leptin and resistin, have pleiotropic effects in reproduction (Tersigni et al., 2011). Therefore, it will also be of interest to determine whether combining the circulating level data of other adipokines with those of adiponectin would improve the accuracy of diagnosis tools.

Finally, most of the above studies reporting results for adiponectin receptors in reproductive tissues focus on AdipoR1 and AdipoR2. The fact that T-cadherin (*CDH-13* gene) binds hexameric and HMW adiponectin isoforms, support angiogenesis (Hebbard et al., 2008) and is upregulated by estradiol, progesterone, and the epidermal growth factor (Bromhead, Miller, & McDonald, 2006), suggest a role for this receptor in mediating adiponectin signal transduction in reproductive tissues. It will thus be of interest to see whether this receptor plays an essential role in reproduction.

ACKNOWLEDGMENTS

We thank Marie-Ève Fortin and Josée Lemay-Couchesne for their valuable help with bibliographical research.

REFERENCES

Agarwal, N., Rice, S. P. L., Bolusani, H., Luzio, S. D., Dunseath, G., Ludgate, M., et al. (2010). Metformin reduces arterial stiffness and improves endothelial function in young women with polycystic ovary syndrome: A randomized, placebo-controlled, crossover trial. *Journal of Clinical Endocrinology and Metabolism*, *95*, 722.

Andersen, K. K., Frystyk, J., Wolthers, O. D., Heuck, C., & Flyvbjerg, A. (2007). Gender differences of oligomers and total adiponectin during puberty: A cross-sectional study of 859 Danish school children. *Journal of Clinical Endocrinology and Metabolism*, *92*, 1857.

Archanco, M., Gomez-Ambrosi, J., Tena-Sempere, M., Fruhbeck, G., & Burrell, M. A. (2007). Expression of leptin and adiponectin in the rat oviduct. *Journal of Histochemistry and Cytochemistry*, *55*, 1027.

Ardawi, M. S., & Rouzi, A. A. (2005). Plasma adiponectin and insulin resistance in women with polycystic ovary syndrome. *Fertility and Sterility*, *83*, 1708.

Arita, Y., Kihara, S., Ouchi, N., Takahashi, M., Maeda, K., Miyagawa, J., et al. (1999). Paradoxical decrease of an adipose-specific protein, adiponectin, in obesity. *Biochemical and Biophysical Research Communications*, *257*, 79.

Aroda, V., Ciaraldi, T. P., Chang, S. A., Dahan, M. H., Chang, R. J., & Henry, R. R. (2008). Circulating and cellular adiponectin in polycystic ovary syndrome: Relationship to glucose tolerance and insulin action. *Fertility and Sterility*, *89*, 1200.

Asimakopoulos, B., Milousis, A., Gioka, T., Kabouromiti, G., Gianisslis, G., Troussa, A., et al. (2009). Serum pattern of circulating adipokines throughout the physiological menstrual cycle. *Endocrine Journal*, *56*, 425.

Barber, T. M., Hazell, M., Christodoulides, C., Golding, S. J., Alvey, C., Burling, K., et al. (2008). Serum levels of retinol-binding protein 4 and adiponectin in women with polycystic ovary syndrome: Associations with visceral fat but no evidence for fat mass-independent effects on pathogenesis in this condition. *Journal of Clinical Endocrinology and Metabolism*, *93*, 2859.

Barber, T. M., McCarthy, M. I., Wass, J. A., & Franks, S. (2006). Obesity and polycystic ovary syndrome. *Clinical Endocrinology, 65*, 137.

Benaitreau, D., Dieudonné, M. N., Dos Santos, E., Leneveu, M. C., de Mazancourt, P., & Pecquery, R. (2009). Antiproliferative effects of adiponectin on human trophoblastic cell lines JEG-3 and BeWo. *Biology of Reproduction, 80*, 1107.

Benaitreau, D., Dos Santos, E., Leneveu, M. C., Alfaidy, N., Feige, J. J., de Mazancourt, P., et al. (2010). Effects of adiponectin on human trophoblast invasion. *Journal of Endocrinology, 207*, 45.

Benaitreau, D., Dos Santos, E., Leneveu, M. C., de Mazancourt, P., Pecquery, R., & Dieudonné, M. N. (2010). Adiponectin promotes syncytialisation of BeWo cell line and primary trophoblast cells. *Reproductive Biology and Endocrinology, 8*, 128.

Ben-Ziv, R. G., & Hod, M. (2008). Gestational diabetes mellitus. *Fetal and Maternal Medicine Review, 19*, 245.

Bersinger, N. A., Birkhauser, M. H., & Wunder, D. M. (2006). Adiponectin as a marker of success in intracytoplasmic sperm injection/embryo transfer cycles. *Gynecological Endocrinology, 22*, 479.

Bik, W., Baranowska-Bik, A., Wolinska-Witort, E., Chmielowska, M., Martynska, L., & Baranowska, B. (2007). The relationship between metabolic status and levels of adiponectin and ghrelin in lean women with polycystic ovary syndrome. *Gynecological Endocrinology, 23*, 325.

Bjursell, M., Ahnmark, A., Bohlooly-Y, M., William-Olsson, L., Rhedin, M., Peng, X. R., et al. (2007). Opposing effects of adiponectin receptors 1 and 2 on energy metabolism. *Diabetes, 56*, 583.

Bottner, A., Kratzsch, J., Muller, G., Kapellen, T. M., Bluher, S., Keller, E., et al. (2004). Gender differences of adiponectin levels develop during the progression of puberty and are related to serum androgen levels. *The Journal of Clinical Endocrinology and Metabolism, 89*, 4053.

Bouchard, L., Guay, S. P., St-Pierre, J., Hivert, M. F., Perrron, P., & Brisson, D. (2011). Placental adiponectin gene DNA methylation levels is correlated with maternal blood glucose concentration. *Canadian Journal of Diabetes, 35*(2), 30.

Brassard, M., AinMelk, Y., & Baillargeon, J. P. (2008). Basic infertility including polycystic ovary syndrome. *The Medical Clinics of North America, 92*, 1163.

Brewer, C. J., & Balen, A. H. (2010). The adverse effects of obesity on conception and implantation. *Reproduction, 140*, 347.

Briana, D. D., & Malamitsi-Puchner, A. (2009). Reviews: Adipocytokines in normal and complicated pregnancies. *Reproductive Sciences, 16*, 921.

Brochu-Gaudreau, K., Rehfeldt, C., Blouin, R., Bordignon, V., Murphy, B. D., & Palin, M. F. (2010). Adiponectin action from head to toe. *Endocrine, 37*, 11.

Bromhead, C., Miller, J. H., & McDonald, F. J. (2006). Regulation of T-cadherin by hormones, glucocorticoid and EGF. *Gene, 374*, 58.

Brufani, C., Tozzi, A., Fintini, D., Ciampalini, P., Grossi, A., Fiori, R., et al. (2009). Sexual dimorphism of body composition and insulin sensitivity across pubertal development in obese Caucasian subjects. *European Journal of Endocrinology, 160*, 769.

Budak, E., Fernandez Sanchez, M., Bellver, J., Cervero, A., Simon, C., & Pellicer, A. (2006). Interactions of the hormones leptin, ghrelin, adiponectin, resistin, and PYY3-36 with the reproductive system. *Fertility and Sterility, 85*, 1563.

Caja, S., Torrente, M., Martinez, I., Abelenda, M., & Puerta, M. (2005). Adiponectin values are unchanged during pregnancy in rats. *Journal of Endocrinological Investigation, 28*, 609.

Caminos, J. E., Nogueiras, R., Gallego, R., Bravo, S., Tovar, S., Garcia-Caballero, T., et al. (2005). Expression and regulation of adiponectin and receptor in human and rat placenta. *Journal of Clinical Endocrinology and Metabolism, 90*, 4276.

Campos, D. B., Palin, M. F., Bordignon, V., & Murphy, B. D. (2008). The 'beneficial' adipokines in reproduction and fertility. *International Journal of Obesity*, *32*, 223.

Carmina, E., Orio, F., Palomba, S., Cascella, T., Longo, R. A., Colao, A. M., et al. (2005). Evidence for altered adipocyte function in polycystic ovary syndrome. *European Journal of Endocrinology*, *152*, 389.

Carmina, E., Orio, F., Palomba, S., Longo, R. A., Cascella, T., Colao, A., et al. (2006). Endothelial dysfunction in PCOS: Role of obesity and adipose hormones. *American Journal of Medicine*, *119*, 356 e1.

Chabrolle, C., Tosca, L., Crochet, S., Tesseraud, S., & Dupont, J. (2007). Expression of adiponectin and its receptors (AdipoR1 and AdipoR2) in chicken ovary: Potential role in ovarian steroidogenesis. *Domestic Animal Endocrinology*, *33*, 480.

Chabrolle, C., Tosca, L., & Dupont, J. (2007). Regulation of adiponectin and its receptors in rat ovary by human chorionic gonadotrophin treatment and potential involvement of adiponectin in granulosa cell steroidogenesis. *Reproduction*, *133*, 719.

Chabrolle, C., Tosca, L., Ramé, C., Lecomte, P., Royère, D., & Dupont, J. (2009). Adiponectin increases insulin-like growth factor I-induced progesterone and estradiol secretion in human granulosa cells. *Fertility and Sterility*, *92*, 1988.

Chalvatzas, N., Dafopoulos, K., Kosmas, G., Kallitsaris, A., Pournaras, S., & Messinis, I. E. (2009). Effect of ovarian hormones on serum adiponectin and resistin concentrations. *Fertility and Sterility*, *91*, 1189.

Chandrasekar, B., Boylston, W. H., Venkatachalam, K., Webster, N. J. G., Prabhu, S. D., & Valente, A. J. (2008). Adiponectin blocks interleukin-18-mediated endothelial cell death via APPL1-dependent AMP-activated protein kinase (AMPK) activation and IKK/NF-kappaB/PTEN suppression. *Journal of Biological Chemistry*, *283*, 24889.

Chappaz, E., Albornoz, M. S., Campos, D., Che, L., Palin, M. F., Murphy, B. D., et al. (2008). Adiponectin enhances in vitro development of swine embryos. *Domestic Animal Endocrinology*, *35*, 198.

Chen, J., Tan, B., Karteris, E., Zervou, S., Digby, J., Hillhouse, E. W., et al. (2006). Secretion of adiponectin by human placenta: Differential modulation of adiponectin and its receptors by cytokines. *Diabetologia*, *49*, 1292.

Chu, M. C., Cosper, P., Nakhuda, G. S., & Lobo, R. A. (2006). A comparison of oral and transdermal short-term estrogen therapy in postmenopausal women with metabolic syndrome. *Fertility and Sterility*, *86*, 1669.

Chu, M. C., Cosper, P., Orio, F., Carmina, E., & Lobo, R. A. (2006). Insulin resistance in postmenopausal women with metabolic syndrome and the measurements of adiponectin, leptin, resistin, and ghrelin. *American Journal of Obstetrics and Gynecology*, *194*, 100.

Cikos, S., Burkus, J., Bukovska, A., Fabian, D., Rehak, P., & Koppel, J. (2010). Expression of adiponectin receptors and effects of adiponectin isoforms in mouse preimplantation embryos. *Human Reproduction*, *25*, 2247.

Cnop, M., Havel, P. J., Utzschneider, K. M., Carr, D. B., Sinha, M. K., Boyko, E. J., et al. (2003). Relationship of adiponectin to body fat distribution, insulin sensitivity and plasma lipoproteins: Evidence for independent roles of age and sex. *Diabetologia*, *46*, 459.

Colombel, A., & Charbonnel, B. (1997). Weight gain and cardiovascular risk factors in the post-menopausal women. *Human Reproduction*, *12*, 134.

Combs, T. P., Berg, A. H., Rajala, M. W., Klebanov, S., Iyengar, P., Jimenez-Chillaron, J. C., et al. (2003). Sexual differentiation, pregnancy, calorie restriction, and aging affect the adipocyte-specific secretory protein adiponectin. *Diabetes*, *52*, 268.

Corbetta, S., Bulfamante, G., Cortelazzi, D., Barresi, V., Cetin, I., Mantovani, G., et al. (2005). Adiponectin expression in human fetal tissues during mid- and late gestation. *Journal of Clinical Endocrinology and Metabolism*, *90*, 2397.

Cortelazzi, D., Corbetta, S., Ronzoni, S., Pelle, F., Marconi, A., Cozzi, V., et al. (2007). Maternal and foetal resistin and adiponectin concentrations in normal and complicated pregnancies. *Clinical Endocrinology, 66*, 447.

Crosignani, P. G., Colombo, M., Vegetti, W., Somigliana, E., Gessati, A., & Ragni, G. (2003). Overweight and obese anovulatory patients with polycystic ovaries: Parallel improvements in anthropometric indices, ovarian physiology and fertility rate induced by diet. *Human Reproduction, 18*, 1928.

Crosignani, P. G., & Nicolosi, A. E. (2001). Polycystic ovarian disease: Heritability and heterogeneity. *Human Reproduction Update, 7*, 3.

Cseh, K., Baranyi, E., Melczer, Z., Kaszas, E., Palik, E., & Winkler, G. (2004). Plasma adiponectin and pregnancy-induced insulin resistance. *Diabetes Care, 27*, 274.

D'Anna, R., Baviera, G., Corrado, F., Giordano, D., De Vivo, A., Nicocia, G., et al. (2006). Adiponectin and insulin resistance in early- and late-onset pre-eclampsia. *BJOG: An International Journal of Obstetrics and Gynaecology, 113*, 1264.

Dafopoulos, K., Sourlas, D., Kallitsaris, A., Pournaras, S., & Messinis, I. E. (2009). Blood ghrelin, resistin, and adiponectin concentrations during the normal menstrual cycle. *Fertility and Sterility, 92*, 1389.

Dalamaga, M., Srinivas, S. K., Elovitz, M. A., Chamberland, J., & Mantzoros, C. S. (2011). Serum adiponectin and leptin in relation to risk for preeclampsia: Results from a large case-control study. *Metabolism, 60*, 1539.

Daoud, G., Amyot, M., Rassart, E., Masse, A., Simoneau, L., & Lafond, J. (2005). ERK1/2 and p38 regulate trophoblasts differentiation in human term placenta. *The Journal of Physiology, 566*. 2, 409.

Deepa, S. S., & Dong, L. Q. (2009). APPL1: Role in adiponectin signaling and beyond. *American Journal of Physiology, Endocrinology and Metabolism, 296*, E22.

Diamanti-Kandarakis, E., Papavassiliou, A. G., Kandarakis, S. A., & Chrousos, G. P. (2007). Pathophysiology and types of dyslipidemia in PCOS. *Trends in Endocrinology and Metabolism, 18*, 280.

Douchi, T., Ijuin, H., Nakamura, S., Oki, T., Yamamoto, S., & Nagata, Y. (1995). Body fat distribution in women with polycystic ovary syndrome. *Obstetrics and Gynecology, 86*, 516.

Dupont, J., Chabrolle, C., Rame´, C., Tosca, L., & Coyral-Castel, S. (2008). Role of the peroxisome proliferator-activated receptors, adenosine monophosphate-activated kinase, and adiponectin in the ovary. *PPAR Research, 2008*, 176275.

Ebinuma, H., Miida, T., Yamauchi, T., Hada, Y., Hara, K., Kubota, N., et al. (2007). Improved ELISA for selective measurement of adiponectin multimers and identification of adiponectin in human cerebrospinal fluid. *Clinical Chemistry, 53*, 1541.

Elkind-Hirsch, K., Marrioneaux, O., Bhushan, M., Vernor, D., & Bhushan, R. (2008). Comparison of single and combined treatment with exenatide and metformin on menstrual cyclicity in overweight women with polycystic ovary syndrome. *Journal of Clinical Endocrinology and Metabolism, 93*, 2670.

Escobar-Morreale, H. F., Villuendas, G., Botella-Carretero, J. I., Alvarez-Blasco, F., Sanchon, R., Luque-Ramirez, M., et al. (2006). Adiponectin and resistin in PCOS: A clinical, biochemical and molecular genetic study. *Human Reproduction, 21*, 2257.

Fasshauer, M., Bluher, M., Stumvoll, M., Tonessen, P., Faber, R., & Stepan, H. (2007). Differential regulation of visfatin and adiponectin in pregnancies with normal and abnormal placental function. *Clinical Endocrinology, 66*, 434.

Fasshauer, M., & Paschke, R. (2003). Regulation of adipocytokines and insulin resistance. *Diabetologia, 46*, 1594.

Fasshauer, M., Waldeyer, T., Seeger, J., Schrey, S., Ebert, T., Kratzsch, J., et al. (2008). Circulating high-molecular-weight adiponectin is upregulated in preeclampsia and is related to insulin sensitivity and renal function. *European Journal of Endocrinology, 158*, 197.

Filippi, E., Sentinelli, F., Trischitta, V., Romeo, S., Arca, M., Leonetti, F., et al. (2004). Association of the human adiponectin gene and insulin resistance. *European Journal of Human Genetics, 12*, 199.

Fischer, S., Santos, A. N., Thieme, R., Ramin, N., & Fischer, B. (2010). Adiponectin stimulates glucose uptake in rabbit blastocysts. *Biology of Reproduction, 83*, 859.

Frish, R. E. (1984). Body fat, puberty and fertility. *Biological Reviews, 59*, 161.

Fry, M., Smith, P. M., Hoyda, T. D., Duncan, M., Ahima, R. S., Sharkey, K. A., et al. (2006). Area postrema neurons are modulated by the adipocyte hormone adiponectin. *Journal of Neuroscience, 26*, 9695.

Gavrila, A., Chan, J. L., Yiannakouris, N., Kontogianni, M., Miller, L. C., Orlova, C., et al. (2003). Serum adiponectin levels are inversely associated with overall and central fat distribution but are not directly regulated by acute fasting or leptin administration in humans: Cross-sectional and interventional studies. *Journal of Clinical Endocrinology and Metabolism, 88*, 4823.

Georgiou, H. M., Lappas, M., Georgiou, G. M., Marita, A., Bryant, V. J., Hiscock, R., et al. (2008). Screening for biomarkers predictive of gestational diabetes mellitus. *Acta Diabetologica, 45*, 157.

Glintborg, D., Andersen, M., Hagen, C., Frystyk, J., Hulstrom, V., Flyvbjerg, A., et al. (2006). Evaluation of metabolic risk markers in polycystic ovary syndrome (PCOS). Adiponectin, ghrelin, leptin and body composition in hirsute PCOS patients and controls. *European Journal of Endocrinology, 155*, 337.

Glintborg, D., Frystyk, J., Hojlund, K., Andersen, K. K., Henriksen, J. E., Hermann, A. P., et al. (2008). Total and high molecular weight (HMW) adiponectin levels and measures of glucose and lipid metabolism following pioglitazone treatment in a randomized placebo-controlled study in polycystic ovary syndrome. *Clinical Endocrinology, 68*, 165.

Goran, M. I., & Gower, B. A. (2001). Longitudinal study on pubertal insulin resistance. *Diabetes, 50*, 2444.

Gui, Y., Silha, J. V., & Murphy, L. J. (2004). Sexual dimorphism and regulation of resistin, adiponectin, and leptin expression in the mouse. *Obesity Research, 12*, 1481.

Guillod-Maximin, E., Roy, A. F., Vacher, C. M., Aubourg, A., Bailleux, V., Lorsignol, A., et al. (2009). Adiponectin receptors are expressed in hypothalamus and colocalized with proopiomelanocortin and neuropeptide Y in rodent arcuate neurons. *The Journal of Endocrinology, 200*, 93.

Gulcelik, N. E., Aral, Y., Serter, R., Demir, Y., & Culha, C. (2006). Adiponectin is an independent determinant of insulin resistance in women with polycystic ovary syndrome. *Gynecological Endocrinology, 22*, 511.

Gulcelik, N. E., Aral, Y., Serter, R., & Koc, G. (2008). Association of hypoadiponectinemia with metabolic syndrome in patients with polycystic ovary syndrome. *Journal of the National Medical Association, 100*, 64.

Gutman, G., Barak, V., Maslovitz, S., Amit, A., Lessing, J. B., & Geva, E. (2009). Recombinant luteinizing hormone induces increased production of ovarian follicular adiponectin in vivo: Implications for enhanced insulin sensitivity. *Fertility and Sterility, 91*, 1837.

Haap, M., Machicao, F., Stefan, N., Thamer, C., Tschritter, O., Schnuck, F., et al. (2005). Genetic determinants of insulin action in polycystic ovary syndrome. *Experimental and Clinical Endocrinology & Diabetes, 113*, 275.

Hall, N., White, C., & O'Sullivan, A. J. (2009). The relationship between adiponectin, progesterone, and temperature across the menstrual cycle. *Journal of Endocrinological Investigation, 32*, 279.

Hara, K., Boutin, P., Mori, Y., Tobe, K., Dina, C., Yasuda, K., et al. (2002). Genetic variation in the gene encoding adiponectin is associated with an increased risk of type 2 diabetes in the Japanese population. *Diabetes, 51*, 536.

Haugen, F., Ranheim, T., Harsem, N. K., Lips, E., Staff, A. C., & Drevon, C. A. (2006). Increased plasma levels of adipokines in preeclampsia: Relationship to placenta and adipose tissue gene expression. *American Journal of Physiology, Endocrinology and Metabolism, 290*, E326.

Hebbard, L. W., Garlatti, M., Young, L. J., Cardiff, R. D., Oshima, R. G., & Ranscht, B. (2008). T-cadherin supports angiogenesis and adiponectin association with the vasculature in a mouse mammary tumor model. *Cancer Research, 68*, 1407.

Heinonen, S., Korhonen, S., Helisalmi, S., Koivunen, R., Tapanainen, J., Hippelainen, M., et al. (2005). Associations between two single nucleotide polymorphisms in the adiponectin gene and polycystic ovary syndrome. *Gynecological Endocrinology, 21*, 165.

Hendler, I., Blackwell, S. C., Mehta, S. H., Whitty, J. E., Russell, E., Sorokin, Y., et al. (2005). The levels of leptin, adiponectin, and resistin in normal weight, overweight, and obese pregnant women with and without preeclampsia. *American Journal of Obstetrics and Gynecology, 193*, 979.

Herse, F., Youpeng, B., Staff, A. C., Yong-Meid, J., Dechend, R., & Rong, Z. (2009). Circulating and uteroplacental adipocytokine concentrations in preeclampsia. *Reproductive Sciences, 16*, 584.

Hong, S. C., Yoo, S. W., Cho, G. J., Kim, T., Hur, J. Y., Park, Y. K., et al. (2007). Correlation between estrogens and serum adipocytokines in premenopausal and postmenopausal women. *Menopause, 14*, 835.

Horenburg, S., Fischer-Posovszky, P., Debatin, K. M., & Wabitsch, M. (2008). Influence of sex hormones on adiponectin expression in human adipocytes. *Hormone and Metabolic Research, 40*, 779.

Hoyda, T. D., Fry, M., Ahima, R. S., & Ferguson, A. V. (2007). Adiponectin selectively inhibits oxytocin neurons of the paraventricular nucleus of the hypothalamus. *The Journal of Physiology, 585*, 805.

Hu, E., Liang, P., & Spiegelman, B. M. (1996). AdipoQ is a novel adipose-specific gene dysregulated in obesity. *The Journal of Biological Chemistry, 271*, 10697.

Huang, K. C., Lue, B. H., Yen, R. F., Shen, C. G., Ho, S. R., Tai, T. Y., et al. (2004). Plasma adiponectin levels and metabolic factors in nondiabetic adolescents. *Obesity Research, 12*, 119.

Hug, C., Wang, J., Ahmad, N. S., Bogan, J. S., Tsao, T. S., & Lodish, H. F. (2004). T-cadherin is a receptor for hexameric and high-molecular-weight forms of Acrp30/adiponectin. *Proceedings of the National Academy of Sciences of the United States of America, 101*, 10308.

Ibanez, L., Sebastiani, G., Lopez-Bermejo, A., Diaz, M., Gomez-Roig, M. D., & de Zegher, F. (2008). Gender specificity of body adiposity and circulating adiponectin, visfatin, insulin, and insulin growth factor-I at term birth: Relation to prenatal growth. *The Journal of Clinical Endocrinology and Metabolism, 93*, 2774.

Ichida, K., Moriyama, T., Morita, H., Kondo, T., Yoshida, S., Ohara, N., et al. (2007). Plasma adiponectin concentrations and placental adiponectin expression in preeclamptic women. *Gynecological Endocrinology, 23*, 238.

Im, J. A., Lee, J. W., Lee, H. R., & Lee, D. C. (2006). Plasma adiponectin levels in postmenopausal women with or without long-term hormone therapy. *Maturitas, 54*, 65.

Iwaki, M., Matsuda, M., Maeda, N., Funahashi, T., Matsuzawa, Y., Makishima, M., et al. (2003). Induction of adiponectin, a fat-derived antidiabetic and antiatherogenic factor, by nuclear receptors. *Diabetes, 52*, 1655.

Jakubowska, J., Bohdanowicz-Pawlak, A., Milewicz, A., Szymczak, J., Bednarek-Tupikowska, G., & Demissie, M. (2008). Plasma cytokines in obese women with polycystic ovary syndrome, before and after metformin treatment. *Gynecological Endocrinology, 24*, 378.

Jarvenpaa, J., Vuoristo, J. T., Santaniemi, M., Ukkola, O., Savolainen, E. R., Jaaskelainen, M., et al. (2009). Adiponectin induced placental cell apoptosis could be mediated via the ADIPOR1-receptor in pre-eclampsia with IUGR. *Journal of Perinatal Medicine, 37*, 257.

Kaaja, R. J. (2008). Metabolic syndrome and the menopause. *Menopause International, 14*, 21.

Kajantie, E., Kaaja, R., Ylikorkala, O., Andersson, S., & Laivuori, H. (2005). Adiponectin concentrations in maternal serum: Elevated in preeclampsia but unrelated to insulin sensitivity. *Journal of the Society for Gynecologic Investigation, 12*, 433.

Kamoda, T., Saitoh, H., Saito, M., Sugiura, M., & Matsui, A. (2004). Serum adiponectin concentrations in newborn infants in early postnatal life. *Pediatric Research, 56*, 690.

Kanaley, J. A., Sames, C., Swisher, L., Swick, A. G., Ploutz-Snyder, L. L., Steppan, C. M., et al. (2001). Abdominal fat distribution in pre- and postmenopausal women: The impact of physical activity, age, and menopausal status. *Metabolism, 50*, 976.

Kaplowitz, P. B., Slora, E. J., Wasserman, R. C., & Pedlow, S. E. (2001). Earlier onset of puberty in girls: Relation to increased body mass index and race. *Pediatrics, 108*, 347.

Katz, A., Nambi, S. S., Mather, K., Baron, A. D., Follmann, D. A., Sullivan, G., et al. (2000). Quantitative insulin sensitivity check index: A simple, accurate method for assessing insulin sensitivity in humans. *The Journal of Clinical Endocrinology and Metabolism, 85*, 2402.

Kim, S. T., Marquard, K., Stephens, S., Louden, E., Allsworth, J., & Moley, K. H. (2011). Adiponectin and adiponectin receptors in the mouse preimplantation embryo and uterus. *Human Reproduction, 26*, 82.

Kirchengast, S., & Huber, J. (2001). Body composition characteristics and body fat distribution in lean women with polycystic ovary syndrome. *Human Reproduction, 16*, 1255.

Kleiblova, P., Springer, D., & Haluzik, M. (2006). The influence of hormonal changes during menstrual cycle on serum adiponectin concentrations in healthy women. *Physiological Research, 55*, 661.

Koga, K., Osuga, Y., Yoshino, O., Hirota, Y., Ruimeng, X., Hirata, T., et al. (2003). Elevated serum soluble vascular endothelial growth factor receptor 1 (sVEGFR-1) levels in women with preeclampsia. *Journal of Clinical Endocrinology and Metabolism, 88*, 2348.

Komar, C. M. (2005). Peroxisome proliferator-activated receptors (PPARs) and ovarian function–Implications for regulating steroidogenesis, differentiation, and tissue remodeling. *Reproductive Biology and Endocrinology, 3*, 41.

Kos, K., Harte, A. L., da Silva, N. F., Tonchev, A., Chaldakov, G., James, S., et al. (2007). Adiponectin and resistin in human cerebrospinal fluid and expression of adiponectin receptors in the human hypothalamus. *Journal of Clinical Endocrinology and Metabolism, 92*, 1129.

Kusminski, C. M., McTernan, P. G., Schraw, T., Kos, K., O'Hare, J. P., Ahima, R., et al. (2007). Adiponectin complexes in human cerebrospinal fluid: Distinct complex distribution from serum. *Diabetologia, 50*, 634.

Kyriakakou, M., Malamitsi-Puchner, A., Militsi, H., Boutsikou, T., Margeli, A., Hassiakos, D., et al. (2008). Leptin and adiponectin concentrations in intrauterine growth restricted and appropriate for gestational age fetuses, neonates, and their mothers. *European Journal of Endocrinology, 158*, 343.

Lagaly, D. V., Aad, P. Y., Grado-Ahuir, J. A., Hulsey, L. B., & Spicer, L. J. (2008). Role of adiponectin in regulating ovarian theca and granulosa cell function. *Molecular and Cellular Endocrinology, 284*, 38.

Lara-Castro, C., Luo, N., Wallace, P., Klein, R. L., & Garvey, W. T. (2006). Adiponectin multimeric complexes and the metabolic syndrome trait cluster. *Diabetes, 55*, 249.

Ledoux, S., Campos, D. B., Lopes, F. L., Dobias-Goff, M., Palin, M. F., & Murphy, B. D. (2006). Adiponectin induces periovulatory changes in ovarian follicular cells. *Endocrinology, 147*, 5178.

Lee, C. G., Carr, M. C., Murdoch, S. J., Mitchell, E., Woods, N. F., Wener, M. H., et al. (2009). Adipokines, inflammation, and visceral adiposity across the menopausal transition: A prospective study. *Journal of Clinical Endocrinology and Metabolism*, *94*, 1104.

Lee, M. H., Klein, R. L., El-Shewy, H. M., Luttrell, D. K., & Luttrell, L. M. (2008). The adiponectin receptors AdipoR1 and AdipoR2 activate ERK1/2 through a Src/Ras-dependent pathway and stimulate cell growth. *Biochemistry*, *47*, 11682.

Leff, T. (2003). AMP-activated protein kinase regulates gene expression by direct phosphorylation of nuclear proteins. *Biochemical Society Transactions*, *31*, 224.

Leung, K. C., Xu, A., Craig, M. E., Martin, A., Lam, K. S. L., & O'Sullivan, A. J. (2009). Adiponectin isoform distribution in women–relationship to female sex steroids and insulin sensitivity. *Metabolism*, *58*, 239.

Levy-Marchal, C., & Jaquet, D. (2004). Long-term metabolic consequences of being born small for gestational age. *Pediatric Diabetes*, *5*, 147.

Li, L., Yun, J. H., Lee, J. H., Song, S., Choi, B. C., & Baek, K. H. (2011). Association study of +45G15G(T/G) and +276(G/T) polymorphisms in the adiponectin gene in patients with polycystic ovary syndrome. *International Journal of Molecular Medicine*, *27*, 283.

Liu, Y. H., Tsai, E. M., Wu, L. C., Chen, S. Y., Chang, Y. H., Jong, S. B., et al. (2005). Higher basal adiponectin levels are associated with better ovarian response to gonadotropin stimulation during in vitro fertilization. *Gynecologic and Obstetric Investigation*, *60*, 167.

Lobo, R. A. (2008). Metabolic syndrome after menopause and the role of hormones. *Maturitas*, *60*, 10.

Lord, E., Ledoux, S., Murphy, B. D., Beaudry, D., & Palin, M. F. (2005). Expression of adiponectin and its receptors in swine. *Journal of Animal Science*, *83*, 565.

Low, C. F., Tohit, E. R. M., Chong, P. P., & Idris, F. (2011). Adiponectin SNP45TG is associated with gestational diabetes mellitus. *Archives of Gynecology and Obstetrics*, *283*, 1255.

Lu, M., Tang, Q., Olefsky, J. M., Mellon, P. L., & Webster, N. J. G. (2008). Adiponectin activates adenosine monophosphate-activated protein kinase and decreases luteinizing hormone secretion in LbetaT2 gonadotropes. *Molecular Endocrinology*, *22*, 760.

Lu, D., Yang, X., Wu, Y., Wang, H., Huang, H., & Dong, M. (2006). Serum adiponectin, leptin and soluble leptin receptor in pre-eclampsia. *International Journal of Gynaecology and Obstetrics*, *95*, 121.

Luque-Ramirez, M., Alvarez-Blasco, F., & Escobar-Morreale, H. F. (2009). Antiandrogenic contraceptives increase serum adiponectin in obese polycystic ovary syndrome patients. *Obesity*, *17*, 3.

Maddineni, S., Metzger, S., Ocon, O., Hendricks, G., 3rd, & Ramachandran, R. (2005). Adiponectin gene is expressed in multiple tissues in the chicken: Food deprivation influences adiponectin messenger ribonucleic acid expression. *Endocrinology*, *146*, 4250.

Maeda, K., Okubo, K., Shimomura, I., Funahashi, T., Matsuzawa, Y., & Matsubara, K. (1996). cDNA cloning and expression of a novel adipose specific collagen-like factor, apM1 (AdiPose Most abundant Gene transcript 1). *Biochemical and Biophysical Research Communications*, *221*, 286.

Maeda, N., Shimomura, I., Kishida, K., Nishizawa, H., Matsuda, M., Nagaretani, H., et al. (2002). Diet-induced insulin resistance in mice lacking adiponectin/ACRP30. *Nature Medicine*, *8*, 731.

Maillard, V., Uzbekova, S., Guignot, F., Perreau, C., Rame, C., Coyral-Castel, S., et al. (2010). Effect of adiponectin on bovine granulosa cell steroidogenesis, oocyte maturation and embryo development. *Reproductive Biology and Endocrinology*, *8*, 23.

Majuri, A., Santaniemi, M., Rautio, K., Kunnari, A., Vartiainen, J., Ruokonen, A., et al. (2007). Rosiglitazone treatment increases plasma levels of adiponectin and decreases levels of resistin in overweight women with PCOS: A randomized placebo-controlled study. *European Journal of Endocrinology*, *156*, 263.

Mao, X., Kikani, C. K., Riojas, R. A., Langlais, P., Wang, L., Ramos, F. J., et al. (2006). APPL1 binds to adiponectin receptors and mediates adiponectin signalling and function. *Nature Cell Biology*, *8*, 516.
Martinez-Cordero, C., Amador-Licona, N., Guizar-Mendoza, J. M., Hernandez-Mendez, J., & Ruelas-Orozco, G. (2006). Body fat at birth and cord blood levels of insulin, adiponectin, leptin, and insulin-like growth factor-I in small-for-gestational-age infants. *Archives of Medical Research*, *37*, 490.
Martos-Moreno, G. A., Barrios, V., & Argente, J. (2006). Normative data for adiponectin, resistin, interleukin 6, and leptin/receptor ratio in a healthy Spanish pediatric population: Relationship with sex steroids. *European Journal of Endocrinology*, *155*, 429.
Martos-Moreno, G. A., Barrios, V., de Pipaon, M. S., Pozo, J., Dorronsoro, I., Martinez-Biarge, M., et al. (2009). Influence of prematurity and growth restriction on the adipokine profile, IGF1, and ghrelin levels in cord blood: Relationship with glucose metabolism. *European Journal of Endocrinology*, *161*, 381.
Masuyama, H., & Hiramatsu, Y. (2010). Angiogenic proteins and adipocytokines as markers for prediction of preeclampsia. *Expert Review of Obstetrics and Gynecology*, *5*, 717.
Masuyama, H., Segawa, T., Sumida, Y., Masumoto, A., Inoue, S., Akahori, Y., et al. (2010). Different profiles of circulating angiogenic factors and adipocytokines between early- and late-onset pre-eclampsia. *BJOG: An International Journal of Obstetrics and Gynaecology*, *117*, 314.
Matsubara, M., Maruoka, S., & Katayose, S. (2002). Inverse relationship between plasma adiponectin and leptin concentrations in normal-weight and obese women. *European Journal of Endocrinology*, *147*, 173.
Maury, E., Ehala-Aleksejev, K., Guiot, Y., Detry, R., Vandenhooft, A., & Brichard, S. M. (2007). Adipokines oversecreted by omental adipose tissue in human obesity. *American Journal of Physiology, Endocrinology and Metabolism*, *293*, E656.
Maynard, S. E., & Karumanchi, S. A. (2011). Angiogenic factors and preeclampsia. *Seminars in Nephrology*, *31*, 33.
Maynard, S. E., Min, J. Y., Merchan, J., Lim, K. H., Li, J., Mondal, S., et al. (2003). Excess placental soluble fms-like tyrosine kinase 1 (sFlt1) may contribute to endothelial dysfunction, hypertension, and proteinuria in preeclampsia. *The Journal of Clinical Investigation*, *111*, 649.
Mazaki-Tovi, S., Romero, R., Vaisbuch, E., Erez, O., Mittal, P., Chaiworapongsa, T., et al. (2009). Maternal serum adiponectin multimers in gestational diabetes. *Journal of Perinatal Medicine*, *37*, 637.
Mazaki-Tovi, S., Romero, R., Vaisbuch, E., Kusanovic, J. P., Erez, O., Gotsch, F., et al. (2009). Maternal serum adiponectin multimers in preeclampsia. *Journal of Perinatal Medicine*, *37*, 349.
McDonald, E. A., & Wolfe, M. W. (2009). Adiponectin attenuation of endocrine function within human term trophoblast cells. *Endocrinology*, *150*, 4358.
Merki-Feld, G. S., Imthurn, B., Rosselli, M., & Spanaus, K. (2011). Serum concentrations of high-molecular weight adiponectin and their association with sex steroids in premenopausal women. *Metabolism*, *60*, 180.
Michalakis, K. G., & Segars, J. H. (2010). The role of adiponectin in reproduction: From polycystic ovary syndrome to assisted reproduction. *Fertility and Sterility*, *94*, 1949.
Moore, R. K., Otsuka, F., & Shimasaki, S. (2001). Role of ERK1/2 in the differential synthesis of progesterone and estradiol by granulosa cells. *Biochemical and Biophysical Research Communications*, *289*, 796.
Moran, A., Jacobs, D. R., Steinberger, J., Hong, C. P., Prineas, R., Luepker, R. V., et al. (1999). Insulin resistance during puberty: Results from clamp studies in 357 children. *Diabetes*, *48*, 2039.

Moran, L. J., Misso, M. L., Wild, R. A., & Norman, R. J. (2010). Impaired glucose tolerance, type 2 diabetes and metabolic syndrome in polycystic ovary syndrome: A systematic review and meta-analysis. *Human Reproduction Update, 16*, 347.

Mori, T., Shinohara, K., Wakatsuki, A., Watanabe, K., & Fujimaki, A. (2010). Adipocytokines and endothelial function in preeclamptic women. *Hypertension Research, 33*, 250.

Nakano, Y., Tobe, T., Choi-Miura, N. H., Mazda, T., & Tomita, M. (1996). Isolation and characterization of GBP28, a novel gelatin-binding protein purified from human plasma. *Journal of Biochemistry, 120*, 803.

Nakashima, R., Kamei, N., Yamane, K., Nakanishi, S., Nakashima, A., & Kohno, N. (2006). Decreased total and high molecular weight adiponectin are independent risk factors for the development of type 2 diabetes in Japanese-Americans. *Journal of Clinical Endocrinology and Metabolism, 91*, 3873.

Naruse, K., Yamasaki, M., Umekage, H., Sado, T., Sakamoto, Y., & Morikawa, H. (2005). Peripheral blood concentrations of adiponectin, an adipocyte-specific plasma protein, in normal pregnancy and preeclampsia. *Journal of Reproductive Immunology, 65*, 65.

Nawrocki, A. R., Rajala, M. W., Tomas, E., Pajvani, U. B., Saha, A. K., Trumbauer, M. E., et al. (2006). Mice lacking adiponectin show decreased hepatic insulin sensitivity and reduced responsiveness to peroxisome proliferator-activated receptor gamma agonists. *Journal of Biological Chemistry, 281*, 2654.

Neumeier, M., Weigert, J., Buettner, R., Wanninger, J., Schaffler, A., Muller, A. M., et al. (2007). Detection of adiponectin in cerebrospinal fluid in humans. *American Journal of Physiology, Endocrinology and Metabolism, 293*, E965.

Nien, J. K., Mazaki-Tovi, S., Romero, R., Erez, O., Kusanovic, J. P., Gotsch, F., et al. (2007a). Adiponectin in severe preeclampsia. *Journal of Perinatal Medicine, 35*, 503.

Nien, J. K., Mazaki-Tovi, S., Romero, R., Erez, O., Kusanovic, J. P., Gotsch, F., et al. (2007b). Plasma adiponectin concentrations in non-pregnant, normal and overweight pregnant women. *Journal of Perinatal Medicine, 35*, 522.

Nishizawa, H., Shimomura, I., Kishida, K., Maeda, N., Kuriyama, H., Nagaretani, H., et al. (2002). Androgens decrease plasma adiponectin, an insulin-sensitizing adipocyte-derived protein. *Diabetes, 51*, 2734.

Norman, R. J., Dewailly, D., Legro, R. S., & Hicket, T. E. (2007). Polycystic ovary syndrome. *Lancet, 370*, 685.

O'Brien, T. E., Ray, J. G., & Chan, W. S. (2003). Maternal body mass index and the risk of preeclampsia: A systematic overview. *Epidemiology, 14*, 368.

O'Connor, A., Phelan, N., Kyaw Tun, T., Boran, G., Gibney, J., & Roche, H. M. (2010). High-molecular-weight adiponectin is selectively reduced in women with polycystic ovary syndrome independent of body mass index and severity of insulin resistance. *Journal of Clinical Endocrinology and Metabolism, 95*, 1378.

Odden, N., & Morkrid, L. (2007). High molecular weight adiponectin dominates in cord blood of newborns but is unaffected by pre-eclamptic pregnancies. *Clinical Endocrinology, 67*, 891.

Ogland, B., Romundstad, P. R., Vefring, H., Forman, M. R., Nilsen, S. T., & Vatten, L. J. (2010). Preeclampsia and adiponectin in cord blood. *Hormone Research in Paediatrics, 74*, 92.

Ong, K. K., Petry, C. J., Emmett, P. M., Sandhu, M. S., Kiess, W., Hales, C. N., et al. (2004). Insulin sensitivity and secretion in normal children related to size at birth, postnatal growth, and plasma insulin-like growth factor-I levels. *Diabetologia, 47*, 1064.

Orio, F., Palomba, S., Cascella, T., Milan, G., Mioni, R., Pagano, C., et al. (2003). Adiponectin levels in women with polycystic ovary syndrome. *Journal of Clinical Endocrinology and Metabolism, 88*, 2619.

Ortega, H. H., Rey, F., Velazquez, M. M. L., & Padmanabhan, V. (2010). Developmental programming: Effect of prenatal steroid excess on intraovarian components of insulin signaling pathway and related proteins in sheep. *Biology of Reproduction, 82*, 1065.

Ouchi, N., Kobayashi, H., Kihara, S., Kumada, M., Sato, K., Inoue, T., et al. (2004). Adiponectin stimulates angiogenesis by promoting cross-talk between AMP-activated protein kinase and Akt signaling in endothelial cells. *Journal of Biological Chemistry, 279*, 1304.

Ouyang, Y., Chen, H., & Chen, H. (2007). Reduced plasma adiponectin and elevated leptin in pre-eclampsia. *International Journal of Gynecology & Obstetrics, 98*, 110.

Pajvani, U. B., Du, X., Combs, T. P., Berg, A. H., Rajala, M. W., Schulthess, T., et al. (2003). Structure-function studies of the adipocyte-secreted hormone Acrp30/ adiponectin. Implications fpr metabolic regulation and bioactivity. *Journal of Biological Chemistry, 278*, 9073.

Pajvani, U. B., Hawkins, M., Combs, T. P., Rajala, M. W., Doebber, T., Berger, J. P., et al. (2004). Complex distribution, not absolute amount of adiponectin, correlates with thiazolidinedione-mediated improvement in insulin sensitivity. *Journal of Biological Chemistry, 279*, 12152.

Pan, W., Tu, H., & Kastin, A. J. (2006). Differential BBB interactions of three ingestive peptides: Obestatin, ghrelin, and adiponectin. *Peptides, 27*, 911.

Panidis, D., Kourtis, A., Farmakiotis, D., Mouslech, T., Rousso, D., & Koliakos, G. (2003). Serum adiponectin levels in women with polycystic ovary syndrome. *Human Reproduction, 18*, 1790.

Panidis, D., Kourtis, A., Kukuvitis, A., Farmakiotis, D., Xita, N., Georgiou, I., et al. (2004). Association of the T45G polymorphism in exon 2 of the adiponectin gene with polycystic ovary syndrome: Role of Delta4-androstenedione. *Human Reproduction, 19*, 1728.

Paradisi, G., Ianniello, F., Tomei, C., Bracaglia, M., Carducci, B., Gualano, M. R., et al. (2010). Longitudinal changes of adiponectin, carbohydrate and lipid metabolism in pregnant women at high risk for gestational diabetes. *Gynecological Endocrinology, 26*, 539.

Peake, P. W., Kriketos, A. D., Campbell, L. V., Shen, Y., & Charlesworth, J. A. (2005). The metabolism of isoforms of human adiponectin: Studies in human subjects and in experimental animals. *European Journal of Endocrinology, 153*, 409.

Petersen, S., Gotfredsen, A., & Knudsen, F. U. (1988). Lean body mass in small for gestational age and appropriate for gestational age infants. *Journal of Pediatrics, 113*, 886.

Peterson, J. M., Wei, Z., & Wong, G. W. (2010). C1q/TNF-related protein-3 (CTRP3), a novel adipokine that regulates hepatic glucose output. *Journal of Biological Chemistry, 285*, 39691.

Petrone, A., Zavarella, S., Caiazzo, A., Leto, G., Spoletini, M., Potenziani, S., et al. (2006). The promoter region of the adiponectin gene is a determinant in modulating insulin sensitivity in childhood obesity. *Obesity, 14*, 1498.

Pierre, P., Froment, P., Nègre, D., Ramé, C., Barateau, V., Chabrolle, C., et al. (2009). Role of adiponectin receptors, AdipoR1 and AdipoR2, in the steroidogenesis of the human granulosa tumor cell line, KGN. *Human Reproduction, 24*, 2890.

Punthakee, Z., Delvin, E. E., O'Loughlin, J., Paradis, G., Levy, E., Platt, R. W., et al. (2006). Adiponectin, adiposity, and insulin resistance in children and adolescents. *Journal of Clinical Endocrinology and Metabolism, 91*, 2119.

Qi, Y., Takahashi, N., Hileman, S. M., Patel, H. R., Berg, A. H., Pajvani, U. B., et al. (2004). Adiponectin acts in the brain to decrease body weight. *Nature Medicine, 10*, 524.

Ramachandran, R., Ocon-Grove, O. M., & Metzger, S. L. (2007). Molecular cloning and tissue expression of chicken AdipoR1 and AdipoR2 complementary deoxyribonucleic acids. *Domestic Animal Endocrinology, 33*, 19.

Ramsay, J. E., Jamieson, N., Greer, I. A., & Sattar, N. (2003). Paradoxical elevation in adiponectin concentrations in women with preeclampsia. *Hypertension, 42*, 891.

Ranheim, T., Haugen, F., Staff, A. C., Braekke, K., Harsem, N., & Drevon, C. A. (2004). Adiponectin is reduced in gestational diabetes mellitus in normal weight women. *Acta Obstetricia et Gynecologica Scandinavica, 83*, 341.

Ranjzad, F., Mahmoudi, T., Shemirani, A. I., Mahban, A., Nikzamir, A., Vahedi, M., et al. (2011). A common variant in the adiponectin gene and polycystic ovary syndrome risk. *Molecular Biology Reports, 39*, 2313.

Retnakaran, R., Connelly, P. W., Maguire, G., Sermer, M., Zinman, B., & Hanley, A. J. G. (2007). Decreased high-molecular-weight adiponectin in gestational diabetes: Implications for the pathophysiology of Type 2 diabetes. *Diabetic Medicine, 24*, 245.

Robinson, R. S., Woad, K. J., Hammond, A. J., Laird, M., Hunter, M. G., & Mann, G. E. (2009). Angiogenesis and vascular function in the ovary. *Reproduction, 138*, 869.

Robitaille, J., & Grant, A. M. (2008). The genetics of gestational diabetes mellitus: Evidence for relationship with type 2 diabetes mellitus. *Genetics in Medicine, 10*, 240.

Rodriguez-Pacheco, F., Martinez-Fuentes, A. J., Tovar, S., Pinilla, L., Tena-Sempere, M., Dieguez, C., et al. (2007). Regulation of pituitary cell function by adiponectin. *Endocrinology, 148*, 401.

Rouen, P. A., Lukacs, J. L., & Reame, N. E. (2010). Adipokine concentrations in nonobese women: A study of reproductive aging, body mass index, and menstrual cycle effects. *Biological Research for Nursing, 12*, 54.

Ruchat, S. M., Loos, R. J., Rankinen, T., Vohl, M. C., Weisnagel, S. J., Despres, J. P., et al. (2008). Associations between glucose tolerance, insulin sensitivity and insulin secretion phenotypes and polymorphisms in adiponectin and adiponectin receptor genes in the Quebec Family Study. *Diabetic Medicine, 25*, 400.

Sawidou, M. D., Nelson, S. M., Makgoba, M., Messow, C. M., Sattar, N., & Nicolaides, K. (2010). First-trimester prediction of gestational diabetes mellitus: Examining the potential of combining maternal characteristics and laboratory measures. *Diabetes, 59*, 3017.

Sawidou, M. D., Sotiriadis, A., Kaihura, C., Nicolaides, K. H., & Sattar, N. (2008). Circulating levels of adiponectin and leptin at 23-25 weeks of pregnancy in women with impaired placentation and in those with established fetal growth restriction. *Clinical Science, 115*, 219.

Scherer, P. E., Williams, S., Fogliano, M., Baldini, G., & Lodish, H. F. (1995). A novel serum protein similar to C1q, produced exclusively in adipocytes. *Journal of Biological Chemistry, 270*, 26746.

Schmidt, T., Fischer, S., Tsikolia, N., Navarrete Santos, A., Rohrbach, S., Ramin, N., et al. (2008). Expression of adipokines in preimplantation rabbit and mice embryos. *Histochemistry and Cell Biology, 129*, 817.

Schnatz, P. F., & Schnatz, J. D. (2006). Dyslipidemia in menopause: Mechanisms and management. *Obstetrical and Gynecological Survey, 61*, 608.

Schoppen, S., Riestra, P., Garcia-Anguita, A., Lopez-Simon, L., Cano, B., de Oya, I., et al. (2010). Leptin and adiponectin levels in pubertal children: Relationship with anthropometric variables and body composition. *Clinical Chemistry and Laboratory Medicine, 48*, 707.

Sekar, N., Garmey, J. C., & Veldhuis, J. D. (2000). Mechanisms underlying the steroidogenic synergy of insulin and luteinizing hormone in porcine granulosa cells: Joint amplification of pivotal sterol-regulatory genes encoding the low-density lipoprotein (LDL) receptor, steroidogenic acute regulatory (stAR) protein and cytochrome P450 side-chain cleavage (P450scc) enzyme. *Molecular and Cellular Endocrinology, 159*, 25.

Sepilian, V., & Nagamani, M. (2005). Adiponectin levels in women with polycystic ovary syndrome and severe insulin resistance. *Journal of the Society for Gynecologic Investigation, 12*, 129.

Shi, Q. J., Lei, Z. M., Rao, Ch. V., & Lin, J. (1993). Novel role of human chorionic gonadotropin in differentiation of human cytotrophoblasts. *Endocrinology, 132*, 1387.

Shroff, R., Kerchner, A., Maifeld, M., Van Beek, E. J., Jagasia, D., & Dokras, A. (2007). Young obese women with polycystic ovary syndrome have evidence of early coronary atherosclerosis. *Journal of Clinical Endocrinology and Metabolism*, *92*, 4609.

Sibai, B., Dekker, G., & Kupferminc, M. (2005). Pre-eclampsia. *Lancet*, *365*, 785.

Sieminska, L., Marek, B., Kos-Kudla, B., Niedziolka, D., Kajdaniuk, D., Nowak, M., et al. (2004). Serum adiponectin in women with polycystic ovarian syndrome and its relation to clinical, metabolic and endocrine parameters. *Journal of Endocrinological Investigation*, *27*, 528.

Sieminska, L., Wojciechowska, C., Niedziolka, D., Marek, B., Kos-Kudla, B., Kajdaniuk, D., et al. (2005). Effect of postmenopause and hormone replacement therapy on serum adiponectin levels. *Metabolism*, *54*, 1610.

Sir-Petermann, T., Maliqueo, M., Codner, E., Echiburu, B., Crisosto, N., Perez, V., et al. (2007). Early metabolic derangements in daughters of women with polycystic ovary syndrome. *Journal of Clinical Endocrinology and Metabolism*, *92*, 4637.

Soheilykhah, S., Mohammadi, M., Mojibian, M., Rahimi-Saghand, S., Rashidi, M., Hadinedoushan, H., et al. (2009). Maternal serum adiponectin concentration in gestational diabetes. *Gynecological Endocrinology*, *25*, 593.

Soni, A. C., Conroy, M. B., Mackey, R. H., & Kuller, L. H. (2011). Ghrelin, leptin, adiponectin, and insulin levels and concurrent and future weight change in overweight, postmenopausal women. *Menopause*, *18*, 296.

Soto, N., Bazaes, R. A., Pena, V., Salazar, T., Avila, A., Iniguez, G., et al. (2003). Insulin sensitivity and secretion are related to catch-up growth in small-for-gestational-age infants at age 1 year: Results from a prospective cohort. *Journal of Clinical Endocrinology and Metabolism*, *88*, 3645.

Sowers, M. F., Wildman, R. P., Mancuso, P., Eyvazzadeh, A. D., Karvonen-Gutierrez, C. A., Rillamas-Sun, E., et al. (2008). Change in adipocytokines and ghrelin with menopause. *Maturitas*, *59*, 149.

Spranger, J., Mohlig, M., Wegewitz, U., Ristow, M., Pfeiffer, A. F. H., Schill, T., et al. (2004). Adiponectin is independently associated with insulin sensitivity in women with polycystic ovary syndrome. *Clinical Endocrinology*, *61*, 738.

Spranger, J., Verma, S., Gohring, I., Bobbert, T., Seifert, J., Sindler, A. L., et al. (2006). Adiponectin does not cross the blood-brain barrier but modifies cytokine expression of brain endothelial cells. *Diabetes*, *55*, 141.

Steiner, C. A., Janez, A., Jensterle, M., Reisinger, K., Forst, T., & Pfutzner, A. (2007). Impact of treatment with rosiglitazone or metformin on biomarkers for insulin resistance and metabolic syndrome in patients with polycystic ovary syndrome. *Journal of Diabetes Science and Technology*, *1*, 211.

Street, M. E., Volta, C., Ziveri, M. A., Viani, I., & Bernasconi, S. (2009). Markers of insulin sensitivity in placentas and cord serum of intrauterine growth-restricted newborns. *Clinical Endocrinology*, *71*, 394.

Stumvoll, M., Tschritter, O., Fritsche, A., Staiger, H., Renn, W., Weisser, M., et al. (2002). Association of the T-G polymorphism in adiponectin (exon 2) with obesity and insulin sensitivity: Interaction with family history of type 2 diabetes. *Diabetes*, *51*, 37.

Sumino, H., Takahashi, T., Itoh, T., Kusaka, K., Yamakawa, J., Ichikawa, S., et al. (2004). Plasma adiponectin levels in post-menopausal women receiving hormone replacement therapy. *The Journal of International Medical Research*, *32*, 639.

Sun, X., He, J., Mao, C., Han, R., Wang, Z., Liu, Y., et al. (2008). Negative regulation of adiponectin receptor 1 promoter by insulin via a repressive nuclear inhibitory protein element. *FEBS Letters*, *582*, 3401.

Tabandeh, M. R., Hosseini, A., Saeb, M., Kafi, M., & Saeb, S. (2010). Changes in the gene expression of adiponectin and adiponectin receptors (AdipoR1 and AdipoR2) in ovarian follicular cells of dairy cow at different stages of development. *Theriogenology*, *73*, 659.

Takaya, J., Yamato, F., Higashino, H., & Kaneko, K. (2007). Intracellular magnesium and adipokines in umbilical cord plasma and infant birth size. *Pediatric Research*, *62*, 700.

Takemura, Y., Osuga, Y., Yamauchi, T., Kobayashi, M., Harada, M., Hirata, T., et al. (2006). Expression of adiponectin receptors and its possible implication in the human endometrium. *Endocrinology*, *147*, 3203.

Tang, Y. T., Hu, T., Arterburn, M., Boyle, B., Bright, J. M., Emtage, P. C., et al. (2005). PAQR proteins: A novel membrane receptor family defined by an ancient 7-transmembrane pass motif. *Journal of Molecular Evolution*, *61*, 372.

Tersigni, C., Di Nicuolo, F., D'Ippolito, S., Veglia, M., Castellucci, M., & Di Simone, N. (2011). Adipokines: New emerging roles in fertility and reproduction. *Obstetrical and Gynecological Survey*, *66*, 47.

The Rotterdam ESHRE/ASRM-Sponsored PCOS Consensus Workshop Group, (2004). Revised 2003 consensus on diagnostic criteria and long-term health risks related to polycystic ovary syndrome (PCOS). *Human Reproduction*, *19*, 41.

The ESHRE Capri Workshop Group, (2006). Nutrition and reproduction in women. *Human Reproduction Update*, *12*, 193.

Tomlinson, J., Millward, A., Stenhouse, E., & Pinkey, J. (2010). Type 2 diabetes and cardiovascular disease in polycystic ovary syndrome: What are the risks and can they be reduced? *Diabetic Medicine*, *27*, 498.

Tosca, L., Froment, P., Solnais, P., Ferre, P., Foufelle, F., & Dupont, J. (2005). Adenosine 5'-monophosphate-activated protein kinase regulates progesterone secretion in rat granulosa cells. *Endocrinology*, *146*, 4500.

Toulis, K. A., Goulis, D. G., Farmakiotis, D., Georgopoulos, N. A., Katsikis, I., Tarlatzis, B. C., et al. (2009). Adiponectin levels in women with polycystic ovary syndrome: A systematic review and a meta-analysis. *Human Reproduction Update*, *15*, 297.

Travers, S. H., Jeffers, B. W., Bloch, C. A., Hill, J. O., & Eckel, R. H. (1995). Gender and Tanner stage differences in body composition and insulin sensitivity in early pubertal children. *Journal of Clinical Endocrinology and Metabolism*, *80*, 172.

Trayhurn, P., & Wood, I. S. (2004). Adipokines: Inflammation and the pleiotropic role of white adipose tissue. *British Journal of Nutrition*, *92*, 347.

Trolle, B., Lauszus, F. F., Frystyk, J., & Flyvbjerg, A. (2010). Adiponectin levels in women with polycystic ovary syndrome: Impact of metformin treatment in a randomized controlled study. *Fertility and Sterility*, *94*, 2234.

Trujillo, M. E., & Scherer, P. E. (2005). Adiponectin–journey from an adipocyte secretory protein to biomarker of the metabolic syndrome. *Journal of Internal Medicine*, *257*, 167.

Tsai, P. J., Yu, C. H., Hsu, S. P., Lee, Y. H., Huang, I. T., Ho, S. C., et al. (2005). Maternal plasma adiponectin concentrations at 24 to 31 weeks of gestation: Negative association with gestational diabetes mellitus. *Nutrition*, *21*, 1095.

Tsuchiya, K., Akaza, I., Yoshimoto, T., & Hirata, Y. (2009). Pioglitazone improves endothelial function with increased adiponectin and high-density lipoprotein cholesterol levels in type 2 diabetes. *Endocrine Journal*, *56*, 691.

Tuuli, M. G., & Odibo, A. O. (2010). First- and second-trimester screening for preeclampsia and intrauterine growth restriction. *Clinics in Laboratory Medicine*, *30*, 727.

Unluturk, U., Harmanci, A., Kocaefe, C., & Yildiz, B. O. (2007). The genetic basis of the polycystic ovary syndrome: A literature review including discussion of PPAR-gamma. *PPAR Research*, *2007*, 49109.

Van Binsbergen, C. J., Voelingh Bennink, H. J., Odink, J., Haspels, A. A., & Koppeschaar, H. P. (1990). A comparative and longitudinal study on endocrine changes related to ovarian function in patients with anorexia nervosa. *The Journal of Clinical Endocrinology and Metabolism*, *71*, 705.

Vink, J. M., Sadrzadeh, S., Lambalk, C. B., & Boomsma, D. I. (2006). Heritability of polycystic ovary syndrome in a Dutch twin-family study. *Journal of Clinical Endocrinology and Metabolism*, *91*, 2100.

Vitoratos, N., Deliveliotou, A., Vlahos, N. F., Mastorakos, G., Papadias, K., Botsis, D., et al. (2008). Serum adiponectin during pregnancy and postpartum in women with gestational diabetes and normal controls. *Gynecological Endocrinology*, *24*, 614.

Vrbikova, J., Dvorakova, K., Hill, M., Vcelak, J., Stanicka, S., Vankova, M., et al. (2005). Determinants of circulating adiponectin in women with polycystic ovary syndrome. *Gynecologic and Obstetric Investigation*, *60*, 155.

Wang, J., Shang, L. X., Dong, X., Wang, X., Wu, N., Wang, S. H., et al. (2010). Relationship of adiponectin and resistin levels in umbilical serum, maternal serum and placenta with neonatal birth weight. *The Australian and New Zealand Journal of Obstetrics and Gynaecology*, *50*, 432.

Wen, J. P., Lv, W. S., Yang, J., Nie, A. F., Cheng, X. B., Yang, Y., et al. (2008). Globular adiponectin inhibits GnRH secretion from GT1-7 hypothalamic GnRH neurons by induction of hyperpolarization of membrane potential. *Biochemical and Biophysical Research Communications*, *371*, 756.

Weyer, C., Funahashi, T., Tanaka, S., Hotta, K., Matsuzawa, Y., Pratley, R. E., et al. (2001). Hypoadiponectinemia in obesity and type 2 diabetes: Close association with insulin resistance and hyperinsulinemia. *Journal of Clinical Endocrinology and Metabolism*, *86*, 1930.

Whitworth, K. M., Agca, C., Kim, J. G., Patel, R. V., Springer, G. K., Bivens, N. J., et al. (2005). Transcriptional profiling of pig embryogenesis by using a 15-K member unigene set specific for pig reproductive tissues and embryos. *Biology of Reproduction*, *72*, 1437.

WHO, (2005). Attending to 136 million births, every year. World health report: Make every mother and child count. Geneva: WHO. p. 41

Wickham, E. P., Cheang, K. I., Clore, J. N., Baillargeon, J. P., & Nestler, J. E. (2011). Total and high-molecular weight adiponectin in women with the polycystic ovary syndrome. *Metabolism*, *60*, 366.

Wong, G. W., Krawczyk, S. A., Kitidis-Mitrokostas, C., Ge, G., Spooner, E., Hug, C., et al. (2009). Identification and characterization of CTRP9, a novel secreted glycoprotein, from adipose tissue that reduces serum glucose in mice and forms heterotrimers with adiponectin. *The FASEB Journal*, *23*, 241.

Wong, G. W., Krawczyk, S. A., Kitidis-Mitrokostas, C., Revett, T., Gimeno, R., & Lodish, H. F. (2008). Molecular, biochemical and functional characterizations of C1q/TNF family members: Adipose-tissue-selective expression patterns, regulation by PPAR-gamma agonist, cysteine-mediated oligomerizations, combinatorial associations and metabolic functions. *Biochemical Journal*, *416*, 161.

Wong, G. W., Wang, J., Hug, C., Tsao, T. S., & Lodish, H. F. (2004). A family of Acrp30/adiponectin structural and functional paralogs. *Proceedings of the National Academy of Sciences of the United States of America*, *101*, 10302.

Woo, J. G., Dolan, L. M., Daniels, S. R., Goodman, E., & Martin, L. J. (2005). Adolescent sex differences in adiponectin are conditional on pubertal development and adiposity. *Obesity Research*, *13*, 2095.

Worda, C., Leipold, H., Gruber, C., Kautzky-Willer, A., Knofler, M., & Bancher-Todesca, D. (2004). Decreased plasma adiponectin concentrations in women with gestational diabetes mellitus. *American Journal of Obstetrics and Gynecology*, *191*, 2120.

Xita, N., Georgiou, I., Chatzikyriakidou, A., Vounatsou, M., Papassotiriou, G. P., Papassotiriou, I., et al. (2005). Effect of adiponectin gene polymorphisms on circulating adiponectin and insulin resistance indexes in women with polycystic ovary syndrome. *Clinical Chemistry*, *51*, 416.

Yajnik, C. S., Lubree, H. G., Rege, S. S., Naik, S. S., Deshpande, J. A., Deshpande, S. S., et al. (2002). Adiposity and hyperinsulinemia in Indians are present at birth. *Journal of Clinical Endocrinology and Metabolism*, *87*, 5575.
Yamauchi, T., Kamon, J., Ito, Y., Tsuchida, A., Yokomizo, T., Kita, S., et al. (2003). Cloning of adiponectin receptors that mediate antidiabetic metabolic effects. *Nature*, *423*, 762.
Yang, W. S., Lee, W. J., Funahashi, T., Tanaka, S., Matsuzawa, Y., Chao, C. L., et al. (2001). Weight reduction increases plasma levels of an adipose-derived anti-inflammatory protein, adiponectin. *Journal of Clinical Endocrinology and Metabolism*, *86*, 3815.
Yoshihara, K., Yahata, T., Kashima, K., Mikami, T., & Tanaka, K. (2009). Association of single nucleotide polymorphisms in adiponectin and its receptor genes with polycystic ovary syndrome. *The Journal of Reproductive Medicine*, *54*, 669.
Zhang, N., Shi, Y. H., Hao, C. F., Gu, H. F., Li, Y., Zhao, Y. R., et al. (2008). Association of 45G15G(T/G) and +276(G/T) polymorphisms in the ADIPOQ gene with polycystic ovary syndrome among Han Chinese women. *European Journal of Endocrinology*, *158*, 255.

CHAPTER ELEVEN

Adiponectin in the Heart and Vascular System

Min Ding[*,†,1], **Eva M. Rzucidlo**[‡,1], **Jennifer C. Davey**[‡], **Yi Xie**[*,†], **Renjing Liu**[*,†], **Yu Jin**[*,†], **Lindsey Stavola**[*,†], **Kathleen A. Martin**[*,†,‡,2]

*Department of Medicine, Yale School of Medicine, New Haven, Connecticut, USA
†Department of Pharmacology, Yale School of Medicine, New Haven, Connecticut, USA
‡Section of Vascular Surgery, Department of Surgery, Dartmouth Medical School, Lebanon, New Hampshire, USA
[1]These authors contributed equally to this work.
[2]Corresponding author: e-mail address: kathleen.martin@yale.edu

Contents

Abstract

Adipose tissue is not only a storage depot for energy, but also an active endocrine tissue. Adipokines, hormones and cytokines secreted from adipocytes, relay information about energy stores to peripheral tissues throughout the body. Most adipokines are produced in direct proportion to fat mass, and many have proinflammatory or otherwise adverse

Vitamins and Hormones, Volume 90
ISSN 0083-6729
http://dx.doi.org/10.1016/B978-0-12-398313-8.00011-7

effects on the cardiovascular system. The notable exception is the cardioprotective adipokine adiponectin, which is secreted in inverse proportion to fat mass. Circulating adiponectin levels are highest in lean individuals and inversely correlate with fat mass. Low levels of serum adiponectin are now appreciated as a risk factor in a variety of cardiovascular diseases including coronary artery disease and restenosis, type 2 diabetes mellitus, and hypertension. In this chapter, we provide an introduction to adiponectin and review the extensive evidence in humans and in mouse and *in vitro* models for adiponectin's cardioprotective effects.

1. INTRODUCTION TO ADIPONECTIN

1.1. Adiponectin is an adipokine produced in inverse proportion to fat mass

In recent years, adipose tissue has been recognized to function not only as an energy storage depot, but also as an endocrine tissue, secreting bioactive substances into circulation (Walsh, 2009). The discovery of leptin in 1994 gave rise to the new and exciting field of adipokines, a variety of hormones and cytokines secreted from adipocytes. These adipokines relay information about energy stores to peripheral tissues throughout the body. The majority of adipokines are produced in direct proportion to fat mass, and many of these have proinflammatory or otherwise adverse effects on the cardiovascular system, including IL-6, IL-8, TNF-α, IL-1β, PAI-1, Angiotensin II (AngII), and resistin. The notable exception is adiponectin (also known as Acrp30, apM1, GBP28, and AdipoQ), the most abundant adipokine produced by adipocytes, which is secreted in inverse proportion to fat mass (Brochu-Gaudreau et al., 2010). Circulating adiponectin levels are highest in lean individuals and inversely correlate with body mass index (BMI; Matsubara, Maruoka, & Katayose, 2002). Plasma concentrations of adiponectin in normal individuals range from 3 to 30 μg/ml. Low levels of serum adiponectin are now recognized as a risk factor in obesity and a variety of cardiovascular diseases (CVDs) including coronary artery disease, type 2 diabetes mellitus, and hypertension (Hopkins, Ouchi, Shibata, & Walsh, 2007). These clinical correlations will be discussed extensively in Section 2.

The best appreciated function of adiponectin to date is its role in promoting insulin sensitivity. Adiponectin influences systemic metabolism by altering key functions in liver and skeletal muscle. Adiponectin promotes insulin sensitivity and fatty acid oxidation, and inhibits fatty acid synthesis in the liver (Brochu-Gaudreau et al., 2010). In skeletal muscle, adiponectin enhances insulin sensitivity, promoting GLUT4-dependent glucose uptake, increased glucose utilization, and fatty acid oxidation (Brochu-Gaudreau et al., 2010). Recent studies revealed that antidiabetic thiazolidinedione drugs (TZDs), including pioglitazone and rosiglitazone, act mainly by inducing expression of adiponectin. Adiponectin was shown to be required for TZD-induced improvements in insulin sensitivity (Nawrocki et al., 2006), cardiomyocyte hypertrophy (Amin, Mathews, Alli, & Leff, 2010), and endothelial function (Wong et al., 2011).

1.2. Oligomeric forms of adiponectin

The adiponectin gene (*Adipoq*) encodes a 30-kDa monomeric protein, which is assembled into oligomers in the endoplasmic reticulum by chaperone proteins including Erp44, Ero1-Lα, and DsbL-A, prior to secretion from the cell (Simpson & Whitehead, 2010). Circulating adiponectin is found in three major oligomeric forms: trimeric, hexameric (low-molecular weight), and high-molecular weight (HMW). These oligomers are highly stable and do not interconvert once secreted. The HMW complexes, generally considered the most potent, can be 12–36 mers (Schraw, Wang, Halberg, Hawkins, & Scherer, 2008), with a prominent form consisting of 18 monomers, arranged as hexamers of trimers (Simpson & Whitehead, 2010). This HMW form accounts for the majority of the reduction in total adiponectin levels seen with obesity (Schraw et al., 2008). A decrease in circulating HMW adiponectin is common in patients with coronary artery disease and type 2 diabetes (Wang, Lam, Yau, & Xu, 2008). Adiponectin contains two major domains: a collagenous domain and a globular domain. While there have been many studies documenting biological effects of recombinant adiponectin encoding only the globular domain (generated in bacterial systems), experts in the field report no evidence that this globular-only form exists *in vivo* under normal physiological conditions (Holland et al., 2011). Recombinant oligomeric adiponectin can be properly processed, but only in mammalian cells.

1.3. Adiponectin receptors and signaling

Adiponectin has been shown to function via the two seven-transmembrane domain receptors AdipoR1 and AdipoR2 (Yamauchi et al., 2003). Adiponectin receptors are widely distributed in various organs, with AdipoR1 highly expressed in skeletal muscle. AdipoR2 is the predominant receptor in the liver (Shetty, Kusminski, & Scherer, 2009; Yamauchi et al., 2003). Tissues of obese or type 2 diabetes mellitus patients express lower levels of AdipoR1 and AdipoR2 mRNA compared to lean healthy subjects. More recently, T-cadherin was found to be a receptor for adiponectin based on its ability to bind the HMW and hexameric isoforms of adiponectin in smooth muscle cells, endothelial cells (ECs), and cardiac myocytes (Denzel et al., 2010). As T-cadherin lacks an intracellular domain, it has been proposed that AdipoR1 or R2 work in concert with T-cadherin as a coreceptor, in adiponectin binding and signaling (Turer & Scherer, 2012). T-cadherin is ubiquitously expressed, but is present at high levels in the heart and arteries (Ivanov, Philippova, & Tkachuk, 2001). Adiponectin protection in models of cardiac hypertrophy and ischemia–reperfusion (I/R) injury (see Sections 3.1 and 3.2) is greatly attenuated in mice lacking T-cadherin (Denzel et al., 2010).

The insulin-sensitizing effects of adiponectin have been attributed to its activation of AMPK, p38MAPK, and peroxisome proliferator-activated receptor alpha (PPAR-α) (Brochu-Gaudreau et al., 2010). Although AdipoR1 and AdipoR2 have a seven-transmembrane topology, they are not G protein-coupled receptors. The mechanism by which they transduce signals to downstream effectors was unknown until a recent groundbreaking study revealed that these receptors exhibit ceramidase activity which is required for many of the actions of adiponectin, including insulin sensitization in liver, and protection from apoptosis in cardiac myocytes and pancreatic beta cells (Holland et al., 2011). AMPK activation appears to be secondary to adiponectin-induced deacylation of ceramide, leading to generation and release of sphingosine-1-phosphate (S1P). This model is consistent with previous evidence that S1P acting on its own receptor can induce calcium transient that then lead to AMPK activation (Holland et al., 2011). Characterization of the ceramidase activity of the adiponectin receptors may suggest a mechanism for activation of other effectors reportedly activated by adiponectin, including p38, PPAR-α, PKA, and Akt.

1.4. New evidence that adiponectin is produced in other cell types

The "dogma" has been that adiponectin is exclusively expressed within white adipose tissue (Kitta et al., 2008; Scherer, Williams, Fogliano, Baldini, & Lodish, 1995). However, multiple recent publications suggested that adiponectin is also expressed in several other cells and tissues, including brown adipose tissue (Viengchareun, Zennaro, Pascual-Le Tallec, & Lombes, 2002), cardiomyocytes (Pineiro et al., 2005), skeletal muscle (Krause et al., 2008), vascular smooth muscle cells (VSMCs) (Ding et al., 2012), bone-forming cells (Berner et al., 2004), placenta (Caminos et al.,

2005), and pituitary cells (Psilopanagioti, Papadaki, Kranioti, Alexandrides, & Varakis, 2009). This locally produced adiponectin, together with circulating adiponectin, seems to play important roles via autocrine and paracrine mechanisms. It was reported that cardiomyocyte-derived adiponectin may regulate cardiac metabolism (Pineiro et al., 2005) and protect heart against I/R injury (Wang et al., 2010, 2011) as well as cardiac hypertrophy (Amin, Mathews, Alli, et al., 2010). In patients with dilated cardiomyopathy, cardiac adiponectin protein expression is downregulated (Skurk et al., 2008). Adiponectin secreted by skeletal muscle maintains muscle contractile function (Krause et al., 2008), is involved in rosiglitazone-induced insulin sensitization (Amin, Mathews, Camp, Ding, & Leff, 2010; Liu et al., 2009), and has local anti-inflammatory function (Delaigle, Jonas, Bauche, Cornu, & Brichard, 2004). Our lab recently reported that VSMC-derived adiponectin is necessary for maintaining the differentiated VSMC phenotype (Ding et al., 2012). While the amount of locally produced adiponectin is far lower than circulating adiponectin levels, the above studies confirm biological roles for this locally synthesized adipokine.

2. ADIPONECTIN AND CVD IN HUMANS

Recently, the correlation between adiponectin and cardiovascular health has received a great deal of attention based on the many cardioprotective functions attributed to this adipokine in mouse and *in vitro* studies (reviewed in Section 3). In light of the current epidemic of obesity, and the clear association between obesity and CVD, the question has been raised whether adipokines mediate this connection at the molecular level. Whether low adiponectin levels predict CVD in humans has been a subject of controversy, but the bulk of the evidence now supports this notion. The overwhelming experimental evidence from animal research supports the hypothesis that adiponectin has a cardioprotective effect through its vasodilator, antiapoptotic, anti-inflammatory, and antioxidative activities in both cardiac and vascular cells (Hui, Lam, Vanhoutte, & Xu, 2012).

2.1. Epicardial and pericardial fat and disease

Most blood vessels and the heart are surrounded by adipose tissue. The heart is covered by epicardial and pericardial fat (Ouwens, Sell, Greulich, & Eckel, 2010). Pericardial adipose tissue covers 80% of the heart and accounts for 20–50% of the total heart weight (Ouwens et al., 2010). These fat deposits are used as a local energy source for the heart and help to maintain contractile activity by releasing fatty acids through lipolysis. However, expanded epicardial and pericardial fat are important risk factors for obesity-related cardiac dysfunction. Increased amounts of fat around the heart are an independent predictor for CVD (Ouwens et al., 2010; Rosito et al., 2008; Tadros et al., 2010). Both the thickness and the volume of epicardial fat were noted to be increased in patients with CVD and unstable angina, and epicardial fat thickness was related to severity of CVD (Eroglu et al., 2009) and subclinical markers of atherosclerosis (Natale et al., 2009). In morbidly obese individuals, epicardial fat thickness is associated with diastolic dysfunction (Iacobellis, Corradi, & Sharma, 2005). Importantly, both plasma adiponectin and epicardial adiponectin levels were decreased in patients with coronary artery disease compared to those without coronary artery disease (Iacobellis et al., 2009).

2.2. Adiponectin sexual dimorphism and cardiovascular risk

Women have about 40% higher circulating levels of adiponectin than men (Arita et al., 1999). The reason is not completely elucidated; however, sex hormones, including estrogen, progesterone and androgen, may affect plasma concentrations of adiponectin. Estrogen, in particular, potentially has multiple beneficial effects on CVD (Gilligan, Quyyumi, & Cannon, 1994; Williams, Adams, Herrington, & Clarkson, 1992). *In vitro* studies have suggested that testosterone may sequester a coactivator ARA70, common to the androgen receptor and PPAR-γ. Such competition may result in the reduced expression of PPAR-γ-regulated genes such as adiponectin (Heinlein, Ting, Yeh, & Chang, 1999). There is also a sexual dimorphism in adiponectin concentrations and complex distribution. The vast majority of adiponectin in men is in the lower molecular weight forms, whereas in women, the complexes are more equally distributed (Pajvani et al., 2003).

Decreased plasma adiponectin levels are observed in patients with obesity, type 2 diabetes, hypertension, metabolic syndrome, and coronary artery disease (Hotta et al., 2000; Iwashima et al., 2004; Kumada et al., 2003; Ouchi et al., 1999; Salmenniemi et al., 2004). Hypoadiponectinemia correlates significantly and independently with CVD (Kumada et al., 2003). In a long-term study, male subjects with adiponectin levels in the highest quintile had decreased risk of myocardial infarction (MI) compared with the lowest quintile (Pischon et al., 2004). This effect was also seen in men with type 2 diabetes (Schulze et al., 2005). Male patients with hypoadiponectinemia have a twofold increase in CVD prevalence, independent of well-known risk factors (Kumada et al., 2003; Ouchi et al., 1999). Adiponectin levels, however, have not been as predictive of coronary heart events in women (Lawlor, Davey Smith, Ebrahim, Thompson, & Sattar, 2005). Further study is required to determine if there is a true gender difference in the effect of adiponectin on CVD.

Adiponectin has been shown to be lower in patients with CVD even in age-, gender-, and BMI-matched controls (Ouchi et al., 1999). Adiponectin levels in patients with acute coronary syndrome are significantly lower than those patients with stable angina and in normal control patients (Nakamura et al., 2004). Plasma levels of adiponectin are significantly associated with coronary lesion complexity in men with CVD (Otsuka et al., 2006). Progression of coronary artery calcification in type 1 diabetic and nondiabetic subjects was correlated independently with low adiponectin levels (Maahs et al., 2005). Several studies have noted that patients undergoing percutaneous coronary intervention (PCI) who have hypoadiponectinemia have an increased risk of in-stent restenosis (Kitta et al., 2008; Moldoveanu et al., 2008). In patients with end-stage renal disease and stroke, adiponectin levels are inverse predictor of cardiovascular outcomes (Chen et al., 2005; Efstathiou et al., 2005; Zoccali et al., 2002). These results indicate that hypoadiponectinemia is an independent risk factor for CVD and that adiponectin may directly protect against abnormal vascular remodeling.

2.3. Adiponectin single nucleotide polymorphisms, CVD, and hypertension

The adiponectin gene *Adipoq* is localized to chromosome 3q27, a major locus associated with metabolic syndrome (Kissebah et al., 2000) and type 2 diabetes (Vionnet et al., 2000). Genetic variants of *Adipoq* have been associated with low levels of adiponectin and type 2 diabetes (Hivert et al., 2008; Patel et al., 2008). The single nucleotide polymorphism (SNP) rs266729 (−11377C to G) in *Adipoq* has been associated with CVD in Caucasian men and in Asians (Hoefle et al., 2007; Oguri et al., 2009). The *Adipoq* SNP rs1501299 (+276G to T) has been associated with CVD in men with diabetes, but not in women with diabetes (Qi et al., 2006, 2005). It has also been shown to be associated with low adiponectin levels in the

Framingham Offspring Study and in healthy individuals (Heid et al., 2006; Hivert et al., 2008).

Hypertension is a risk factor for CVD including stroke, MI, and aneurysm. Several studies have demonstrated the relationship of plasma adiponectin concentration with hypertension (Adamczak et al., 2003; Iwashima et al., 2004; Kazumi, Kawaguchi, Sakai, Hirano, & Yoshino, 2002; Murakami et al., 2003). An inverse correlation is observed between adiponectin and mean systolic and diastolic blood pressure, including patients with type 2 diabetes and metabolic syndrome (Choi et al., 2004). Interestingly, studies have shown that hypoadiponectinemia is a risk factor for hypertension independent of insulin resistance and diabetes (Iwashima et al., 2004). Adiponectin has also been shown to be associated with the progression of hypertension in a prospective study (Chow et al., 2007). Hypoadiponectinemia at baseline was a strong predictor of future hypertension even after adjusting for mean blood pressure, C-reactive protein, BMI, and waist circumference. Patients with hypoadiponectinemia had three times higher morbidity of future hypertension than those with normal adiponectin levels (Chow et al., 2007). Another prospective study confirmed that men with significantly lower baseline adiponectin levels developed hypertension more often than those with normal or high baseline levels (Imatoh, Miyazaki, Momose, Tanihara, & Une, 2008). A larger prospective study in the Netherlands found no correlation between adiponectin level and new-onset hypertension. However, their subjects were quite lean and mildly hypertensive which may have contributed to the result (Asferg et al., 2010). Numerous other clinical studies have explored the relationship between hypertension and adiponectin with mixed conclusions, and the variability is likely due to the ethnicity of the patients, size, and type of study, medication subjects took, and the fact that adiponectin levels are influenced by obesity which is also a high risk factor for hypertension (Adamczak et al., 2003; Cesari et al., 2006; Iwashima et al., 2004; Ohashi et al., 2006; Patel, Lim, Dubb, Hughes, & Lip, 2009; Thomopoulos et al., 2011).

Mutations in *Adipoq* provide further evidence of its association with hypertension. A SNP at position 164 (T164C) has been associated with hypoadiponectinemia and high blood pressure in the Japanese population (Iwashima et al., 2004). This mutation has also been associated with metabolic syndrome and CVD (Kondo et al., 2002; Ohashi et al., 2004). Another study did not find an association between hypertension and adiponectin with SNP T164C, but did find a distinct SNP in adiponectin that linked hypertension and metabolic syndrome (Leu et al., 2011). The rs266729 SNP in the adiponectin gene is associated with hypertension in a large study of normotensive and hypertensive Chinese subjects (Ong et al., 2010).

2.4. Adiponectin in hypertrophy and heart failure

Hypoadiponectinemia is associated with the progression of left ventricular hypertrophy (LVH) with diastolic dysfunction among patients with essential hypertension (Hong, Park, Seo, Oh, & Ro, 2004). In healthy patients, adiponectin levels are inversely and independently associated with LVH diagnosed by electrocardiography (Mitsuhashi et al., 2007). Another study using echocardiography showed that circulating total adiponectin and HMW adiponectin were related to left ventricular wall thickness and diastolic function independent of age and metabolic factors (Kozakova et al., 2008). All these studies suggest that adiponectin could be involved in hypertrophic progression of cardiomyocytes.

The role of adiponectin in heart failure is controversial. Several studies have noted high levels of adiponectin in patients with chronic heart failure (CHF). Strikingly, these high levels of adiponectin are associated with CHF severity and mortality despite the protective effect of adiponectin on CHF in mice (Kistorp et al., 2005; Nakamura et al., 2006; Shibata et al., 2007; Tamura et al., 2007). Also, high adiponectin levels are associated with mortality, independent of risk markers for CHF severity among patients with chronic kidney disease (Kistorp et al., 2005; Menon et al., 2006). Therefore, the cardioprotective role of adiponectin in congestive heart failure and chronic kidney disease remains controversial.

The exact mechanism for the upregulation is unknown; however, several hypotheses are possible. Increased expression of adiponectin may be a compensatory response to the stress of heart failure, similar to the mechanism described for B-type natriuretic peptide secretion (Daniels & Maisel, 2007). Elevated plasma levels of adiponectin among patients with CHF can be a reflection of accompanying renal dysfunction, or "adiponectin resistance" including impaired adiponectin signal transduction in the myocardium (Bruce, Mertz, Heigenhauser, & Dyck, 2005; Chen et al., 2005). Last, increased levels of adiponectin may result as a compensatory response to aberrant expression of adiponectin receptors. This would disrupt adiponectin signaling in the target tissues and may act as a compensatory response and partially explain the observed increase in adiponectin levels in these disease states. Further careful studies may be required to clarify the relationship between adiponectin and heart failure.

3. PROTECTIVE EFFECTS OF ADIPONECTIN IN THE CARDIOVASCULAR SYSTEM: EVIDENCE FROM RODENT AND *IN VITRO* STUDIES

3.1. Adiponectin and cardiac hypertrophy

Cardiac hypertrophy is an adaptive stress response to both intrinsic and extrinsic stimuli, initially used to compensate for increases in mechanical load. However, these beneficial effects can become maladaptive, leading to contractile dysfunction and subsequent heart failure. Increased demand on the heart leads to both remodeling and metabolic adaptations. Because of its role in sensing cellular energy levels and modulating metabolic responses, AMPK has been implicated as a key molecule in cardiac responses to both overload and ischemia (Kim, Miller, & Young, 2009). As AMPK is a major signaling effector of adiponectin, it is likely that both circulating adiponectin and adiponectin expressed in cardiac myocytes contribute to regulation of cardiomyocyte hypertrophy (Ding et al., 2007).

Mice lacking adiponectin exhibit enhanced concentric cardiac hypertrophy and increased mortality in a pressure overload model (Liao et al., 2005; Shibata, Ouchi, Ito, et al., 2004), which can be effectively rescued with adenovirus-mediated delivery of adiponectin. AngII has long been recognized to modulate cardiac contractility, metabolism, and structure as well as stimulate cardiac myocyte hypertrophy through multiple signaling pathways (Hein et al., 1997; Kudoh et al., 1997; Sadoshima & Izumo, 1993; Takano et al., 1996). Treatment with adiponectin was able to attenuate AngII-induced cardiac hypertrophy and fibrosis in mice by stimulating AMPK and repressing ERK signaling (Shibata, Ouchi, Ito, et al., 2004). AngII has also been reported to increase reactive oxygen species (ROS) production by activating the NAD(P)H oxidase, which can potentially modulate the signaling mechanisms involved in cardiac hypertrophy (Das, Otani, Maulik, & Das, 2006; Kimura et al., 2005). Recently, it has been shown that adiponectin protects cardiomyocytes against pathological cardiac remodeling associated with oxidative stress. Essick and colleagues demonstrated protection against fibrosis and significantly reduced MMP-2 and MMP-9 matrix remodeling activities with adiponectin treatment in AngII-infused adiponectin knockout (APN-KO) hearts, linking ROS production, cardiac hypertrophy, and MMP activity (Essick et al., 2011). In agreement with previously published data, these authors confirmed the action of adiponectin is through the upregulation of AMPK signaling with concomitant inhibition of the ERK and NF-κB activity.

Pioglitazone has been shown to attenuate hypertension-related cardiac remodeling and fibrosis in rats (Kato et al., 2008; Nakamoto et al., 2008) and prevent pressure overload-induced cardiac hypertrophy in mice (Asakawa et al., 2002). Using an AngII-induced cardiac hypertrophy model, Li and colleagues have shown that pioglitazone exerts its protective effects by increasing circulating levels of adiponectin (Li et al., 2010). Transfection of siRNA-targeting adiponectin receptors AdipoR1 or AdipoR2 was able to attenuate the cardioprotective effects of adiponectin on cultured cardiomyocytes induced to undergo hypertrophy with endothelin-1 treatment (Fujioka et al., 2006), raising the possibility that adiponectin may act as a cardiac endocrine inhibitor for endothelin-1 (Yin et al., 2011).

Collectively, these studies underscore the complexity of the hypertrophic signaling pathway in cardiomyocytes. It appears that adiponectin exerts cardioprotective effects by inhibiting the ERK pathway, while simultaneously activating AMPK signaling. As described in Section 2.4, adiponectin levels are decreased in patients with ischemic heart disease and coronary artery disease, but are paradoxically increased in patients with heart failure (George et al., 2006; Kistorp et al., 2005). Additional studies will be required to elucidate the interconnectivity between adiponectin and other pathological cardiac conditions.

3.2. Adiponectin and I/R

The cardioprotective role of adiponectin in cardiac I/R injury has been suggested by a number of studies. Serum adiponectin levels rapidly decreased in patients after acute MI (Kojima et al., 2003), and higher plasma adiponectin levels were associated with a lower risk of MI (Pischon et al., 2004). Notably, APN-KO mice displayed increased MI size, myocardial apoptosis, and TNF-α expression as compared with wild-type (WT) mice after I/R injury (Shibata et al., 2005), all of which were reversed by administration of adenoviral adiponectin. Also, heart tissue obtained from APN-KO mice had significantly higher I/R-induced iNOS and gp91(phox) protein expression, as well as NO, superoxide, and peroxynitrite formation. These were all reversed by preadministration of the globular domain of adiponectin (Tao et al., 2007). In addition, adiponectin significantly increased left ventricular function and coronary flow while reducing infarct size in isolated

I/R-injured rat hearts, and coadministration of the NO-synthase inhibitor nitro-L-arginine abrogated these effects (Gonon et al., 2008). All these studies supported that systemic administration of adiponectin has protective effects during myocardial I/R injury.

Interestingly, as mentioned in Section 1.4, local cardiomyocyte-derived adiponectin is also bioactive in protecting against myocardial I/R injury. It was reported that rosiglitazone upregulated expression of adiponectin and its receptors AdipoR1 and AdipoR2, reduced I/R-induced apoptosis, and decreased LDH release in WT cardiomyocytes. However, the cell-protective effects of rosiglitazone were completely lost in APN-KO cardiomyocytes, despite of the upregulated AdipoR1/AdipoR2 expression, suggesting that rosiglitazone protects against I/R injury via induction of adiponectin expression. Moreover, in AdipoR1-knock-down/AdipoR2-KO cardiomyocytes, rosiglitazone further induced adiponectin expression, while the cellular protective effects of rosiglitazone were significantly blunted in these cells (Wang et al., 2010). This data suggested that locally produced adiponectin achieves its protective effect against I/R damage primarily through paracrine/autocrine activation of its receptors. Further, a recent study showed that a combination of the antioxidant reagents *N*-acetylcysteine and allopurinol decreased postischemic MI size and myocardial I/R injury, which were associated with increased cardiac adiponectin content, AdipoR2 expression, and its downstream phosphorylation of Akt, STAT3, and eNOS in diabetic rats (Wang et al., 2011). This finding suggested that local cardiomyocyte-expressed adiponectin may protect against myocardial I/R injury via its receptor-mediated downstream eNOS activation.

Several mechanisms could be responsible for the beneficial effects of adiponectin against I/R injury. *In vitro*, adiponectin inhibited cardiomyocyte apoptosis via activation of AMPK (Shibata et al., 2005). Adiponectin also induced cyclooxygenase (COX)-2-dependent synthesis of prostaglandin E (2) in cardiac cells, which was involved in the inhibitory effects of adiponectin on TNF-α production and infarct size (Shibata et al., 2005). In addition, adiponectin inhibited iNOS and NADPH-oxidase protein expression, and resulted in decreased oxidative/nitrative stress during the I/R injury (Tao et al., 2007). Further, adiponectin was reported to protect against cerebral I/R injury through its anti-inflammatory action (Chen et al., 2009). In summary, adiponectin displays significant cardioprotective functions during cardiac I/R injury process and may be a therapeutic target in the treatment of MI.

3.3. Adiponectin and heart transplant rejection

Heart transplant rejection can be acute or chronic, with acute rejection defined as days to weeks, and chronic rejection as months to years. Pathologically, acute rejection involves damage by T cells and other inflammatory cells to ECs in capillaries of the graft. Chronic rejection is similarly initiated by alloimmune damage to the endothelium, but results in arteriosclerotic remodeling of the coronary arteries, often referred to as cardiac allograft vasculopathy (CAV). CAV is characterized by VSMC proliferation and can be diagnosed angiographically by the formation of intimal hyperplasia and luminal narrowing in coronaries (Wehner & Baldwin, 2010). The annual incidence of CAV is about 5–10% and is progressive, with no available effective treatment or cure. Metabolic risk factors including hyperlipidemia and hyperglycemia may contribute to CAV as well.

Postoperative obesity increases the risk of morbidity and mortality after cardiac transplantation (Grady et al., 2005). Since perivascular adipose tissue modulates several other forms of vascular remodeling (Miao & Li, 2012), it may also play important roles in the development of CAV (Wehner & Baldwin, 2010). It is well known that adiponectin levels are lower in obese patients. In heart transplant recipients, plasma adiponectin levels were found to be normal or higher, probably due to renal failure (Ambrosi et al., 2009). However, higher plasma adiponectin levels may predict better outcome among recipients of heart transplant, as adiponectin inhibited allograft rejection in a murine cardiac transplantation model (Okamoto et al., 2009). As compared with WT mice, MHC class II incompatible allografts transplanted into APN-KO mice showed reduced contraction and increased accumulation of CD4- and CD8-positive T lymphocytes and Mac3-positive macrophages, indicating more severe acute rejection. Injection with adiponectin-expressing adenovirus was able to reverse these acute rejection responses in the APN-KO mice. The rejected allografts in APN-KO mice expressed higher levels of inflammatory cytokines, chemokines, and chemokine receptors, including TNF-α, IFN-γ, RANTES, MCP-1, and IP-10. *In vitro*, physiological concentrations of adiponectin inhibited anti-CD3-induced mouse T-lymphocyte proliferation and production of inflammatory cytokines and chemokines (Okamoto et al., 2009). These data suggested that adiponectin reduces allograft rejection by blocking lymphocyte proliferation and recruitment, therefore providing beneficial effects in inhibiting alloimmune responses. This pathway may represent a new therapeutic target for allograft rejection after heart transplantation.

3.4. Adiponectin and restenosis/vascular remodeling

Restenosis following PCI is an overreaction of the normal vascular healing process. This is characterized by endothelial denudation, inflammation, granulation, VSMC proliferation and migration, as well as extracellular matrix remodeling, leading to neointimal hyperplasia formation and constrictive remodeling of the injured arterial segments (Liu, Roubin, & King, 1989). In addition, vessel recoil and plaque prolapse after balloon-only angioplasty and additional inflammatory mechanisms associated with stents also contribute to this process (Welt & Rogers, 2002). Notably, it was reported that low serum adiponectin level predicted late-in-stent restenosis after PCI with bare-metal stents (Kitta et al., 2008), which was independent of inflammation markers (Moldoveanu et al., 2008). Moreover, hyperadiponectinemia (plasma adiponectin levels $\geq$6.0 μg/ml) was independently associated with a lower risk of restenosis in patients with stable angina after PCI (Mito et al., 2011). These data all suggested that plasma adiponectin level may be useful in predicting vascular restenosis after angioplasty.

Recently, it was reported that periadventitial fat protected against neointimal formation after vascular injury in mice under physiological conditions and that the inflammatory changes in the periadventitial fat may have a direct role in the pathogenesis of obesity-accelerated vascular disease (Takaoka et al., 2009). Adiponectin seems to be the key adipokine in periadventitial fat that plays anti-inflammatory and antirestenotic functions in vessel remodeling. The antirestenotic effect of adiponectin was first identified using APN-KO mice. These mice displayed severe neointimal thickening and increased VSMC proliferation in mechanically injured arteries, which were rescued by adenoviral delivery of adiponectin (Matsuda et al., 2002). Interestingly, decreased adiponectin expression in the periadventitial adipose tissue under high-fat, high-sucrose feeding conditions was also associated with enhanced neointima formation after endovascular injury, and local delivery but not systemic administration of recombinant adiponectin to the periadventitial area reversed the markedly enhanced lesion formation in APN-KO mice (Takaoka et al., 2009). This study suggested that increasing local adiponectin level may be a therapeutic strategy to treat restenosis after PCI.

In vitro, adiponectin attenuates VSMC DNA synthesis induced by growth factors including platelet-derived growth factor, heparin-binding epidermal growth factor (EGF)-like growth factor (HB-EGF), basic fibroblast growth factor and EGF, as well as VSMC proliferation and migration induced by HB-EGF (Matsuda et al., 2002). This study suggested that the antiproliferative and antimigratory effects of adiponectin in VSMC may be due in part to its ability to directly bind growth factors, thus inhibiting growth factor-induced ERK signaling (Arita et al., 2002). Interestingly, VSMC themselves also express and secrete functional adiponectin, which is necessary for maintenance of contractile phenotype *in vitro* via autocrine and paracrine actions (Ding et al., 2012). Our lab reported that HMW adiponectin was as efficacious as rapamycin, the gold standard for drug-eluting stents, in promoting differentiation of VSMC *in vitro*. Trimeric adiponectin was effective to a lesser extent, but globular adiponectin had no effect in this assay. We showed that HMW adiponectin induces VSMC differentiation via AMPK inhibition of mTORC1 and FoxO4 (Ding et al., 2011). These studies suggest an active signaling role of adiponectin, as opposed to merely binding growth factors (as suggested by Matsuda et al., 2002). While rapamycin potently inhibits intimal hyperplasia, it also inhibits reendothelialization. Notably, adiponectin has similar beneficial effects on VSMC as rapamycin, but also promotes EC survival and function (Ding et al., 2011), suggesting that targeting the adiponectin pathway could have advantages over rapamycin for potential new drug-eluting stent therapies.

3.5. Adiponectin and hypertension

Animal studies have shown a clear relationship between hypertension and adiponectin, with a negative correlation between plasma adiponectin levels and blood pressure. APN-KO mice develop hypertension, without insulin resistance, when fed a high salt diet, and adiponectin therapy lowers their induced high blood pressure (Ohashi et al., 2006). The KKAy strain of mice is severely obese, hyperlipidemic, insulin resistant, and have high blood pressure. Adenoviral delivery of adiponectin significantly decreased their high blood pressure (Ohashi et al., 2006).

As AngII promotes vasoconstriction and increased blood pressure, this pathway is a common target for antihypertensive therapy. Interestingly, AngII type 1 receptor blocker (ARB) drugs have been shown to induce levels of adiponectin while concurrently lowering blood pressure in hypertensive patients (Nakamura et al., 2009; Uchida et al., 2008; Watanabe et al., 2006). Interestingly, antihypertensive drugs that lower blood pressure by blocking calcium channels do not affect adiponectin levels, indicating that AngII affects adiponectin levels through a mechanism other than simply lowering blood pressure (Watanabe et al., 2006). Also, AngII fed directly to rats on a high-fructose diet lowered their adiponectin levels (Ran et al., 2006). A putative mechanism by which ARBs induce adiponectin is through their ability to act as partial agonists for PPAR-γ. PPAR-γ has been shown to induce adiponectin in mice, humans, and adipocyte cell culture (Ohashi, Ouchi, & Matsuzawa, 2011).The actions of adiponectin in these *in vivo* studies have been further studied at the cellular level in order to understand the mechanisms by which lack of adiponectin may promote hypertension and the CVD that ensues.

At the cellular level, adiponectin has been shown to affect the actions of ECs. Low adiponectin levels are linked to vascular endothelial dysfunction (Shimabukuro et al., 2003), and many studies have shown that adiponectin is beneficial to the healthy activities of the endothelium (Ohashi et al., 2011). One pathway implicated in the beneficial effects of adiponectin on the vascular endothelium is the AMPK–eNOS–NO signaling pathway which is stimulated by adiponectin. NO is a key regulator of endothelial actions and vessel relaxation and its inhibition can lead to hypertension. Adiponectin-mediated AMPK phosphorylation of eNOS promotes its activity. In addition, APN-KO mice have lower levels of eNOS transcripts in their aortas and plasma NO metabolites compared to WT mice when fed a high salt diet. Exogenous adiponectin rescues this phenotype and restores the level of eNOS (Ohashi et al., 2006, 2011).

Pulmonary arterial hypertension (PAH) is characterized by elevation in pulmonary arterial pressure. It leads to right ventricular failure and is associated with congenital heart disease, liver disease, HIV infection, schistosomiasis, and other inflammatory conditions, though it can be a primary disorder. There are no clinical studies to date regarding association between adiponectin levels and PAH, but animal studies suggest adiponectin plays a role (Hansmann & Rabinovitch, 2010; Medoff et al., 2009; Sam et al., 2010; Summer et al., 2009; Weng, Baron, et al., 2011; Weng, Raher, et al., 2011). PAH is characterized by abnormal proliferation of pulmonary artery smooth muscle cells (PASMCs) and inflammation in the pulmonary artery wall. Aged APN-KO mice develop inflammation and increased blood pressure in the pulmonary artery and symptoms of PAH (Summer et al., 2009). The ADP-KO mice show an increase in E-selectin, an inflammatory marker, in the pulmonary perivascular region and *in vitro*, adiponectin was able to repress TNF-α-induction of E-selectin. This points to adiponectin acting as an anti-inflammatory factor (Summer et al., 2009). Other evidence suggests that adiponectin can play a direct or indirect role in pulmonary artery remodeling and helps confirm its role in the recruitment of inflammatory cells. Transgenic mice that overexpress adiponectin by two- to threefold resist eosinophilic inflammation-induced PAH via exposure to ovalbumin. Though they show an increase in inflammatory eosinophils, they resist remodeling of the PASMCs indicating a direct effect of adiponectin on remodeling (Weng, Raher, et al., 2011). APN-KO mice that develop eosinophilic-inflammation-induced PAH via exposure to ovalbumin develop more inflammation and remodeling than WT mice. The increase in disease severity is attributed to an increase in macrophage signaling which leads to recruitment of more eosinophils and therefore more PASMC proliferation. In this case, adiponectin inhibits eosinophil recruitment through an indirect mechanism and the signaling involved in this is not known. The specificity of the role of eosinophils was demonstrated by inhibiting their recruitment with an IL-5 antibody and thereby inhibiting disease severity (Weng, Baron, et al., 2011).

Adiponectin plays a role in various forms of hypertension and may well play a role in forms not mentioned here, for instance preeclampsia, but the evidence is not yet clear. It acts as an anti-inflammatory factor and as an inhibitor of proliferation in PAH and more research is needed to understand the pathways by which it signals.

3.6. Adiponectin and atherosclerosis

Endothelial inflammation plays a major role in atherosclerosis. Early studies showed that adiponectin, through a cAMP/PKA pathway, inhibits TNF-α-induced NF-κB signaling and expression of adhesion molecules including VCAM-1, E-selectin, and ICAM-1 in ECs, leading to reduced monocyte binding (Ouchi et al., 1999, 2000). Similarly, adiponectin deficiency promoted the TNF-α-induced leukocyte–EC interaction *in vivo*, and eNOS–NO signaling was required (Ouedraogo et al., 2007). Further, adiponectin deficiency exacerbates endothelial activation due to upregulated chemoattractant cytokines and VCAM-1 and ICAM-1 expression in a peritonitis model (Teoh et al., 2008). This is also supported by studies of adiponectin receptor function, where overexpression of adiponectin receptor potentiates the anti-inflammatory effect of a subeffective dose of globular adiponectin and reduces the TNF-α-induced NF-κB activation and ICAM-1 expression in EC (Zhang, Wang, Fan, Tang, & Wang, 2009). In addition, adiponectin was shown to suppress the synthesis and/or secretion of the proinflammatory factors IL-8 (Kobashi et al., 2005), CD40/CD40L (Natal et al., 2008), and C-reactive protein

(Devaraj, Torok, Dasu, Samols, & Jialal, 2008), which are normally upregulated in CVD. Here, the inhibition of NF-κB activity via AMPK or PKA pathways was suggested to be the mechanism (Kobashi et al., 2005; Natal et al., 2008; Wu et al., 2007). Adiponectin also shows a positive effect on maintaining endothelium barrier function, as adiponectin could inhibit AngII- or TNF-α-induced endothelial monolayer hyperpermeability by regulating actin cytoskeleton and microtubule stability through PKA activation (Xu et al., 2008).

While the anti-inflammatory effects of adiponectin on EC are clear, studies on atherosclerosis in mouse models have been less conclusive. One study found that globular adiponectin reduced accumulation of T-lymphocytes in the ApoE KO atherosclerosis model, as well as inhibiting LPS-induced cytokine release from macrophages (Okamoto et al., 2008). In contrast, a very thorough study of atherogenesis in response to high-fat diet in LDL-R KO mice revealed no exacerbation of atherosclerosis in APN-KO mice and no protective effect in adiponectin transgenic mice (Nawrocki et al., 2010). The authors attribute their lack of an effect despite expected changes in insulin sensitivity to the limitations of using mouse models to study atherosclerosis. In marked contrast to humans, mice are not normally susceptible to atherosclerosis with high-fat diet alone and require substantial genetic manipulations in cholesterol handling to observe this phenotype. Mice are also resistant to plaque rupture and MI, two major complications of atherosclerosis in humans. The bulk of the clinical literature, as well as some *in vitro* mechanistic findings, suggests that adiponectin levels likely do influence atherosclerosis and response to therapy in humans (see Section 2).

3.7. Adiponectin and revascularization

Adiponectin plays an important role in angiogenesis, arteriogenesis, and revascularization. Early studies revealed that adiponectin stimulates the differentiation and migration of human umbilical vein EC *in vitro* and stimulated blood vessel growth *in vivo* using mouse Matrigel plug implantation and rat corneal models (Ouchi et al., 2004). Notably, revascularization of ischemic hindlimbs was impaired in APN-KO mice as evaluated by laser Doppler flow and capillary density analysis, and this could be rescued by adenovirus-mediated adiponectin supplement (Shibata, Ouchi, Kihara, et al., 2004). Caloric restriction was shown to improve revascularization in the hindlimb ischemia model by elevating circulating adiponectin. This effect required AMPK and eNOS, and caloric restriction had no effect in APN-KO mice

(Kondo et al., 2009). Pressure overload by aortic banding induces cardiac hypertrophy and myocardial angiogenesis. APN-KO mice had more severe cardiac dysfunction following pressure overload due to reduced VEGF levels and inhibition of myocardial angiogenesis through the AdipoR1–LKB1–AMPK axis (Shimano et al., 2010).

Many studies have investigated the proangiogenic mechanisms of adiponectin. AMPK, a major downstream effector of adiponectin, promotes angiogenesis both *in vitro* and *in vivo* in response to hypoxia by phosphorylating eNOS at Ser-1177, which elevates NO production (Chen et al., 1999; Nagata, Mogi, & Walsh, 2003). AMPKα1 and α2 knockout mice exhibit impaired ischemic angiogenesis (Xu et al., 2011). AMPK has also been reported to lead to VEGF mRNA stabilization, thus upregulating VEGF levels and promoting angiogenesis in prostate carcinoma (Yun, Lee, Kim, & Ha, 2005) and skeletal muscle (Ouchi, Shibata, & Walsh, 2005). This may be an AMPKα2 isoform-specific effect (Neurath, Keough, Mikkelsen, & Claffey, 2006).

Besides AMPK, signaling via COX-2 downstream of adiponectin has also been implicated in revascularization. Adiponectin upregulates COX-2 expression through a pathway that requires calreticulin/CD91 and PI3K/Akt. Endothelial-specific knockout of COX-2 suppressed adiponectin-induced revascularization in ischemic hindlimb and adiponectin-induced EC migration, differentiation, and survival (Ohashi et al., 2009). Adiponectin can also inhibit EC apoptosis (Kobayashi et al., 2004), which may also favor angiogenesis. More specifically, adiponectin could oppose IL-18-induced EC death via AMPK signaling (Chandrasekar et al., 2008). Adiponectin also inhibited palmitate-induced ROS production and endothelial apoptosis in an AMPK- and PKA-dependent manner (Kim et al., 2010).

Studies in APN-KO mice clearly support a proangiogenic role for adiponectin. In tumor tissues, lack of adiponectin (Landskroner-Eiger et al., 2009) or T-cadherin (Hebbard et al., 2008), a membrane coreceptor for adiponectin, suppresses mammary tumor neovascularization and tumor growth, suggesting that endogenous adiponectin is required for tumor angiogenesis. However, several studies suggest that high doses of exogenously delivered adiponectin may have an antiangiogenic effect. Most of these studies employ the recombinant globular form of adiponectin, which is not thought to be present *in vivo* (Holland et al., 2011). Recombinant adiponectin could inhibit the tumor angiogenesis, thus suppressing tumor growth by inducing the caspase-mediated EC apoptosis (Bråkenhielm et al., 2004). Injection of high levels of an 18-amino acid adiponectin peptide inhibited angiogenesis in a mouse model of laser-induced choroidal neovascularization by reducing VEGF levels (Bora et al., 2007). Adenoviral delivery of adiponectin suppressed liver tumor growth and metastasis by inhibiting tumor angiogenesis and lamellipodial formation (Man et al., 2010). These data suggest that recombinant adiponectin may have therapeutic potential for tumor antiangiogenesis.

4. CONCLUSIONS/FUTURE QUESTIONS

Herein, we have reviewed the evidence for the protective effects of adiponectin in the heart and vascular system. Adiponectin may be an important link between obesity, metabolic syndrome, and CVD, and as such, may be an important pathway for future therapeutics. There is still much to be learned about the actions of adiponectin. The signaling mechanisms utilized by adiponectin's unconventional receptors and coreceptors are only beginning to be deciphered and are an area of active research. Because oligomeric adiponectin is such a large protein ligand, it would not likely be commercially viable to use oligomeric adiponectin itself as a therapeutic, but whether globular domain-only adiponectin protein, or small molecule activators of adiponectin receptors might be viable therapeutics remains to be determined. Distinct effects of different oligomeric versus globular adiponectin are still being investigated in different target tissues. Once thought to be derived exclusively from adipocytes, the realization that adiponectin produced locally by other cell types in tissues has distinct biological effects indicates the need for further study of the actions of circulating versus paracrine-acting adiponectin. The knowledge that adiponectin is the key effector of TZDs suggests other strategies for treatment of diabetes and CVD. The continued search for answers to these outstanding issues will lead to better understanding of how to increase endogenous expression of adiponectin in patients, and how to effectively target this pathway for optimal prevention and treatment of CVD.

REFERENCES

Adamczak, M., Wiecek, A., Funahashi, T., Chudek, J., Kokot, F., & Matsuzawa, Y. (2003). Decreased plasma adiponectin concentration in patients with essential hypertension. *American Journal of Hypertension*, *16*, 72–75.

Ambrosi, P., Oddoze, C., Nicolay, A., Penet-Lorec, A. M., Riberi, A., Metras, D., et al. (2009). Plasma adiponectin in heart transplant recipients. *Clinical Transplantation*, *23*, 83–88.

Amin, R. H., Mathews, S. T., Alli, A., & Leff, T. (2010). Endogenously produced adiponectin protects cardiomyocytes from hypertrophy by a PPARgamma-dependent autocrine mechanism. *American Journal of Physiology. Heart and Circulatory Physiology*, *299*, H690–H698.

Amin, R. H., Mathews, S. T., Camp, H. S., Ding, L., & Leff, T. (2010). Selective activation of PPARgamma in skeletal muscle induces endogenous production of adiponectin and protects mice from diet-induced insulin resistance. *American Journal of Physiology. Endocrinology and Metabolism*, *298*, E28–E37.

Arita, Y., Kihara, S., Ouchi, N., Maeda, K., Kuriyama, H., Okamoto, Y., et al. (2002). Adipocyte-derived plasma protein adiponectin acts as a platelet-derived growth factor-BB-binding protein and regulates growth factor-induced common postreceptor signal in vascular smooth muscle cell. *Circulation*, *105*, 2893–2898.

Arita, Y., Kihara, S., Ouchi, N., Takahashi, M., Maeda, K., Miyagawa, J., et al. (1999). Paradoxical decrease of an adipose-specific protein, adiponectin, in obesity. *Biochemical and Biophysical Research Communications*, *257*, 79–83.

Asakawa, M., Takano, H., Nagai, T., Uozumi, H., Hasegawa, H., Kubota, N., et al. (2002). Peroxisome proliferator-activated receptor gamma plays a critical role in inhibition of cardiac hypertrophy in vitro and in vivo. *Circulation*, *105*, 1240–1246.

Asferg, C., Mogelvang, R., Flyvbjerg, A., Frystyk, J., Jensen, J. S., Marott, J. L., et al. (2010). Leptin, not adiponectin, predicts hypertension in the Copenhagen City Heart Study. *American Journal of Hypertension*, *23*, 327–333.

Berner, H. S., Lyngstadaas, S. P., Spahr, A., Monjo, M., Thommesen, L., Drevon, C. A., et al. (2004). Adiponectin and its receptors are expressed in bone-forming cells. *Bone*, *35*, 842–849.

Bora, P. S., Kaliappan, S., Lyzogubov, V. V., Tytarenko, R. G., Thotakura, S., Viswanathan, T., et al. (2007). Expression of adiponectin in choroidal tissue and inhibition of laser induced choroidal neovascularization by adiponectin. *FEBS Letters*, *581*, 1977–1982.

Bråkenhielm, E., Veitonmäki, N., Cao, R., Kihara, S., Matsuzawa, Y., Zhivotovsky, B., et al. (2004). Adiponectin-induced antiangiogenesis and antitumor activity involve caspase-mediated endothelial cell apoptosis. *Proceedings of the National Academy of Sciences of the United States of America*, *101*, 2476–2481.

Brochu-Gaudreau, K., Rehfeldt, C., Blouin, R., Bordignon, V., Murphy, B. D., & Palin, M. F. (2010). Adiponectin action from head to toe. *Endocrine*, *37*, 11–32.

Bruce, C. R., Mertz, V. A., Heigenhauser, G. J., & Dyck, D. J. (2005). The stimulatory effect of globular adiponectin on insulin-stimulated glucose uptake and fatty acid oxidation is impaired in skeletal muscle from obese subjects. *Diabetes*, *54*, 3154–3160.

Caminos, J. E., Nogueiras, R., Gallego, R., Bravo, S., Tovar, S., Garcia-Caballero, T., et al. (2005). Expression and regulation of adiponectin and receptor in human and rat placenta. *Journal of Clinical Endocrinology and Metabolism*, *90*, 4276–4286.

Cesari, M., Pessina, A. C., Zanchetta, M., De Toni, R., Avogaro, A., Pedon, L., et al. (2006). Low plasma adiponectin is associated with coronary artery disease but not with hypertension in high-risk nondiabetic patients. *Journal of Internal Medicine*, *260*, 474–483.

Chandrasekar, B., Boylston, W. H., Venkatachalam, K., Webster, N. J., Prabhu, S. D., & Valente, A. J. (2008). Adiponectin blocks interleukin-18-mediated endothelial cell death via APPL1-dependent AMP-activated protein kinase (AMPK) activation and IKK/NF-kappaB/PTEN suppression. *Journal of Biological Chemistry*, *283*, 24889–24898.

Chen, B., Liao, W. Q., Xu, N., Xu, H., Wen, J. Y., Yu, C. A., et al. (2009). Adiponectin protects against cerebral ischemia-reperfusion injury through anti-inflammatory action. *Brain Research*, *1273*, 129–137.

Chen, M. B., McAinch, A. J., Macaulay, S. L., Castelli, L. A., O'Brien, P. E., Dixon, J. B., et al. (2005). Impaired activation of AMP-kinase and fatty acid oxidation by globular adiponectin in cultured human skeletal muscle of obese type 2 diabetics. *Journal of Clinical Endocrinology and Metabolism*, *90*, 3665–3672.

Chen, Z. P., Mitchelhill, K. I., Michell, B. J., Stapleton, D., Rodriguez-Crespo, I., Witters, L. A., et al. (1999). AMP-activated protein kinase phosphorylation of endothelial NO synthase. *FEBS Letters, 443*, 285–289.

Chen, M. P., Tsai, J. C., Chung, F. M., Yang, S. S., Hsing, L. L., Shin, S. J., et al. (2005). Hypoadiponectinemia is associated with ischemic cerebrovascular disease. *Arteriosclerosis, Thrombosis, and Vascular Biology, 25*, 821–826.

Choi, K. M., Lee, J., Lee, K. W., Seo, J. A., Oh, J. H., Kim, S. G., et al. (2004). Serum adiponectin concentrations predict the developments of type 2 diabetes and the metabolic syndrome in elderly Koreans. *Clinical Endocrinology, 61*, 75–80.

Chow, W. S., Cheung, B. M., Tso, A. W., Xu, A., Wat, N. M., Fong, C. H., et al. (2007). Hypoadiponectinemia as a predictor for the development of hypertension: A 5-year prospective study. *Hypertension, 49*, 1455–1461.

Daniels, L. B., & Maisel, A. S. (2007). Natriuretic peptides. *Journal of the American College of Cardiology, 50*, 2357–2368.

Das, S., Otani, H., Maulik, N., & Das, D. K. (2006). Redox regulation of angiotensin II preconditioning of the myocardium requires MAP kinase signaling. *Journal of Molecular and Cellular Cardiology, 41*, 248–255.

Delaigle, A. M., Jonas, J. C., Bauche, I. B., Cornu, O., & Brichard, S. M. (2004). Induction of adiponectin in skeletal muscle by inflammatory cytokines: in vivo and in vitro studies. *Endocrinology, 145*, 5589–5597.

Denzel, M. S., Scimia, M. C., Zumstein, P. M., Walsh, K., Ruiz-Lozano, P., & Ranscht, B. (2010). T-cadherin is critical for adiponectin-mediated cardioprotection in mice. *The Journal of Clinical Investigation, 120*, 4342–4352.

Devaraj, S., Torok, N., Dasu, M. R., Samols, D., & Jialal, I. (2008). Adiponectin decreases C-reactive protein synthesis and secretion from endothelial cells: Evidence for an adipose tissue-vascular loop. *Arteriosclerosis, Thrombosis, and Vascular Biology, 28*, 1368–1374.

Ding, G., Qin, Q., He, N., Francis-David, S. C., Hou, J., Liu, J., et al. (2007). Adiponectin and its receptors are expressed in adult ventricular cardiomyocytes and upregulated by activation of peroxisome proliferator-activated receptor gamma. *Journal of Molecular and Cellular Cardiology, 43*, 73–84.

Ding, M., Xie, Y., Wagner, R. J., Jin, Y., Carrao, A. C., Liu, L. S., et al. (2011). Adiponectin induces vascular smooth muscle cell differentiation via repression of mammalian target of rapamycin complex 1 and FoxO4. *Arteriosclerosis, Thrombosis, and Vascular Biology, 31*, 1403–1410.

Ding, M., Carrao, A. C., Wagner, R. J., Xie, Y., Jin, Y., Rzucidlo, E. M., et al. (2012). Vascular smooth muscle cell-derived adiponectin: A paracrine regulator of contractile phenotype. *Journal of Molecular and Cellular Cardiology, 52*, 474–484.

Efstathiou, S. P., Tsioulos, D. I., Tsiakou, A. G., Gratsias, Y. E., Pefanis, A. V., & Mountokalakis, T. D. (2005). Plasma adiponectin levels and five-year survival after first-ever ischemic stroke. *Stroke, 36*, 1915–1919.

Eroglu, S., Sade, L. E., Yildirir, A., Bal, U., Ozbicer, S., Ozgul, A. S., et al. (2009). Epicardial adipose tissue thickness by echocardiography is a marker for the presence and severity of coronary artery disease. *Nutrition, Metabolism, and Cardiovascular Diseases, 19*, 211–217.

Essick, E. E., Ouchi, N., Wilson, R. M., Ohashi, K., Ghobrial, J., Shibata, R., et al. (2011). Adiponectin mediates cardioprotection in oxidative stress-induced cardiac myocyte remodeling. *American Journal of Physiology. Heart and Circulatory Physiology, 301*, H984–H993.

Fujioka, D., Kawabata, K., Saito, Y., Kobayashi, T., Nakamura, T., Kodama, Y., et al. (2006). Role of adiponectin receptors in endothelin-induced cellular hypertrophy in cultured cardiomyocytes and their expression in infarcted heart. *American Journal of Physiology. Heart and Circulatory Physiology, 290*, H2409–H2416.

George, J., Patal, S., Wexler, D., Sharabi, Y., Peleg, E., Kamari, Y., et al. (2006). Circulating adiponectin concentrations in patients with congestive heart failure. *Heart, 92*, 1420–1424.
Gilligan, D. M., Quyyumi, A. A., & Cannon, R. O., 3rd (1994). Effects of physiological levels of estrogen on coronary vasomotor function in postmenopausal women. *Circulation, 89*, 2545–2551.
Gonon, A. T., Widegren, U., Bulhak, A., Salehzadeh, F., Persson, J., Sjoquist, P. O., et al. (2008). Adiponectin protects against myocardial ischaemia-reperfusion injury via AMP-activated protein kinase, Akt, and nitric oxide. *Cardiovascular Research, 78*, 116–122.
Grady, K. L., Naftel, D., Pamboukian, S. V., Frazier, O. H., Hauptman, P., Herre, J., et al. (2005). Post-operative obesity and cachexia are risk factors for morbidity and mortality after heart transplant: Multi-institutional study of post-operative weight change. *The Journal of Heart and Lung Transplantation, 24*, 1424–1430.
Hansmann, G., & Rabinovitch, M. (2010). The protective role of adiponectin in pulmonary vascular disease. *American Journal of Physiology. Lung Cellular and Molecular Physiology, 298*, L1–L2.
Hebbard, L. W., Garlatti, M., Young, L. J., Cardiff, R. D., Oshima, R. G., & Ranscht, B. (2008). T-cadherin supports angiogenesis and adiponectin association with the vasculature in a mouse mammary tumor model. *Cancer Research, 68*, 1407–1416.
Heid, I. M., Wagner, S. A., Gohlke, H., Iglseder, B., Mueller, J. C., Cip, P., et al. (2006). Genetic architecture of the APM1 gene and its influence on adiponectin plasma levels and parameters of the metabolic syndrome in 1,727 healthy Caucasians. *Diabetes, 55*, 375–384.
Hein, L., Stevens, M. E., Barsh, G. S., Pratt, R. E., Kobilka, B. K., & Dzau, V. J. (1997). Overexpression of angiotensin AT1 receptor transgene in the mouse myocardium produces a lethal phenotype associated with myocyte hyperplasia and heart block. *Proceedings of the National Academy of Sciences of the United States of America, 94*, 6391–6396.
Heinlein, C. A., Ting, H. J., Yeh, S., & Chang, C. (1999). Identification of ARA70 as a ligand-enhanced coactivator for the peroxisome proliferator-activated receptor gamma. *Journal of Biological Chemistry, 274*, 16147–16152.
Hivert, M. F., Manning, A. K., McAteer, J. B., Florez, J. C., Dupuis, J., Fox, C. S., et al. (2008). Common variants in the adiponectin gene (ADIPOQ) associated with plasma adiponectin levels, type 2 diabetes, and diabetes-related quantitative traits: The Framingham Offspring Study. *Diabetes, 57*, 3353–3359.
Hoefle, G., Muendlein, A., Saely, C. H., Risch, L., Rein, P., Koch, L., et al. (2007). The -11377 C>G promoter variant of the adiponectin gene, prevalence of coronary atherosclerosis, and incidence of vascular events in men. *Thrombosis and Haemostasis, 97*, 451–457.
Holland, W. L., Miller, R. A., Wang, Z. V., Sun, K., Barth, B. M., Bui, H. H., et al. (2011). Receptor-mediated activation of ceramidase activity initiates the pleiotropic actions of adiponectin. *Nature Medicine, 17*, 55–63.
Hong, S. J., Park, C. G., Seo, H. S., Oh, D. J., & Ro, Y. M. (2004). Associations among plasma adiponectin, hypertension, left ventricular diastolic function and left ventricular mass index. *Blood Pressure, 13*, 236–242.
Hopkins, T. A., Ouchi, N., Shibata, R., & Walsh, K. (2007). Adiponectin actions in the cardiovascular system. *Cardiovascular Research, 74*, 11–18.
Hotta, K., Funahashi, T., Arita, Y., Takahashi, M., Matsuda, M., Okamoto, Y., et al. (2000). Plasma concentrations of a novel, adipose-specific protein, adiponectin, in type 2 diabetic patients. *Arteriosclerosis, Thrombosis, and Vascular Biology, 20*, 1595–1599.
Hui, X., Lam, K. S., Vanhoutte, P. M., & Xu, A. (2012). Adiponectin and cardiovascular health: An update. *British Journal of Pharmacology, 165*, 574–590.

Iacobellis, G., Corradi, D., & Sharma, A. M. (2005). Epicardial adipose tissue: Anatomic, biomolecular and clinical relationships with the heart. *Nature Clinical Practice. Cardiovascular Medicine*, *2*, 536–543.

Iacobellis, G., di Gioia, C. R., Cotesta, D., Petramala, L., Travaglini, C., De Santis, V., et al. (2009). Epicardial adipose tissue adiponectin expression is related to intracoronary adiponectin levels. *Hormone and Metabolic Research—Hormon- und Stoffwechselforschung—Hormones et metabolisme*, *41*, 227–231.

Imatoh, T., Miyazaki, M., Momose, Y., Tanihara, S., & Une, H. (2008). Adiponectin levels associated with the development of hypertension: A prospective study. *Hypertension Research*, *31*, 229–233.

Ivanov, D. B., Philippova, M. P., & Tkachuk, V. A. (2001). Structure and functions of classical cadherins. *Biochemistry (Biokhimiia)*, *66*, 1174–1186.

Iwashima, Y., Katsuya, T., Ishikawa, K., Ouchi, N., Ohishi, M., Sugimoto, K., et al. (2004). Hypoadiponectinemia is an independent risk factor for hypertension. *Hypertension*, *43*, 1318–1323.

Kato, M. F., Shibata, R., Obata, K., Miyachi, M., Yazawa, H., Tsuboi, K., et al. (2008). Pioglitazone attenuates cardiac hypertrophy in rats with salt-sensitive hypertension: Role of activation of AMP-activated protein kinase and inhibition of Akt. *Journal of Hypertension*, *26*, 1669–1676.

Kazumi, T., Kawaguchi, A., Sakai, K., Hirano, T., & Yoshino, G. (2002). Young men with high-normal blood pressure have lower serum adiponectin, smaller LDL size, and higher elevated heart rate than those with optimal blood pressure. *Diabetes Care*, *25*, 971–976.

Kim, A. S., Miller, E. J., & Young, L. H. (2009). AMP-activated protein kinase: A core signalling pathway in the heart. *Acta Physiologica*, *196*, 37–53.

Kim, J. E., Song, S. E., Kim, Y. W., Kim, J. Y., Park, S. C., Park, Y. K., et al. (2010). Adiponectin inhibits palmitate-induced apoptosis through suppression of reactive oxygen species in endothelial cells: Involvement of cAMP/protein kinase A and AMP-activated protein kinase. *Journal of Endocrinology*, *207*, 35–44.

Kimura, S., Zhang, G. X., Nishiyama, A., Shokoji, T., Yao, L., Fan, Y. Y., et al. (2005). Role of NAD(P)H oxidase- and mitochondria-derived reactive oxygen species in cardioprotection of ischemic reperfusion injury by angiotensin II. *Hypertension*, *45*, 860–866.

Kissebah, A. H., Sonnenberg, G. E., Myklebust, J., Goldstein, M., Broman, K., James, R. G., et al. (2000). Quantitative trait loci on chromosomes 3 and 17 influence phenotypes of the metabolic syndrome. *Proceedings of the National Academy of Sciences of the United States of America*, *97*, 14478–14483.

Kistorp, C., Faber, J., Galatius, S., Gustafsson, F., Frystyk, J., Flyvbjerg, A., et al. (2005). Plasma adiponectin, body mass index, and mortality in patients with chronic heart failure. *Circulation*, *112*, 1756–1762.

Kitta, Y., Takano, H., Nakamura, T., Kodama, Y., Umetani, K., Fujioka, D., et al. (2008). Low adiponectin levels predict late in-stent restenosis after bare metal stenting in native coronary arteries. *International Journal of Cardiology*, *131*, 78–82.

Kobashi, C., Urakaze, M., Kishida, M., Kibayashi, E., Kobayashi, H., Kihara, S., et al. (2005). Adiponectin inhibits endothelial synthesis of interleukin-8. *Circulation Research*, *97*, 1245–1252.

Kobayashi, H., Ouchi, N., Kihara, S., Walsh, K., Kumada, M., Abe, Y., et al. (2004). Selective suppression of endothelial cell apoptosis by the high molecular weight form of adiponectin. *Circulation Research*, *94*, e27–e31.

Kojima, S., Funahashi, T., Sakamoto, T., Miyamoto, S., Soejima, H., Hokamaki, J., et al. (2003). The variation of plasma concentrations of a novel, adipocyte derived protein, adiponectin, in patients with acute myocardial infarction. *Heart*, *89*, 667.

Kondo, M., Shibata, R., Miura, R., Shimano, M., Kondo, K., Li, P., et al. (2009). Caloric restriction stimulates revascularization in response to ischemia via adiponectin-mediated

activation of endothelial nitric-oxide synthase. *Journal of Biological Chemistry*, *284*, 1718–1724.
Kondo, H., Shimomura, I., Matsukawa, Y., Kumada, M., Takahashi, M., Matsuda, M., et al. (2002). Association of adiponectin mutation with type 2 diabetes: A candidate gene for the insulin resistance syndrome. *Diabetes*, *51*, 2325–2328.
Kozakova, M., Muscelli, E., Flyvbjerg, A., Frystyk, J., Morizzo, C., Palombo, C., et al. (2008). Adiponectin and left ventricular structure and function in healthy adults. *Journal of Clinical Endocrinology and Metabolism*, *93*, 2811–2818.
Krause, M. P., Liu, Y., Vu, V., Chan, L., Xu, A., Riddell, M. C., et al. (2008). Adiponectin is expressed by skeletal muscle fibers and influences muscle phenotype and function. *American Journal of Physiology. Cell Physiology*, *295*, C203–C212.
Kudoh, S., Komuro, I., Mizuno, T., Yamazaki, T., Zou, Y., Shiojima, I., et al. (1997). Angiotensin II stimulates c-Jun NH2-terminal kinase in cultured cardiac myocytes of neonatal rats. *Circulation Research*, *80*, 139–146.
Kumada, M., Kihara, S., Sumitsuji, S., Kawamoto, T., Matsumoto, S., Ouchi, N., et al. (2003). Association of hypoadiponectinemia with coronary artery disease in men. *Arteriosclerosis, Thrombosis, and Vascular Biology*, *23*, 85–89.
Landskroner-Eiger, S., Qian, B., Muise, E. S., Nawrocki, A. R., Berger, J. P., Fine, E. J., et al. (2009). Proangiogenic contribution of adiponectin toward mammary tumor growth in vivo. *Clinical Cancer Research*, *15*, 3265–3276.
Lawlor, D. A., Davey Smith, G., Ebrahim, S., Thompson, C., & Sattar, N. (2005). Plasma adiponectin levels are associated with insulin resistance, but do not predict future risk of coronary heart disease in women. *Journal of Clinical Endocrinology and Metabolism*, *90*, 5677–5683.
Leu, H. B., Chung, C. M., Lin, S. J., Jong, Y. S., Pan, W. H., & Chen, J. W. (2011). Adiponectin gene polymorphism is selectively associated with the concomitant presence of metabolic syndrome and essential hypertension. *PLoS One*, *6*, e19999.
Li, P., Shibata, R., Unno, K., Shimano, M., Furukawa, M., Ohashi, T., et al. (2010). Evidence for the importance of adiponectin in the cardioprotective effects of pioglitazone. *Hypertension*, *55*, 69–75.
Liao, Y., Takashima, S., Maeda, N., Ouchi, N., Komamura, K., Shimomura, I., et al. (2005). Exacerbation of heart failure in adiponectin-deficient mice due to impaired regulation of AMPK and glucose metabolism. *Cardiovascular Research*, *67*, 705–713.
Liu, Y., Chewchuk, S., Lavigne, C., Brule, S., Pilon, G., Houde, V., et al. (2009). Functional significance of skeletal muscle adiponectin production, changes in animal models of obesity and diabetes, and regulation by rosiglitazone treatment. *American Journal of Physiology. Endocrinology and Metabolism*, *297*, E657–E664.
Liu, M. W., Roubin, G. S., & King, S. B., 3rd (1989). Restenosis after coronary angioplasty. Potential biologic determinants and role of intimal hyperplasia. *Circulation*, *79*, 1374–1387.
Maahs, D. M., Ogden, L. G., Kinney, G. L., Wadwa, P., Snell-Bergeon, J. K., Dabelea, D., et al. (2005). Low plasma adiponectin levels predict progression of coronary artery calcification. *Circulation*, *111*, 747–753.
Man, K., Ng, K. T., Xu, A., Cheng, Q., Lo, C. M., Xiao, J. W., et al. (2010). Suppression of liver tumor growth and metastasis by adiponectin in nude mice through inhibition of tumor angiogenesis and downregulation of Rho kinase/IFN-inducible protein 10/matrix metalloproteinase 9 signaling. *Clinical Cancer Research*, *16*, 967–977.
Matsubara, M., Maruoka, S., & Katayose, S. (2002). Inverse relationship between plasma adiponectin and leptin concentrations in normal-weight and obese women. *European Journal of Endocrinology*, *147*, 173–180.
Matsuda, M., Shimomura, I., Sata, M., Arita, Y., Nishida, M., Maeda, N., et al. (2002). Role of adiponectin in preventing vascular stenosis. The missing link of adipo-vascular axis. *Journal of Biological Chemistry*, *277*, 37487–37491.

Medoff, B. D., Okamoto, Y., Leyton, P., Weng, M., Sandall, B. P., Raher, M. J., et al. (2009). Adiponectin deficiency increases allergic airway inflammation and pulmonary vascular remodeling. *American Journal of Respiratory Cell and Molecular Biology, 41*, 397–406.

Menon, V., Li, L., Wang, X., Greene, T., Balakrishnan, V., Madero, M., et al. (2006). Adiponectin and mortality in patients with chronic kidney disease. *Journal of the American Society of Nephrology, 17*, 2599–2606.

Miao, C. Y., & Li, Z. Y. (2012). The role of perivascular adipose tissue in vascular smooth muscle cell growth. *British Journal of Pharmacology, 165*, 643–658.

Mito, T., Miura, S., Iwata, A., Morii, J., Sugihara, M., Ike, A., et al. (2011). Determination of the cut-off plasma adiponectin level associated with a lower risk of restenosis in patients with stable angina. *Coronary Artery Disease, 22*, 451–457.

Mitsuhashi, H., Yatsuya, H., Tamakoshi, K., Matsushita, K., Otsuka, R., Wada, K., et al. (2007). Adiponectin level and left ventricular hypertrophy in Japanese men. *Hypertension, 49*, 1448–1454.

Moldoveanu, E., Mut-Vitcu, B., Tanaseanu, G. R., Marta, D. S., Manea, G., Kosaka, T., et al. (2008). Low basal levels of circulating adiponectin in patients undergoing coronary stenting predict in-stent restenosis, independently of basal levels of inflammatory markers: Lipoprotein associated phospholipase A2, and myeloperoxidase. *Clinical Biochemistry, 41*, 1429–1433.

Murakami, H., Ura, N., Furuhashi, M., Higashiura, K., Miura, T., & Shimamoto, K. (2003). Role of adiponectin in insulin-resistant hypertension and atherosclerosis. *Hypertension Research, 26*, 705–710.

Nagata, D., Mogi, M., & Walsh, K. (2003). AMP-activated protein kinase (AMPK) signaling in endothelial cells is essential for angiogenesis in response to hypoxic stress. *Journal of Biological Chemistry, 278*, 31000–31006.

Nakamoto, M., Ohya, Y., Shinzato, T., Mano, R., Yamazato, M., Sakima, A., et al. (2008). Pioglitazone, a thiazolidinedione derivative, attenuates left ventricular hypertrophy and fibrosis in salt-sensitive hypertension. *Hypertension Research, 31*, 353–361.

Nakamura, T., Funayama, H., Kubo, N., Yasu, T., Kawakami, M., Saito, M., et al. (2006). Association of hyperadiponectinemia with severity of ventricular dysfunction in congestive heart failure. *Circulation Journal, 70*, 1557–1562.

Nakamura, T., Kawachi, K., Saito, Y., Saito, T., Morishita, K., Hoshino, J., et al. (2009). Effects of ARB or ACE-inhibitor administration on plasma levels of aldosterone and adiponectin in hypertension. *International Heart Journal, 50*, 501–512.

Nakamura, Y., Shimada, K., Fukuda, D., Shimada, Y., Ehara, S., Hirose, M., et al. (2004). Implications of plasma concentrations of adiponectin in patients with coronary artery disease. *Heart, 90*, 528–533.

Natal, C., Restituto, P., Iñigo, C., Colina, I., Díez, J., & Varo, N. (2008). The proinflammatory mediator CD40 ligand is increased in the metabolic syndrome and modulated by adiponectin. *Journal of Clinical Endocrinology and Metabolism, 93*, 2319–2327.

Natale, F., Tedesco, M. A., Mocerino, R., de Simone, V., Di Marco, G. M., Aronne, L., et al. (2009). Visceral adiposity and arterial stiffness: Echocardiographic epicardial fat thickness reflects, better than waist circumference, carotid arterial stiffness in a large population of hypertensives. *European Journal of Echocardiography, 10*, 549–555.

Nawrocki, A. R., Hofmann, S. M., Teupser, D., Basford, J. E., Durand, J. L., Jelicks, L. A., et al. (2010). Lack of association between adiponectin levels and atherosclerosis in mice. *Arteriosclerosis, Thrombosis, and Vascular Biology, 30*, 1159–1165.

Nawrocki, A. R., Rajala, M. W., Tomas, E., Pajvani, U. B., Saha, A. K., Trumbauer, M. E., et al. (2006). Mice lacking adiponectin show decreased hepatic insulin sensitivity and reduced responsiveness to peroxisome proliferator-activated receptor gamma agonists. *Journal of Biological Chemistry, 281*, 2654–2660.

Neurath, K. M., Keough, M. P., Mikkelsen, T., & Claffey, K. P. (2006). AMP-dependent protein kinase alpha 2 isoform promotes hypoxia-induced VEGF expression in human glioblastoma. *Glia, 53*, 733–743.

Oguri, M., Kato, K., Yokoi, K., Itoh, T., Yoshida, T., Watanabe, S., et al. (2009). Association of genetic variants with myocardial infarction in Japanese individuals with metabolic syndrome. *Atherosclerosis, 206*, 486–493.

Ohashi, K., Kihara, S., Ouchi, N., Kumada, M., Fujita, K., Hiuge, A., et al. (2006). Adiponectin replenishment ameliorates obesity-related hypertension. *Hypertension, 47*, 1108–1116.

Ohashi, K., Ouchi, N., Kihara, S., Funahashi, T., Nakamura, T., Sumitsuji, S., et al. (2004). Adiponectin I164T mutation is associated with the metabolic syndrome and coronary artery disease. *Journal of the American College of Cardiology, 43*, 1195–1200.

Ohashi, K., Ouchi, N., & Matsuzawa, Y. (2011). Adiponectin and hypertension. *American Journal of Hypertension, 24*, 263–269.

Ohashi, K., Ouchi, N., Sato, K., Higuchi, A., Ishikawa, T. O., Herschman, H. R., et al. (2009). Adiponectin promotes revascularization of ischemic muscle through a cyclooxygenase 2-dependent mechanism. *Molecular and Cellular Biology, 29*, 3487–3499.

Okamoto, Y., Christen, T., Shimizu, K., Asano, K., Kihara, S., Mitchell, R. N., et al. (2009). Adiponectin inhibits allograft rejection in murine cardiac transplantation. *Transplantation, 88*, 879–883.

Okamoto, Y., Folco, E. J., Minami, M., Wara, A. K., Feinberg, M. W., Sukhova, G. K., et al. (2008). Adiponectin inhibits the production of CXC receptor 3 chemokine ligands in macrophages and reduces T-lymphocyte recruitment in atherogenesis. *Circulation Research, 102*, 218–225.

Ong, K. L., Li, M., Tso, A. W., Xu, A., Cherny, S. S., Sham, P. C., et al. (2010). Association of genetic variants in the adiponectin gene with adiponectin level and hypertension in Hong Kong Chinese. *European Journal of Endocrinology, 163*, 251–257.

Otsuka, F., Sugiyama, S., Kojima, S., Maruyoshi, H., Funahashi, T., Matsui, K., et al. (2006). Plasma adiponectin levels are associated with coronary lesion complexity in men with coronary artery disease. *Journal of the American College of Cardiology, 48*, 1155–1162.

Ouchi, N., Kihara, S., Arita, Y., Maeda, K., Kuriyama, H., Okamoto, Y., et al. (1999). Novel modulator for endothelial adhesion molecules: Adipocyte-derived plasma protein adiponectin. *Circulation, 100*, 2473–2476.

Ouchi, N., Kihara, S., Arita, Y., Okamoto, Y., Maeda, K., Kuriyama, H., et al. (2000). Adiponectin, an adipocyte-derived plasma protein, inhibits endothelial NF-kappaB signaling through a cAMP-dependent pathway. *Circulation, 102*, 1296–1301.

Ouchi, N., Kobayashi, H., Kihara, S., Kumada, M., Sato, K., Inoue, T., et al. (2004). Adiponectin stimulates angiogenesis by promoting cross-talk between AMP-activated protein kinase and Akt signaling in endothelial cells. *Journal of Biological Chemistry, 279*, 1304–1309.

Ouchi, N., Shibata, R., & Walsh, K. (2005). AMP-activated protein kinase signaling stimulates VEGF expression and angiogenesis in skeletal muscle. *Circulation Research, 96*, 838–846.

Ouedraogo, R., Gong, Y., Berzins, B., Wu, X., Mahadev, K., Hough, K., et al. (2007). Adiponectin deficiency increases leukocyte-endothelium interactions via upregulation of endothelial cell adhesion molecules in vivo. *The Journal of Clinical Investigation, 117*, 1718–1726.

Ouwens, D. M., Sell, H., Greulich, S., & Eckel, J. (2010). The role of epicardial and perivascular adipose tissue in the pathophysiology of cardiovascular disease. *Journal of Cellular and Molecular Medicine, 14*, 2223–2234.

Pajvani, U. B., Du, X., Combs, T. P., Berg, A. H., Rajala, M. W., Schulthess, T., et al. (2003). Structure-function studies of the adipocyte-secreted hormone Acrp30/adiponectin. Implications for metabolic regulation and bioactivity. *Journal of Biological Chemistry, 278*, 9073–9085.

Patel, S., Flyvbjerg, A., Kozakova, M., Frystyk, J., Ibrahim, I. M., Petrie, J. R., et al. (2008). Variation in the ADIPOQ gene promoter is associated with carotid intima media thickness independent of plasma adiponectin levels in healthy subjects. *European Heart Journal*, *29*, 386–393.

Patel, J. V., Lim, H. S., Dubb, K., Hughes, E. A., & Lip, G. Y. (2009). Circulating levels of adiponectin, leptin, and tumour necrosis factor alpha in hypertension. *Annals of Medicine*, *41*, 291–300.

Pineiro, R., Iglesias, M. J., Gallego, R., Raghay, K., Eiras, S., Rubio, J., et al. (2005). Adiponectin is synthesized and secreted by human and murine cardiomyocytes. *FEBS Letters*, *579*, 5163–5169.

Pischon, T., Girman, C. J., Hotamisligil, G. S., Rifai, N., Hu, F. B., & Rimm, E. B. (2004). Plasma adiponectin levels and risk of myocardial infarction in men. *JAMA: The Journal of the American Medical Association*, *291*, 1730–1737.

Psilopanagioti, A., Papadaki, H., Kranioti, E. F., Alexandrides, T. K., & Varakis, J. N. (2009). Expression of adiponectin and adiponectin receptors in human pituitary gland and brain. *Neuroendocrinology*, *89*, 38–47.

Qi, L., Doria, A., Manson, J. E., Meigs, J. B., Hunter, D., Mantzoros, C. S., et al. (2006). Adiponectin genetic variability, plasma adiponectin, and cardiovascular risk in patients with type 2 diabetes. *Diabetes*, *55*, 1512–1516.

Qi, L., Li, T., Rimm, E., Zhang, C., Rifai, N., Hunter, D., et al. (2005). The +276 polymorphism of the APM1 gene, plasma adiponectin concentration, and cardiovascular risk in diabetic men. *Diabetes*, *54*, 1607–1610.

Ran, J., Hirano, T., Fukui, T., Saito, K., Kageyama, H., Okada, K., et al. (2006). Angiotensin II infusion decreases plasma adiponectin level via its type 1 receptor in rats: An implication for hypertension-related insulin resistance. *Metabolism*, *55*, 478–488.

Rosito, G. A., Massaro, J. M., Hoffmann, U., Ruberg, F. L., Mahabadi, A. A., Vasan, R. S., et al. (2008). Pericardial fat, visceral abdominal fat, cardiovascular disease risk factors, and vascular calcification in a community-based sample: The Framingham Heart Study. *Circulation*, *117*, 605–613.

Sadoshima, J., & Izumo, S. (1993). Molecular characterization of angiotensin II–induced hypertrophy of cardiac myocytes and hyperplasia of cardiac fibroblasts. Critical role of the AT1 receptor subtype. *Circulation Research*, *73*, 413–423.

Salmenniemi, U., Ruotsalainen, E., Pihlajamaki, J., Vauhkonen, I., Kainulainen, S., Punnonen, K., et al. (2004). Multiple abnormalities in glucose and energy metabolism and coordinated changes in levels of adiponectin, cytokines, and adhesion molecules in subjects with metabolic syndrome. *Circulation*, *110*, 3842–3848.

Sam, F., Duhaney, T. A., Sato, K., Wilson, R. M., Ohashi, K., Sono-Romanelli, S., et al. (2010). Adiponectin deficiency, diastolic dysfunction, and diastolic heart failure. *Endocrinology*, *151*, 322–331.

Scherer, P. E., Williams, S., Fogliano, M., Baldini, G., & Lodish, H. F. (1995). A novel serum protein similar to C1q, produced exclusively in adipocytes. *Journal of Biological Chemistry*, *270*, 26746–26749.

Schraw, T., Wang, Z. V., Halberg, N., Hawkins, M., & Scherer, P. E. (2008). Plasma adiponectin complexes have distinct biochemical characteristics. *Endocrinology*, *149*, 2270–2282.

Schulze, M. B., Shai, I., Rimm, E. B., Li, T., Rifai, N., & Hu, F. B. (2005). Adiponectin and future coronary heart disease events among men with type 2 diabetes. *Diabetes*, *54*, 534–539.

Shetty, S., Kusminski, C. M., & Scherer, P. E. (2009). Adiponectin in health and disease: Evaluation of adiponectin-targeted drug development strategies. *Trends in Pharmacological Sciences*, *30*, 234–239.

Shibata, R., Izumiya, Y., Sato, K., Papanicolaou, K., Kihara, S., Colucci, W. S., et al. (2007). Adiponectin protects against the development of systolic dysfunction following myocardial infarction. *Journal of Molecular and Cellular Cardiology*, *42*, 1065–1074.

Shibata, R., Ouchi, N., Ito, M., Kihara, S., Shiojima, I., Pimentel, D. R., et al. (2004). Adiponectin-mediated modulation of hypertrophic signals in the heart. *Nature Medicine*, *10*, 1384–1389.

Shibata, R., Ouchi, N., Kihara, S., Sato, K., Funahashi, T., & Walsh, K. (2004). Adiponectin stimulates angiogenesis in response to tissue ischemia through stimulation of amp-activated protein kinase signaling. *Journal of Biological Chemistry*, *279*, 28670–28674.

Shibata, R., Sato, K., Pimentel, D. R., Takemura, Y., Kihara, S., Ohashi, K., et al. (2005). Adiponectin protects against myocardial ischemia-reperfusion injury through AMPK- and COX-2-dependent mechanisms. *Nature Medicine*, *11*, 1096–1103.

Shimabukuro, M., Higa, N., Asahi, T., Oshiro, Y., Takasu, N., Tagawa, T., et al. (2003). Hypoadiponectinemia is closely linked to endothelial dysfunction in man. *Journal of Clinical Endocrinology and Metabolism*, *88*, 3236–3240.

Shimano, M., Ouchi, N., Shibata, R., Ohashi, K., Pimentel, D. R., Murohara, T., et al. (2010). Adiponectin deficiency exacerbates cardiac dysfunction following pressure overload through disruption of an AMPK-dependent angiogenic response. *Journal of Molecular and Cellular Cardiology*, *49*, 210–220.

Simpson, F., & Whitehead, J. P. (2010). Adiponectin—It's all about the modifications. *The International Journal of Biochemistry & Cell Biology*, *42*, 785–788.

Skurk, C., Wittchen, F., Suckau, L., Witt, H., Noutsias, M., Fechner, H., et al. (2008). Description of a local cardiac adiponectin system and its deregulation in dilated cardiomyopathy. *European Heart Journal*, *29*, 1168–1180.

Summer, R., Fiack, C. A., Ikeda, Y., Sato, K., Dwyer, D., Ouchi, N., et al. (2009). Adiponectin deficiency: A model of pulmonary hypertension associated with pulmonary vascular disease. *American Journal of Physiology. Lung Cellular and Molecular Physiology*, *297*, L432–L438.

Tadros, T. M., Massaro, J. M., Rosito, G. A., Hoffmann, U., Vasan, R. S., Larson, M. G., et al. (2010). Pericardial fat volume correlates with inflammatory markers: The Framingham Heart Study. *Obesity*, *18*, 1039–1045.

Takano, H., Komuro, I., Zou, Y., Kudoh, S., Yamazaki, T., & Yazaki, Y. (1996). Activation of p70 S6 protein kinase is necessary for angiotensin II-induced hypertrophy in neonatal rat cardiac myocytes. *FEBS Letters*, *379*, 255–259.

Takaoka, M., Nagata, D., Kihara, S., Shimomura, I., Kimura, Y., Tabata, Y., et al. (2009). Periadventitial adipose tissue plays a critical role in vascular remodeling. *Circulation Research*, *105*, 906–911.

Tamura, T., Furukawa, Y., Taniguchi, R., Sato, Y., Ono, K., Horiuchi, H., et al. (2007). Serum adiponectin level as an independent predictor of mortality in patients with congestive heart failure. *Circulation Journal*, *71*, 623–630.

Tao, L., Gao, E., Jiao, X., Yuan, Y., Li, S., Christopher, T. A., et al. (2007). Adiponectin cardioprotection after myocardial ischemia/reperfusion involves the reduction of oxidative/nitrative stress. *Circulation*, *115*, 1408–1416.

Teoh, H., Quan, A., Bang, K. W., Wang, G., Lovren, F., Vu, V., et al. (2008). Adiponectin deficiency promotes endothelial activation and profoundly exacerbates sepsis-related mortality. *American Journal of Physiology. Endocrinology and Metabolism*, *295*, E658–E664.

Turer, A. T., & Scherer, P. E. (2012). Adiponectin: mechanistic insights and clinical implications, Diabetologia, June 12, 2012. [Epub ahead of print].

Thomopoulos, C., Daskalaki, M., Papazachou, O., Rodolakis, N., Bratsas, A., Papadopoulos, D. P., et al. (2011). Association of resistin and adiponectin with different clinical blood pressure phenotypes. *Journal of Human Hypertension*, *25*, 38–46.

Uchida, T., Shimizu, M., Sakai, Y., Nakano, T., Hara, K., Takebayashi, K., et al. (2008). Effects of losartan on serum total and high-molecular weight adiponectin concentrations in hypertensive patients with metabolic syndrome. *Metabolism*, *57*, 1278–1285.

Viengchareun, S., Zennaro, M. C., Pascual-Le Tallec, L., & Lombes, M. (2002). Brown adipocytes are novel sites of expression and regulation of adiponectin and resistin. *FEBS Letters, 532*, 345–350.

Vionnet, N., Hani, E. H., Dupont, S., Gallina, S., Francke, S., Dotte, S., et al. (2000). Genomewide search for type 2 diabetes-susceptibility genes in French whites: Evidence for a novel susceptibility locus for early-onset diabetes on chromosome 3q27-qter and independent replication of a type 2-diabetes locus on chromosome 1q21-q24. *American Journal of Human Genetics, 67*, 1470–1480.

Walsh, K. (2009). Adipokines, myokines and cardiovascular disease. *Circulation Journal, 73*, 13–18.

Wang, Y., Lam, K. S., Yau, M. H., & Xu, A. (2008). Post-translational modifications of adiponectin: Mechanisms and functional implications. *The Biochemical Journal, 409*, 623–633.

Wang, Y., Lau, W. B., Gao, E., Tao, L., Yuan, Y., Li, R., et al. (2010). Cardiomyocyte-derived adiponectin is biologically active in protecting against myocardial ischemia-reperfusion injury. *American Journal of Physiology. Endocrinology and Metabolism, 298*, E663–E670.

Wang, T., Qiao, S., Lei, S., Liu, Y., Ng, K. F., Xu, A., et al. (2011). N-acetylcysteine and allopurinol synergistically enhance cardiac adiponectin content and reduce myocardial reperfusion injury in diabetic rats. *PLoS One, 6*, e23967.

Watanabe, S., Okura, T., Kurata, M., Irita, J., Manabe, S., Miyoshi, K., et al. (2006). The effect of losartan and amlodipine on serum adiponectin in Japanese adults with essential hypertension. *Clinical Therapeutics, 28*, 1677–1685.

Wehner, J. R., & Baldwin, W. M., 3rd (2010). Cardiac allograft vasculopathy: Do adipocytes bridge alloimmune and metabolic risk factors? *Current Opinion in Organ Transplantation, 15*, 639–644.

Welt, F. G., & Rogers, C. (2002). Inflammation and restenosis in the stent era. *Arteriosclerosis, Thrombosis, and Vascular Biology, 22*, 1769–1776.

Weng, M., Baron, D. M., Bloch, K. D., Luster, A. D., Lee, J. J., & Medoff, B. D. (2011). Eosinophils are necessary for pulmonary arterial remodeling in a mouse model of eosinophilic inflammation-induced pulmonary hypertension. *American Journal of Physiology. Lung Cellular and Molecular Physiology, 301*, L927–L936.

Weng, M., Raher, M. J., Leyton, P., Combs, T. P., Scherer, P. E., Bloch, K. D., et al. (2011). Adiponectin decreases pulmonary arterial remodeling in murine models of pulmonary hypertension. *American Journal of Respiratory Cell and Molecular Biology, 45*, 340–347.

Williams, J. K., Adams, M. R., Herrington, D. M., & Clarkson, T. B. (1992). Short-term administration of estrogen and vascular responses of atherosclerotic coronary arteries. *Journal of the American College of Cardiology, 20*, 452–457.

Wong, W. T., Tian, X. Y., Xu, A., Yu, J., Lau, C. W., Hoo, R. L., et al. (2011). Adiponectin is required for PPARgamma-mediated improvement of endothelial function in diabetic mice. *Cell Metabolism, 14*, 104–115.

Wu, X., Mahadev, K., Fuchsel, L., Ouedraogo, R., Xu, S. Q., & Goldstein, B. J. (2007). Adiponectin suppresses IkappaB kinase activation induced by tumor necrosis factor-alpha or high glucose in endothelial cells: Role of cAMP and AMP kinase signaling. *American Journal of Physiology. Endocrinology and Metabolism, 293*, E1836–E1844.

Xu, S. Q., Mahadev, K., Wu, X., Fuchsel, L., Donnelly, S., Scalia, R. G., et al. (2008). Adiponectin protects against angiotensin II or tumor necrosis factor alpha-induced endothelial cell monolayer hyperpermeability: Role of cAMP/PKA signaling. *Arteriosclerosis, Thrombosis, and Vascular Biology, 28*, 899–905.

Xu, M. J., Song, P., Shirwany, N., Liang, B., Xing, J., Viollet, B., et al. (2011). Impaired expression of uncoupling protein 2 causes defective postischemic angiogenesis in mice deficient in AMP-activated protein kinase α subunits. *Arteriosclerosis, Thrombosis, and Vascular Biology, 31*, 1757–1765.

Yamauchi, T., Kamon, J., Ito, Y., Tsuchida, A., Yokomizo, T., Kita, S., et al. (2003). Cloning of adiponectin receptors that mediate antidiabetic metabolic effects. *Nature*, *423*, 762–769.

Yin, W. H., Chen, Y. H., Wei, J., Jen, H. L., Huang, W. P., Young, M. S., et al. (2011). Associations between endothelin-1 and adiponectin in chronic heart failure. *Cardiology*, *118*, 207–216.

Yun, H., Lee, M., Kim, S. S., & Ha, J. (2005). Glucose deprivation increases mRNA stability of vascular endothelial growth factor through activation of AMP-activated protein kinase in DU145 prostate carcinoma. *Journal of Biological Chemistry*, *280*, 9963–9972.

Zhang, P., Wang, Y., Fan, Y., Tang, Z., & Wang, N. (2009). Overexpression of adiponectin receptors potentiates the antiinflammatory action of subeffective dose of globular adiponectin in vascular endothelial cells. *Arteriosclerosis, Thrombosis, and Vascular Biology*, *29*, 67–74.

Zoccali, C., Mallamaci, F., Tripepi, G., Benedetto, F. A., Cutrupi, S., Parlongo, S., et al. (2002). Adiponectin, metabolic risk factors, and cardiovascular events among patients with end-stage renal disease. *Journal of the American Society of Nephrology*, *13*, 134–141.

CHAPTER TWELVE

Adiponectin Interactions in Bone and Cartilage Biology and Disease

Massimiliano Ruscica[*,†], **Liliana Steffani**[*,†], **Paolo Magni**[*,†,1]
[*]Department of Pharmacological and Biomolecular Sciences, Università degli Studi di Milano, Milan, Italy
[†]Centro Universitario Dislipidemie, Ospedale Niguarda Cà Granda, Milan, Italy
[1]Corresponding author: e-mail address: paolo.magni@unimi.it

Contents

Abstract

The adipokine adiponectin promotes insulin sensitivity and fat β-oxidation. In addition to its metabolic effects, adiponectin is an important local and systemic modulator of bone remodeling and cartilage biology, involving direct and indirect mechanisms and a large set of downstream molecular signals. Moreover, data suggest that changes in adiponectin signaling may be associated with bone and cartilage diseases. Adiponectin seems to exert a negative net effect on bone mass and to be an independent predictor of lower bone mass, whereas available data about actions on cartilage are more controversial, showing both pro- and anti-inflammatory actions. Adiponectin-bone cross talk seems to be reciprocal, as osteocalcin, produced by osteoblasts, has been shown to stimulate adiponectin expression and to improve glucose tolerance. Adiponectin-related signaling in bone and cartilage should be considered within the network of hormonal and nutritional signals that may influence skeleton biology, together with body homeostasis and adipose mass changes.

ABBREVIATIONS

AdipoR adiponectin receptor
AMPK AMP-activated kinase
BMD bone mineral density
BMI body mass index
IL interleukin
MAPK mitogen-activated protein kinase

Vitamins and Hormones, Volume 90
ISSN 0083-6729
http://dx.doi.org/10.1016/B978-0-12-398313-8.00012-9

MMP matrix metalloproteinase
NFAT nuclear factor-activated T cell
NO nitric oxide
OA osteoarthritis
OPG osteoprotegerin
RA rheumatoid arthritis
RANKL receptor activator of nuclear factor κB ligand
SF synovial fluid
TNF-α tumor necrosis factor-alpha

1. INTRODUCTION

The growing information on the novel functions of the adipose tissue obtained since the discovery of leptin in 1994 (Zhang et al., 1994) contributed to consolidate the concept that this organ works as a true endocrine gland, producing several active molecules (Mohamed-Ali, Pinkney, & Coppack, 1998). These secretory products, called adipokines, show multiple actions, such as modulation of eating behavior, energy expenditure, fatty acid oxidation, insulin sensitivity, and glucose metabolism, as well as regulation of cardiovascular and reproductive functions (Scherer, 2006).

Recent observations have shown that some adipokines also participate in the complex network regulating bone metabolism, suggesting peculiar adipose–bone molecular interactions (reviewed in Magni, Dozio, Galliera, Ruscica, & Corsi, 2010). In particular, some adipose–bone molecules appear to modulate bone metabolism directly, also through fat infiltration in the medullary space, as well as indirectly, by central regulation of both metabolism and bone mass (Rosen & Bouxsein, 2006). In this field, the relationship between adipokines and cartilage is another important emerging aspect (reviewed in Dozio et al., 2011).

The biology of different components of the skeleton, like bone and joint cartilage, and its derangement, which leads to the development of pathological states, like osteoporosis and osteoarthritis (OA), are known to be controlled by a wide range of local (paracrine) and systemic (endocrine, nervous, nutritional, biophysical, etc.) factors (Cohen, 2006; Parfitt, 1996), which thus includes different adipokines and, specifically, adiponectin.

As the circulating levels of adipokines vary according to body weight and fat mass, the relationships between these hormones and the biology and pathophysiology of bone and cartilage need to be placed within the broader area of obesity and weight reduction strategies. Indeed, weight increase and obesity generally appear to represent important protective factors for osteoporotic fractures, while thinness seems a risk factor for them (Reid, 2008). In most studies, for example, obese postmenopausal women appeared protected against the development of osteoporosis, despite the lack of ovarian estrogens and the associated trophic actions on bone (Zhao et al., 2008). On the contrary, obesity seems to exert a negative impact on articular cartilage, not only because of increased mechanical stress on joint tissues but also through the action of soluble factors, including adipokines, that might play an important role in the onset and progression of the degradative process of cartilage associated to obesity (Anderson & Felson, 1988; Cicuttini, Baker, & Spector, 1996).

The adipokine adiponectin is secreted by the adipose tissue and circulates in high concentrations in the blood in normal weight subjects, whereas in obese and diabetic patients, it is markedly reduced (Arita et al., 1999; Hotta et al., 2000; Matsuzawa, 2005; Tsao et al., 2003; Vilarrasa et al., 2005). However, after marked weight loss, plasma adiponectin levels are increased (Yang et al., 2001). Adiponectin promotes insulin sensitivity and fatty acid oxidation and shows anti-inflammatory properties (Berg, Combs, & Scherer, 2002; Nawrocki & Scherer, 2004). In addition, it exerts relevant actions in bone and cartilage biology and is potentially involved in some related diseases, as discussed in this chapter.

2. ADIPOSE, BONE, AND CARTILAGE RELATIONSHIPS: OBSERVATIONS IN HUMANS

The current information about the role of adiponectin and of the other adipokines in skeletal physiology and pathophysiology has been preceded over the past decades by several studies on the relationship between adipose mass and bone and cartilage health and disease. Excess adipose tissue has been generally found associated with increased bone mass (Felson, Zhang, Hannan, & Anderson, 1993) and reduced bone turnover (Cifuentes et al., 2003; Papakitsou et al., 2004) and loss (Nguyen, Sambrook, & Eisman, 1998; Reid, 2010; Reid, Ames, Evans, Sharpe, & Gamble, 1994). Interestingly, high bone density tended to be reduced upon weight and fat loss (Ensrud et al., 2005; Langlois et al., 2001). In particular, a total 10% weight loss resulted in about 1–2% bone loss at various bone sites (Compston, Laskey, Croucher, Coxon, & Kreitzman, 1992; Salamone et al., 1999; Svendsen, Hassager, & Christiansen, 1993).

Moreover, bone loss was greater when weight loss occurred in normal weight subjects in comparison to overweight or obese individuals (Nguyen et al., 1998). Conversely, low body weight was mainly associated with low bone mass and increased fracture risk (Espallargues et al., 2001).

Cross-sectional and longitudinal studies in adults and children have shown that bone mass is positively related to body weight and, to a lesser extent, to body mass index (BMI) (Ravn et al., 1999; Reid, 2008, 2010; Reid et al., 1994; Robbins, Schott, Azari, & Kronmal, 2006; Wu et al., 2002). In women, body weight is actually the strongest predictor of bone mineral density (BMD) at various skeletal sites, independently of previous estrogen use (Grinspoon et al., 2000), although it is controversial whether lean mass or fat mass is the more important determinant of BMD (Wang, Bachrach, et al., 2005).

Different possible mechanisms may explain the positive association between body weight and bone mass: (1) a greater body weight might increase mechanical load on the skeleton, particularly on cortical elements, and (2) excess fat mass is associated with altered secretion of hormones active on bone and produced by adipocytes or by inflamed adipose tissue (i.e., estrogen, adipokines, inflammatory cytokines) and pancreatic beta cells (insulin, amylin) (Reid, 2008, 2010). Interestingly, the precise fat–BMD relationships found are dependent on gender, the fat–bone relationship being weaker in men (Goulding, Jones, Taylor, Williams, & Manning, 2001); on menopausal status, as larger fat effect are found in the postmenopause; on exercise status, as the relationship is stronger in sedentary populations (Elefteriou et al., 2005); and on ethnicity (Reid, 2008; Wang, Bachrach, et al., 2005).

Body weight, in addition to BMD, has also been negatively associated with fracture risk by many authors (Brown & Rosen, 2003), showing that high BMI is protective against total fractures, osteoporotic fractures, and hip fractures both in men and women (De Laet et al., 2005).

In addition to a systemic connection, a local bone–fat relationship within the bone itself has also been proposed, based on the observation of a greater amount of adipocytes in bone-marrow samples from women with osteoporosis, compared to healthy young subjects (Justesen et al., 2001; Meunier, Aaron, Edouard, & Vignon, 1971; Rozman et al., 1989; Verma, Rajaratnam, Denton, Hoyland, & Byers, 2002). Indeed, increased bone-marrow fat in the vertebral bodies of older women with low bone mass increased the risk for compression fracture in addition to that associated with low BMD (Wehrli et al., 2000). Moreover, women with low BMD had significantly greater bone-marrow adipose than those with normal BMD (Yeung et al., 2005). Interestingly, marrow stromal cells isolated from postmenopausal osteoporotic patients express more adipocytic differentiation markers than those with normal bone mass (Sekiya, Larson, Vuoristo, Cui, & Prockop, 2004) and more likely enter an adipocyte differentiation program than an osteoblast one (Rodriguez, Montecinos, Rios, Reyes, & Martinez, 2000). Fat in bone marrow might not only suppress osteoblastogenesis but might also promote bone resorption, as also marrow adipocytes secrete adipokines and inflammatory cytokines capable of recruiting osteoclasts (Weisberg et al., 2003). Fat infiltration in the bone marrow may then represent an example of lipotoxicity (Duque, 2008) and is associated with skeletal fragility.

Obesity has also been identified as an important risk factor for joint degeneration and OA. For weight-bearing joints, the combination of increased load and altered joint biomechanics could be regarded as the main underlying principle for this relationship. Obese subjects adjust their movement strategy of everyday movements and show altered biomechanics, which could be related to the initiation of OA by a modification in the load-bearing regions of the articular cartilage in the weight-bearing joints (Runhaar, Koes, Clockaerts, & Bierma-Zeinstra, 2011). Weight loss has been shown to improve pain and function in hip and knee joints with OA. Interestingly, bariatric surgery, a relatively common surgical approach to high grade obesity, along with marked weight loss and metabolic improvement, is generally associated to improved hip and knee OA (Gill et al., 2011). Recent evidence suggest that the link between obesity and articular cartilage disease may not just be due to increased mechanical stress on joint tissue, which is maintained by chondrocytes through a biologically complex process that undergoes metabolic changes as a result of aging, disease, and injury. A multifactorial process then leads to damage of cartilage, deposition of subchondral bone matrix, and release of proinflammatory cytokines. Moreover, adipokines, including adiponectin, may also play an important role in the development of cartilage degradative process in obese subjects (Anderson & Felson, 1988; Berry, Jones, Cicuttini, Wluka, & Maciewicz, 2011; Cicuttini et al., 1996). It has been shown that osteoblasts and chondrocytes, in addition to adipocytes, are able to produce and release adipokines, suggesting that local adipokine production, including that by infrapatellar fat, may be of great importance in bone and cartilage biology (Presle et al., 2006; Teichtahl, Wluka, Proietto, & Cicuttini, 2005). Moreover, there is presently no evidence that adipokines cross the joint capsule to enter the joint and this further suggests that local effects of adipokines are mainly dependent on their local production. It is possible that only the increased permeability of inflamed synovial membrane might be a condition allowing adipokines to enter the joint. Anyway, the correlation between synovial fluid (SF) adipokine levels and BMI seems to suggest that, in addition to the mechanical stress associated to weight gain, increased levels of adipokines may be a potential link between obesity and cartilage disease.

3. ADIPONECTIN, BONE, AND CARTILAGE: INTERACTIONS IN HUMANS

The potential involvement of adiponectin in the regulation of bone metabolism is suggested by clinical studies reporting its association with BMD. An inverse relationship between adiponectin plasma concentration and total body BMD was found in adult men and women, also after adjustment for fat mass (Lenchik et al., 2003; Richards, Valdes, Burling, Perks, & Spector, 2007), as also confirmed by a recent meta-analysis (Biver et al., 2011). This inverse relationship was also present in adolescent women but was no longer observed after adjustment for fat mass and Tanner stage (Huang et al., 2004). A negative association of adiponectin with total BMD was generally found in healthy premenopausal (Jurimae & Jurimae, 2007) and perimenopausal women, even after adjusting for body composition, insulin, and leptin (Jurimae, Rembel, Jurimae, & Rehand, 2005). In a group of healthy elderly women, adiponectin levels were found to explain up to about 12% of total body, volumetric, and femoral neck BMD variance (Zoico et al., 2008). The adiponectin–BMD correlation was also confirmed in large cohorts of women with a wide range of age and BMI (Napoli et al., 2010; Richards et al., 2007), showing a negative correlation between plasma adiponectin levels and BMD, also at non-load-bearing sites, after adjustment for total fat mass. However, when the analysis was limited to premenopausal women or men, this association was lost. Other studies found, however, a negative association between adiponectin levels and BMD in men over 60 years, which was even stronger when BMI exceeded 27 kg/m^2 (Basurto et al., 2009), and also in young healthy men at the time of peak bone mass (Frost et al., 2010). In addition, total and high-molecular-weight adiponectin levels were found associated with vertebral fractures in men with type 2 diabetes (Kanazawa, Yamaguchi, Yamamoto, Yamauchi,

Yano, et al., 2009), but not in elderly men (Michaelsson et al., 2008). Interestingly, after 1-year treatment of patients with type 2 diabetes mellitus, serum total adiponectin, but not high-molecular-weight adiponectin, was significantly and positively correlated with percentage change in femoral neck BMD (Kanazawa, Yamaguchi, & Sugimoto, 2010).

A potential determinant of adiponectin–BMD relationship may also be ethnicity, as adiponectin was inversely correlated with BMD in Caucasian, but not in Hispanic women (King, Deemer, & Thompson, 2010). Gastric bypass surgery, which results in rapid phenotypic and hormonal changes in obese patients, was associated with significant BMD loss in female subjects, in association with increased adiponectin levels (Carrasco et al., 2009). Increased serum adiponectin was also associated with decreased BMD in male hemodialysis patients, suggesting a potential role of this adipokine in the mineral and bone disorder observed in patients with chronic renal failure (Okuno et al., 2011).

The existence of at least three different actions of adiponectin on bone has been proposed: (a) a positive action through an autocrine/paracrine pathway, (b) a negative direct endocrine action, and (c) a positive indirect endocrine action through enhancement of insulin osteogenic signaling (Shinoda et al., 2006). Therefore, the discrepancies among different studies may derive from the relative prevalence of negative versus positive effects of adiponectin on bone formation in each study cohort. As a general consideration, circulating adiponectin appears to exert a dominant negative effect on bone mass in humans and to be an independent predictor of lower bone mass.

The well-known protective biological properties of adiponectin (anti-inflammatory, insulin sensitizer, activator of fatty acid oxidation; Berg et al., 2002; Nawrocki & Scherer, 2004) do not appear to be univocally present in the context of articular cartilage biology, as this adipokine has been found to act both as an anti- and a proinflammatory agent involved in matrix degradation. Serum and SF adiponectin levels have been shown to correlate with anthropometric parameters, such as BMI, although there is presently no evidence that adipokines may cross the joint capsule to enter the joint itself (Matsubara, Katayose, & Maruoka, 2003; Matsubara, Namioka, & Katayose, 2003). The amount of adiponectin present in the SF has thus been shown to derive from synovium, infrapatellar fat pad, osteophyte, and cartilage, more than from the circulation (Presle et al., 2006). Adiponectin receptors (AdipoRs), and mainly the AdipoR1, have been identified on the surface of human chondrocytes (Chen et al., 2006; Lago et al., 2008).

Although several studies support the role of adiponectin on inflammation and cartilage destruction in OA, its actual actions in the pathogenesis of OA remain controversial.

In women affected by OA, adiponectin levels in the SF were found to be significantly lower compared to paired plasma levels and to correlate with degradation markers of cartilage matrix (Hao et al., 2011). Evaluation of multimeric adiponectin levels in SF of patients with OA revealed that the amount of high molecular weight relative to total adiponectin in OA SF was lower than in OA plasma, while that of the hexamer form was similar and that of the trimer form was higher (Ebina et al., 2009). Plasma adiponectin levels and adiponectin released from OA cartilage were found to be greater in patients with radiologically most severe OA than in patients with less severe disease. Plasma adiponectin concentrations correlated positively with biomarkers of OA, like matrix metalloproteinase (MMP)-3. Adiponectin enhanced production of nitric oxide (NO), interleukin (IL)-6, MMP-1, and MMP-3 in OA cartilage and in primary chondrocytes *in vitro* by a mitogen-activated protein kinase (MAPK)-dependent manner. This study thus suggests that adiponectin is associated with and possibly mediates cartilage destruction in OA (Koskinen et al., 2011). On the contrary, in another study, adiponectin concentrations in plasma and SF showed significant inverse correlation with knee OA severity. Plasma adiponectin levels positively correlated with SF adiponectin levels and both significantly decreased as the severity of OA increased, suggesting a protective role of adiponectin in OA (Honsawek & Chayanupatkul, 2010).

Patients affected by rheumatoid arthritis (RA) showed SF adiponectin levels greater than subjects with OA (Otero et al., 2006; Schaffler et al., 2003). These elevated adiponectin levels in patients with RA have been interpreted as an attempt to overcome the well-known proinflammatory effect of leptin, for example, by counteracting the proinflammatory effects of tumor necrosis factor-alpha (TNF-α) and reducing the production of IL-6 and C reactive protein in RA. Interestingly, etanercept (a soluble fusion protein that specifically binds to TNF-α) treatment in RA patients was able to increase adiponectin levels in women with RA (Lewicki, Kotyla, & Kucharz, 2008). This finding has also been suggested using infliximab, another anti-TNF-α treatment (Nagashima et al., 2008). Other studies, however, indicated that adiponectin may have a proinflammatory role through increased production of NO synthase type II, MMP-3 and -9 (Ehling et al., 2006; Lago et al., 2008), IL-6, IL-8, prostaglandin E2, and vascular endothelial growth factor (Choi et al., 2009).

Some studies in different biomedical research areas (cardiovascular, diabetes, etc.) (Kotani, Sakane, Saiga, & Kurozawa, 2005; Norata et al., 2007) suggested that integrating different adipokine values may represent a better predictor of disease progression. The serum leptin/adiponectin ratio has been shown to be an important predictor of inflammatory diseases, as it may reflect the balance between the proinflammatory effect of leptin and the anti-inflammatory role of adiponectin. In OA patients, a lower SF leptin/adiponectin ratio predicted lower knee OA pain as measured by the MPQ-SF, but not the WOMAC pain scale, with a greater significance than each individual adipokine level alone (Gandhi, Takahashi, Smith, Rizek, & Mahomed, 2010; Gandhi, Takahashi, Syed, Davey, & Mahomed, 2010).

In summary, the role of adiponectin in synovial joint disease is still not clear, as some authors have suggested that leptin may have a clear-cut proinflammatory role, whereas both pro- and anti-inflammatory actions have been described for adiponectin. Therefore, also in this context, the ratio of different adipokine values may better describe the net effect of this complex networking of signals on tissue biology.

4. ADIPONECTIN, BONE, AND CARTILAGE: EXPERIMENTAL OBSERVATIONS

The current knowledge on adipokine–bone relationships at the cellular and molecular levels has been improved by several experimental studies, which, however, contributed to identify a very complex picture. For example, the prototypical adipokine leptin seems to modulate the activity of bone cells both directly and indirectly influencing skeletal metabolism through its actions on the central nervous system (Karsenty, 2006; Takeda et al., 2002). A functional role of adiponectin in bone biology has been supported by some studies (Oshima et al., 2005; Shinoda et al., 2006), showing that this hormone and its receptors are expressed in osteoblasts (Berner et al., 2004), which also differentiate in response to this hormone. Adiponectin has also been found to stimulate the proliferation and mineralization of osteoblasts via the AdipoR1- and AMP-activated kinase (AMPK) signaling pathway in autocrine and/or paracrine fashions (Kanazawa et al., 2007). Adiponectin seems to stimulate the receptor activator of nuclear factor κB ligand (RANKL) pathway, to inhibit the production of osteoprotegerin (OPG) in human osteoblasts, and to indirectly promote osteoclastogenesis (Luo et al., 2006). Adiponectin was able to inhibit the induction of TNF-α/RANKL-stimulated nuclear factor-activated T cell

(NFAT) c1 (Yamaguchi et al., 2008). However, the effects of adiponectin on bone metabolism are not univocal among different experimental systems and between *in vivo* and *in vitro* studies (Luo et al., 2005; Oshima et al., 2005; Shinoda et al., 2006). Adiponectin has been shown to inhibit osteoclastogenesis and osteoclast activity *in vitro*, to reduce bone resorption, and to increase bone mass in mice *in vivo* (Oshima et al., 2005). On the contrary, adiponectin was also found to modulate osteoblast production of both RANKL and OPG, with reduced osteoclastogenesis (Luo et al., 2006). Moreover, adiponectin has been shown to bind some growth factors (Wang, Lam, et al., 2005) and to reduce circulating insulin concentrations, which would counteract any anabolic effects. Adiponectin was also found to enhance bone morphogenetic protein 2 expression in osteoblastic cells, through the involvement of AdipoR1, AMPK, p38, and NF-κB signaling pathways (Huang et al., 2010). In another study using primary human and rat osteoblasts, adiponectin stimulated osteoblast growth but inhibited osteoclastogenesis, probably via an effect on stromal cells (Williams et al., 2009). Indeed, the sustained release of adiponectin by a matrigel controlled-release system suppressed osteoclastic activity both *in vitro* and *in vivo* and improved the peri-implant osteogenesis in ovariectomized rabbits, while the short-term adiponectin treatment did not (Luo et al., 2012). Interestingly, the adiponectin knockout mouse displayed increased bone mass, suggesting that adiponectin may also have indirect effects on bone, possibly through modulating growth factor action or insulin sensitivity (Williams et al., 2009).

Another study emphasized the role of leptin, but not of adiponectin, in regulating the expression of cartilage-specific markers in chondrocytes obtained from patients with OA (Francin et al., 2011). In the mouse ATDC5 cell line, an *in vitro* model of chondrogenesis, adiponectin increased chondrocyte proliferation, proteoglycan synthesis, and matrix mineralization, through the upregulation of the expression of type II collagen, aggrecan, Runx2, and type X collagen, and of alkaline phosphatase and MMP-9 activity, suggesting the possibility of a direct role in chondrocyte proliferation and differentiation (Challa, Rais, & Ornan, 2010).

Experimental studies also showed contrasting effects of adiponectin on joint inflammation. Recent reports showed that adenovirus-mediated gene transfer of adiponectin reduced the severity of collagen-induced arthritis in mice, thus preventing inflammation and joint destruction (Ebina et al., 2009). However, adiponectin has been found to exert proinflammatory actions by inducing human synovial fibroblasts to specifically release IL-6, MMP-1, and MMP-3, which are main mediators of RA pathophysiology, via AdipoR1 and the p38 MAPK and AMPK pathways (Ehling et al., 2006; Tong et al., 2011). Notably, these effects could be blocked by TNF-α inhibitors.

As mentioned before, adiponectin has been implicated in the pathogenesis of OA. The expression levels of both AdipoR1 and AdipoR2 were significantly higher in lesional than in nonlesional areas of OA cartilage. The increased rate of AdipoR1-positive chondrocytes was twice that of AdipoR2-positive chondrocytes when compared between nonlesional and lesional areas. Adiponectin-stimulated OA chondrocytes showed increased total NO and MMP-1, -3, and -13 levels compared with nonstimulated cells. Adiponectin thus was found to cause matrix degradation in OA cartilage and to increase MMPs and inducible NO synthase expression via the AMPK and JNK pathways in human OA chondrocytes (Kang et al., 2010).

The findings of some studies indicate that adiponectin induces gene expression and protein synthesis in human RA SF cells, lymphocytes, endothelial cells, and chondrocytes, supporting the concept of adiponectin being involved in the pathophysiologic modulation of RA effector cells (Frommer et al., 2010). Adiponectin may promote inflammation through cytokine synthesis, attraction of inflammatory cells to the synovium, and recruitment of prodestructive cells via chemokines, thus activating matrix destruction at sites of cartilage invasion (Frommer et al., 2010).

5. CONCLUSIONS

In summary, adiponectin seems to be an important local and systemic modulator of bone remodeling and cartilage biology, involving direct and indirect mechanisms and a large set of downstream molecular signals. Moreover, several data suggest that changes in adiponectin signaling may be associated with bone and cartilage disease states. Being adiponectin plasma levels and cellular receptor expression and function regulated in part by adipose mass variations, such effects of this adipokine should be put in the context of obesity and thinness and may be in balance with those exerted by other adipose–bone signals. The increase of plasma adiponectin concentrations is regarded as positive, in a cardiometabolic perspective, upon weight loss and optimal metabolic control in obese subjects and in diabetics, respectively. In consideration of the evidence presented herein, it thus appears important, within the above-mentioned context, to take into consideration also the consequent positive or negative changes in bone and cartilage biology and disease.

However, it should be emphasized that the extent of local production of adiponectin by bone and cartilage cells also appears to be important, if not dominant, in modulating tissue health or disease progression. Adiponectin-related signaling in bone and cartilage should be considered within the whole network of hormonal and nutritional signals that may influence skeleton biology, together with body homeostasis (Conde et al., 2011; Gomez et al., 2011). Moreover, adiponectin-bone cross talk seems to be reciprocal, as osteocalcin, produced by osteoblasts, has been shown to stimulate cyclinD1 and insulin expression in beta cells and adiponectin expression in adipocytes and to improve glucose tolerance *in vivo* (Confavreux, Levine, & Karsenty, 2009; Lee et al., 2007). Serum osteocalcin is also associated with glucose and total adiponectin levels, fat mass, and atherosclerosis parameters in patients with type 2 diabetes, indicating its important role also for glucose and fat metabolism (Kanazawa, Yamaguchi, Yamamoto, Yamauchi, Kurioka, et al., 2009).

In conclusion, adiponectin participates in the endocrine and local/paracrine regulation of bone and cartilage biology and may show relevant actions in the pathophysiology of some related diseases. Adiponectin seems to exert a negative net effect on bone mass and to be an independent predictor of lower bone mass, whereas available data about actions on cartilage are more controversial, showing both pro- and anti-inflammatory actions, depending on the model system or the biological context. Several aspects still remain to be elucidated at the clinical, genetic, and molecular levels, indicating the need for new model systems and more extensive studies.

REFERENCES

Anderson, J. J., & Felson, D. T. (1988). Factors associated with osteoarthritis of the knee in the first national Health and Nutrition Examination Survey (HANES I). Evidence for an association with overweight, race, and physical demands of work. *American Journal of Epidemiology*, *128*, 179–189.

Arita, Y., Kihara, S., Ouchi, N., Takahashi, M., Maeda, K., Miyagawa, J., et al. (1999). Paradoxical decrease of an adipose-specific protein, adiponectin, in obesity. *Biochemical and Biophysical Research Communications*, *257*, 79–83.

Basurto, L., Galvan, R., Cordova, N., Saucedo, R., Vargas, C., Campos, S., et al. (2009). Adiponectin is associated with low bone mineral density in elderly men. *European Journal of Endocrinology*, *160*, 289–293.

Berg, A. H., Combs, T. P., & Scherer, P. E. (2002). ACRP30/adiponectin: An adipokine regulating glucose and lipid metabolism. *Trends in Endocrinology and Metabolism*, *13*, 84–89.

Berner, H. S., Lyngstadaas, S. P., Spahr, A., Monjo, M., Thommesen, L., Drevon, C. A., et al. (2004). Adiponectin and its receptors are expressed in bone-forming cells. *Bone*, *35*, 842–849.

Berry, P. A., Jones, S. W., Cicuttini, F. M., Wluka, A. E., & Maciewicz, R. A. (2011). Temporal relationship between serum adipokines, biomarkers of bone and cartilage turnover, and cartilage volume loss in a population with clinical knee osteoarthritis. *Arthritis and Rheumatism*, *63*, 700–707.

Biver, E., Salliot, C., Combescure, C., Gossec, L., Hardouin, P., Legroux-Gerot, I., et al. (2011). Influence of adipokines and ghrelin on bone mineral density and fracture risk: A systematic review and meta-analysis. *The Journal of Clinical Endocrinology and Metabolism*, *96*, 2703–2713.

Brown, S. A., & Rosen, C. J. (2003). Osteoporosis. *The Medical Clinics of North America*, *87*, 1039–1063.

Carrasco, F., Ruz, M., Rojas, P., Csendes, A., Rebolledo, A., Codoceo, J., et al. (2009). Changes in bone mineral density, body composition and adiponectin levels in morbidly obese patients after bariatric surgery. *Obesity Surgery*, *19*, 41–46.

Challa, T. D., Rais, Y., & Ornan, E. M. (2010). Effect of adiponectin on ATDC5 proliferation, differentiation and signaling pathways. *Molecular and Cellular Endocrinology*, *323*, 282–291.

Chen, T. H., Chen, L., Hsieh, M. S., Chang, C. P., Chou, D. T., & Tsai, S. H. (2006). Evidence for a protective role for adiponectin in osteoarthritis. *Biochimica et Biophysica Acta*, *1762*, 711–718.

Choi, H. M., Lee, Y. A., Lee, S. H., Hong, S. J., Hahm, D. H., Choi, S. Y., et al. (2009). Adiponectin may contribute to synovitis and joint destruction in rheumatoid arthritis by stimulating vascular endothelial growth factor, matrix metalloproteinase-1, and matrix metalloproteinase-13 expression in fibroblast-like synoviocytes more than proinflammatory mediators. *Arthritis Research & Therapy*, *11*, R161.

Cicuttini, F. M., Baker, J. R., & Spector, T. D. (1996). The association of obesity with osteoarthritis of the hand and knee in women: A twin study. *The Journal of Rheumatology*, *23*, 1221–1226.

Cifuentes, M., Johnson, M. A., Lewis, R. D., Heymsfield, S. B., Chowdhury, H. A., Modlesky, C. M., et al. (2003). Bone turnover and body weight relationships differ in normal-weight compared with heavier postmenopausal women. *Osteoporosis International*, *14*, 116–122.

Cohen, M. M., Jr. (2006). The new bone biology: Pathologic, molecular, and clinical correlates. *American Journal of Medical Genetics. Part A*, *140*, 2646–2706.

Compston, J. E., Laskey, M. A., Croucher, P. I., Coxon, A., & Kreitzman, S. (1992). Effect of diet-induced weight loss on total body bone mass. *Clinical Science (London, England)*, *82*, 429–432.

Conde, J., Gomez, R., Bianco, G., Scotece, M., Lear, P., Dieguez, C., et al. (2011). Expanding the adipokine network in cartilage: Identification and regulation of novel factors in human and murine chondrocytes. *Annals of the Rheumatic Diseases*, *70*, 551–559.

Confavreux, C. B., Levine, R. L., & Karsenty, G. (2009). A paradigm of integrative physiology, the crosstalk between bone and energy metabolisms. *Molecular and Cellular Endocrinology*, *310*, 21–29.

De Laet, C., Kanis, J. A., Oden, A., Johanson, H., Johnell, O., Delmas, P., et al. (2005). Body mass index as a predictor of fracture risk: A meta-analysis. *Osteoporosis International*, *16*, 1330–1338.

Dozio, E., Corsi, M. M., Ruscica, M., Passafaro, L., Steffani, L., Banfi, G., et al. (2011). Adipokine actions on cartilage homeostasis. *Advances in Clinical Chemistry*, *55*, 61–79.

Duque, G. (2008). Bone and fat connection in aging bone. *Current Opinion in Rheumatology*, *20*, 429–434.

Ebina, K., Fukuhara, A., Ando, W., Hirao, M., Koga, T., Oshima, K., et al. (2009). Serum adiponectin concentrations correlate with severity of rheumatoid arthritis evaluated by extent of joint destruction. *Clinical Rheumatology*, *28*, 445–451.

Ehling, A., Schaffler, A., Herfarth, H., Tarner, I. H., Anders, S., Distler, O., et al. (2006). The potential of adiponectin in driving arthritis. *The Journal of Immunology*, *176*, 4468–4478.
Elefteriou, F., Ahn, J. D., Takeda, S., Starbuck, M., Yang, X., Liu, X., et al. (2005). Leptin regulation of bone resorption by the sympathetic nervous system and CART. *Nature*, *434*, 514–520.
Ensrud, K. E., Fullman, R. L., Barrett-Connor, E., Cauley, J. A., Stefanick, M. L., Fink, H. A., et al. (2005). Voluntary weight reduction in older men increases hip bone loss: The osteoporotic fractures in men study. *The Journal of Clinical Endocrinology and Metabolism*, *90*, 1998–2004.
Espallargues, M., Sampietro-Colom, L., Estrada, M. D., Sola, M., del Rio, L., Setoain, J., et al. (2001). Identifying bone-mass-related risk factors for fracture to guide bone densitometry measurements: A systematic review of the literature. *Osteoporosis International*, *12*, 811–822.
Felson, D. T., Zhang, Y., Hannan, M. T., & Anderson, J. J. (1993). Effects of weight and body mass index on bone mineral density in men and women: The Framingham study. *Journal of Bone and Mineral Research*, *8*, 567–573.
Francin, P. J., Guillaume, C., Humbert, A. C., Pottie, P., Netter, P., Mainard, D., et al. (2011). Association between the chondrocyte phenotype and the expression of adipokines and their receptors: Evidence for a role of leptin but not adiponectin in the expression of cartilage-specific markers. *Journal of Cellular Physiology*, *226*, 2790–2797.
Frommer, K. W., Zimmermann, B., Meier, F. M., Schroder, D., Heil, M., Schaffler, A., et al. (2010). Adiponectin-mediated changes in effector cells involved in the pathophysiology of rheumatoid arthritis. *Arthritis and Rheumatism*, *62*, 2886–2899.
Frost, M., Abrahamsen, B., Nielsen, T. L., Frystyk, J., Flyvbjerg, A., Hagen, C., et al. (2010). Adiponectin and peak bone mass in men: A cross-sectional, population-based study. *Calcified Tissue International*, *87*, 36–43.
Gandhi, R., Takahashi, M., Smith, H., Rizek, R., & Mahomed, N. N. (2010). The synovial fluid adiponectin-leptin ratio predicts pain with knee osteoarthritis. *Clinical Rheumatology*, *29*, 1223–1228.
Gandhi, R., Takahashi, M., Syed, K., Davey, J. R., & Mahomed, N. N. (2010). Relationship between body habitus and joint leptin levels in a knee osteoarthritis population. *Journal of Orthopaedic Research*, *28*, 329–333.
Gill, R. S., Al-Adra, D. P., Shi, X., Sharma, A. M., Birch, D. W., & Karmali, S. (2011). The benefits of bariatric surgery in obese patients with hip and knee osteoarthritis: A systematic review. *Obesity Reviews*, *12*, 1083–1089.
Gomez, R., Conde, J., Scotece, M., Gomez-Reino, J. J., Lago, F., & Gualillo, O. (2011). What's new in our understanding of the role of adipokines in rheumatic diseases? *Nature Reviews. Rheumatology*, 7, 528–536.
Goulding, A., Jones, I. E., Taylor, R. W., Williams, S. M., & Manning, P. J. (2001). Bone mineral density and body composition in boys with distal forearm fractures: A dual-energy x-ray absorptiometry study. *The Journal of Pediatrics*, *139*, 509–515.
Grinspoon, S., Thomas, E., Pitts, S., Gross, E., Mickley, D., Miller, K., et al. (2000). Prevalence and predictive factors for regional osteopenia in women with anorexia nervosa. *Annals of Internal Medicine*, *133*, 790–794.
Hao, D., Li, M., Wu, Z., Duan, Y., Li, D., & Qiu, G. (2011). Synovial fluid level of adiponectin correlated with levels of aggrecan degradation markers in osteoarthritis. *Rheumatology International*, *31*(11), 1433–1437.
Honsawek, S., & Chayanupatkul, M. (2010). Correlation of plasma and synovial fluid adiponectin with knee osteoarthritis severity. *Archives of Medical Research*, *41*, 593–598.
Hotta, K., Funahashi, T., Arita, Y., Takahashi, M., Matsuda, M., Okamoto, Y., et al. (2000). Plasma concentrations of a novel, adipose-specific protein, adiponectin, in type 2 diabetic patients. *Arteriosclerosis, Thrombosis, and Vascular Biology*, *20*, 1595–1599.

Huang, K. C., Cheng, W. C., Yen, R. F., Tsai, K. S., Tai, T. Y., & Yang, W. S. (2004). Lack of independent relationship between plasma adiponectin, leptin levels and bone density in nondiabetic female adolescents. *Clinical Endocrinology, 61*, 204–208.
Huang, C. Y., Lee, C. Y., Chen, M. Y., Tsai, H. C., Hsu, H. C., & Tang, C. H. (2010). Adiponectin increases BMP-2 expression in osteoblasts via AdipoR receptor signaling pathway. *Journal of Cellular Physiology, 224*, 475–483.
Jurimae, J., & Jurimae, T. (2007). Adiponectin is a predictor of bone mineral density in middle-aged premenopausal women. *Osteoporosis International, 18*, 1253–1259.
Jurimae, J., Rembel, K., Jurimae, T., & Rehand, M. (2005). Adiponectin is associated with bone mineral density in perimenopausal women. *Hormone and Metabolic Research, 37*, 297–302.
Justesen, J., Stenderup, K., Ebbesen, E. N., Mosekilde, L., Steiniche, T., & Kassem, M. (2001). Adipocyte tissue volume in bone marrow is increased with aging and in patients with osteoporosis. *Biogerontology, 2*, 165–171.
Kanazawa, I., Yamaguchi, T., & Sugimoto, T. (2010). Baseline serum total adiponectin level is positively associated with changes in bone mineral density after 1-year treatment of type 2 diabetes mellitus. *Metabolism, 59*, 1252–1256.
Kanazawa, I., Yamaguchi, T., Yamamoto, M., Yamauchi, M., Kurioka, S., Yano, S., et al. (2009). Serum osteocalcin level is associated with glucose metabolism and atherosclerosis parameters in type 2 diabetes mellitus. *The Journal of Clinical Endocrinology and Metabolism, 94*, 45–49.
Kanazawa, I., Yamaguchi, T., Yamamoto, M., Yamauchi, M., Yano, S., & Sugimoto, T. (2009). Relationships between serum adiponectin levels versus bone mineral density, bone metabolic markers, and vertebral fractures in type 2 diabetes mellitus. *European Journal of Endocrinology, 160*, 265–273.
Kanazawa, I., Yamaguchi, T., Yano, S., Yamauchi, M., Yamamoto, M., & Sugimoto, T. (2007). Adiponectin and AMP kinase activator stimulate proliferation, differentiation, and mineralization of osteoblastic MC3T3-E1 cells. *BMC Cell Biology, 8*, 51.
Kang, E. H., Lee, Y. J., Kim, T. K., Chang, C. B., Chung, J. H., Shin, K., et al. (2010). Adiponectin is a potential catabolic mediator in osteoarthritis cartilage. *Arthritis Research & Therapy, 12*, R231.
Karsenty, G. (2006). Convergence between bone and energy homeostases: Leptin regulation of bone mass. *Cell Metabolism, 4*, 341–348.
King, G. A., Deemer, S. E., & Thompson, D. L. (2010). Relationship between leptin, adiponectin, bone mineral density, and measures of adiposity among pre-menopausal Hispanic and Caucasian women. *Endocrine Research, 35*, 106–117.
Koskinen, A., Juslin, S., Nieminen, R., Moilanen, T., Vuolteenaho, K., & Moilanen, E. (2011). Adiponectin associates with markers of cartilage degradation in osteoarthritis and induces production of proinflammatory and catabolic factors through mitogen-activated protein kinase pathways. *Arthritis Research & Therapy, 13*, R184.
Kotani, K., Sakane, N., Saiga, K., & Kurozawa, Y. (2005). Leptin: Adiponectin ratio as an atherosclerotic index in patients with type 2 diabetes: Relationship of the index to carotid intima-media thickness. *Diabetologia, 48*, 2684–2686.
Lago, R., Gomez, R., Otero, M., Lago, F., Gallego, R., Dieguez, C., et al. (2008). A new player in cartilage homeostasis: Adiponectin induces nitric oxide synthase type II and pro-inflammatory cytokines in chondrocytes. *Osteoarthritis and Cartilage, 16*, 1101–1109.
Langlois, J. A., Mussolino, M. E., Visser, M., Looker, A. C., Harris, T., & Madans, J. (2001). Weight loss from maximum body weight among middle-aged and older white women and the risk of hip fracture: The NHANES I epidemiologic follow-up study. *Osteoporosis International, 12*, 763–768.
Lee, N. K., Sowa, H., Hinoi, E., Ferron, M., Ahn, J. D., Confavreux, C., et al. (2007). Endocrine regulation of energy metabolism by the skeleton. *Cell, 130*, 456–469.

Lenchik, L., Register, T. C., Hsu, F. C., Lohman, K., Nicklas, B. J., Freedman, B. I., et al. (2003). Adiponectin as a novel determinant of bone mineral density and visceral fat. *Bone, 33*, 646–651.
Lewicki, M., Kotyla, P., & Kucharz, E. (2008). Etanercept increases adiponectin level in woman with rheumatoid arthritis. *Clinical Rheumatology, 27*, 1337–1338.
Luo, X. H., Guo, L. J., Xie, H., Yuan, L. Q., Wu, X. P., Zhou, H. D., et al. (2006). Adiponectin stimulates RANKL and inhibits OPG expression in human osteoblasts through the MAPK signaling pathway. *Journal of Bone and Mineral Research, 21*, 1648–1656.
Luo, X. H., Guo, L. J., Yuan, L. Q., Xie, H., Zhou, H. D., Wu, X. P., et al. (2005). Adiponectin stimulates human osteoblasts proliferation and differentiation via the MAPK signaling pathway. *Experimental Cell Research, 309*, 99–109.
Luo, E., Hu, J., Bao, C., Li, Y., Tu, Q., Murray, D., et al. (2012). Sustained release of adiponectin improves osteogenesis around hydroxyapatite implants by suppressing osteoclast activity in ovariectomized rabbits. *Acta Biomaterialia, 8*, 734–743.
Magni, P., Dozio, E., Galliera, E., Ruscica, M., & Corsi, M. M. (2010). Molecular aspects of adipokine-bone interactions. *Current Molecular Medicine, 10*, 522–532.
Matsubara, M., Katayose, S., & Maruoka, S. (2003). Decreased plasma adiponectin concentrations in nondiabetic women with elevated homeostasis model assessment ratios. *European Journal of Endocrinology, 148*, 343–350.
Matsubara, M., Namioka, K., & Katayose, S. (2003). Decreased plasma adiponectin concentrations in women with low-grade C-reactive protein elevation. *European Journal of Endocrinology, 148*, 657–662.
Matsuzawa, Y. (2005). White adipose tissue and cardiovascular disease. *Best Practice & Research. Clinical Endocrinology & Metabolism, 19*, 637–647.
Meunier, P., Aaron, J., Edouard, C., & Vignon, G. (1971). Osteoporosis and the replacement of cell populations of the marrow by adipose tissue. A quantitative study of 84 iliac bone biopsies. *Clinical Orthopaedics and Related Research, 80*, 147–154.
Michaelsson, K., Lind, L., Frystyk, J., Flyvbjerg, A., Gedeborg, R., Berne, C., et al. (2008). Serum adiponectin in elderly men does not correlate with fracture risk. *The Journal of Clinical Endocrinology and Metabolism, 93*, 4041–4047.
Mohamed-Ali, V., Pinkney, J. H., & Coppack, S. W. (1998). Adipose tissue as an endocrine and paracrine organ. *International Journal of Obesity and Related Metabolic Disorders, 22*, 1145–1158.
Nagashima, T., Okubo-Fornbacher, H., Aoki, Y., Kamata, Y., Kimura, H., Kamimura, T., et al. (2008). Increase in plasma levels of adiponectin after administration of anti-tumor necrosis factor agents in patients with rheumatoid arthritis. *The Journal of Rheumatology, 35*, 936–938.
Napoli, N., Pedone, C., Pozzilli, P., Lauretani, F., Ferrucci, L., & Incalzi, R. A. (2010). Adiponectin and bone mass density: The InCHIANTI study. *Bone, 47*, 1001–1005.
Nawrocki, A. R., & Scherer, P. E. (2004). The delicate balance between fat and muscle: Adipokines in metabolic disease and musculoskeletal inflammation. *Current Opinion in Pharmacology, 4*, 281–289.
Nguyen, T. V., Sambrook, P. N., & Eisman, J. A. (1998). Bone loss, physical activity, and weight change in elderly women: The Dubbo Osteoporosis Epidemiology Study. *Journal of Bone and Mineral Research, 13*, 1458–1467.
Norata, G. D., Raselli, S., Grigore, L., Garlaschelli, K., Dozio, E., Magni, P., et al. (2007). Leptin: Adiponectin ratio is an independent predictor of intima media thickness of the common carotid artery. *Stroke, 38*, 2844–2846.
Okuno, S., Ishimura, E., Norimine, K., Tsuboniwa, N., Kagitani, S., Yamakawa, K., et al. (2012). Serum adiponectin and bone mineral density in male hemodialysis patients. *Osteoporosis International, 23*(7), 2027–2035.

Oshima, K., Nampei, A., Matsuda, M., Iwaki, M., Fukuhara, A., Hashimoto, J., et al. (2005). Adiponectin increases bone mass by suppressing osteoclast and activating osteoblast. *Biochemical and Biophysical Research Communications*, *331*, 520–526.

Otero, M., Lago, R., Gomez, R., Lago, F., Dieguez, C., Gomez-Reino, J. J., et al. (2006). Changes in plasma levels of fat-derived hormones adiponectin, leptin, resistin and visfatin in patients with rheumatoid arthritis. *Annals of the Rheumatic Diseases*, *65*, 1198–1201.

Papakitsou, E. F., Margioris, A. N., Dretakis, K. E., Trovas, G., Zoras, U., Lyritis, G., et al. (2004). Body mass index (BMI) and parameters of bone formation and resorption in postmenopausal women. *Maturitas*, *47*, 185–193.

Parfitt, A. M. (1996). Skeletal heterogeneity and the purposes of bone remodeling: implications for the understanding of osteoporosis. In R. Marcus, D. Feldman & J. Kelsey (Eds.), *Osteoporosis* (pp. 315–339). San Diego: Academic Press.

Presle, N., Pottie, P., Dumond, H., Guillaume, C., Lapicque, F., Pallu, S., et al. (2006). Differential distribution of adipokines between serum and synovial fluid in patients with osteoarthritis. Contribution of joint tissues to their articular production. *Osteoarthritis and Cartilage*, *14*, 690–695.

Ravn, P., Cizza, G., Bjarnason, N. H., Thompson, D., Daley, M., Wasnich, R. D., et al. (1999). Low body mass index is an important risk factor for low bone mass and increased bone loss in early postmenopausal women. Early Postmenopausal Intervention Cohort (EPIC) study group. *Journal of Bone and Mineral Research*, *14*, 1622–1627.

Reid, I. R. (2008). Relationships between fat and bone. *Osteoporosis International*, *19*, 595–606.

Reid, I. R. (2010). Fat and bone. *Archives of Biochemistry and Biophysics*, *503*, 20–27.

Reid, I. R., Ames, R. W., Evans, M. C., Sharpe, S. J., & Gamble, G. D. (1994). Determinants of the rate of bone loss in normal postmenopausal women. *The Journal of Clinical Endocrinology and Metabolism*, *79*, 950–954.

Richards, J. B., Valdes, A. M., Burling, K., Perks, U. C., & Spector, T. D. (2007). Serum adiponectin and bone mineral density in women. *The Journal of Clinical Endocrinology and Metabolism*, *92*, 1517–1523.

Robbins, J., Schott, A. M., Azari, R., & Kronmal, R. (2006). Body mass index is not a good predictor of bone density: Results from WHI, CHS, and EPIDOS. *Journal of Clinical Densitometry*, *9*, 329–334.

Rodriguez, J. P., Montecinos, L., Rios, S., Reyes, P., & Martinez, J. (2000). Mesenchymal stem cells from osteoporotic patients produce a type I collagen-deficient extracellular matrix favoring adipogenic differentiation. *Journal of Cellular Biochemistry*, *79*, 557–565.

Rosen, C. J., & Bouxsein, M. L. (2006). Mechanisms of disease: Is osteoporosis the obesity of bone? *Nature Clinical Practice. Rheumatology*, *2*, 35–43.

Rozman, C., Feliu, E., Berga, L., Reverter, J. C., Climent, C., & Ferran, M. J. (1989). Age-related variations of fat tissue fraction in normal human bone marrow depend both on size and number of adipocytes: A stereological study. *Experimental Hematology*, *17*, 34–37.

Runhaar, J., Koes, B. W., Clockaerts, S., & Bierma-Zeinstra, S. M. (2011). A systematic review on changed biomechanics of lower extremities in obese individuals: A possible role in development of osteoarthritis. *Obesity Reviews*, *12*, 1071–1082.

Salamone, L. M., Cauley, J. A., Black, D. M., Simkin-Silverman, L., Lang, W., Gregg, E., et al. (1999). Effect of a lifestyle intervention on bone mineral density in premenopausal women: A randomized trial. *The American Journal of Clinical Nutrition*, *70*, 97–103.

Schaffler, A., Ehling, A., Neumann, E., Herfarth, H., Tarner, I., Scholmerich, J., et al. (2003). Adipocytokines in synovial fluid. *JAMA: The Journal of the American Medical Association*, *290*, 1709–1710.

Scherer, P. E. (2006). Adipose tissue: From lipid storage compartment to endocrine organ. *Diabetes*, *55*, 1537–1545.

Sekiya, I., Larson, B. L., Vuoristo, J. T., Cui, J. G., & Prockop, D. J. (2004). Adipogenic differentiation of human adult stem cells from bone marrow stroma (MSCs). *Journal of Bone and Mineral Research, 19*, 256–264.

Shinoda, Y., Yamaguchi, M., Ogata, N., Akune, T., Kubota, N., Yamauchi, T., et al. (2006). Regulation of bone formation by adiponectin through autocrine/paracrine and endocrine pathways. *Journal of Cellular Biochemistry, 99*, 196–208.

Svendsen, O. L., Hassager, C., & Christiansen, C. (1993). Effect of an energy-restrictive diet, with or without exercise, on lean tissue mass, resting metabolic rate, cardiovascular risk factors, and bone in overweight postmenopausal women. *The American Journal of Medicine, 95*, 131–140.

Takeda, S., Elefteriou, F., Levasseur, R., Liu, X., Zhao, L., Parker, K. L., et al. (2002). Leptin regulates bone formation via the sympathetic nervous system. *Cell, 111*, 305–317.

Teichtahl, A. J., Wluka, A. E., Proietto, J., & Cicuttini, F. M. (2005). Obesity and the female sex, risk factors for knee osteoarthritis that may be attributable to systemic or local leptin biosynthesis and its cellular effects. *Medical Hypotheses, 65*, 312–315.

Tong, K. M., Chen, C. P., Huang, K. C., Shieh, D. C., Cheng, H. C., Tzeng, C. Y., et al. (2011). Adiponectin increases MMP-3 expression in human chondrocytes through AdipoR1 signaling pathway. *Journal of Cellular Biochemistry, 112*, 1431–1440.

Tsao, T. S., Tomas, E., Murrey, H. E., Hug, C., Lee, D. H., Ruderman, N. B., et al. (2003). Role of disulfide bonds in Acrp30/adiponectin structure and signaling specificity. Different oligomers activate different signal transduction pathways. *The Journal of Biological Chemistry, 278*, 50810–50817.

Verma, S., Rajaratnam, J. H., Denton, J., Hoyland, J. A., & Byers, R. J. (2002). Adipocytic proportion of bone marrow is inversely related to bone formation in osteoporosis. *Journal of Clinical Pathology, 55*, 693–698.

Vilarrasa, N., Vendrell, J., Maravall, J., Broch, M., Estepa, A., Megia, A., et al. (2005). Distribution and determinants of adiponectin, resistin and ghrelin in a randomly selected healthy population. *Clinical Endocrinology, 63*, 329–335.

Wang, M. C., Bachrach, L. K., Van Loan, M., Hudes, M., Flegal, K. M., & Crawford, P. B. (2005). The relative contributions of lean tissue mass and fat mass to bone density in young women. *Bone, 37*, 474–481.

Wang, Y., Lam, K. S., Xu, J. Y., Lu, G., Xu, L. Y., Cooper, G. J., et al. (2005). Adiponectin inhibits cell proliferation by interacting with several growth factors in an oligomerization-dependent manner. *The Journal of Biological Chemistry, 280*, 18341–18347.

Wehrli, F. W., Hopkins, J. A., Hwang, S. N., Song, H. K., Snyder, P. J., & Haddad, J. G. (2000). Cross-sectional study of osteopenia with quantitative MR imaging and bone densitometry. *Radiology, 217*, 527–538.

Weisberg, S. P., McCann, D., Desai, M., Rosenbaum, M., Leibel, R. L., & Ferrante, A. W., Jr. (2003). Obesity is associated with macrophage accumulation in adipose tissue. *The Journal of Clinical Investigation, 112*, 1796–1808.

Williams, G. A., Wang, Y., Callon, K. E., Watson, M., Lin, J. M., Lam, J. B., et al. (2009). In vitro and in vivo effects of adiponectin on bone. *Endocrinology, 150*(8), 3603–3610.

Wu, F., Ames, R., Clearwater, J., Evans, M. C., Gamble, G., & Reid, I. R. (2002). Prospective 10-year study of the determinants of bone density and bone loss in normal postmenopausal women, including the effect of hormone replacement therapy. *Clinical Endocrinology, 56*, 703–711.

Yamaguchi, N., Kukita, T., Li, Y. J., Kamio, N., Fukumoto, S., Nonaka, K., et al. (2008). Adiponectin inhibits induction of TNF-alpha/RANKL-stimulated NFATc1 via the AMPK signaling. *FEBS Letters, 582*, 451–456.

Yang, W. S., Lee, W. J., Funahashi, T., Tanaka, S., Matsuzawa, Y., Chao, C. L., et al. (2001). Weight reduction increases plasma levels of an adipose-derived anti-inflammatory protein, adiponectin. *The Journal of Clinical Endocrinology and Metabolism, 86*, 3815–3819.

Yeung, D. K., Griffith, J. F., Antonio, G. E., Lee, F. K., Woo, J., & Leung, P. C. (2005). Osteoporosis is associated with increased marrow fat content and decreased marrow fat unsaturation: A proton MR spectroscopy study. *Journal of Magnetic Resonance Imaging*, *22*, 279–285.

Zhang, Y., Proenca, R., Maffei, M., Barone, M., Leopold, L., & Friedman, J. M. (1994). Positional cloning of the mouse obese gene and its human homologue. *Nature*, *372*, 425–432.

Zhao, L. J., Jiang, H., Papasian, C. J., Maulik, D., Drees, B., Hamilton, J., et al. (2008). Correlation of obesity and osteoporosis: Effect of fat mass on the determination of osteoporosis. *Journal of Bone and Mineral Research*, *23*, 17–29.

Zoico, E., Zamboni, M., Di Francesco, V., Mazzali, G., Fantin, F., De Pergola, G., et al. (2008). Relation between adiponectin and bone mineral density in elderly postmenopausal women: Role of body composition, leptin, insulin resistance, and dehydroepiandrosterone sulfate. *Journal of Endocrinological Investigation*, *31*, 297–302.

CHAPTER THIRTEEN

Lipid-Lowering Drugs and Circulating Adiponectin

Desiree Wanders*, Eric P. Plaisance†, Robert L. Judd*,[1]
*Department of Anatomy, Physiology and Pharmacology, College of Veterinary Medicine, Auburn University, Auburn, Alabama, USA
†Laboratory of Nutrient Sensing and Adipocyte Signaling, Pennington Biomedical Research Center, Louisiana State University System, Baton Rouge, Louisiana, USA
[1]Corresponding author: e-mail address: juddrob@auburn.edu

Contents

Abstract

Pharmacological agents used to treat primary and combined hyperlipidemia reduce cardiovascular disease morbidity and mortality. Risk reduction has been attributed to improvements in blood lipid and lipoprotein characteristics. However, each class of available lipid-lowering drugs has been shown to exhibit pleiotropic effects that broaden their anticipated actions. Indeed, the results of a growing number of available studies suggest that a strong relationship exists between pharmacological reductions in blood lipids and circulating concentrations of the adipose tissue derived protein, adiponectin. Adiponectin is the most abundantly secreted protein from adipose tissue and has been shown to decrease hepatic glucose production, increase fatty acid oxidation in liver and skeletal muscle, and decrease vascular inflammation. In this chapter, we present a comprehensive analysis of the effects of the available classes of lipid-lowering drugs (statins, fibrates, niacin, and omega-3-fatty acids) on circulating adiponectin and the known mechanisms which produce these important events.

Vitamins and Hormones, Volume 90
ISSN 0083-6729
http://dx.doi.org/10.1016/B978-0-12-398313-8.00013-0

1. INTRODUCTION

Cardiovascular disease (CVD) remains one of the leading causes of morbidity and mortality in the United States accounting for 33% of all deaths annually (Roger et al., 2011). Derangements in blood lipid and lipoprotein metabolism are among the most frequently observed risk factors for the development of CVD. Over the past 40 years, epidemiological studies have consistently demonstrated a strong positive association between total serum cholesterol and low-density lipoprotein-cholesterol (LDL-C) with CVD. In contrast, the relationship between serum triglycerides (TG) and CVD is more controversial due to the inverse metabolic relationship between HDL-C and TG-rich lipoproteins (Gotto, 1998). Nonetheless, elevated TGs are increasingly recognized as part of a cluster of metabolic abnormalities referred to as atherogenic dyslipidemia which is characterized by moderately elevated LDL-C (130–159 mg/dL) and TGs (>150 mg/dL), small LDL particles, and low high-density lipoprotein-cholesterol (HDL-C; <35 mg/dL) (Grundy, 1997). Defining the independent contribution of serum TGs to atherogenesis is further compounded by the fact that the lipid and lipoprotein changes of atherogenic dyslipidemia aggregate with other CVD risk factors such as obesity, insulin resistance, and hypertension.

Pharmacological treatment strategies for primary elevations in LDL-C or atherogenic dyslipidemia mainly consist of the 3-hydroxy-3-methylglutaryl coenzyme A (HMG-CoA) reductase inhibitors (statins). HMG-CoA inhibition reduces the synthesis of cholesterol and subsequently increases the expression of LDL receptors which dramatically reduces serum cholesterol content and the number of circulating LDL particles. The effects of statins on tissue and circulating TGs are more modest and along with reductions in HDL-C, render them less than optimal as monotherapy for the treatment of obesity-related atherogenic dyslipidemia. Therefore, other pharmacological agents such as fibrates, niacin, and omega-3 fatty acids are routinely used in combination with statins or alone, and each has proven to independently reduce atherogenic dyslipidemia and CVD morbidity and mortality. Although their primary action is to improve blood lipid and lipoprotein characteristics, it is becoming increasingly clear that each of the available classes of lipid-lowering agents possess pleiotropic benefits in addition to their known therapeutic effects. Indeed, the reported cardioprotective benefits of lipid-lowering drugs from each available class include improvements in endothelial function, reduced oxidative stress, decreased platelet adhesion, and increased atherosclerotic plaque stability. Emerging evidence also suggests that lipid-lowering agents modulate the synthesis and secretion of certain adipokines, including adiponectin.

2. ADIPONECTIN

Although adipose tissue is a well-known storage depot for excess energy, its role as an endocrine organ was not appreciated until the discoveries of leptin and adiponectin in the mid-1990s. Of over 100 adipokines identified, adiponectin is the most abundantly secreted protein from adipose tissue, accounting for up to 0.05% of total plasma proteins (Scherer, Williams, Fogliano, Baldini, & Lodish, 1995). Adiponectin is constitutively expressed as a 30 kDa monomer and undergoes posttranslational modification to form higher order structures including: (1) homotrimers of 90 kDa, (2) hexamers of 180 kDa [low-molecular weight (LMW)], and (3) 12–18mers of greater than 300 kDa [high-molecular weight (HMW)] (Trujillo & Scherer, 2005). The expression and circulating concentrations of adiponectin are reduced in obesity (Hu, Liang, & Spiegelman, 1996; Yamauchi, Kamon, Waki, Terauchi, et al., 2001), and the reduction in adiponectin appears to precede the development of insulin resistance (Hotta et al., 2001). The most compelling evidence for the role of adiponectin in the development of insulin resistance was made in mice with a targeted deletion of adiponectin. Adiponectin null mice had delayed clearance of free fatty acids in the plasma, low mRNA levels of fatty acid transport protein in skeletal muscle, and elevated adipose tissue and circulating TNF-α concentrations (Maeda et al., 2002). Despite having normal glucose tolerance on a chow diet, adiponectin null mice experienced more severe high-fat diet-induced obesity and glucose intolerance than wild-type mice which was reversed with adenoviral-mediated expression of adiponectin (Maeda et al., 2002).

Multiple studies have shown that adiponectin lowers blood glucose acutely through a reduction in hepatic glucose production (HGP) with little or no effect on glucose disposal (Berg, Combs, Du, Brownlee, & Scherer, 2001; Combs, Berg, Obici, Scherer, & Rossetti, 2001; Qi et al., 2004). Reductions in circulating adiponectin impairs glucose tolerance and increases HGP providing further support for the role of adiponectin in normal hepatic metabolism (Nawrocki et al., 2006). The glucose-lowering action of adiponectin in the liver has been attributed to AMPK activation, providing a mechanistic link to a signal transduction pathway already established as an antagonist of HGP and lipogenesis (Viollet et al., 2009).

Adiponectin has also been shown to be actively engaged in the control of hepatic and serum TGs and has been shown to prevent alcoholic and non-alcoholic steatosis in the liver (Xu et al., 2003; Yamauchi, Kamon, Waki, Terauchi, et al., 2001). The TG lowering effects of adiponectin appear to be mediated through activation of the AdipoR1 and AdipoR2 receptors which activate AMPK and PPARα, respectively, which are both well known to increase fatty acid oxidation (Yamauchi et al., 2007). Targeted disruption of both receptors abolishes adiponectin binding, resulting in increased tissue TG content, inflammation, and marked glucose intolerance (Yamauchi et al., 2007).

Adiponectin has also been shown to activate AMPK in the arcuate nucleus of the hypothalamus via activation of the AdipoR1 receptor (Kubota et al., 2007). Receptor activation was shown to elicit an increase in food intake and decrease in energy expenditure. Other studies show that i.c.v. infusion of adiponectin produced no effect on food intake but increased energy expenditure (Qi et al., 2004).

Finally, adiponectin directly improves endothelial dysfunction by stimulating production of nitric oxide (Deng, Long, Yu, & Li, 2010), inhibiting proliferation of vascular smooth muscle cells (Arita et al., 2002), and inhibiting the conversion of macrophages to foam cells (Ouchi et al., 2001). Based on transgenic and adenoviral overexpression studies and the predicted functions of adiponectin, it is reasonable to consider a variety of approaches to raise circulating adiponectin concentrations. Interestingly, each of the available lipid-lowering drugs has been shown to increase circulating adiponectin, suggesting a role for adiponectin as part of the pleiotropic benefits of each class of agents.

Thus, the purpose of this chapter will be to present our current understanding of the capacity and potential mechanisms by which the most commonly used lipid-lowering drugs (statins, fibrates, niacin, and omega-3 fatty acids) alter circulating adiponectin concentrations. A more thorough appreciation of the capacity of lipid-lowering agents to alter the production of adiponectin has the potential to improve our understanding of the metabolic pathways that regulate glucose and lipid metabolism and could lead to the development of superior treatments in the future.

3. STATINS

Statins account for over 90% of the medications dispensed for the treatment of blood lipids (Ma, Sehgal, Ayanian, & Stafford, 2005). Meta-analysis of randomized clinical trials show that statin therapy reduces short- and long-term incidence of major coronary events, coronary revascularization, and stroke by about 20% for every 40 mg/dL reduction in LDL-C, irrespective of the initial lipid profile or other presenting characteristics (Baigent et al., 2005). Statins may have the greatest impact of all lipid-lowering drugs on CVD mortality risk as a result of their superior capacity to reduce serum cholesterol (Gould, Rossouw, Santanello, Heyse, & Furberg, 1998). The primary mechanism of cholesterol reduction is by inhibition of the rate-limiting enzyme in cholesterol biosynthesis, HMG-CoA reductase. The reduction in cholesterol synthesis also reduces the synthesis of apolipoprotein B100, leading to a reduction in the formation of very low-density lipoprotein (VLDL), which results in a reduction in cholesterol and triglyceride secretion from the liver. The reduction in circulating cholesterol increases the expression of LDL receptors which binds apolipoprotein B100 arising from peripheral tissues along with apolipoprotein B48 entering the circulation from the gut. The culmination of increased LDL receptor expression and the uptake of apolipoprotein B lipoproteins serve to reduce circulating cholesterol.

Although it is clear that a significant portion of the benefit of statins involve reductions in LDL-C, studies conducted over the past 10 years suggest that statins possess pleiotropic benefits which include reductions in vascular inflammation and improvements in insulin sensitivity. The improvements in insulin sensitivity may be associated with increases in circulating adiponectin. Serum adiponectin concentrations increase in response to atorvastatin, pravastatin, and pitavastatin treatment (Ando et al., 2008; Blanco-Colio et al., 2008; Miyagishima et al., 2007; Nakamura et al., 2007; Nomura, Inami, et al., 2009; Nomura, Shouzu, et al., 2009; Sakamoto, Sakamoto, & Ogawa, 2006; Sugiyama et al., 2007; Takagi et al., 2008; Table 13.1). In contrast, other studies using simvastatin demonstrate that circulating adiponectin concentrations were either unchanged (Devaraj et al., 2007; Koh, Quon, Han, Ahn, et al., 2005; Koh et al., 2004; Nomura, Shouzu, et al., 2009; Table 13.1) or reduced (Forst et al., 2007; Koh, Quon, Sakuma, et al., 2011; Koh et al., 2008, 2009; Pfützner et al., 2009; Table 13.1). Indeed, Nomura, Shouzu, et al. (2009) tested the effects of either pitavastatin or simvastatin in humans with hypercholesterolemia and found that total plasma adiponectin concentrations increased after 6 months of pitavastatin treatment but were unaffected by simvastatin. Furthermore, a comparison of pravastatin and simvastatin therapy in hypercholesterolemic patients showed that pravastatin increased while simvastatin decreased plasma adiponectin concentrations (Koh et al., 2009). Pravastatin treatment has

Table 13.1 The effects of statins on adiponectin

References	Subjects	Treatment	Effects on adiponectin
Yin, Tu, and Yang (2007)	3T3-L1 adipocytes	16 h of simvastatin (0.1, 1, and 10 mM) then LPS[a] (100 ng/mL) was added for 8 h	Simvastatin increased LPS-induced secretion and mRNA expression of adiponectin
Takagi et al. (2008)	C57BL/6J mice	0–15 weeks pravastatin (0.06% of diet)	Adiponectin increased
Ando et al. (2008)	Humans with hypercholesterolemia	4 months of pravastatin or atorvastatin (10 mg/day)	Adiponectin increased from 10.7 ± 4.7 to 11.0 ± 5.1 μg/mL in response to atorvastatin
Nomura, Shouzu, et al. (2009)	Humans with hyperlipidemia	6 months of simvastatin (10 mg/day) or pitavastatin (2 mg/day)	Adiponectin increased in response to pitavastatin but not in response to simvastatin
Blanco-Colio et al. (2008)	Humans at high cardiovascular risk	12 weeks atorvastatin (10–80 mg/day)	Adiponectin increased by 10% with maximal increase (25%) observed at 80 mg/day
Sakamoto et al. (2006)	Humans with hypercholesterolemia and CAD[b]	6 months pravastatin (10–20 mg/day)	Adiponectin increased by 16.8% from 7.2 to 7.8 μg/mL
Miyagishima et al. (2007)	Humans with hypercholesterolemia and ischemic heart disease	3 months atorvastatin (10 mg/day)	Adiponectin increased from 9.7 ± 7.4 to 13.9 ± 9.98 μg/mL
Takagi et al. (2008)	Men with hypercholesterolemia	1 year pravastatin (40 mg/day)	Adiponectin increased by 1.47 ± 0.33 μg/mL
Sugiyama et al. (2007)	Humans with CAD and IGT[c]	6 months pravastatin (20 mg/day)	Adiponectin increased by 35% from 5.2 to 6.1 μg/mL

Nakamura et al. (2007)	Humans with stable CAD, hypercholesterolemia, and hypertriglyceridemia	6 months of atorvastatin (10 mg/day)	Adiponectin increased by 4 weeks, with further increases by 6 months
Koh, Han, et al. (2005)	Humans with combined hyperlipidemia	2 months of atorvastatin (10 mg/day) plus placebo or atorvastatin plus fenofibrate (200 mg/day) or fenofibrate plus placebo	Atorvastatin alone had no effect on adiponectin. Atorvastatin combined with fenofibrate increased adiponectin from 3.4 to 3.5, and fenofibrate alone increased adiponectin from 3.2 to 3.6 μg/mL
Koh, Quon, Han, et al. (2005)	Humans with T2DM[d]	2 months of simvastatin (20 mg/day) plus placebo or simvastatin plus ramipril (10 mg/day) or losartan plus placebo	Simvastatin had no effect on adiponectin
Koh et al. (2004)	Humans with htn[a] and hypercholesterolemia	2 months of simvastatin (20 mg/day) plus placebo or simvastatin plus losartan (100 mg/day) or losartan plus placebo	Simvastatin had no effect on adiponectin
Thongtang et al. (2011)	Humans with hyperlipidemia	6 weeks of atorvastatin (80 mg/day) or rosuvastatin (40 mg/day)	No change in adiponectin
Devaraj, Siegel, and Jialal (2007)	Humans with MetS[e]	8 weeks of simvastatin (40 mg/day)	No change in adiponectin
Gannage-Yared et al. (2005)	Healthy humans	12 weeks of pravastatin (40 mg/day)	No change in adiponectin
Anagnostis et al. (2011)	Humans with dyslipidemia without T2DM	12 weeks of atorvastatin (20 mg/day) or rosuvastatin (10 mg/day)	No change in adiponectin

Continued

Table 13.1 The effects of statins on adiponectin—cont'd

References	Subjects	Treatment	Effects on adiponectin
Shetty, Economides, Horton, Mantzoros, and Veves (2004)	Humans with T2DM or at high risk for T2DM	12 weeks of atorvastatin (20 mg/day)	No change in adiponectin
Bayes et al. (2005)	Kidney transplant recipients	12 weeks of atorvastatin (10 mg/day)	No change in adiponectin
Otto et al. (2007)	Humans with T2DM and mixed hyperlipoproteinemia	6 weeks of atorvastatin (10 mg/day)	No change in adiponectin
Sonmez et al. (2006)	Humans with hypercholesterolemia on fluvastatin plus TLC[f] and humans with normal cholesterol on TLC alone	12 weeks of fluvastatin (80 mg/day) plus TLC ($n=24$) or 12 weeks of TLC alone ($n=19$)	Adiponectin increased from 5.3 ± 1.5 to 6.2 ± 2.2 μg/mL in response to TLC but was unchanged in response to fluvastatin
von et al. (2009)	Humans with T2DM	8 weeks of atorvastatin (40 mg/day)	HMW[g] increased by 42.3% while MMW[h] and LMW[i] decreased by 21% and 23%, respectively. Total adiponectin did not change
Koh et al. (2010)	Humans with hypercholesterolemia	2 months of atorvastatin (10–80 mg/day) or placebo	Atorvastatin decreased adiponectin when compared to baseline, but when compared to placebo, there were no effects of atorvastatin on adiponectin
Matafome et al. (2011)	Male Goto-Kakizaki rats on HFD[j]	1 month of atorvastatin (10 mg/kg/day dissolved in water) or metformin or atorvastatin plus metformin	Atorvastatin alone decreased adiponectin; metformin alone had no effect on adiponectin; atorvastatin plus metformin increased adiponectin

Koh et al. (2009)	Humans with hypercholesterolemia	2 months of simvastatin (20 mg/day) or pravastatin (40 mg/day)	Simvastatin decreased adiponectin (5.8 ± 0.8–5.2 ± 0.6 μg/mL); pravastatin increased adiponectin (5.6 ± 0.6– 6.1 ± 0.6 μg/mL)
Koh, Quon, Han, et al. (2011)	Humans with htn	2 months of atorvastatin (20 mg/day)	Adiponectin decreased from 3.3 ± 0.3 to 3.0 ± 0.3 μg/mL
Forst et al. (2007)	Humans at cardiovascular risk	3 months of simvastatin (40 mg/day)	Adiponectin decreased from 15.5 ± 12.7 to 11.6 ± 7.0 μg/mL
Pfutzner et al. (2009)	Nondiabetic humans with increased cardiovascular risk	3 months of simvastatin (40 mg/day)	Adiponectin decreased
Koh et al. (2008)	Humans with hypercholesterolemia	2 months of simvastatin (10–80 mg/day)	Adiponectin decreased with maximal decrease (10%) observed at 80 mg/day
Koh, Quon, Sakuma, et al. (2011)	Humans with hypercholesterolemia and/or T2DM	2 months of simvastatin (20 or 40 mg/day)	Both 20 and 40 mg doses decreased adiponectin by 9% and 8%, respectively

[a]Hypertension.
[b]Coronary artery disease.
[c]Impaired glucose tolerance.
[d]Type 2 diabetes mellitus.
[e]Metabolic syndrome.
[f]Therapeutic lifestyle changes.
[g]High-molecular weight.
[h]Middle molecular weight.
[i]Low-molecular weight.
[j]High-fat diet.

also been shown to increase plasma adiponectin levels in men with hypercholesterolemia and increases adiponectin production and secretion in 3T3-L1 adipocytes (Takagi et al., 2008).

The difference in adiponectin responses to simvastatin versus other statins may be due to lipophilicity. It has been noted that simvastatin, the most lipophilic of the statins, may cause undesirable pleiotropic effects such as reductions in insulin secretion and exacerbation of insulin resistance (Chamberlain, 2001; Kanda, Satoh, & Ichihara, 2003; Koh et al., 2009; Yada, Nakata, Shiraishi, & Kakei, 1999). However, pravastatin, the most hydrophilic, may cause favorable effects such as improvement in insulin sensitivity (Sugiyama et al., 2007). Furthermore, pravastatin reduces the risk of new onset type 2 diabetes (Freeman et al., 2001), while other statins increase the risk (Koh, Sakuma, & Quon, 2011). The greater the lipophilicity, the greater the levels of permeation in nonhepatic tissues; the greater the hydrophilicity, the more hepatoselective the statin (Hamelin & Turgeon, 1998; Serajuddin, Ranadive, & Mahoney, 1991). Therefore, it is possible that the difference in lipophilicity and hydrophilicity between statins may play a role in the varying effects of different statins on adiponectin.

Although many studies showed an increase in circulating adiponectin after statin therapy, several studies examining simvastatin (Nomura, Shouzu, et al., 2009), pravastatin (Gannage-Yared et al., 2005), fluvastatin (Sonmez et al., 2006), rosuvastatin (Anagnostis et al., 2011), or atorvastatin (Anagnostis et al., 2011; Bayes et al., 2005; Koh, Quon, Han, et al., 2011; Otto et al., 2007; Shetty et al., 2004; Thongtang et al., 2011) in a variety of subjects (healthy humans, humans with hypercholesterolemia or hyperlipoproteinemia, or kidney transplant recipients) all found no change or a reduction (Matafome et al., 2011) in circulating adiponectin concentrations in response to statin treatment. A possible explanation for this discrepancy may involve the analysis of the different multimers of adiponectin. One study conducted in humans with type 2 diabetes found no change in total adiponectin but observed a 42% increase in the biologically active HMW adiponectin and a 21% and 23% decrease in the middle- and low-molecular weight forms of adiponectin, respectively, in response to 8 weeks of atorvastatin (von et al., 2009). A recent study demonstrated that simvastatin caused a 40% reduction in the HMW form of adiponectin, whereas total adiponectin levels remained unchanged (Khan, Hamilton, Mundy, Chua, & Scherer, 2009). Therefore, simvastatin caused a redistribution of complexes, that is, a reduction in percentage of HMW levels (Khan et al., 2009). Patients with type 2 diabetes experience a cardioprotective effect from low dose atorvastatin therapy (Colhoun et al., 2004); however, the magnitude of protection may be greater than what is seen in nondiabetic individuals (von et al., 2005). This may be due in part to statin-induced elevations in pathologically reduced adiponectin levels seen in obesity and type 2 diabetes.

The mechanism(s) by which statins increase adiponectin is largely unknown. It has been reported that statins activate PPARγ via extracellular signal-regulated kinase 1/2 and p38 mitogen-activated protein kinase activation (Yano et al., 2007). PPARγ activation using thiazolidinediones (TZDs) produces well-known increases in adiponectin gene expression and secretion *in vitro* as well as serum adiponectin concentrations *in vivo* suggesting that statins may increase adiponectin expression in a similar fashion. In light of recent studies showing that TZDs acutely increase the release of adiponectin in rodents, it will be interesting to see if statins perform similarly and to determine if these effects are modulated in humans to a similar extent.

4. FIBRATES

Fibrates are primarily indicated for the treatment of hypertriglyceridemia and low HDL-C and are used alone or in combination with statins. Systematic review and meta-analysis of the literature concluded that fibrates reduce the risk of major cardiovascular events primarily by reducing serum triglyceride concentrations (Jun et al., 2010). Fibrates lower TG by limiting substrate availability for triglyceride synthesis in the liver (Kissebah, Adams, Harrigan, & Wynn, 1974; Maragoudakis, 1970), promoting the action of lipoprotein lipase (Lithell, Boberg, Hellsing, Lundqvist, & Vessby, 1978), enhancing LDL–receptor/ligand interaction (Kleinman et al., 1985), promoting cholesterol excretion via bile (Palmer, 1987), and stimulating reverse cholesterol transport (Shepherd, 1993). The ultimate result is a reduction in TG, modest reductions in LDL-C and inconsistent effects on HDL-C. A meta-analysis of randomized clinical trials indicates that fibrates increase HDL-C in approximately 50% of all trials (Birjmohun, Hutten, Kastelein, & Stroes, 2005). However, fibrates are particularly effective as combination therapy to attenuate commonly observed reductions in circulating HDL-C associated with statins (Vega et al., 2003). However, the combination of high dose statins and fibrates can increase risk of myopathy, so statin doses should be kept relatively low when used in combination with a fibrate (Smith et al., 2006). Of the fibrates, gemfibrozil and fenofibrate are available in the United States and Canada, bezafibrate is only available in Canada, and fenofibric acid, a relatively new fibrate, is only available in the United States (Jackevicius et al., 2011).

In addition to improving blood lipids, fibrate therapy often results in an improvement in insulin sensitivity, and a reduction in the incidence, and delayed onset of type 2 diabetes (Tenenbaum et al., 2004). The mechanisms by which fibrates improve insulin sensitivity are unclear, yet studies suggest that fibrates bind to the nuclear receptor, PPARα, and increase hepatic fatty acid oxidation and reduce the expression of lipogenic genes. An additional mechanism by which fibrates, particularly bezafibrate, may improve insulin sensitivity is through increasing circulating concentrations of adiponectin.

There are no reports of gemfibrozil impacting adiponectin gene expression or plasma concentrations. However, fenofibric acid, alone and in combination with atorvastatin, increases adiponectin release from adipose tissue of patients with a normal lipid profile and mixed dyslipidemia (Krysiak, Labuzek, & Okopien, 2009). Both fenofibrate and bezafibrate have been shown to increase adiponectin gene expression and circulating adiponectin concentrations. Furthermore, both fibrates increase the expression of the adiponectin receptor, AdipoR2 (Chinetti, Zawadski, Fruchart, & Staels, 2004). Adiponectin mRNA levels measured in epididymal fat isolated from rats on a high-fat, high-glucose diet increased as a result of fenofibrate treatment (Naderali, Fatani, Telles, & Hunter, 2008). Several studies showed increases in circulating total adiponectin (Hiuge et al., 2007; Koh, Han, et al. 2005; Koh, Quon, Han, Chung, et al., 2005; Koh, Quon, Lim, et al., 2011; Nakamura et al., 2007; Rosenson, 2009; Wu et al., 2009; Table 13.2) with fibrate therapy, while others showed an increase in HMW adiponectin but not in total adiponectin (Oki et al., 2007; Table 13.2). However, some studies showed no change in adiponectin in response to fibrate therapy (Anderlova et al., 2007; Choi et al., 2005; Naderali et al., 2008; Otto et al., 2007; Table 13.2). Interestingly, a recent study demonstrated that fenofibrate promotes ischemia-induced revascularization in an adiponectin-dependent manner (Li et al., 2010). The investigators found that 4 weeks of fenofibrate treatment increased circulating adiponectin concentrations and promoted revascularization following hindlimb ischemia in wild-type mice but not their adiponectin-deficient littermates.

Interestingly, Nakano et al. demonstrated that bezafibrate increased, while fenofibrate decreased plasma adiponectin concentrations in *db/db* mice (Nakano et al., 2007). Further, they confirmed in 3T3-L1 adipocytes that bezafibrate increased adiponectin mRNA and secretion into the media,

Table 13.2 The effects of fibrates on adiponectin

References	Subjects	Treatment	Effects on adiponectin
Nakano et al. (2007)	3T3-L1 Adipocytes	Bezafibrate (100 or 300 μM) or fenofibrate (300 μM)	100 and 300 μM bezafibrate increased adiponectin secretion into media, while fenofibrate had no effect Bezafibrate increased mRNA levels of adiponectin at 16 and 24 h posttreatment, while fenofibrate reduced adiponectin mRNA at 16 h posttreatment
Nakano et al. (2007)	*db/db* mice	6 weeks of bezafibrate (100 or 300 mg/kg/day) or fenofibrate (300 mg/kg/day)	300 mg/kg/day bezafibrate increased adiponectin, while 100 mg/kg/day bezafibrate had no effect on adiponectin and fenofibrate reduced adiponectin
Naderali et al. (2008)	Obese male Wistar rats	8 weeks of fenofibrate (50 mg/kg/day)	Fenofibrate increased plasma adiponectin concentrations and adiponectin mRNA in epididymal fat pads
Mori, Oana, Matsuzawa, Akahane, and Tajima (2004)	Male Otsuka Long-Evans Tokushima Fatty rats	14 days of bezafibrate (10 or 100 mg/kg/day)	Circulating adiponectin increased on day 3, 7, and 14 (all days measured) in response to 100 mg/kg/day bezafibrate, but not to 10 mg/kg/day Adiponectin mRNA in retroperitoneal white adipose tissue increased only by day 14 of rats treated with 100 mg/kg/day bezafibrate
Koh, Quon, Lim, et al. (2011)	Humans with hypertriglyceridemia	8 weeks of fenofibrate (160 mg/day)	Adiponectin increased by 17%

Continued

Table 13.2 The effects of fibrates on adiponectin—cont'd

References	Subjects	Treatment	Effects on adiponectin
Rosenson (2009)	Humans with hypertriglyceridemia, central obesity, other characteristics of MetS[a]	12 weeks of fenofibrate (160 mg/day)	Adiponectin increased 7.7% from 4.10 to 4.50 μg/mL
Nakamura et al. (2007)	Humans with stable CAD[b], hypercholesterolemia and hypertriglyceridemia	6 months of bezafibrate (400 mg/day)	Adiponectin increased by 4 weeks, with further increases by 6 months
Wu et al. (2009)	Nondiabetic men with insulin resistance and dyslipidemia; Male Sprague-Dawley rats on HFD[c]; 3T3-L1 adipocytes	8 weeks of fenofibrate (200 mg/day) in humans 2 weeks of fenofibrate (200 mg/kg/day) in rats 24 h of fenofibrate (10, 50, and 100 μM) in 3T3-L1 adipocytes	Fenofibrate increased adiponectin from 11.3 ± 6.7 to 16.5 ± 7.2 μg/mL in humans Fenofibrate attenuated the HFD-induced decrease in adiponectin mRNA in adipose tissue of rats Fenofibrate increased adiponectin mRNA levels by 46% at highest concentration in 3T3-L1 adipocytes
Hiuge et al. (2007)	Humans with MetS	Longitudinal study over a period of ~6.2 years of 400 mg/day bezafibrate	After 2 years of treatment, adiponectin increased, with the median percentage change being +9.8%
Koh, Han, et al. (2005)	Humans with hypertriglyceridemia; some with MetS	8 weeks of fenofibrate (200 mg/day)	Adiponectin increased by 14 ± 5%
Oki, Koide, Nakanishi, Nakashima, and Yamane (2007)	Men with hypertriglyceridemia	12 weeks of 150 mg/day ($n=7$) or 4 weeks of 150 mg/day then 8 weeks of 300 mg/day ($n=4$) fenofibrate	HMW[d] adiponectin increased from 3.0 ± 1.5 to 3.4 ± 1.7 μg/mL No change in total adiponectin

Koh, Han, et al. (2005)	Humans with combined hyperlipidemia	2 months of atorvastatin (10 mg/day) plus placebo or atorvastatin plus fenofibrate (200 mg/day) or fenofibrate plus placebo	Fenofibrate alone or combined with atorvastatin increased adiponectin from 3.4 to 3.5 and 3.2 to 3.6 μg/mL
Koh, Quon, et al. (2005)	Humans with combined hyperlipidemia	2 months of atorvastatin (10 mg/day), atorvastatin and fenofibrate (10 and 200 mg/day, respectively), or fenofibrate 200 mg/day	Combined therapy or fenofibrate alone increased plasma adiponectin from 3.4 to 3.5 μg/mL ($p=0.001$) and 3.2 to 3.6 μg/mL ($p=0.004$), respectively
Otto et al. (2007)	Humans with T2DM[e] and mixed hyperlipoproteinemia	6 weeks of fenofibrate (200 mg/day)	No change in adiponectin
Anderlova et al. (2007)	Obese women with T2DM	3 months of fenofibrate (200 mg/day)	No change in adiponectin
Choi et al. (2005)	Male Long-Evans Tokushima Otsuka rats	40 weeks of fenofibrate (100 mg/kg/day)	No effect on plasma adiponectin, yet adiponectin mRNA increased in visceral fat depots

[a]Metabolic syndrome.
[b]Coronary artery disease.
[c]High-fat diet.
[d]High-molecular weight.
[e]Type 2 diabetes mellitus.

while fenofibrate reduced adiponectin mRNA and no detectable effect on adiponectin secretion. This resulted in improved glycemic control in response to bezafibrate, but not to fenofibrate. Similarly, Mori et al. (2004) also found that bezafibrate increased circulating adiponectin concentrations and adiponectin mRNA in adipose tissue of rats. These studies could point to a direct effect of bezafibrate on adiponectin production, resulting in the increased circulating adiponectin concentrations.

Whether or not the increase in adiponectin is a result of fibrate-mediated PPARα activation remains unclear. For instance, bezafibrate and fenofibrate, both PPARα agonists, have been shown to have differing effects on adiponectin production and secretion. Further, WY-14,643, a selective PPARα agonist, had no effect on adiponectin mRNA levels or secretion from 3T3-L1 adipocytes (Nakano et al., 2007). Although one study has shown that fenofibrate increases adiponectin mRNA in adipose tissue, this resulted in no change in circulating adiponectin concentrations (Choi et al., 2005). While all fibrates have some binding ability at PPARγ, bezafibrate activates PPARα, PPARγ, and PPARδ at comparable doses (Willson, Brown, Sternbach, & Henke, 2000). It is possible that some of the effects of fibrates, particularly bezafibrate, to increase adiponectin are mediated through PPARγ, rather than PPARα. Additional studies will be required to determine if these changes in PPARα activation and insulin sensitivity are responsible in any way for the fibrate-induced changes in plasma adiponectin concentrations.

The discrepancy of whether adiponectin is unchanged or increased in response to fibrate treatment is somewhat puzzling considering that the conditions and participants included in the studies were similar. Analysis of the different adiponectin complexes may be necessary to fully assess the impact of fibrates on adiponectin levels since there can be a shift in adiponectin multimer distribution without altering total adiponectin concentrations.

5. NIACIN

Niacin is a collective term used for nicotinic acid and nicotinamide. While both compounds are equally effective as nutrients or cofactors in glucose and lipid metabolic pathways, pharmacological doses of the acid form of niacin (nicotinic acid) have proven lipid-lowering qualities that do not exist with niacin in the amide form (Dipalma & Thayer, 1991). The pharmacological properties of nicotinic acid were initially discovered in 1955 by Rudulf Altschul and his colleague Abram Hoffer during their quest to identify pharmacological agents for the treatment of hypercholesterolemia and schizophrenia. Nicotinic acid and nicotinamide produced no beneficial effects on the symptoms of schizophrenia, but the authors did find that pharmacological doses (3000 mg/day) of nicotinic acid decreased plasma cholesterol in normal and hypercholesterolemic subjects (Altschul, Hoffer, & Stephen, 1955).

Prospective studies provide convincing evidence that niacin reduces CVD risk. The Coronary Drug Project (1973) (Canner et al., 1986) was a nationwide, double-blind, placebo-controlled study designed to evaluate the long-term effects of niacin, and other lipid-lowering agents on the primary endpoint of all-cause mortality in men with previous myocardial infarction. Niacin (3000 mg/day) reduced total cholesterol by 10% and TG by 26% during the 6-year follow-up period. These changes were associated with reductions in nonfatal myocardial infarction and cerebrovascular accident by 26 and 24% compared to those treated with placebo. The combination of niacin and colestipol, a bile acid sequestrant, in patients with known CVD increased HDL-C, decreased 2-year mortality rates (Brown et al., 1990; Carlson & Rosenhamer, 1988), and reduced the progression of atherosclerotic plaque formation in 61% of patients treated (Blankenhorn et al., 1987). In contrast, the recently completed AIM-HIGH study was terminated approximately 18 months earlier than planned because the addition of niacin produced no further benefits on cardiovascular events than statins alone and actually showed an unexplainable increase in the risk of ischemic cerebrovascular accidents (Dolgin, 2011). The results of these studies bring in to question the safety and efficacy of niacin, but it is important to note that the majority of studies using niacin in high-risk populations impart reductions in overall CVD morbidity and mortality. Furthermore, the results of this trial may be difficult to infer to other populations for primary prevention of heart disease and in individuals with higher LDL-C or atherogenic dyslipidemia (average LDL-C was 71 mg/dL in AIM-HIGH).

The majority of studies conducted to date clearly show that both immediate and extended-release formulations of niacin possess equivalent benefits on blood lipid and lipoprotein characteristics at doses ranging from 500 to 2500 mg/day (Goldberg, 1998; Knopp et al., 1998). Niacin's primary effects are to lower serum TG and raise HDL-C. In fact, niacin is the most effective lipid-lowering agent to increase HDL-C (Walldius & Wahlberg, 1985) and is the only agent known to reduce the proatherogenic lipoprotein, Lp(a). Recent studies from our lab also show that extended-release niacin increases HDL and LDL particle size (E. Plaisance et al, unpublished observations) which would be expected to elicit important beneficial effects on reverse cholesterol transport and vascular lipid deposition, respectively. In agreement with the significant improvement in tissue and circulating lipid metabolism produced by niacin, our group and others have shown that niacin produces remarkable changes in circulating adiponectin (Linke et al., 2009; Plaisance, Grandjean, Brunson, & Judd, 2008; Plaisance, Lukasova, et al., 2009; Westphal, Borucki, Taneva, Makarova, & Luley, 2006, 2007; Westphal & Luley, 2008; Table 13.3).

In our initial studies, we demonstrated that treating primary rat adipocytes and 3T3-L1 adipocytes stably expressing the niacin receptor with niacin resulted in a significant increase in adiponectin secretion (Plaisance, Lukasova, et al., 2009). Additional studies showed that a single dose of niacin acutely elevates serum adiponectin concentrations in rats (Plaisance, Lukasova, et al., 2009). Westphal et al. (2006) reported that serum total adiponectin concentrations increased by 54% in humans after 4 weeks of extended-release niacin (1000 mg/day) and by 94% after 6 weeks (1500 mg/day). Additional studies reported that 6 weeks of extended-release niacin (1500 mg/day) increased serum total adiponectin concentrations in men with metabolic syndrome (Plaisance et al., 2008; Westphal et al., 2007). Niacin treatment significantly increased serum low- and medium-molecular weight adiponectin but produced a dramatic (63–88%) increase in the most biologically active HMW form of adiponectin in men with metabolic syndrome (Plaisance et al., 2008; Westphal & Luley, 2008).

The mechanism by which niacin increases adiponectin remains unclear, but results from our laboratory show that niacin has direct effects on adipose tissue secretion of adiponectin that appear to be dependent on the HCA_2 receptor. HCA_2 is a G_i-protein coupled receptor primarily located on adipocytes which binds niacin with high affinity. Receptor activation produces a reduction in the formation of cAMP from adenylate cyclase which subsequently reduces the downstream activation of protein kinase A, adipose triglyceride lipase, and hormone-sensitive lipase which ultimately reduces triglyceride hydrolysis. Results from our laboratory using G-protein uncoupling agents and HCA_2 loss-of-function studies in 3T3-L1 adipocytes show that the increase in niacin is dependent on receptor activation (Plaisance, Lukasova, et al., 2009). The observation that HCA_2 activation reduces triglyceride hydrolysis and increases adiponectin secretion suggests that these events are coordinately regulated. Although the exact mechanisms are poorly understood, it is possible that niacin increases adiponectin secretion by reducing endoplasmic retention time in a manner reminiscent to that

Table 13.3 The effects of niacin on adiponectin

References	Subjects	Treatment	Effects on adiponectin
Plaisance, Lukasova, et al. (2009)	Male Sprague–Dawley rats	Single dose IR[a] niacin (30 mg/kg)	Adiponectin increased by 37% at 10 min, peaked at 1 h and remained elevated for 24 h
Plaisance et al. (2008)	Men with MetS[b]	6 weeks of ER[c] niacin (1500 mg/day)	Adiponectin increased by 46% from 5.7 ± 0.5 to 8.4 ± 0.7 μg/mL
Westphal et al. (2007)	Men with MetS	6 weeks of ER niacin (1500 mg/day)	Adiponectin increased by 56% from 6.1 ± 2.3 to 10.1 ± 4.0 μg/mL
Westphal et al. (2008)	Men with MetS	6 weeks of ER niacin (1500 mg/day)	LMW[d] and MMW[e] adiponectin increased by 35% and 33%, respectively, but HMW[f] adiponectin increased by 88%
Westphal et al. (2006)	Humans	6 weeks of ER niacin (1500 mg/day)	Adiponectin increased by 94% from 4.83 ± 2.39 to 9.35 ± 6.06 μg/mL
Westphal et al. (2006)	Humans	4 weeks of ER niacin (1000 mg/day)	Adiponectin increased by 54% from 4.83 ± 2.39 to 7.45 ± 5.71 μg/mL
Linke et al. (2009)	Humans with IGT[g] and low HDL-C concentrations	6 months of extended-release niacin (1000 mg/day)	Serum adiponectin increased from 3.7 ± 0.2 to 5.1 ± 0.4 μg/mL Adiponectin mRNA increased fourfold in abdominal subcutaneous adipose tissue

[a]Immediate-release.
[b]Metabolic syndrome.
[c]Extended release.
[d]Low-molecular weight.
[e]Middle molecular weight.
[f]High-molecular weight.
[g]Impaired glucose tolerance.

produced by PPARγ agonists (Wang et al., 2007). Although niacin can acutely (within 10 min) increase circulating adiponectin in rodents (Plaisance, Lukasova, et al., 2009), long-term niacin treatment in patients with impaired glucose tolerance increased adiponectin mRNA expression by fourfold in adipose tissue and was associated with an increased circulating adiponectin (Linke et al., 2009). In support of these findings, direct treatment of 3T3-L1 adipocytes with niacin also increases adiponectin gene expression (Digby et al., 2010). Taken together, the data suggest that niacin produces short-term effects that increase the vesicular secretion of adiponectin into the circulation along with long-term effects that appear to modulate adiponectin gene expression. Future studies will be required to characterize the short- and long-term effects of niacin on adiponectin secretion and gene expression with exploration of the downstream nutrient sensing events that lead to the increase in circulating adiponectin.

6. OMEGA-3 FATTY ACIDS

Omega-3 fatty acids are polyunsaturated fatty acids which are derived from plants [α-linolenic acid (ALA)] and deep water fish [eicosopentanoic acid (EPA) and docosahexanoic acid (DHA)]. The majority of investigations which have examined the effects of ALA on blood lipids and CVD suggest moderate benefits of ALA consumption. In contrast, DHA and EPA obtained from the dietary consumption of deep water fish or over-the-counter or prescription forms indicate that DHA and EPA produce robust effects on CVD events and mortality. In the Nurse's Health Study, the risk of mortality was 34% lower in individuals who consumed deep water fish five times per week compared to those who consumed fish less than 1 day per week (Hu et al., 2002). The JELIS (Japan EPA Lipid Intervention Study) trial consisted of over 18,000 individuals with hyperlipidemia who were followed for 5 years on statin alone or in combination with 1.8 g/day omega-3 fatty acids. Omega-3 fatty acid administration produced a significant reduction in death, revascularization, myocardial infarction, and unstable angina compared to statin alone (Yokoyama et al., 2007). The greatest risk reduction was observed in patients with hypertriglyceridemia and low HDL-C and occurred in the absence of changes in LDL-C.

Recommendations commonly include the intake of 1 g/day of EPA and DHA (Mori & Beilin, 2004; von & Harris, 2007), but a dose–response reduction in TG of 25–45% can be observed up to 4000 mg/day for the treatment of hypertriglyceridemia (Harris, 1997; Plaisance, Grandjean, &

Mahurin, 2009). The greatest benefit of omega-3 fatty acids is a reduction of serum triglyceride concentrations which occurs most frequently in individuals with hypertriglyceridemia (Harris, 1997).

Omega-3 fatty acids suppress the hepatic secretion of VLDL particles which primarily consist of TG, and to a lesser extent, cholesterol, which is thought to explain the significant reduction in circulating TG and less extensive reduction in LDL-C. Although the effects of omega-3 fatty acids on circulating HDL-C are not powerful, they shift the distribution of HDL-C particles from the smaller HDL_3-C to larger HDL_2-C particles which are more active in reverse cholesterol transport.

In addition, omega-3 fatty acids have been shown to alter the circulating adipokine profile in a way that would be expected to yield health-related benefits. Rossi et al. (2005) established that dietary fish oil reversed the inhibitory effects of a long-term sucrose-rich diet on serum leptin and total adiponectin levels in insulin-resistant rats. The increase in adiponectin levels due to dietary fish oil was accompanied by a reversal of insulin resistance and dyslipidemia (Rossi et al., 2005). Since then, additional studies have confirmed that dietary fish oil and omega-3 fatty acid supplements increase total adiponectin concentrations (Duda et al., 2007; Flachs et al., 2006; Neschen et al., 2006; Nomura, Inami, et al., 2009; Perez-Matute, Perez-Echarri, Martinez, Marti, & Moreno-Aliaga, 2007; Saraswathi et al., 2007; Table 13.4).

Sneddon et al. (2008) examined the effects of conjugated linoleic acid (CLA) and omega-3 fatty acids on plasma leptin and adiponectin concentrations. While there was no effect on other adipokines examined, they demonstrated that plasma adiponectin concentrations increased by 12% with CLA plus omega-3 fatty acids in young, obese subjects, but no change was observed in the older obese subjects or in the treated lean subjects (Sneddon et al., 2008). Furthermore, a study conducted on young, nonobese Japanese subjects found that an 8-week diet high in fish intake increased serum adiponectin concentrations in females, but not in males (Kondo et al., 2010). Interestingly, serum levels of omega-3 fatty acids increased more in females than males, suggesting that the increased circulating omega-3 fatty acids may explain the different responses.

Some studies with omega-3 fatty acids have observed no effect on plasma total or HMW adiponectin levels (Kratz et al., 2008), while others have demonstrated that in cell culture conditions, EPA actually reduces adiponectin gene expression and secretion (Lorente-Cebrian et al., 2006). The authors attributed this to reduced PPARγ mRNA levels which resulted

Table 13.4 The effects of omega-3 fatty acids on adiponectin

References	Subjects	Treatment	Effects on adiponectin
Rossi et al. (2005)	Male Wistar rats on SRD[a]	2 months of fish oil	Adiponectin increased from 1.74 ± 0.15 to 3.20 ± 0.20 μg/mL
Flachs et al. (2006)	Male C57BL/6J mice on HFD[b]	5 weeks of EPA/DHA[c] concentrate	Adiponectin increased from 9.0 ± 0.3 to 12.1 ± 0.5 μg/mL with free access to food Adiponectin increased from 9.2 ± 1.1 to 11.2 ± 0.5 μg/mL with 30% food restriction
Neschen et al. (2006)	Male 129 Sv mice	15 days of fish oil in food	Adiponectin increased two- to threefold in a dose-dependent manner, and the concentrations remained approximately twofold higher for 7 days when the fish oil diet was replaced by a safflower oil diet
Itoh et al. (2007)	Male WT, *ob/ob*, and DIO[d] C57BL/6 mice	4 weeks of EPA in food	Adiponectin tended to increase in WT mice and significantly increased in *ob/ob* and DIO mice
Duda et al. (2007)	Male Wistar rats	10 weeks of fish oil diet	Plasma adiponectin and adiponectin mRNA in epididymal adipose tissue increased
Perez-Matute et al. (2007)	Lean and obese male Wistar rats	5 weeks of EPA (1 g/kg/day)	When correcting for adiposity, EPA treatment increased circulating adiponectin concentrations. EPA treatment increased adiponectin gene expression in white adipose tissue
Saraswathi et al. (2007)	C57BL/6 $LDLR^{-/-}$ mice	12 weeks of fish oil diet	All multimers of adiponectin were increased in plasma Adiponectin protein levels in white adipose tissue were increased to 90% higher than controls

Itoh et al. (2007)	Obese humans	3 months of EPA	Adiponectin increased from 4.7 ± 0.6 to 7.4 ± 0.5 μg/mL
Kondo et al. (2010)	Healthy humans	8 weeks fish-diet (3.0 g n-3 PUFA[e]/day)	Fish-diet intervention increased adiponectin in women (13.5 ± 4.6–15.8 ± 5.2 μg/mL), but not in men
Nomura, Inami, et al. (2009)	Humans with hyperlipidemia and T2DM[f]	6 months of EPA (1800 mg/day)	Adiponectin increased after 6 months, but not 3 months of EPA therapy
Sneddon et al. (2008)	Lean and obese young men (23–38 years) and older men (50–64 years)	12 weeks of supplements containing CLA[g] (2280 mg) plus n-3 PUFA (900 mg EPA and 630 mg DHA)	Circulating adiponectin increased by 12% in the young obese group Overall in the younger group, adiponectin increased by 9% Adiponectin did not change in the older group
Kratz et al. (2008)	Moderately obese men and women	14 weeks of a diet rich in n-3 PUFA and supplements containing 8.7 g fish oil/day	There was no effect on plasma total or HMW adiponectin
Lorente-Cebrian, Perez-Matute, Martinez, Marti, and Moreno-Aliaga (2006)	Rat adipocytes	EPA (100 and 200 μM) for 96 h	EPA decreased adiponectin gene expression and protein secretion

[a]Sucrose-rich diet.
[b]High-fat diet.
[c]Eicosapentaenoic acid/docosahexaenoic acid.
[d]Diet-induced obese.
[e]Polyunsaturated fatty acids.
[f]Type 2 diabetes mellitus.
[g]Conjugated linoleic acid.

from EPA treatment. PPARγ activation stimulates adiponectin gene expression and secretion (Yamauchi, Kamon, Waki, Murakami, et al., 2001). The reduction in adiponectin gene expression, synthesis, and secretion in the presence of EPA, which increases fatty acid oxidation, may be a decompensatory mechanism in adipose tissue since omega-3 fatty acids are working directly on the liver and adiponectin increases fatty acid oxidation in liver (Yamauchi et al., 2002).

A 2006 study showed that replacing vegetable oils with EPA and DHA for 5 weeks in the high-fat diets of mice resulted in increased plasma adiponectin concentrations while adiponectin gene expression in dorsolumbar and epididymal white adipose tissue was unchanged (Flachs et al., 2006). Similarly, others have shown that 4 weeks of EPA treatment increased adiponectin serum concentrations in *ob/ob* mice, and diet-induced obese mice without altering adiponectin mRNA in epididymal white adipose tissue (Itoh et al., 2007). In contrast, another study found that within 15 days, incorporation of fish oil into the diets of mice increased plasma total and HMW adiponectin concentrations, as well as adiponectin gene expression in epididymal fat (Neschen et al., 2006).

While many studies have demonstrated the ability of omega-3 fatty acids to increase plasma adiponectin concentrations, the mechanism by which this occurs remains unclear. Studies have suggested that the increase in adiponectin following omega-3 fatty acid administration appears to be due to PPARγ activation. PPARγ is known to be critical in adiponectin production (Combs et al., 2002; Yang et al., 2002), and omega-3 fatty acids and their derivatives are capable of activating PPARγ (Krey et al., 1997; Yamamoto et al., 2005). Indeed, Neschen et al. showed that the increase in plasma adiponectin concentrations by dietary fish oil was abolished when a PPARγ inhibitor was used, signifying the important role of PPARγ in the mechanism of action of fish oil to increase adiponectin concentrations. They also demonstrated that the ability of omega-3 fatty acids, which have been shown to be PPARα ligands (Forman, Chen, & Evans, 1997; Kliewer et al., 1997; Krey et al., 1997), to increase circulating adiponectin is not mediated through PPARα, as evidenced through studies conducted with PPARα null mice. Furthermore, Neschen et al. (2006) showed that a single dose of fish oil delivered by oral gavage had no acute effects on plasma adiponectin concentrations at 24 h but increased adiponectin gene expression in epididymal fat. Another mechanism explored to explain the ability of omega-3 fatty acids to increase serum adiponectin concentrations involves the inflammatory macrophages often found in adipose tissue of obese subjects. Adiponectin mRNA expression and secretion are reduced when adipocytes are cocultured with macrophages (Suganami, Nishida, & Ogawa, 2005). Itoh et al. (2007) discovered that treatment with EPA reversed the coculture-induced decrease in adiponectin secretion without altering adiponectin mRNA expression. TNF-α treatment of adipocytes reduces adiponectin secretion, and the authors found that EPA treatment suppressed the coculture-induced increase in TNF-α expression, suggesting a possible mechanism by which EPA increases serum adiponectin concentrations.

7. CONCLUSIONS

With the increased occurrence of obesity comes a greater incidence of cardiovascular and metabolic diseases, often accompanied by hyperlipidemia. The use of lipid-lowering drugs is critical in the treatment of hyperlipidemia, but they may also affect circulating adipokines in ways that could benefit health. Indeed, the majority of lipid-lowering agents also beneficially increase circulating adiponectin concentrations. High circulating adiponectin levels are known to have many protective effects in the body including antidiabetic, antiatherosclerotic, and anticancer properties. With the exception of simvastatin, statins, particularly pravastatin, increase total and HMW adiponectin. Additionally, fibrates increase adiponectin gene expression and circulating adiponectin concentrations. Niacin also increases circulating adiponectin concentrations, primarily the biologically active HMW multimers both acutely and chronically. In addition to their effects on circulating TG, omega-3 fatty acids also increase circulating adiponectin levels. These effects could be due to the direct effects of these lipid-lowering agents on adiponectin gene expression and/or secretion. However, these effects on adiponectin could also be indirectly produced by the reduction in blood lipids.

Some of the protective effects of lipid-lowering drugs may be explained by the modulation of the synthesis and secretion of adiponectin. However, more studies are needed to fully evaluate the effects of some classes of lipid-lowering drugs on adiponectin. Furthermore, other effects of these drugs (in addition to their effects on adipokines) might play a role in the supplementary health benefits observed aside from the reduction in blood lipids. Most importantly, studies will be required to determine if the increase in adiponectin produced by lipid-lowering agents produces beneficial reductions in metabolic and CVD risk.

ACKNOWLEDGMENT

Boshell Diabetes and Metabolic Diseases Research Program.

REFERENCES

Altschul, R., Hoffer, A., & Stephen, J. D. (1955). Influence of nicotinic acid on serum cholesterol in man. *Archives of Biochemistry and Biophysics, 54*, 558–559.

Anagnostis, P., Selalmatzidou, D., Polyzos, S. A., Panagiotou, A., Slavakis, A., Panagiotidou, A., et al. (2011). Comparative effects of rosuvastatin and atorvastatin on glucose metabolism and adipokine levels in non-diabetic patients with dyslipidaemia: A prospective randomised open-label study. *International Journal of Clinical Practice, 65*, 679–683.

Anderlova, K., Dolezalova, R., Housova, J., Bosanska, L., Haluzikova, D., Kremen, J., et al. (2007). Influence of PPAR-alpha agonist fenofibrate on insulin sensitivity and selected adipose tissue-derived hormones in obese women with type 2 diabetes. *Physiological Research, 56*, 579–586.

Ando, H., Sugimoto, K., Yanagihara, H., Tsuruoka, S., Saito, T., Takamura, T., et al. (2008). Effects of atorvastatin and pravastatin on glucose tolerance, adipokine levels and inflammatory markers in hypercholesterolaemic patients. *Clinical and Experimental Pharmacology & Physiology, 35*, 1012–1017.

Arita, Y., Kihara, S., Ouchi, N., Maeda, K., Kuriyama, H., Okamoto, Y., et al. (2002). Adipocyte-derived plasma protein adiponectin acts as a platelet-derived growth factor-BB-binding protein and regulates growth factor-induced common postreceptor signal in vascular smooth muscle cell. *Circulation, 105*, 2893–2898.

Baigent, C., Keech, A., Kearney, P. M., Blackwell, L., Buck, G., Pollicino, C., et al. (2005). Efficacy and safety of cholesterol-lowering treatment: Prospective meta-analysis of data from 90,056 participants in 14 randomised trials of statins. *The Lancet, 366*, 1267–1278.

Bayes, B., Granada, M. L., Lauzurica, R., Pastor, M. C., Navarro, M., Bonet, J., et al. (2005). Effect of low doses of atorvastatin on adiponectin, glucose homeostasis, and clinical inflammatory markers in kidney transplant recipients. *Transplantation Proceedings, 37*, 3808–3812.

Berg, A. H., Combs, T. P., Du, X., Brownlee, M., & Scherer, P. E. (2001). The adipocyte-secreted protein Acrp30 enhances hepatic insulin action. *Nature Medicine*, 7, 947–953.

Birjmohun, R. S., Hutten, B. A., Kastelein, J. J., & Stroes, E. S. (2005). Efficacy and safety of high-density lipoprotein cholesterol-increasing compounds: A meta-analysis of randomized controlled trials. *Journal of the American College of Cardiology, 45*, 185–197.

Blanco-Colio, L. M., Martin-Ventura, J. L., Gomez-Guerrero, C., Masramon, X., de, T. E., Farsang, C., et al. (2008). Adiponectin plasma levels are increased by atorvastatin treatment in subjects at high cardiovascular risk. *European Journal of Pharmacology, 586*, 259–265.

Blankenhorn, D. H., Nessim, S. A., Johnson, R. L., Sanmarco, M. E., Azen, S. P., & Cashin-Hemphill, L. (1987). Beneficial effects of combined colestipol-niacin therapy on coronary atherosclerosis and coronary venous bypass grafts. *JAMA: The Journal of the American Medical Association, 257*, 3233–3240.

Brown, G., Albers, J. J., Fisher, L. D., Schaefer, S. M., Lin, J. T., Kaplan, C., et al. (1990). Regression of coronary artery disease as a result of intensive lipid-lowering therapy in men with high levels of apolipoprotein B. *The New England Journal of Medicine, 323*, 1289–1298.

Canner, P. L., Berge, K. G., Wenger, N. K., Stamler, J., Friedman, L., Prineas, R. J., et al. (1986). Fifteen year mortality in Coronary Drug Project patients: Long-term benefit with niacin. *Journal of the American College of Cardiology, 8*, 1245–1255.

Carlson, L. A., & Rosenhamer, G. (1988). Reduction of mortality in the Stockholm Ischaemic Heart Disease Secondary Prevention Study by combined treatment with clofibrate and nicotinic acid. *Acta Medica Scandinavica*, *223*, 405–418.

Chamberlain, L. H. (2001). Inhibition of isoprenoid biosynthesis causes insulin resistance in 3T3-L1 adipocytes. *FEBS Letters*, *507*, 357–361.

Chinetti, G., Zawadski, C., Fruchart, J. C., & Staels, B. (2004). Expression of adiponectin receptors in human macrophages and regulation by agonists of the nuclear receptors PPARalpha, PPARgamma, and LXR. *Biochemical and Biophysical Research Communications*, *314*, 151–158.

Choi, K. C., Ryu, O. H., Lee, K. W., Kim, H. Y., Seo, J. A., Kim, S. G., et al. (2005). Effect of PPAR-alpha and -gamma agonist on the expression of visfatin, adiponectin, and TNF-alpha in visceral fat of OLETF rats. *Biochemical and Biophysical Research Communications*, *336*, 747–753.

Colhoun, H. M., Betteridge, D. J., Durrington, P. N., Hitman, G. A., Neil, H. A., Livingstone, S. J., et al. (2004). Primary prevention of cardiovascular disease with atorvastatin in type 2 diabetes in the Collaborative Atorvastatin Diabetes Study (CARDS): Multicentre randomised placebo-controlled trial. *The Lancet*, *364*, 685–696.

Combs, T. P., Berg, A. H., Obici, S., Scherer, P. E., & Rossetti, L. (2001). Endogenous glucose production is inhibited by the adipose-derived protein Acrp30. *The Journal of Clinical Investigation*, *108*, 1875–1881.

Combs, T. P., Wagner, J. A., Berger, J., Doebber, T., Wang, W. J., Zhang, B. B., et al. (2002). Induction of adipocyte complement-related protein of 30 kilodaltons by PPARgamma agonists: A potential mechanism of insulin sensitization. *Endocrinology*, *143*, 998–1007.

Deng, G., Long, Y., Yu, Y. R., & Li, M. R. (2010). Adiponectin directly improves endothelial dysfunction in obese rats through the AMPK-eNOS Pathway. *International Journal of Obesity*, *34*, 165–171.

Devaraj, S., Siegel, D., & Jialal, I. (2007). Simvastatin (40 mg/day), adiponectin levels, and insulin sensitivity in subjects with the metabolic syndrome. *The American Journal of Cardiology*, *100*, 1397–1399.

Digby, J. E., McNeill, E., Dyar, O. J., Lam, V., Greaves, D. R., & Choudhury, R. P. (2010). Anti-inflammatory effects of nicotinic acid in adipocytes demonstrated by suppression of fractalkine, RANTES, and MCP-1 and upregulation of adiponectin. *Atherosclerosis*, *209*, 89–95.

Dipalma, J. R., & Thayer, W. S. (1991). Use of niacin as a drug. *Annual Review of Nutrition*, *11*, 169–187.

Dolgin, E. (2011). Trial puts niacin—and cholesterol dogma—in the line of fire. *Nature Medicine*, *17*, 756.

Duda, M. K., O'Shea, K. M., Lei, B., Barrows, B. R., Azimzadeh, A. M., McElfresh, T. E., et al. (2007). Dietary supplementation with omega-3 PUFA increases adiponectin and attenuates ventricular remodeling and dysfunction with pressure overload. *Cardiovascular Research*, *76*, 303–310.

Flachs, P., Mohamed-Ali, V., Horakova, O., Rossmeisl, M., Hosseinzadeh-Attar, M. J., Hensler, M., et al. (2006). Polyunsaturated fatty acids of marine origin induce adiponectin in mice fed a high-fat diet. *Diabetologia*, *49*, 394–397.

Forman, B. M., Chen, J., & Evans, R. M. (1997). Hypolipidemic drugs, polyunsaturated fatty acids, and eicosanoids are ligands for peroxisome proliferator-activated receptors alpha and delta. *Proceedings of the National Academy of Sciences of the United States of America*, *94*, 4312–4317.

Forst, T., Pfutzner, A., Lubben, G., Weber, M., Marx, N., Karagiannis, E., et al. (2007). Effect of simvastatin and/or pioglitazone on insulin resistance, insulin secretion, adiponectin, and proinsulin levels in nondiabetic patients at cardiovascular risk–the PIOSTAT study. *Metabolism, 56*, 491–496.

Freeman, D. J., Norrie, J., Sattar, N., Neely, R. D., Cobbe, S. M., Ford, I., et al. (2001). Pravastatin and the development of diabetes mellitus: Evidence for a protective treatment effect in the West of Scotland Coronary Prevention Study. *Circulation, 103*, 357–362.

Gannage-Yared, M. H., Azar, R. R., Amm-Azar, M., Khalife, S., Germanos-Haddad, M., Neemtallah, R., et al. (2005). Pravastatin does not affect insulin sensitivity and adipocytokines levels in healthy nondiabetic patients. *Metabolism, 54*, 947–951.

Goldberg, A. C. (1998). Clinical trial experience with extended-release niacin (Niaspan): Dose-escalation study. *The American Journal of Cardiology, 82*, 35U–38U.

Gotto, A. M., Jr. (1998). Triglyceride as a risk factor for coronary artery disease. *The American Journal of Cardiology, 82*, 22Q–25Q.

Gould, A. L., Rossouw, J. E., Santanello, N. C., Heyse, J. F., & Furberg, C. D. (1998). Cholesterol reduction yields clinical benefit: Impact of statin trials. *Circulation, 97*, 946–952.

Grundy, S. M. (1997). Small LDL, atherogenic dyslipidemia, and the metabolic syndrome. *Circulation, 95*, 1–4.

Hamelin, B. A., & Turgeon, J. (1998). Hydrophilicity/lipophilicity: Relevance for the pharmacology and clinical effects of HMG-CoA reductase inhibitors. *Trends in Pharmacological Sciences, 19*, 26–37.

Harris, W. S. (1997). n-3 fatty acids and serum lipoproteins: Human studies. *The American Journal of Clinical Nutrition, 65*, 1645S–1654S.

Hiuge, A., Tenenbaum, A., Maeda, N., Benderly, M., Kumada, M., Fisman, E. Z., et al. (2007). Effects of peroxisome proliferator-activated receptor ligands, bezafibrate and fenofibrate, on adiponectin level. *Arteriosclerosis, Thrombosis, and Vascular Biology, 27*, 635–641.

Hotta, K., Funahashi, T., Bodkin, N. L., Ortmeyer, H. K., Arita, Y., Hansen, B. C., et al. (2001). Circulating concentrations of the adipocyte protein adiponectin are decreased in parallel with reduced insulin sensitivity during the progression to type 2 diabetes in rhesus monkeys. *Diabetes, 50*, 1126–1133.

Hu, E., Liang, P., & Spiegelman, B. M. (1996). AdipoQ is a novel adipose-specific gene dysregulated in obesity. *The Journal of Biological Chemistry, 271*, 10697–10703.

Hu, F. B., Bronner, L., Willett, W. C., Stampfer, M. J., Rexrode, K. M., Albert, C. M., et al. (2002). Fish and omega-3 fatty acid intake and risk of coronary heart disease in women. *JAMA: The Journal of the American Medical Association, 287*, 1815–1821.

Itoh, M., Suganami, T., Satoh, N., Tanimoto-Koyama, K., Yuan, X., Tanaka, M., et al. (2007). Increased adiponectin secretion by highly purified eicosapentaenoic acid in rodent models of obesity and human obese subjects. *Arteriosclerosis, Thrombosis, and Vascular Biology, 27*, 1918–1925.

Jackevicius, C. A., Tu, J. V., Ross, J. S., Ko, D. T., Carreon, D., & Krumholz, H. M. (2011). Use of fibrates in the United States and Canada. *JAMA: The Journal of the American Medical Association, 305*, 1217–1224.

Jun, M., Foote, C., Lv, J., Neal, B., Patel, A., Nicholls, S. J., et al. (2010). Effects of fibrates on cardiovascular outcomes: A systematic review and meta-analysis. *The Lancet, 375*, 1875–1884.

Kanda, M., Satoh, K., & Ichihara, K. (2003). Effects of atorvastatin and pravastatin on glucose tolerance in diabetic rats mildly induced by streptozotocin. *Biological & Pharmaceutical Bulletin, 26*, 1681–1684.

Khan, T., Hamilton, M. P., Mundy, D. I., Chua, S. C., & Scherer, P. E. (2009). Impact of simvastatin on adipose tissue: Pleiotropic effects in vivo. *Endocrinology, 150*, 5262–5272.

Kissebah, A. H., Adams, P. W., Harrigan, P., & Wynn, V. (1974). The mechanism of action of clofibrate and tetranicotinoylfructose (Bradilan) on the kinetics of plasma free fatty acid and triglyceride transport in type IV and type V hypertriglyceridaemia. *European Journal of Clinical Investigation*, *4*, 163–174.

Kleinman, Y., Eisenberg, S., Oschry, Y., Gavish, D., Stein, O., & Stein, Y. (1985). Defective metabolism of hypertriglyceridemic low density lipoprotein in cultured human skin fibroblasts. Normalization with bezafibrate therapy. *The Journal of Clinical Investigation*, *75*, 1796–1803.

Kliewer, S. A., Sundseth, S. S., Jones, S. A., Brown, P. J., Wisely, G. B., Koble, C. S., et al. (1997). Fatty acids and eicosanoids regulate gene expression through direct interactions with peroxisome proliferator-activated receptors alpha and gamma. *Proceedings of the National Academy of Sciences of the United States of America*, *94*, 4318–4323.

Knopp, R. H., Alagona, P., Davidson, M., Goldberg, A. C., Kafonek, S. D., Kashyap, M., et al. (1998). Equivalent efficacy of a time-release form of niacin (Niaspan) given once-a-night versus plain niacin in the management of hyperlipidemia. *Metabolism*, *47*, 1097–1104.

Koh, K. K., Quon, M. J., Han, S. H., Chung, W. J., Ahn, J. Y., Seo, Y. H., et al. (2004). Additive beneficial effects of losartan combined with simvastatin in the treatment of hypercholesterolemic, hypertensive patients. *Circulation*, *110*, 3687–3692.

Koh, K. K., Han, S. H., Quon, M. J., Yeal, A. J., & Shin, E. K. (2005). Beneficial effects of fenofibrate to improve endothelial dysfunction and raise adiponectin levels in patients with primary hypertriglyceridemia. *Diabetes Care*, *28*, 1419–1424.

Koh, K. K., Quon, M. J., Han, S. H., Ahn, J. Y., Jin, D. K., Kim, H. S., et al. (2005). Vascular and metabolic effects of combined therapy with ramipril and simvastatin in patients with type 2 diabetes. *Hypertension*, *45*, 1088–1093.

Koh, K. K., Quon, M. J., Han, S. H., Chung, W. J., Ahn, J. Y., Seo, Y. H., et al. (2005). Additive beneficial effects of fenofibrate combined with atorvastatin in the treatment of combined hyperlipidemia. *Journal of the American College of Cardiology*, *45*, 1649–1653.

Koh, K. K., Quon, M. J., Han, S. H., Lee, Y., Ahn, J. Y., Kim, S. J., et al. (2008). Simvastatin improves flow-mediated dilation but reduces adiponectin levels and insulin sensitivity in hypercholesterolemic patients. *Diabetes Care*, *31*, 776–782.

Koh, K. K., Quon, M. J., Han, S. H., Lee, Y., Kim, S. J., Park, J. B., et al. (2009). Differential metabolic effects of pravastatin and simvastatin in hypercholesterolemic patients. *Atherosclerosis*, *204*, 483–490.

Koh, K. K., Quon, M. J., Han, S. H., Lee, Y., Kim, S. J., & Shin, E. K. (2010). Atorvastatin causes insulin resistance and increases ambient glycemia in hypercholesterolemic patients. *Journal of the American College of Cardiology*, *55*, 1209–1216.

Koh, K. K., Quon, M. J., Han, S. H., Lee, Y., Park, J. B., Kim, S. J., et al. (2011). Additive beneficial effects of atorvastatin combined with amlodipine in patients with mild-to-moderate hypertension. *International Journal of Cardiology*, *146*, 319–325.

Koh, K. K., Quon, M. J., Lim, S., Lee, Y., Sakuma, I., Lee, Y. H., et al. (2011). Effects of fenofibrate therapy on circulating adipocytokines in patients with primary hypertriglyceridemia. *Atherosclerosis*, *214*, 144–147.

Koh, K. K., Quon, M. J., Sakuma, I., Lee, Y., Lim, S., Han, S. H., et al. (2011). Effects of simvastatin therapy on circulating adipocytokines in patients with hypercholesterolemia. *International Journal of Cardiology*, *146*, 434–437.

Koh, K. K., Sakuma, I., & Quon, M. J. (2011). Differential metabolic effects of distinct statins. *Atherosclerosis*, *215*, 1–8.

Kondo, K., Morino, K., Nishio, Y., Kondo, M., Fuke, T., Ugi, S., et al. (2010). Effects of a fish-based diet on the serum adiponectin concentration in young, non-obese, healthy Japanese subjects. *Journal of Atherosclerosis and Thrombosis*, *17*, 628–637.

Kratz, M., Swarbrick, M. M., Callahan, H. S., Matthys, C. C., Havel, P. J., & Weigle, D. S. (2008). Effect of dietary n-3 polyunsaturated fatty acids on plasma total and high-molecular-weight adiponectin concentrations in overweight to moderately obese men and women. *The American Journal of Clinical Nutrition*, *87*, 347–353.

Krey, G., Braissant, O., L'Horset, F., Kalkhoven, E., Perroud, M., Parker, M. G., et al. (1997). Fatty acids, eicosanoids, and hypolipidemic agents identified as ligands of peroxisome proliferator-activated receptors by coactivator-dependent receptor ligand assay. *Molecular Endocrinology*, *11*, 779–791.

Krysiak, R., Labuzek, K., & Okopien, B. (2009). Effect of atorvastatin and fenofibric acid on adipokine release from visceral and subcutaneous adipose tissue of patients with mixed dyslipidemia and normolipidemic subjects. *Pharmacological Reports*, *61*, 1134–1145.

Kubota, N., Yano, W., Kubota, T., Yamauchi, T., Itoh, S., Kumagai, H., et al. (2007). Adiponectin stimulates AMP-activated protein kinase in the hypothalamus and increases food intake. *Cell Metabolism*, *6*, 55–68.

Li, P., Shibata, R., Maruyama, S., Kondo, M., Ohashi, K., Ouchi, N., et al. (2010). Fenofibrate promotes ischemia-induced revascularization through the adiponectin-dependent pathway. *American Journal of Physiology. Endocrinology and Metabolism*, *299*, E560–E566.

Linke, A., Sonnabend, M., Fasshauer, M., Hollriegel, R., Schuler, G., Niebauer, J., et al. (2009). Effects of extended-release niacin on lipid profile and adipocyte biology in patients with impaired glucose tolerance. *Atherosclerosis*, *205*, 207–213.

Lithell, H., Boberg, J., Hellsing, K., Lundqvist, G., & Vessby, B. (1978). Increase of the lipoprotein-lipase activity in human skeletal muscle during clofibrate administration. *European Journal of Clinical Investigation*, *8*, 67–74.

Lorente-Cebrian, S., Perez-Matute, P., Martinez, J. A., Marti, A., & Moreno-Aliaga, M. J. (2006). Effects of eicosapentaenoic acid (EPA) on adiponectin gene expression and secretion in primary cultured rat adipocytes. *Journal of Physiology and Biochemistry*, *62*, 61–69.

Ma, J., Sehgal, N. L., Ayanian, J. Z., & Stafford, R. S. (2005). National trends in statin use by coronary heart disease risk category. *PLoS Medicine*, *2*, e123.

Maeda, N., Shimomura, I., Kishida, K., Nishizawa, H., Matsuda, M., Nagaretani, H., et al. (2002). Diet-induced insulin resistance in mice lacking adiponectin/ACRP30. *Nature Medicine*, *8*, 731–737.

Maragoudakis, M. E. (1970). On the mode of action of lipid-lowering agents. II. in vitro inhibition of acetyl coenzyme A carboxylase by a hypolipidemic drug. *Biochemistry*, *9*, 413–417.

Matafome, P., Louro, T., Rodrigues, L., Crisostomo, J., Nunes, E., Amaral, C., et al. (2011). Metformin and atorvastatin combination further protect the liver in type 2 diabetes with hyperlipidaemia. *Diabetes/Metabolism Research and Reviews*, *27*, 54–62.

Miyagishima, K., Hiramitsu, S., Kato, S., Kato, Y., Kitagawa, F., Teradaira, R., et al. (2007). Efficacy of atorvastatin therapy in ischaemic heart disease—Effects on oxidized low-density lipoprotein and adiponectin. *The Journal of International Medical Research*, *35*, 534–539.

Mori, T. A., & Beilin, L. J. (2004). Omega-3 fatty acids and inflammation. *Current Atherosclerosis Reports*, *6*, 461–467.

Mori, Y., Oana, F., Matsuzawa, A., Akahane, S., & Tajima, N. (2004). Short-term effect of bezafibrate on the expression of adiponectin mRNA in the adipose tissues: A study in spontaneously type 2 diabetic rats with visceral obesity. *Endocrine*, *25*, 247–251.

Naderali, E. K., Fatani, S., Telles, M., & Hunter, L. (2008). The effects of physiological and pharmacological weight loss on adiponectin and leptin mRNA levels in the rat epididymal adipose tissue. *European Journal of Pharmacology*, *579*, 433–438.

Nakamura, T., Kodama, Y., Takano, H., Umetani, K., Fujioka, D., Saito, Y., et al. (2007). Increase in circulating levels of adiponectin after treatment with statin and fibrate in patients with coronary artery disease and hyperlipidemia. *Atherosclerosis*, *193*, 449–451.

Nakano, S., Inada, Y., Masuzaki, H., Tanaka, T., Yasue, S., Ishii, T., et al. (2007). Bezafibrate regulates the expression and enzyme activity of 11beta-hydroxysteroid dehydrogenase type 1 in murine adipose tissue and 3T3-L1 adipocytes. *American Journal of Physiology. Endocrinology and Metabolism, 292*, E1213–E1222.

Nawrocki, A. R., Rajala, M. W., Tomas, E., Pajvani, U. B., Saha, A. K., Trumbauer, M. E., et al. (2006). Mice lacking adiponectin show decreased hepatic insulin sensitivity and reduced responsiveness to peroxisome proliferator-activated receptor gamma agonists. *The Journal of Biological Chemistry, 281*, 2654–2660.

Neschen, S., Morino, K., Rossbacher, J. C., Pongratz, R. L., Cline, G. W., Sono, S., et al. (2006). Fish oil regulates adiponectin secretion by a peroxisome proliferator-activated receptor-gamma-dependent mechanism in mice. *Diabetes, 55*, 924–928.

Nomura, S., Inami, N., Shouzu, A., Omoto, S., Kimura, Y., Takahashi, N., et al. (2009). The effects of pitavastatin, eicosapentaenoic acid and combined therapy on platelet-derived microparticles and adiponectin in hyperlipidemic, diabetic patients. *Platelets, 20*, 16–22.

Nomura, S., Shouzu, A., Omoto, S., Inami, N., Shimazu, T., Satoh, D., et al. (2009). Effects of pitavastatin on monocyte chemoattractant protein-1 in hyperlipidemic patients. *Blood Coagulation & Fibrinolysis, 20*, 440–447.

Oki, K., Koide, J., Nakanishi, S., Nakashima, R., & Yamane, K. (2007). Fenofibrate increases high molecular weight adiponectin in subjects with hypertriglyceridemia. *Endocrine Journal, 54*, 431–435.

Otto, C., Otto, B., Frost, R. J., Vogeser, M., Pfeiffer, A. F., Spranger, J., et al. (2007). Short-term therapy with atorvastatin or fenofibrate does not affect plasma ghrelin, resistin or adiponectin levels in type 2 diabetic patients with mixed hyperlipoproteinaemia. *Acta Diabetologica, 44*, 65–68.

Ouchi, N., Kihara, S., Arita, Y., Nishida, M., Matsuyama, A., Okamoto, Y., et al. (2001). Adipocyte-derived plasma protein, adiponectin, suppresses lipid accumulation and class A scavenger receptor expression in human monocyte-derived macrophages. *Circulation, 103*, 1057–1063.

Palmer, R. H. (1987). Effects of fibric acid derivatives on biliary lipid composition. *The American Journal of Medicine, 83*, 37–43.

Perez-Matute, P., Perez-Echarri, N., Martinez, J. A., Marti, A., & Moreno-Aliaga, M. J. (2007). Eicosapentaenoic acid actions on adiposity and insulin resistance in control and high-fat-fed rats: Role of apoptosis, adiponectin and tumour necrosis factor-alpha. *The British Journal of Nutrition, 97*, 389–398.

Pfutzner, A., Schondorf, T., Hanefeld, M., Lubben, G., Kann, P. H., Karagiannis, E., et al. (2009). Changes in insulin resistance and cardiovascular risk induced by PPARgamma activation have no impact on RBP4 plasma concentrations in nondiabetic patients. *Hormone and Metabolic Research, 41*, 202–206.

Plaisance, E. P., Grandjean, P. W., Brunson, B. L., & Judd, R. L. (2008). Increased total and high-molecular weight adiponectin after extended-release niacin. *Metabolism, 57*, 404–409.

Plaisance, E. P., Grandjean, P. W., & Mahurin, A. J. (2009). Independent and combined effects of aerobic exercise and pharmacological strategies on serum triglyceride concentrations: A qualitative review. *The Physician and Sportsmedicine, 37*, 11–19.

Plaisance, E. P., Lukasova, M., Offermanns, S., Zhang, Y., Cao, G., & Judd, R. L. (2009). Niacin stimulates adiponectin secretion through the GPR109A receptor. *American Journal of Physiology. Endocrinology and Metabolism, 296*, E549–E558.

Qi, Y., Takahashi, N., Hileman, S. M., Patel, H. R., Berg, A. H., Pajvani, U. B., et al. (2004). Adiponectin acts in the brain to decrease body weight. *Nature Medicine, 10*, 524–529.

Roger, V. L., Go, A. S., Lloyd-Jones, D. M., Adams, R. J., Berry, J. D., Brown, T. M., et al. (2011). Heart disease and stroke statistics–2011 update: A report from the American Heart Association. *Circulation, 123*, e18–e209.

Rosenson, R. S. (2009). Effect of fenofibrate on adiponectin and inflammatory biomarkers in metabolic syndrome patients. *Obesity (Silver Spring), 17*, 504–509.

Rossi, A. S., Lombardo, Y. B., Lacorte, J. M., Chicco, A. G., Rouault, C., Slama, G., et al. (2005). Dietary fish oil positively regulates plasma leptin and adiponectin levels in sucrose-fed, insulin-resistant rats. *American Journal of Physiology. Regulatory, Integrative and Comparative Physiology, 289*, R486–R494.

Sakamoto, K., Sakamoto, T., & Ogawa, H. (2006). The effect of 6 months of treatment with pravastatin on serum adiponection concentrations in Japanese patients with coronary artery disease and hypercholesterolemia: A pilot study. *Clinical Therapeutics, 28*, 1012–1021.

Saraswathi, V., Gao, L., Morrow, J. D., Chait, A., Niswender, K. D., & Hasty, A. H. (2007). Fish oil increases cholesterol storage in white adipose tissue with concomitant decreases in inflammation, hepatic steatosis, and atherosclerosis in mice. *The Journal of Nutrition, 137*, 1776–1782.

Scherer, P. E., Williams, S., Fogliano, M., Baldini, G., & Lodish, H. F. (1995). A novel serum protein similar to C1q, produced exclusively in adipocytes. *The Journal of Biological Chemistry, 270*, 26746–26749.

Serajuddin, A. T., Ranadive, S. A., & Mahoney, E. M. (1991). Relative lipophilicities, solubilities, and structure-pharmacological considerations of 3-hydroxy-3-methylglutaryl-coenzyme A (HMG-CoA) reductase inhibitors pravastatin, lovastatin, mevastatin, and simvastatin. *Journal of Pharmaceutical Sciences, 80*, 830–834.

Shepherd, J. (1993). Mechanism of action of fibrates. *Postgraduate Medical Journal, 69*(Suppl. 1), S34–S41.

Shetty, G. K., Economides, P. A., Horton, E. S., Mantzoros, C. S., & Veves, A. (2004). Circulating adiponectin and resistin levels in relation to metabolic factors, inflammatory markers, and vascular reactivity in diabetic patients and subjects at risk for diabetes. *Diabetes Care, 27*, 2450–2457.

Smith, S. C., Jr., Allen, J., Blair, S. N., Bonow, R. O., Brass, L. M., Fonarow, G. C., et al. (2006). AHA/ACC guidelines for secondary prevention for patients with coronary and other atherosclerotic vascular disease: 2006 update: Endorsed by the National Heart, Lung, and Blood Institute. *Circulation, 113*, 2363–2372.

Sneddon, A. A., Tsofliou, F., Fyfe, C. L., Matheson, I., Jackson, D. M., Horgan, G., et al. (2008). Effect of a conjugated linoleic acid and omega-3 fatty acid mixture on body composition and adiponectin. *Obesity (Silver Spring), 16*, 1019–1024.

Sonmez, A., Dogru, T., Tasci, I., Yilmaz, M. I., Pinar, M., Naharci, I., et al. (2006). The effect of fluvastatin on plasma adiponectin levels in dyslipidaemia. *Clinical Endocrinology, 64*, 567–572.

Suganami, T., Nishida, J., & Ogawa, Y. (2005). A paracrine loop between adipocytes and macrophages aggravates inflammatory changes: Role of free fatty acids and tumor necrosis factor alpha. *Arteriosclerosis, Thrombosis, and Vascular Biology, 25*, 2062–2068.

Sugiyama, S., Fukushima, H., Kugiyama, K., Maruyoshi, H., Kojima, S., Funahashi, T., et al. (2007). Pravastatin improved glucose metabolism associated with increasing plasma adiponectin in patients with impaired glucose tolerance and coronary artery disease. *Atherosclerosis, 194*, e43–e51.

Takagi, T., Matsuda, M., Abe, M., Kobayashi, H., Fukuhara, A., Komuro, R., et al. (2008). Effect of pravastatin on the development of diabetes and adiponectin production. *Atherosclerosis, 196*, 114–121.

Tenenbaum, A., Motro, M., Fisman, E. Z., Schwammenthal, E., Adler, Y., Goldenberg, I., et al. (2004). Peroxisome proliferator-activated receptor ligand bezafibrate for prevention of type 2 diabetes mellitus in patients with coronary artery disease. *Circulation, 109*, 2197–2202.

The Coronary Drug Project, (1973). Findings leading to discontinuation of the 2.5-mg day estrogen group. The Coronary Drug Project Research Group. *JAMA: The Journal of the American Medical Association, 226*, 652–657.
Thongtang, N., Ai, M., Otokozawa, S., Himbergen, T. V., Asztalos, B. F., Nakajima, K., et al. (2011). Effects of maximal atorvastatin and rosuvastatin treatment on markers of glucose homeostasis and inflammation. *The American Journal of Cardiology, 107*, 387–392.
Trujillo, M. E., & Scherer, P. E. (2005). Adiponectin–journey from an adipocyte secretory protein to biomarker of the metabolic syndrome. *Journal of Internal Medicine, 257*, 167–175.
Vega, G. L., Ma, P. T., Cater, N. B., Filipchuk, N., Meguro, S., Garcia-Garcia, A. B., et al. (2003). Effects of adding fenofibrate (200 mg/day) to simvastatin (10 mg/day) in patients with combined hyperlipidemia and metabolic syndrome. *The American Journal of Cardiology, 91*, 956–960.
Viollet, B., Guigas, B., Leclerc, J., Hebrard, S., Lantier, L., Mounier, R., et al. (2009). AMP-activated protein kinase in the regulation of hepatic energy metabolism: From physiology to therapeutic perspectives. *Acta Physiologica (Oxford, England), 196*, 81–98.
von, S. C., & Harris, W. S. (2007). Cardiovascular benefits of omega-3 fatty acids. *Cardiovascular Research, 73*, 310–315.
von, E. M., Schneider, J. G., Hadziselimovic, S., Hamann, A., Bierhaus, A., Nawroth, P. P., et al. (2005). Adipocytokines as a novel target for the anti-inflammatory effect of atorvastatin in patients with type 2 diabetes. *Diabetes Care, 28*, 754–755.
von, E. M., Liu, D., Bluemm, A., Schuster, T., Baumann, M., Lutz, J., et al. (2009). Changes in adiponectin multimer distribution in response to atorvastatin treatment in patients with type 2 diabetes. *Clinical Endocrinology, 71*, 27–32.
Walldius, G., & Wahlberg, G. (1985). Effects of nicotinic acid and its derivatives on lipid metabolism and other metabolic factors related to atherosclerosis. *Advances in Experimental Medicine and Biology, 183*, 281–293.
Wang, Z. V., Schraw, T. D., Kim, J. Y., Khan, T., Rajala, M. W., Follenzi, A., et al. (2007). Secretion of the adipocyte-specific secretory protein adiponectin critically depends on thiol-mediated protein retention. *Molecular and Cellular Biology, 27*, 3716–3731.
Westphal, S., & Luley, C. (2008). Preferential increase in high-molecular weight adiponectin after niacin. *Atherosclerosis, 198*, 179–183.
Westphal, S., Borucki, K., Taneva, E., Makarova, R., & Luley, C. (2006). Adipokines and treatment with niacin. *Metabolism, 55*, 1283–1285.
Westphal, S., Borucki, K., Taneva, E., Makarova, R., & Luley, C. (2007). Extended-release niacin raises adiponectin and leptin. *Atherosclerosis, 193*, 361–365.
Willson, T. M., Brown, P. J., Sternbach, D. D., & Henke, B. R. (2000). The PPARs: From orphan receptors to drug discovery. *Journal of Medicinal Chemistry, 43*, 527–550.
Wu, H., Wei, L., Bao, Y., Lu, J., Huang, P., Liu, Y., et al. (2009). Fenofibrate reduces serum retinol-binding protein-4 by suppressing its expression in adipose tissue. *American Journal of Physiology. Endocrinology and Metabolism, 296*, E628–E634.
Xu, A., Wang, Y., Keshaw, H., Xu, L. Y., Lam, K. S., & Cooper, G. J. (2003). The fat-derived hormone adiponectin alleviates alcoholic and nonalcoholic fatty liver diseases in mice. *The Journal of Clinical Investigation, 112*, 91–100.
Yada, T., Nakata, M., Shiraishi, T., & Kakei, M. (1999). Inhibition by simvastatin, but not pravastatin, of glucose-induced cytosolic Ca^{2+} signalling and insulin secretion due to blockade of L-type Ca^{2+} channels in rat islet beta-cells. *British Journal of Pharmacology, 126*, 1205–1213.
Yamamoto, K., Itoh, T., Abe, D., Shimizu, M., Kanda, T., Koyama, T., et al. (2005). Identification of putative metabolites of docosahexaenoic acid as potent PPARgamma agonists and antidiabetic agents. *Bioorganic & Medicinal Chemistry Letters, 15*, 517–522.

Yamauchi, T., Kamon, J., Waki, H., Murakami, K., Motojima, K., Komeda, K., et al. (2001). The mechanisms by which both heterozygous peroxisome proliferator-activated receptor gamma (PPARgamma) deficiency and PPARgamma agonist improve insulin resistance. *The Journal of Biological Chemistry*, *276*, 41245–41254.

Yamauchi, T., Kamon, J., Waki, H., Terauchi, Y., Kubota, N., Hara, K., et al. (2001). The fat-derived hormone adiponectin reverses insulin resistance associated with both lipoatrophy and obesity. *Nature Medicine*, 7, 941–946.

Yamauchi, T., Kamon, J., Minokoshi, Y., Ito, Y., Waki, H., Uchida, S., et al. (2002). Adiponectin stimulates glucose utilization and fatty-acid oxidation by activating AMP-activated protein kinase. *Nature Medicine*, *8*, 1288–1295.

Yamauchi, T., Nio, Y., Maki, T., Kobayashi, M., Takazawa, T., Iwabu, M., et al. (2007). Targeted disruption of AdipoR1 and AdipoR2 causes abrogation of adiponectin binding and metabolic actions. *Nature Medicine*, *13*, 332–339.

Yang, W. S., Jeng, C. Y., Wu, T. J., Tanaka, S., Funahashi, T., Matsuzawa, Y., et al. (2002). Synthetic peroxisome proliferator-activated receptor-gamma agonist, rosiglitazone, increases plasma levels of adiponectin in type 2 diabetic patients. *Diabetes Care*, *25*, 376–380.

Yano, M., Matsumura, T., Senokuchi, T., Ishii, N., Murata, Y., Taketa, K., et al. (2007). Statins activate peroxisome proliferator-activated receptor gamma through extracellular signal-regulated kinase 1/2 and p38 mitogen-activated protein kinase-dependent cyclooxygenase-2 expression in macrophages. *Circulation Research*, *100*, 1442–1451.

Yin, X., Tu, L., & Yang, H. (2007). Effect of simvastatin on IL-6 and adiponectin secretion and mRNA expression in 3T3-L1 adipocytes. *Journal of Huazhong University of Science and Technology. Medical Sciences*, *27*, 248–251.

Yokoyama, M., Origasa, H., Matsuzaki, M., Matsuzawa, Y., Saito, Y., Ishikawa, Y., et al. (2007). Effects of eicosapentaenoic acid on major coronary events in hypercholesterolaemic patients (JELIS): A randomised open-label, blinded endpoint analysis. *The Lancet*, *369*, 1090–1098.

CHAPTER FOURTEEN

Adiponectin and Interleukin-6 in Inflammation-Associated Disease

Li Li, Li-Ling Wu[1]
Department of Physiology and Pathophysiology, Peking University Health Science Center and Key Laboratory of Molecular Cardiovascular Sciences, Ministry of Education, Beijing, China
[1]Corresponding author: e-mail address: pathophy@bjmu.edu.cn

Contents

Abstract

Numerous lines of evidence implicate a role for adipose tissue in the development of a systemic inflammatory state that contributes to metabolic, cardiovascular, and autoimmune disorders. Serum levels of adiponectin, a cytokine that is mainly produced by adipocytes, are paradoxically decreased in individuals with obesity, type 2 diabetes, and cardiovascular disease compared with healthy individuals. Mounting experimental data have revealed that adiponectin exhibits beneficial effects on energy homeostasis and cardiovascular functions that are attributed to its direct modulation of a proinflammatory factor, interleukin-6. However, some recent studies indicate that adiponectin appears to function as an inducer of proinflammatory factors and the elevated adiponectin level aggravates inflammation response in autoimmune disease. In this review, we focus on the action of adiponectin in chronic inflammation-associated metabolic, cardiovascular, and autoimmune disorders. In particular, we discuss the interaction between adiponectin and interleukin-6 in adipocytes and cardiovascular cells.

ABBREVIATIONS

AMPK adenosine monophosphate-activated protein kinase
CAD coronary artery disease
CRP C-reactive protein

Vitamins and Hormones, Volume 90
ISSN 0083-6729
http://dx.doi.org/10.1016/B978-0-12-398313-8.00014-2

ERK1/2 extracellular signal-regulated kinase 1/2
gAd globular adiponectin
HMW high-molecular weight
IL-6 interleukin-6
NF-κB nuclear factor-κB
PPAR peroxisome proliferator-activated receptor
TNF-α tumor necrosis factor-α

1. INTRODUCTION

Obesity, a major health care issue, is characterized as a state of chronic mild inflammation and is closely related to the development of metabolic and cardiovascular morbidities (Berg & Scherer, 2005; Wellen & Hotamisligil, 2003). Considerable effort has been expended in the past decade to understand the mechanism of obesity-related systemic inflammation in the pathogenesis of metabolic and cardiovascular diseases (Gualillo, González-Juanatey, & Lago, 2007; Lago, Dieguez, Gómez-Reino, & Gualillo, 2007a). Numerous intensive studies have indicated that adipose tissue is more than just an energy storage compartment; it also serves as a major endocrine organ that secretes a variety of bioactive substances known as adipokines (Lago et al., 2007a; Lago, Dieguez, Gómez-Reino, & Gualillo, 2007b). To date, more than 50 cytokines have been identified. They include adipokines, such as leptin, adiponectin, resistin, visfatin, apelin, omentin, chemerin, visceral adipose tissue-derived serpin (vaspin), and hepcidin, which are mainly produced by adipose tissue (Lago et al., 2007a), and other cytokines, such as interleukin (IL)-6, tumor necrosis factor (TNF)-α, serum amyloid A3 (SAA3), monocyte chemoattractant protein 1 (MCP-1), IL-8, plasminogen activator inhibitor 1 (PAI-1), and vascular endothelial growth factor, that are shared with other systems (Gualillo et al., 2007).

Adipokine secretion profiles are altered as obesity develops and they thereby modulate inflammatory processes that contribute to the pathophysiology of obesity-linked metabolic and cardiovascular disorders. In obese individuals, there are increased levels of several proinflammatory adipokines, including IL-6, TNF-α, SAA3, IL-1β, IL-8, PAI-1, leptin, and resistin (Fain, Madan, Hiler, Cheema, & Bahouth, 2004). By contrast, there are decreased levels of adiponectin, an adipokine that exhibits anti-inflammatory actions in a number of animal models and cell types (Ouchi & Walsh, 2007). This review focuses on the interaction between the inflammatory protein,

IL-6, and the anti-inflammatory adipokine, adiponectin, with particular emphasis on their roles as mediators of obesity-associated metabolic, cardiovascular, and autoimmune disorders.

2. BIOLOGICAL ACTIONS AND SIGNALING PATHWAYS ACTIVATED BY ADIPONECTIN

Adiponectin, which is also referred to as adipocyte complement-related protein, 30 kDa (Scherer, Williams, Fegliano, Baldini, & Lodish, 1995), AdipoQ (Hu, Liang, & Spiegelman, 1996), and gelatin-binding protein 28 (Nakano, Tobe, ChoiMiura, Mazda, & Tomita, 1996), is derived mainly from white adipose tissue. Human adiponectin is a 244-amino acid protein that contains a collagen-like domain at its N-terminus and a globular domain at its C-terminus, similar to collagen types VIII and X, and complement factor C1q (Saito et al., 1999). Adiponectin circulates in the bloodstream as polymers including trimers, hexamers, and high-molecular weight (HMW) isoforms. In addition to forming oligomers, full-length adiponectin can also be processed by proteolytic cleavage; a 16-kDa globular domain fragment has been detected in human plasma (Fruebis et al., 2001; Shapiro & Scherer, 1998). These different isoforms of adiponectin may have different binding activity to membrane receptor and share distinct signaling effects in the cardiovascular system (Hada et al., 2007). Adiponectin accounts for as much as 0.01% of total plasma protein; in humans, average levels of plasma adiponectin range from 3 to 30 μg/ml (Arita et al., 1999), and low plasma adiponectin levels (<4 μg/ml) are closely associated with obesity-linked metabolic and cardiovascular disorders, including insulin resistance, type 2 diabetes, hypertension, and coronary artery disease (CAD) (Iwashima et al., 2004; Kumada et al., 2003; Weyer et al., 2001). Adiponectin exerts important effects on metabolic modulation and energy homeostasis (Maeda et al., 2002). Administration of adiponectin *in vivo* and *in vitro* has been shown to increase fatty acid oxidation in the liver and in the skeletal muscle (Yamauchi et al., 2003a), to promote glucose uptake by adipocytes (Wu et al., 2003) and myocytes (Ceddia et al., 2005), to decrease hepatic gluconeogenesis (Combs, Berg, Obici, Scherer, & Rossetti, 2001), and to modulate calcium signaling in skeletal muscles (Iwabu et al., 2010). Adiponectin exerts its biological actions by binding two adiponectin receptors (adipoRs), adipoR1 and R2. AdipoR1, a high-affinity receptor for globular adiponectin (gAd), is most abundantly expressed in skeletal muscle. AdipoR2, a moderate-affinity receptor for both globular and full-length adiponectin, is predominantly expressed in liver (Yamauchi, Kamon, Ito, et al., 2003a). T-cadherin has also been identified as a receptor for the hexameric and HMW forms of adiponectin; it is expressed in the cardiovascular system, nervous system, and muscles (Hug et al., 2004). Interactions between adiponectin and these receptors activate two critical downstream molecules, adenosine monophosphate-activated protein kinase (AMPK) and peroxisome proliferator-activated receptor α (PPARα), thereby affecting many aspects of cellular metabolism including fatty acid oxidation and glucose uptake and utilization (Yamauchi, Kamon, Ito, et al., 2003a) (Fig. 14.1).

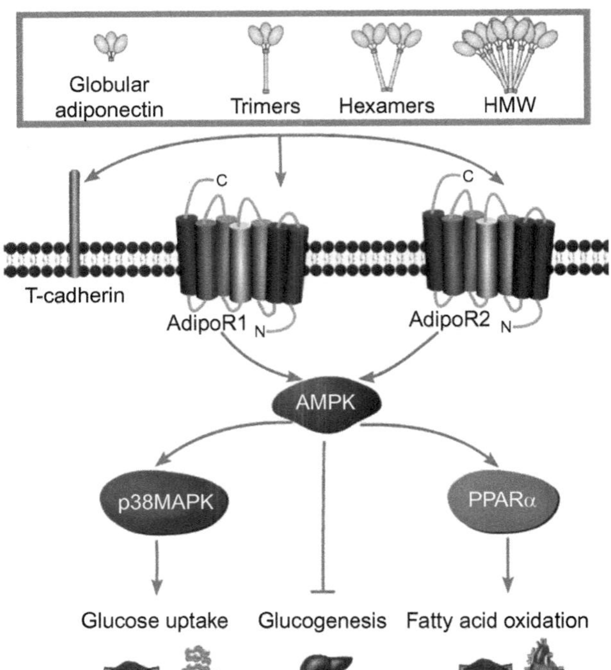

Figure 14.1 Adiponectin-mediated signaling in the regulation of glucose and fatty acid metabolism. Full-length adiponectin (including trimers, hexamers, and HMW forms of adiponectin) has moderate affinity for both adipoR1 and adipoR2 receptors, whereas globular adiponectin has a high affinity for adipoR1 but low affinity for adipoR2. T-cadherin has been identified as another receptor for the hexameric and HMW forms of adiponectin. Interactions between adiponectin and these receptors activate AMPK, PPARα, and p38MAPK, thereby modulating many aspects of cellular metabolism, including the acceleration of glucose uptake in adipocytes and myocytes, the reduction of enzymes involved in gluconeogenesis in hepatocytes, and the promotion of fatty acid oxidation in myocytes.

3. INTERACTION BETWEEN ADIPONECTIN AND IL-6 IN REGULATING DIABETIC STATE

Numerous epidemiological studies have highlighted the association between decreased adiponectin levels and the diabetic state. Levels of adiponectin were significantly reduced in obese subjects compared to non-obese controls (Arita et al., 1999), and a significant negative correlation was found between plasma adiponectin levels and body mass index (BMI), percent body fat, waist-to-hip ratio and intra-abdominal fat (Cnop et al., 2003; Nayak et al., 2010; Ryo et al., 2004). Plasma adiponectin levels were lower in patients with type 2 diabetes than in nondiabetic control subjects with similar BMI (Hotta et al., 2000). A clinical investigation (Lara-Castro, Luo, Wallace, Klein, & Garvey, 2003) indicated that total serum adiponectin was significantly decreased in subjects with insulin resistance and type 2 diabetes; this decrease was attributed to the reduction in HMW adiponectin rather than in trimeric forms of adiponectin. In support of these findings, another study (Blüher et al., 2007) showed that total serum adiponectin levels were associated with insulin sensitivity and low serum adiponectin levels predicted insulin resistance and impaired glucose tolerance. Other clinical studies have also indicated that increased adiponectin levels are a negative predictor of the development of insulin resistance and type 2 diabetes in populations of BMI-adjusted subjects (Snehalatha et al., 2003; Yamamoto, Hirose, Saito, Nishikai, & Saruta, 2004).

Mounting evidence from experimental models has indicated that normal levels of adiponectin play a pivotal role in the regulation of insulin sensitivity and in protection against the development of insulin resistance and type 2 diabetes. In rhesus monkeys, plasma levels of adiponectin were decreased in association with reduced insulin sensitivity during the progression of type 2 diabetes (Hotta et al., 2001). Adiponectin knock-out (APN-KO) mice exhibited diet-induced insulin resistance and the adiponectin supplement protected both APN-KO and ob^{-}/ob^{-} mice from developing diabetes (Yamauchi et al., 2003b). Adiponectin has also been shown to lower plasma glucose levels in normal and diabetic mice, without affecting insulin levels (Berg, Combs, Du, Brownlee, & Scherer, 2001). These data confirm that adiponectin has an insulin-sensitizing activity. The beneficial effect of adiponectin on insulin sensitivity appears to be mediated mainly through its ability to activate AMPK. AMPK causes: (1) glucose transporter-4 (GLUT-4) translocation to the cell surface to accelerate glucose uptake in adipocytes and myocytes (Ceddia et al., 2005; Wu et al., 2003); (2) phosphorylation of acetyl coenzyme A carboxylase and fatty acid oxidation in myocytes (Tomas et al., 2002; Yamauchin, Kamon, & Waki, 2001); and (3) decreased expression and activity of enzymes involved in hepatic gluconeogenesis, leading to reduction of plasma glucose levels (Combs et al., 2001).

Low-grade systemic inflammation has been reported in obese patients. Some studies have linked the proinflammatory cytokine IL-6 to insulin resistance by demonstrating increased levels of circulating IL-6 and of IL-6 secretion from adipocytes in subjects with obesity (Vozarova et al., 2001) or type 2 diabetes (Charles et al., 2011). Basal plasma concentrations of IL-6 have independently predicted future risk of developing type 2 diabetes (Pradhan, Manson, Rifai, Buring, & Ridker, 2001). A polymorphism in the human *IL6* gene that is associated with decreased levels of circulating IL-6 is associated with increased insulin sensitivity (Kubaszek et al., 2003). Plasma concentrations of IL-6 have decreased in parallel with weight loss and improvement of insulin resistance in patients undergoing bariatric surgery (Moschen et al., 2010).

IL-6 may contribute to insulin resistance by mediating gluconeogenesis and impairing insulin receptor signaling in the liver. Administration of recombinant IL-6 into rats induces hepatic gluconeogenesis and in turn leads to hyperglycemia and compensatory hyperinsulinemia (Stith & Luo, 1994). Treating diet-induced obese mice with an IL-6 antibody ameliorates hepatic insulin resistance and increases insulin sensitivity (Klover, Clementi, & Mooney, 2005). Further, chronic treatment with IL-6 impairs autophosphorylation of the insulin receptor and downstream signaling mediators in the liver (Klover, Zimmers, Koniaris, & Mooney, 2003). IL-6 has also been shown to induce expression of suppressor of cytokine signaling-3, a critical inhibitor of insulin receptor signaling, in hepatocytes (Senn et al., 2003).

These data appear to show that adiponectin and IL-6 have opposing roles in the modulation of insulin sensitivity, even though some reports have paradoxically suggested that in some settings adiponectin activates nuclear factor-κB (NF-κB) and induces proinflammatory cytokines (Haugen & Drevon, 2007) and that transient IL-6 upregulation contributes to improved insulin sensitivity (Pedersen & Febbraio, 2008). The exact manner in which adiponectin and IL-6 interact in the regulation of insulin sensitivity is, however, still unknown. IL-6 may interfere with the expression and function of adiponectin during the pathogenesis of insulin resistance and type 2 diabetes. In obese subjects, reduced adiponectin levels may be due to either transcriptional suppression or decreased secretion caused by IL-6. For example, IL-6 and TNF-α levels were positively correlated with visceral fat and negatively correlated with adiponectin levels (Lira et al., 2011). IL-6 expression was highly and persistently upregulated in db^-/db^- mice with continuous downregulation of plasma adiponectin levels (Awazawa et al., 2011). IL-6 KO mice had increased plasma adiponectin levels on both a chow and a high fat/high glucose diet (Di Gregorio, Hensley, Lu, Ranganathan, & Kern, 2004). Treatment of 3T3-L1 adipocytes with IL-6 significantly decreased adiponectin mRNA expression and secretion. This inhibitory effect of IL-6 on adiponectin production is partially mediated by extracellular signal-regulated kinase 1/2 (ERK1/2) (Fasshauer et al., 2003). Therefore, by inhibiting adiponectin production from adipocytes during metabolic disease, the overactivity of proinflammatory cytokines such as IL-6 and TNF-α may lead to low adiponectin levels and to a perpetual state of low-grade chronic inflammation and metabolic disorders.

However, several studies challenge the widely accepted notion that IL-6 signaling promotes the diabetic state. For example, IL-6 KO mice are predisposed to obesity, glucose intolerance, and hypertriglyceridemia. Diminished AMPK activity in their muscle and adipose tissue appears to contribute to their progression to metabolic syndrome (Kelly, Ruderman, & Tomas, 2006). A recent study revealed that IL-6 secreted from macrophages promoted the insulin-sensitizing action of adiponectin in liver. Expression of insulin receptor substrate-2 (IRS-2), a major mediator of insulin signaling in liver that is essential for regulating glucose metabolism in physiological and insulin resistant states, is decreased in the livers of APN-KO mice, db^-/db^- mice, and high fat diet-induced obese mice. Adiponectin injection intraperitoneally increased IRS-2 expression and restored insulin signaling in the livers of these mouse models. However, surprisingly, adiponectin administration caused an acute and transient increase of the plasma IL-6 levels of these animals; further, genetic disruption of IL-6 expression or a neutralizing antibody to IL-6 substantially attenuated adiponectin-induced signal transducer and activator of transcription 3 (STAT3) phosphorylation and IRS-2 upregulation. In mice with hepatocyte-targeted disruption of STAT3, adiponectin-induced STAT3 phosphorylation and IRS-2 upregulation were completely inhibited, whereas transient IL-6 induction by adiponectin was not affected. These data indicate that adiponectin upregulates IRS-2 through STAT3 activation in hepatocytes in an IL-6-dependent manner. A further investigation into the origin of IL-6 induction by adiponectin detected IL-6 expression almost exclusively in stromal vascular cells from white adipose tissue but not in adipocytes. Adiponectin strongly upregulated IL-6 expression in cultured macrophages, but not in fully differentiated 3T3-L1 adipocytes. The transcriptional mechanism involved in adiponectin-induced transient IL-6 expression in macrophages includes increased binding of NF-κB with IL-6 promoter region (Awazawa et al., 2011). Taken together, these data demonstrate that adiponectin can directly increase insulin sensitivity by upregulating IRS-2 in hepatocytes. They also indicate a novel pathway in which adiponectin upregulates IRS-2 via signaling from an unidentified adipoR, and IL-6 can mediate the adiponectin-induced activation of STAT3 and a subsequent increase in IRS-2 expression. Thus, the signaling pathways are complex, but modulating the adiponectin/IL-6 axis may provide a new therapeutic strategy for treating obesity-induced insulin resistance and type 2 diabetes.

4. INTERACTION BETWEEN ADIPONECTIN AND IL-6 IN REGULATING CARDIOVASCULAR DISEASE

It has been reported that plasma adiponectin concentrations are lower in patients with CAD than in age- and BMI-adjusted control subjects (Kumada et al., 2003; Nakamura et al., 2004). A recent study has indicated that overweight patients with CAD have lower plasma levels of adiponectin than overweight patients without CAD (Kollias et al., 2011). High plasma adiponectin levels are associated both with a lower risk of myocardial infarction in men (Pischon et al., 2004) and a moderately decreased risk for CAD in male diabetic patients (Schulze et al., 2005). Collectively, these studies suggest that low serum levels of adiponectin are strongly associated with elevated risk and prevalence of CAD. Because serum HMW adiponectin and the HMW to total adiponectin ratio are significantly lower in men with CAD than in men without CAD, the HMW to total adiponectin ratio appears to be more crucial than the absolute circulating adiponectin concentration in predicting risk of cardiovascular disease (Aso et al., 2006). This inverse correlation between adiponectin levels and cardiovascular disorders also extends to peripheral arterial diseases (Iwashima et al., 2006), essential hypertension (Iwashima et al., 2004), and stroke (Chen et al., 2005). A study performed with a large population sample (Pääkkö et al., 2010) also detected an association between low plasma adiponectin concentration and left ventricular hypertrophy. Carotid intima-media thickness increases with decreased plasma adiponectin (Iglseder et al., 2005).

Several lines of evidence from experimental studies have revealed that adiponectin exerts favorable effects on cardiovascular system through its anti-inflammatory actions. Adiponectin protects against the progression of atherosclerosis and myocardial remodeling through modulating functions of monocytes, vascular smooth muscle cells (VSMCs), endothelial cells, cardiomyocytes, and cardiac fibroblasts.

4.1. Atherosclerosis

Experiments in mice have demonstrated that adiponectin plays a pivotal role in preventing atherosclerotic progression. The cumulative findings have enriched our understanding of the anti-inflammatory mechanism underlying adiponectin-mediated vascular function. Adiponectin regulates many steps and cell types in this atherogenic process, which is characterized by chronic systemic inflammation; often, it does so by interfering with the proinflammatory properties of IL-6. In apolipoprotein E-KO mice, adenovirus-mediated overexpression of adiponectin reduced the formation of atherosclerotic lesions and reduced the levels of class A scavenger receptor (SR-A), TNF-α, and vascular cellular adhesion molecule-1 mRNAs in the vascular wall (Okamoto et al., 2002; Yamauchi, Kamon, Waki, et al., 2003b). In APN-KO mice, mechanical injury-induced neointimal hyperplasia and smooth muscle proliferation were reversed by reexpression of adiponectin (Matsuda et al., 2002).

4.1.1 Monocytes and macrophages

Exposure of primary human monocytes and monocyte-derived macrophages to adiponectin results in the induction of anti-inflammatory mediators including IL-10, IL-1 receptor antagonist, and the tissue inhibitor of metalloproteinase-1 (Wolf, Wolf, Rumpold, Enrich, & Tilg, 2004). Further, adiponectin limits the secretion of TNF-α and IL-6 by activated monocytes and macrophages, in part by inhibiting the activation of NF-κB and ERK1/2 (Neumeier et al., 2006; Wulster-Radcliffe, Ajuwon, Wang, Christian, & Spurlock, 2004). In support of the notion that the anti-inflammatory effects of adiponectin appear to be mediated by suppressing NF-κB activation, adiponectin was found to upregulate expression of PPARγ2, which antagonizes transcriptional activation of NF-κB by lipopolysaccharide (LPS) (Ajuwon & Spurlock, 2005). Adiponectin also inhibits the transformation of macrophages to foam cells and reduces intracellular cholesteryl ester content by suppressing expression of SR-A in human macrophages (Ouchi et al., 2001).

Although it has been suggested that adiponectin exerts anti-inflammatory effects via suppressing the production of proinflammatory cytokines from macrophages, another study provides a novel opinion that the "anti-inflammatory" effect of adiponectin may be due to an induction of macrophage tolerance. gAd powerfully induces TNF-α and IL-6 secretion from primary human peripheral macrophages, the THP-1 human macrophage cell line, and primary mouse peritoneal macrophages. Preexposure of macrophages to gAd induces NF-κB activation and renders them tolerant to further gAd exposure or to other proinflammatory stimuli, such as LPS exposure. These data suggest that the anti-inflammatory effect results from an adiponectin-mediated induction of macrophage tolerance to other proinflammatory stimuli and to adiponectin itself. In this view, high circulating levels of adiponectin in lean subjects are important in maintaining macrophage tolerance, whereas the decrease of adiponectin levels in obesity reduces this effect by rendering macrophages more sensitive to proinflammatory stimuli (Tsatsanis et al., 2005).

4.1.2 Vascular smooth muscle cells

Adiponectin modulates the proliferation and migration of VSMCs during the development and progression of atherosclerotic lesions. *In vitro* studies have shown that adiponectin suppresses both the proliferation and migration of human aortic smooth muscle cells (HASMCs) induced by platelet-derived growth factor (PDGF)-BB and that it inhibits ERK1/2 signaling via direct binding with PDGF-BB (Arita et al., 2002). A recent study also suggested that adiponectin derived from VSMCs is important in maintaining the contractile phenotype of VSMCs, thereby contributing to critical cardioprotective functions in the vascular wall (Ding et al., 2012).

In addition to impeding the development and progression of vascular lesions, a possible mechanism of adiponectin-mediated stabilization of atherosclerotic plaques has been reported (Li et al., 2011). Extracellular matrix components, especially collagen, are thought to be important for maintaining the stability of atherosclerotic plaques, and prolyl-4-hydroxylase (P4H) is a key intracellular enzyme essential for folding procollagen polypeptide chains into stabilized collagen. Experiments in HASMCs treated with adiponectin indicate that adiponectin abrogates the inhibitory effect of IL-6 on collagen III level, suggesting that adiponectin may interfere with IL-6-mediated suppression of collagen III maturation and secretion. Addition of adiponectin substantially increases P4Ha1 mRNA and protein levels in IL-6-stimulated HASMCs. ERK1/2 and Sp1 play a crucial role in adiponectin-mediated increases in P4Ha1 expression. These data suggest that adiponectin attenuates IL-6-inhibited P4Ha1 synthesis and stabilizes collagen formation in HASMCs through a Sp1-ERK1/2-P4Ha1-dependent pathway, and in this way, it may play a major role in the development of the thick fibrous cap of atherosclerotic plaques (Li et al., 2011).

4.1.3 Endothelial cells

Adiponectin induces nitric oxide release from endothelial cells by enhancing the expression and activity of endothelial nitric oxide synthase via an AMPK-dependent pathway (Chen, Montagnani, Funahashi, Shimomura, & Quon, 2003; Ouchi et al., 2004). Adiponectin also inhibits TNF-α-induced NF-κB activation in endothelial cells, thereby effectively attenuating TNF-α-stimulated expression of IL-8 and adhesion molecules including vascular cell adhesion molecule-1, E-selectin, and intracellular adhesion molecule-1, such that TNF-α-stimulated monocyte adhesion to endothelial cells is inhibited. The inhibitory effect of adiponectin on NF-κB is partly mediated by its ability to promote the cyclic AMP/protein kinase A pathway (Ouchi et al., 1999, 2000). Preventing apoptosis in endothelial cells through activation of AMPK signaling may be another reason for adiponectin-mediated anti-inflammatory action because the accumulation of apoptotic debris is known to be able to cause inflammation and immune system dysfunction (Takemura et al., 2007).

4.2. Myocardial ischemia and remodeling

APN-KO mice develop larger infarcts than wild-type mice following ischemia–reperfusion, and these larger infarcts are associated with increased myocardial cell apoptosis and TNF-α expression. Adenovirus-mediated delivery of adiponectin diminishes infarct size, myocardial apoptosis, and TNF-α production in both APN-KO and wild-type mice. The protective action of adiponectin against myocardial ischemia–reperfusion injury appears to be mediated by the activation of two independent pathways involving both AMPK-mediated antiapoptotic actions and cyclooxygenase-2-mediated anti-inflammatory actions (Shibata et al., 2005).

Among various cytokines secreted by cardiac fibroblasts, IL-6 plays an important role in myocardium. The plasma level of IL-6 is elevated in patients with congestive heart failure and increased with the severity of this disease (Tsutamoto et al., 1998). IL-6 mediates the development of cardiac hypertrophy and cardiac fibrosis in chronic cardiac allograft rejection (Diaz et al., 2009). However, the IL-6/soluble IL-6 receptor complex reduces infarct size via inhibiting myocardial apoptosis (Matsushita et al., 2005). IL-6 is obligatory for the development of a cardioprotective phenotype in the late phase of ischemic preconditioning (Smart et al., 2006). IL-6 KO mice display severe cardiac dilation and increased accumulation of interstitial collagen compared with wild-type littermates (Banerjee, Fuseler, Intwala, & Baudino, 2009). Therefore, IL-6 may be a multifunctional cytokine in regulating myocardial structure and function.

Adiponectin induces IL-6 production via adipoR1 signaling in adult mouse cardiac fibroblasts (Liao et al., 2009), and this effect has been confirmed by our team in adult rat cardiac fibroblasts. We further found that activated AMPK, p38 mitogen-activated protein kinase (MAPK), and ERK1/2 serve as signaling molecules to mediate gAd-induced IL-6 production. Moreover, the activated protein-1 (AP-1), nuclear factor interleukin-6 (NF-IL-6 a.k.a. CCAAT-enhancer-binding protein beta [C/EBPβ]), and NF-κB binding sites of the *IL6* gene are responsible for gAd-induced IL-6 transcription in cardiac fibroblasts (Fan et al., 2011) (Fig. 14.2). IL-6 leads to the stimulation of the gp130/Janus-activated kinase signaling pathway and as a result causes STAT3 tyrosine phosphorylation, such that STAT3 dimerizes and translocates from the cytoplasm into the nucleus where it induces the transcription of target genes, leading finally to cardioprotective actions. Therefore, these studies give a new clue to the molecular mechanism of adiponectin regulation of IL-6 expression in cardiac fibroblasts and provide new insights into future therapeutic approaches that might regulate the local myocardial adiponectin/IL-6 system to prevent or reverse cardiac remodeling.

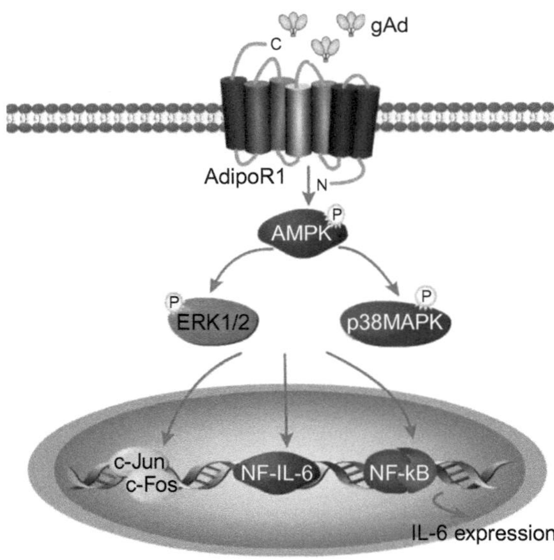

Figure 14.2 Adiponectin signaling and the stimulation of IL-6 production in adult rat cardiac fibroblasts. Globular adiponectin (gAd) enhances IL-6 synthesis and release in adult rat cardiac fibroblasts through adipoR1. Activation of AMPK, p38MAPK, and ERK1/2 mediates the intracellular signal transduction process. The AP-1, NF-IL-6, and NF-kB *cis*-elements are required for gAd-induced IL-6 transcription.

5. INTERACTION BETWEEN ADIPONECTIN AND IL-6 IN REGULATING AUTOIMMUNE DISEASE

In contrast to the protective effect against obesity-related diabetic and cardiovascular disorders, adiponectin appears to function as an inducer of proinflammatory factors in autoimmune disease. Plasma adiponectin levels in patients with various autoimmune diseases, such as rheumatoid arthritis, systemic lupus erythematosus, inflammatory bowel disease, primary Sjögren's syndrome, and type 1 diabetes, are elevated rather than decreased (Otero et al., 2006; Perseghin et al., 2003; Sada et al., 2006; Toussirot et al., 2010). Thus, adiponectin is regulated in the opposite direction and may exert proinflammatory rather than anti-inflammatory activities in autoimmune/inflammatory conditions (Fantuzzi, 2008). Increased levels of adiponectin are observed in the serum and synovial fluid of patients with rheumatoid arthritis, and C-reactive protein (CRP) levels are positively correlated to those of serum adiponectin, the opposite of what is reported for metabolic diseases (Otero et al., 2006; Schäffler et al., 2003; Senolt, Pavelka, Housa, & Haluzík, 2006). Adiponectin enhances IL-6 production in human synovial fibroblasts via activation of AMPK, p38MAPK, and NF-κB (Tang, Chiu, Tan, Yang, & Fu, 2007). Neutralization of TNF-α activity inhibits the stimulatory effect of adiponectin on IL-6 production, indicating that the proinflammatory effect of adiponectin in the synovium appears to be mediated partially by inducing TNF-α (Ehling et al., 2006). In colonic epithelial cells, adiponectin substantially increases the secretion and mRNA levels of IL-8, granulocyte-macrophage colony-stimulating factor and MCP-1, which are all blocked by pharmacological inhibitors of NF-κB, ERK, and p38 MAPK (Ogunwobi & Beales, 2006). In contrast to what has been seen with type 2 diabetes, a significant increase in serum adiponectin levels has been reported for patients with type 1 diabetes (Perseghin et al., 2003). High adiponectin levels in type 1 diabetes are associated with increased levels of IL-6 and CRP, which are negatively correlated with adiponectin in patients with type 2 diabetes (Furuta et al., 2006). Recently, it has been reported that adiponectin rescues salivary gland epithelial cells from spontaneous or interferon-γ-induced apoptosis via activation of AMPK (Katsiougiannis, Tenta, & Skopouli, 2010). Therefore, the modulation of adiponectin production and interaction with IL-6 may depend on the model and tissue investigated and the cause of disease.

6. CONCLUSION

To conclude, plasma levels of adiponectin have been confirmed to be substantially lower in patients with obesity, type 2 diabetes, and CAD than in healthy persons. In addition, hypoadiponectinemia is strongly associated with elevated risk and prevalence of obesity-related diabetic and cardiovascular disorders. Adiponectin exerts favorable effect on energy homeostasis and cardiovascular functions through directly modulating the proinflammatory factor IL-6. However, there are higher than normal levels of adiponectin in patients with inflammatory autoimmune diseases. In those settings, levels of adiponectin expression are positively rather than negatively correlated with levels of IL-6 expression (Fig. 14.3). The interaction between adiponectin

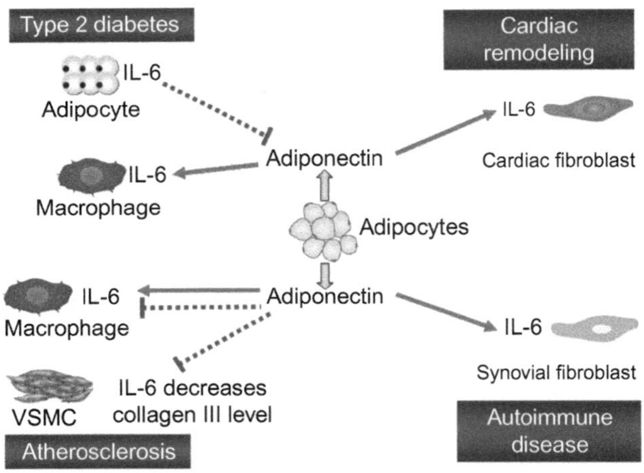

Figure 14.3 The interaction between adiponectin and IL-6 in type 2 diabetes, cardiovascular disease, and autoimmune disorders. IL-6 decreases adiponectin mRNA expression and secretion in adipocytes, suggesting that IL-6 promotes the obesity-related diabetic state by interfering with adiponectin expression. However, some reports suggest that adiponectin induces IL-6 expression in macrophages and transient IL-6 upregulation contributes to improved insulin sensitivity. In obesity-related vascular disorders such as atherosclerosis, IL-6 secretion is found to be inhibited by adiponectin in monocytes and macrophages. However, the fact that globular adiponectin powerfully induces IL-6 secretion in macrophages provides a novel possibility that the "anti-inflammatory" effect of adiponectin may be due to an induction of macrophage tolerance. In vascular smooth muscle cells (VSMCs), adiponectin reverses the inhibitory effect of IL-6 on collagen formation and may play a major role in the development of the thick fibrous cap of atherosclerotic plaques. Adiponectin induces IL-6 production in adult mouse and rat cardiac fibroblasts, which may lead to cardioprotective effects. Plasma adiponectin levels are elevated in autoimmune diseases and adiponectin enhances IL-6 production in human synovial fibroblasts. Arrows, stimulation; dotted lines, inhibition.

and proinflammatory factors is still not completely clear. Moreover, the exact anti-inflammatory or proinflammatory effects of systemic or local adiponectin and IL-6 expression still needs to be further investigated in these different diseases.

REFERENCES

Ajuwon, K. M., & Spurlock, M. E. (2005). Adiponectin inhibits LPS-induced NF-kappaB activation and IL-6 production and increases PPARgamma2 expression in adipocytes. *American Journal of Physiology. Regulatory, Integrative and Comparative Physiology, 288*, R1220–R1225.

Arita, Y., Kihara, S., Ouchi, N., Maeda, K., Kuriyama, H., Okamoto, Y., et al. (2002). Adipocyte-derived plasma protein adiponectin acts as a platelet-derived growth factor-BB binding protein and regulates growth factor-induced common postreceptor signal in vascular smooth muscle cells. *Circulation, 105*, 2893–2898.

Arita, Y., Kihara, S., Ouchi, N., Takahashi, M., Maeda, K., Miyagawa, J., et al. (1999). Paradoxical decrease of an adipose-specific protein, adiponectin, in obesity. *Biochemical and Biophysical Research Communications, 257*, 79–83.

Aso, Y., Yamamoto, R., Wakabayashi, S., Uchida, T., Takayanagi, K., Takebayashi, K., et al. (2006). Comparison of serum high-molecular weight (HMW) adiponectin with total adiponectin concentrations in type 2 diabetic patients with coronary artery disease using a novel enzyme-linked immunosorbent assay to detect HMW adiponectin. *Diabetes, 55*, 1954–1960.

Awazawa, M., Ueki, K., Inabe, K., Yamauchi, T., Kubota, N., Kaneko, K., et al. (2011). Adiponectin enhances insulin sensitivity by increasing hepatic IRS-2 expression via a macrophage-derived IL-6-dependent pathway. *Cell Metabolism, 13*, 401–412.

Banerjee, I., Fuseler, J. W., Intwala, A. R., & Baudino, T. A. (2009). IL-6 loss causes ventricular dysfunction, fibrosis, reduced capillary density, and dramatically alters the cell populations of the developing and adult heart. *American Journal of Physiology. Heart and Circulatory Physiology, 296*, H1694–H1704.

Berg, A. H., Combs, T. P., Du, X. L., Brownlee, M., & Scherer, P. (2001). The adipocyte-secreted protein Acrp30 enhances hepatic insulin action. *Nature Medicine*, 7, 947–953.

Berg, A. H., & Scherer, P. E. (2005). Adipose tissue, inflammation, and cardiovascular disease. *Circulation Research, 96*, 939–949.

Blüher, M., Brennan, A. M., Kelesidis, T., Kratzsch, J., Fasshauer, M., Kralisch, S., et al. (2007). Total and high-molecular weight adiponectin in relation to metabolic variables at baseline and in response to an exercise treatment program: Comparative evaluation of three assays. *Diabetes Care, 30*, 280–285.

Ceddia, R. B., Somwar, R., Maida, A., Fang, X., Bikopoulos, G., & Sweeney, G. (2005). Globular adiponectin increases GLUT4 translocation and glucose uptake but reduces glycogen synthesis in rat skeletal muscle cells. *Diabetologia, 48*, 132–139.

Charles, B. A., Doumatey, A., Huang, H., Zhou, J., Chen, G., Shriner, D., et al. (2011). The roles of IL-6, IL-10, and IL-1RA in obesity and insulin resistance in African-Americans. *The Journal of Clinical Endocrinology and Metabolism, 96*, E2018–E2022.

Chen, H., Montagnani, M., Funahashi, T., Shimomura, I., & Quon, M. J. (2003). Adiponectin stimulates production of nitric oxide in vascular endothelial cells. *The Journal of Biological Chemistry, 278*, 45021–45026.

Chen, M. P., Tsai, J. C., Chung, F. M., Yang, S. S., Hsing, L. L., Shin, S. J., et al. (2005). Hypoadiponectinemia is associated with ischemic cerebrovascular disease. *Arteriosclerosis, Thrombosis, and Vascular Biology, 25*, 821–826.

Cnop, M., Havel, P. J., Utzschneider, K. M., Carr, D. B., Sinha, M. K., Boyko, E. J., et al. (2003). Relationship of adiponectin to body fat distribution, insulin sensitivity and plasma lipoproteins: Evidence for independent roles of age and sex. *Diabetologia, 46*, 459–469.

Combs, T. P., Berg, A. H., Obici, S., Scherer, P. E., & Rossetti, L. (2001). Endogenous glucose production is inhibited by the adipose derived protein Acrp30. *The Journal of Clinical Investigation, 108*, 1875–1881.

Di Gregorio, G. B., Hensley, L., Lu, T., Ranganathan, G., & Kern, P. A. (2004). Lipid and carbohydrate metabolism in mice with a targeted mutation in the IL-6 gene: Absence of development of age-related obesity. *American Journal of Physiology. Endocrinology and Metabolism, 287*, E182–E187.

Diaz, J. A., Booth, A. J., Lu, G., Wood, S. C., Pinsky, D. J., & Bishop, D. K. (2009). Critical role for IL-6 in hypertrophy and fibrosis in chronic cardiac allograft rejection. *American Journal of Transplantation, 9*, 1773–1783.

Ding, M., Carrão, A. C., Wagner, R. J., Xie, Y., Jin, Y., Rzucidlo, E. M., et al. (2012). Vascular smooth muscle cell-derived adiponectin: A paracrine regulator of contractile phenotype. *Journal of Molecular and Cellular Cardiology, 52*, 474–484.

Ehling, A., Schäffler, A., Herfarth, H., Tarner, I. H., Anders, S., Distler, O., et al. (2006). The potential of adiponectin in driving arthritis. *Journal of Immunology, 176*, 4468–4478.

Fain, J. N., Madan, A. K., Hiler, M. L., Cheema, P., & Bahouth, S. W. (2004). Comparison of the release of adipokines by adipose tissue, adipose tissue matrix, and adipocytes from visceral and subcutaneous abdominal adipose tissues of obese humans. *Endocrinology, 145*, 2273–2282.

Fan, D., Li, L., Wang, C., Cui, X. B., Zhou, Y., & Wu, L. L. (2011). Adiponectin induces interleukin-6 production and its underlying mechanism in adult rat cardiac fibroblasts. *Journal of Cellular Physiology, 226*, 1793–1802.

Fantuzzi, G. (2008). Adiponectin and inflammation: Consensus and controversy. *The Journal of Allergy and Clinical Immunology, 121*, 326–330.

Fasshauer, M., Kralisch, S., Klier, M., Lossner, U., Bluher, M., Klein, J., et al. (2003). Adiponectin gene expression and secretion is inhibited by interleukin-6 in 3T3-L1 adipocytes. *Biochemical and Biophysical Research Communications, 301*, 1045–1050.

Fruebis, J., Tsao, T. S., Javorschi, S., Ebbets-Reed, D., Erickson, M. R. S., Yen, F. T., et al. (2001). Proteolytic cleavage product of 30-kDa adipocyte complement-related protein increases fatty acid oxidation in muscle and causes weight loss in mice. *Proceedings of the National Academy of Sciences of the United States of America, 98*, 2005–2010.

Furuta, M., Tamai, M., Hanabusa, T., Yamamoto, Y., Nanjo, K., & Sanke, T. (2006). Serum adiponectin is associated with fasting serum C-peptide in non-obese diabetic patients. *Diabetes Research and Clinical Practice, 72*, 302–307.

Gualillo, O., González-Juanatey, J. R., & Lago, F. (2007). The emerging role of adipokines as mediators of cardiovascular function: Physiologic and clinical perspectives. *Trends in Cardiovascular Medicine, 17*, 275–283.

Hada, Y., Yamauchi, T., Waki, H., Tsuchida, A., Hara, K., Yago, H., et al. (2007). Selective purification and characterization of adiponectin multimer species from human plasma. *Biochemical and Biophysical Research Communications, 356*, 487–493.

Haugen, F., & Drevon, C. A. (2007). Activation of nuclear factor-kappaB by high molecular weight and globular adiponectin. *Endocrinology, 148*, 5478–5486.

Hotta, K., Funahashi, T., Arita, Y., Takahashi, M., Matsuda, M., Okamoto, Y., et al. (2000). Plasma concentrations of a novel, adipose-specific protein, adiponectin, in type 2 diabetic patients. *Arteriosclerosis, Thrombosis, and Vascular Biology, 20*, 1595–1599.

Hotta, K., Funahashi, T., Bodkin, N. L., Ortmeyer, H. K., Arita, Y., Hansen, B. C., et al. (2001). Circulating concentrations of the adipocyte protein adiponectin are decreased in parallel with reduced insulin sensitivity during the progression to type 2 diabetes in rhesus monkeys. *Diabetes, 50*, 1126–1133.

Hu, E., Liang, P., & Spiegelman, B. M. (1996). AdipoQ is a novel adipose-specific gene dysregulated in obesity. *The Journal of Biological Chemistry*, *271*, 10697–10703.

Hug, C., Wang, J., Ahmad, N. S., Bogan, J. S., Tsao, T. S., & Lodish, H. F. (2004). T-cadherin is a receptor for hexameric and high-molecular-weight forms of Acrp30/adiponectin. *Proceedings of the National Academy of Sciences of the United States of America*, *101*, 10308–10313.

Iglseder, B., Mackevics, V., Stadlmayer, A., Tasch, G., Ladurner, G., & Paulweber, B. (2005). Plasma adiponectin levels and sonographic phenotypes of subclinical carotid artery atherosclerosis: Data from the SAPHIR Study. *Stroke*, *36*, 2577–2582.

Iwabu, M., Yamauchi, T., Okada-Iwabu, M., Sato, K., Nakagawa, T., Funata, M., et al. (2010). Adiponectin and AdipoR1 regulate PGC-1alpha and mitochondria by Ca2+ and AMPK/SIRT1. *Nature*, *464*, 1313–1319.

Iwashima, Y., Horio, T., Suzuki, Y., Kihara, S., Rakugi, H., Kangawa, K., et al. (2006). Adiponectin and inflammatory markers in peripheral arterial occlusive disease. *Atherosclerosis*, *188*, 384–390.

Iwashima, Y., Katsuya, T., Ishikawa, K., Ouchi, N., Ohishi, M., Sugimoto, K., et al. (2004). Hypoadiponectinemia is an independent risk factor for hypertension. *Hypertension*, *43*, 1318–1323.

Katsiougiannis, S., Tenta, R., & Skopouli, F. N. (2010). Activation of AMP-activated protein kinase by adiponectin rescues salivary gland epithelial cells from spontaneous and interferon-γ-induced apoptosis. *Arthritis and Rheumatism*, *62*, 414–419.

Kelly, M., Ruderman, N. B., & Tomas, E. (2006). AMP-activated protein kinase and its regulation by adiponectin and interleukin-6. *The Scandinavian Journal of Food and Nutrition*, *50*(Suppl. 2), 85–91.

Klover, P. J., Clementi, A. H., & Mooney, R. A. (2005). Interleukin-6 depletion selectively improves hepatic insulin action in obesity. *Endocrinology*, *146*, 3417–3427.

Klover, P. J., Zimmers, T. A., Koniaris, L. G., & Mooney, R. A. (2003). Chronic exposure to interleukin-6 causes hepatic insulin resistance in mice. *Diabetes*, *52*, 2784–2789.

Kollias, A., Tsiotra, P. C., Ikonomidis, I., Maratou, E., Mitrou, P., Kyriazi, E., et al. (2011). Adiponectin levels and expression of adiponectin receptors in isolated monocytes from overweight patients with coronary artery disease. *Cardiovascular Diabetology*, *10*, 14.

Kubaszek, A., Pihlajamäki, J., Punnonen, K., Karhapää, P., Vauhkonen, I., & Laakso, M. (2003). The C-174G promoter polymorphism of the IL-6 gene affects energy expenditure and insulin sensitivity. *Diabetes*, *52*, 558–561.

Kumada, M., Kihara, S., Sumitsuji, S., Kawamoto, T., Matsumoto, S., Ouchi, N., et al. (2003). Association of hypoadiponectinemia with coronary artery disease in men. *Arteriosclerosis, Thrombosis, and Vascular Biology*, *23*, 85–89.

Lago, F., Dieguez, C., Gómez-Reino, J., & Gualillo, O. (2007a). Adipokines as emerging mediators of immune response and inflammation. *Nature Clinical Practice. Rheumatology*, *3*, 716–724.

Lago, F., Dieguez, C., Gómez-Reino, J., & Gualillo, O. (2007b). The emerging role of adipokines as mediators of inflammation and immune responses. *Cytokine & Growth Factor Reviews*, *18*, 313–325.

Lara-Castro, C., Luo, N., Wallace, P., Klein, R. L., & Garvey, W. T. (2003). Adiponectin multimeric complexes and the metabolic syndrome trait cluster. *Diabetes*, *55*, 249–259.

Li, L., Zhang, K., Cai, X. J., Feng, M., Zhang, Y., & Zhang, M. (2011). Adiponectin upregulates prolyl-4-hydroxylase α1 expression in interleukin 6-stimulated human aortic smooth muscle cells by regulating ERK 1/2 and Sp1. *PloS One*, *6*, e22819.

Liao, W., Yu, C., Wen, J., Jia, W., Li, G., Ke, Y., et al. (2009). Adiponectin induces interleukin-6 production and activates STAT3 in adult mouse cardiac fibroblasts. *Biology of the Cell*, *101*, 263–272.

Lira, F. S., Rosa, J. C., Dos Santos, R. V., Venancio, D. P., Carnier, J., Sanches, P. de.L, et al. (2011). Visceral fat decreased by long-term interdisciplinary lifestyle therapy correlated positively with interleukin-6 and tumor necrosis factor-α and negatively with adiponectin levels in obese adolescents. *Metabolism*, *60*, 359–365.

Maeda, N., Shimomura, I., Kishida, K., Nishizawa, H., Matsuda, M., & Nagaretani, H. (2002). Diet-induced insulin resistance in mice lacking adiponectin/ ACRP30. *Nature Medicine*, *8*, 731–737.

Matsuda, M., Shimomura, I., Sata, M., Arita, Y., Nishida, M., & Maeda, N. (2002). Role of adiponectin in preventing vascular stenosis. The missing link of adipo-vascular axis. *The Journal of Biological Chemistry*, *277*, 37487–37491.

Matsushita, K., Iwanaga, S., Oda, T., Kimura, K., Shimada, M., Sano, M., et al. (2005). Interleukin-6/soluble interleukin-6 receptor complex reduces infarct size via inhibiting myocardial apoptosis. *Laboratory Investigation*, *85*, 1210–1223.

Moschen, A. R., Molnar, C., Geiger, S., Graziadei, I., Ebenbichler, C. F., Weiss, H., et al. (2010). Anti-inflammatory effects of excessive weight loss: Potent suppression of adipose interleukin 6 and tumour necrosis factor alpha expression. *Gut*, *59*, 1259–1264.

Nakamura, Y., Shimada, K., Fukuda, D., Shimada, Y., Ehara, S., Hirose, M., et al. (2004). Implications of plasma concentrations of adiponectin in patients with coronary artery disease. *Heart*, *90*, 528–533.

Nakano, Y., Tobe, T., ChoiMiura, N. H., Mazda, T., & Tomita, M. (1996). Isolation and characterization of GBP28, a novel gelatin-binding protein purified from human plasma. *Journal of Biochemistry*, *120*, 803–812.

Nayak, B. S., Ramsingh, D., Gooding, S., Legall, G., Bissram, S., Mohammed, A., et al. (2010). Plasma adiponectin levels are related to obesity, inflammation, blood lipids and insulin in type 2 diabetic and non-diabetic Trinidadians. *Primary Care Diabetes*, *4*, 187–192.

Neumeier, M., Weigert, J., Schäffler, A., Wehrwein, G., Müller-Ladner, U., & Schölmerich, J. (2006). Different effects of adiponectin isoforms in human monocytic cells. *Journal of Leukocyte Biology*, *79*, 803–808.

Ogunwobi, O. O., & Beales, I. L. (2006). Adiponectin stimulates proliferation and cytokine secretion in colonic epithelial cells. *Regulatory Peptides*, *134*, 105–113.

Okamoto, Y., Kihara, S., Ouchi, N., Nishida, M., Arita, Y., Kumada, M., et al. (2002). Adiponectin reduces atherosclerosis in apolipoprotein E deficient mice. *Circulation*, *106*, 2767–2770.

Otero, M., Lago, R., Gomez, R., Lago, F., Dieguez, C., Gómez-Reino, J. J., et al. (2006). Changes in fat-derived hormones plasma concentrations: Adiponectin, leptin, resistin and visfatin in rheumatoid arthritis subjects. *Annals of the Rheumatic Diseases*, *65*, 1198–1201.

Ouchi, N., Kihara, S., Arita, Y., Maeda, K., Kuriyama, H., Okamoto, Y., et al. (1999). Novel modulator for endothelial adhesion molecules: Adipocyte-derived plasma protein adiponectin. *Circulation*, *100*, 2473–2476.

Ouchi, N., Kihara, S., Arita, Y., Nishida, M., Matsuyama, A., Okamoto, Y., et al. (2001). Adipocyte-derived plasma protein, adiponectin, suppresses lipid accumulation and class A scavenger receptor expression in human monocyte-derived macrophages. *Circulation*, *103*, 1057–1063.

Ouchi, N., Kihara, S., Arita, Y., Okamoto, Y., Maeda, K., Kuriyama, H., et al. (2000). Adiponectin, an adipocyte-derived plasma protein, inhibits endothelial NF-kappa B signaling through a cAMP-dependent pathway. *Circulation*, *102*, 1296–1301.

Ouchi, N., Kobayashi, H., Kihara, S., Kumada, M., Sato, K., Inoue, T., et al. (2004). Adiponectin stimulates angiogenesis by promoting cross-talk between AMP-activated protein kinase and Akt signaling in endothelial cells. *The Journal of Biological Chemistry*, *279*, 1304–1309.

Ouchi, N., & Walsh, K. (2007). Adiponectin as an anti-inflammatory factor. *Clinica Chimica Acta, 380*, 24–30.

Pääkkö, T., Ukkola, O., Ikäheimo, M., & Kesäniemi, Y. A. (2010). Plasma adiponectin levels are associated with left ventricular hypertrophy in a random sample of middle-aged subjects. *Annals of Medicine, 43*, 131–137.

Pedersen, B. K., & Febbraio, M. A. (2008). Muscle as an endocrine organ: Focus on muscle-derived interleukin-6. *Physiological Reviews, 88*, 1379–1406.

Perseghin, G., Lattuada, G., Danna, M., Sereni, L. P., Maffi, P., De Cobelli, F., et al. (2003). Insulin resistance, intramyocellular lipid content, and plasma adiponectin in patients with type 1 diabetes. *American Journal of Physiology. Endocrinology and Metabolism, 285*, E1174–E1181.

Pischon, T., Girman, C. J., Hotamisligil, G. S., Rifai, N., Hu, F. B., & Rimm, E. B. (2004). Plasma adiponectin levels and risk of myocardial infarction in men. *JAMA: The Journal of the American Medical Association, 291*, 1730–1737.

Pradhan, A. D., Manson, J. E., Rifai, N., Buring, J. E., & Ridker, P. M. (2001). C-reactive protein, interleukin 6, and risk of developing type 2 diabetes mellitus. *JAMA: The Journal of the American Medical Association, 286*, 327–334.

Ryo, M., Nakamura, T., Kihara, S., Kumada, M., Shibazaki, S., Takahashi, M., et al. (2004). Adiponectin as a biomarker of the metabolic syndrome. *Circulation Journal, 68*, 975–981.

Sada, K. E., Yamasaki, Y., Maruyama, M., Sugiyama, H., Yamamura, M., Maeshima, Y., et al. (2006). Altered levels of adipocytokines in association with insulin resistance in patients with systemic lupus erythematosus. *The Journal of Rheumatology, 33*, 1545–1552.

Saito, K., Tobe, T., Minoshima, S., Asakawa, S., Sumiya, J., Yoda, M., et al. (1999). Organization of the gene for gelatin binding protein (GBP) 28. *Gene, 229*, 67–73.

Schäffler, A., Ehling, A., Neumann, E., Herfarth, H., Tarner, I., Schölmerich, J., et al. (2003). Adipocytokines in synovial fluid. *JAMA: The Journal of the American Medical Association, 290*, 1709–1710.

Scherer, P. E., Williams, S., Fegliano, M., Baldini, G., & Lodish, H. F. (1995). A novel serum protein similar to C1q produced exclusively in adipocytes. *The Journal of Biological Chemistry, 270*, 26746–26747.

Schulze, M. B., Shai, I., Rimm, E. B., Li, T., Rifai, N., & Hu, F. B. (2005). Adiponectin and future coronary heart disease events among men with type 2 diabetes. *Diabetes, 54*, 534–539.

Senn, J. J., Klover, P. J., Nowak, I. A., Zimmers, T. A., Koniaris, L. G., Furlanetto, R. W., et al. (2003). Suppressor of cytokine signaling-3 (SOCS-3), a potential mediator of interleukin-6-dependent insulin resistance in hepatocytes. *The Journal of Biological Chemistry, 278*, 13740–13746.

Senolt, L., Pavelka, K., Housa, D., & Haluzík, M. (2006). Increased adiponectin is negatively linked to the local inflammatory process in patients with rheumatoid arthritis. *Cytokine, 35*, 247–252.

Shapiro, L., & Scherer, P. E. (1998). The crystal structure of a complement-1q family protein suggests an evolutionary link to tumor necrosis factor. *Current Biology, 8*, 335–340.

Shibata, R., Sato, K., Pimentel, D. R., Takemura, Y., Kihara, S., Ohashi, K., et al. (2005). Adiponectin protects against myocardial ischemia–reperfusion injury through AMPK- and COX-2-dependent mechanisms. *Nature Medicine, 11*, 1096–1103.

Smart, N., Mojet, M. H., Latchman, D. S., Marber, M. S., Duchen, M. R., & Heads, R. J. (2006). IL-6 induces PI 3-kinase and nitric oxide-dependent protection and preserves mitochondrial function in cardiomyocytes. *Cardiovascular Research, 69*, 164–177.

Snehalatha, C., Mukesh, B., Simon, M., Viswanathan, V., Haffner, S. M., & Ramachandran, A. (2003). Plasma adiponectin is an independent predictor of type 2 diabetes in Asian Indians. *Diabetes Care, 26*, 3226–3229.

Stith, R. D., & Luo, J. (1994). Endocrine and carbohydrate responses to interleukin-6 in vivo. *Circulatory Shock*, *44*, 210–215.

Takemura, Y., Ouchi, N., Shibata, R., Aprahamian, T., Kirber, M. T., Summer, R. S., et al. (2007). Adiponectin modulates inflammatory reactions via calreticulin receptor-dependent clearance of early apoptotic bodies. *The Journal of Clinical Investigation*, *117*, 375–386.

Tang, C. H., Chiu, Y. C., Tan, T. W., Yang, R. S., & Fu, W. M. (2007). Adiponectin enhances IL-6 production in human synovial fibroblast via an AdipoR1 receptor, AMPK, p38, and NF-kappa B pathway. *Journal of Immunology*, *179*, 5483–5492.

Tomas, E., Tsao, T. S., Saha, A. K., Murrey, H. E., Zhang, C. C., Itani, S. I., et al. (2002). Enhanced muscle fat oxidation and glucose transport by ACRP30 globular domain: Acetyl-CoA carboxylase inhibition and AMP-activated protein kinase activation. *Proceedings of the National Academy of Sciences of the United States of America*, *99*, 16309–16313.

Toussirot, E., Gaugler, B., Bouhaddi, M., Nguyen, N. U., Saas, P., & Dumoulin, G. (2010). Elevated adiponectin serum levels in women with systemic autoimmune diseases. *Mediators of Inflammation*, *2010*, 938408.

Tsatsanis, C., Zacharioudaki, V., Androulidaki, A., Dermitzaki, E., Charalampopoulos, I., Minas, V., et al. (2005). Adiponectin induces TNF-alpha and IL-6 in macrophages and promotes tolerance to itself and other pro-inflammatory stimuli. *Biochemical and Biophysical Research Communications*, *335*, 1254–1263.

Tsutamoto, T., Hisanaga, T., Wada, A., Maeda, K., Ohnishi, M., Fukai, D., et al. (1998). Interleukin-6 spillover in the peripheral circulation increases with the severity of heart failure, and the high plasma level of interleukin-6 is an important prognostic predictor in patients with congestive heart failure. *Journal of the American College of Cardiology*, *31*, 391–398.

Vozarova, B., Weyer, C., Hanson, K., Tataranni, P. A., Bogardus, C., & Pratley, R. E. (2001). Circulating interleukin-6 in relation to adiposity, insulin action, and insulin secretion. *Obesity Research*, *9*, 414–417.

Wellen, K. E., & Hotamisligil, G. S. (2003). Obesity-induced inflammatory changes in adipose tissue. *The Journal of Clinical Investigation*, *112*, 1785–1788.

Weyer, C., Funahashi, T., Tanaka, S., Hotta, K., Matsuzawa, Y., Pratley, R. E., et al. (2001). Hypoadiponectinemia in obesity and type 2 diabetes: Close association with insulin resistance and hyperinsulinemia. *The Journal of Clinical Endocrinology and Metabolism*, *86*, 1930–1935.

Wolf, A. M., Wolf, D., Rumpold, H., Enrich, B., & Tilg, H. (2004). Adiponectin induces the anti-inflammatory cytokines IL-10 and IL-1RA in human leukocytes. *Biochemical and Biophysical Research Communications*, *323*, 630–635.

Wu, X. D., Motoshima, H., Mahadev, K., Stalker, T. J., Scalia, R., & Goldstein, B. J. (2003). Involvement of AMP-activated protein kinase in glucose uptake stimulated by the globular domain of adiponectin in primary rat adipocyte. *Diabetes*, *52*, 1355–1363.

Wulster-Radcliffe, M. C., Ajuwon, K. M., Wang, J., Christian, J. A., & Spurlock, M. E. (2004). Adiponectin differentially regulates cytokines in porcine macrophages. *Biochemical and Biophysical Research Communications*, *316*, 924–929.

Yamamoto, Y., Hirose, H., Saito, I., Nishikai, K., & Saruta, T. (2004). Adiponectin, an adipocyte-derived protein, predicts future insulin resistance: Two-year follow-up study in Japanese population. *The Journal of Clinical Endocrinology and Metabolism*, *89*, 87–90.

Yamauchi, T., Kamon, J., Ito, Y., Tsuchida, A., Yokomizo, T., Kita, S., et al. (2003). Cloning of adiponectin receptors that mediate antidiabetic metabolic effects. *Nature Medicine*, *423*, 762–769.

Yamauchi, T., Kamon, J., Waki, H., Imai, Y., Shimozawa, N., Hioki, K., et al. (2003). Globular adiponectin protected ob/ob mice from diabetes and ApoE deficient mice from atherosclerosis. *The Journal of Biological Chemistry*, *278*, 2461–2468.
Yamauchin, T., Kamon, J., & Waki, H. (2001). The fat derived hormone adiponectin reverses insulin resistance associated with both lipoatrophy and obesity. *Nature Medicine*, 7, 941–956.

CHAPTER FIFTEEN

New Insights into Anticarcinogenic Properties of Adiponectin: A Potential Therapeutic Approach in Breast Cancer?

Laetitia Delort[*,†,1,2], **Thierry Jardé**[*,†,2], **Virginie Dubois**[*,†], **Marie-Paule Vasson**[*,†,‡], **Florence Caldefie-Chézet**[*,†]

[*]Clermont Université, Université d'Auvergne, UFR Pharmacie, Laboratoire SVFp, 28 place Henri Dunant, F-63000 Clermont-Ferrand

[†]INRA, UMR 1019, ECRIN, CRNH Auvergne, F-63000 Clermont-Ferrand

[‡]Centre Jean Perrin, Unité de Nutrition, F-63003 Clermont-Ferrand

[1]Corresponding author: e-mail address: laetitia.delort@udamail.fr

[2]Both authors contributed equally to the writing of this review.

Contents

Abstract

Obesity is a recognized breast cancer risk factor in postmenopausal women. A recent hypothesis suggests a major role for adipose tissue in carcinogenesis. During many years, the adipose tissue was only considered as a fat storage of energy. This tissue is now described as an endocrine organ secreting a large range of molecules called adipokines. Among these adipokines, adiponectin may play a major role in breast cancer. Plasma adiponectin levels were found to be decreased in cases of breast cancer and in obese patients. Adiponectin may act directly on breast cancer cells by inhibiting

Vitamins and Hormones, Volume 90
ISSN 0083-6729
http://dx.doi.org/10.1016/B978-0-12-398313-8.00015-4

proliferation and angiogenesis or by stimulating apoptosis. Increasing adiponectin levels may be of major importance in the prevention and/or the treatment of breast cancer. This therapeutic approach may be of particular significance for obese patients. The beneficial effects of adiponectin and its possible therapeutic applications will be discussed in this review.

1. INTRODUCTION

Breast cancer is by far the most common cancer among women worldwide, with 1.38 million new cases diagnosed in 2008 and causing 458,000 deaths (Jemal et al., 2011). The incidence of this cancer varies geographically, with a low incidence in developing regions (South America, Africa, and Asia) and a high incidence in developed countries (Unites States and Western Europe) (Ferlay et al., 2010). In parallel, the world overweight population was 937 million (23.2%) in 2005, with 396 million obese patients (9.8%). In many industrialized countries, over one-fifth of the adult population is obese, and this proportion is also increasing in developing countries (Renehan, Soerjomataram, & Leitzmann, 2010). Obesity is now recognized as a breast cancer risk factor in postmenopausal women. The excess of adipose tissue significantly increases the risk of breast cancer by 30–50% (Huang et al., 1997; Trentham-Dietz et al., 1997). Epidemiological studies have shown that obesity is correlated with poor prognosis, larger tumor size, a higher incidence of lymph node metastasis, and a high tumor grade (Calle & Thun, 2004).

Understanding the role of adipose tissue in breast cancer, growth is of major importance particularly in obese patients. Several hypotheses have been suggested to explain the relationship between breast cancer and adiposity. First, the large amount of adipose tissue in obese patients is associated with a high conversion of androgens into estrogens by aromatase and higher circulating levels of insulin and insulin-growth factor (IGF), both signaling pathways being involved in cell proliferation processes (Maccio, Madeddu, & Mantovani, 2009). Second, adipose tissue produces large amounts of factors called adipokines and the excess of fatty tissue strongly alters circulating levels of these hormones. The adipose tissue is now described as an endocrine organ secreting at least 50 adipokines, including adiponectin, that can act by an endocrine, paracrine, or autocrine manner (Trayhurn & Beattie, 2001).

Adiponectin is produced by fat tissue and circulates in the plasma at high concentrations (Ryan et al., 2003). Adiponectin is considered as a protective hormone, as it plays a major role in the regulation of glucose with potent insulin-sensitizing activity affecting the uptake of glucose in the muscle, in lipid homeostasis, and is involved in the pathophysiology of atherosclerosis with anti-inflammatory activity (Waki et al., 2005). Numerous studies have shown an inhibitory activity of adiponectin on the proliferation of various cell types, including aortic smooth muscle cells, endothelial cells, and several types of cancer cells (Arita et al., 2002; Wang et al., 2005; Yokota et al., 2000). The decreased levels of adiponectin observed both in obese and breast cancer patients suggest an inverse correlation between adiponectin levels and breast cancer risk particularly in obese patients, and lead to the hypothesis that this adipokine may be a protective factor against breast cancer.

The aim of this review is to evaluate the anticarcinogenic potential of adiponectin in breast cancer. The mechanisms explaining its protective role, and the therapeutic potential of adiponectin will be considered.

2. ADIPONECTINEMIA

2.1. Structure of adiponectin

Adiponectin, also called Acrp30 (adipocyte complement-related protein of 30 kDa), adipoQ, ApM1, (adipose most abundant gene transcript 1) and GBP28 (28 kDa gelatin binding protein), is encoded by a gene located on chromosome 3q27 which contains three exons and two introns (Saito et al., 1999). Adiponectin, initially identified as expressed and secreted by differentiated murine adipocytes, is a protein of 247 amino acids with a predictive mass of 28 kDa. Adiponectin is mainly produced by adipocytes and is induced over 100-fold during adipocyte differentiation. This is the most abundant circulating protein produced by the adipose tissue and its secretion is enhanced by insulin (Scherer, Williams, Fogliano, Baldini, & Lodish, 1995).

Adiponectin belongs to the complement 1q family and forms a characteristic homomultimer. Adiponectin can be found into five different configurations which have different biological effects: the globular adiponectin (gAPN), full-length adiponectin (fAPN), low-molecular-weight adiponectin, medium-molecular-weight adiponectin, and high-molecular-weight adiponectin (HMW). The globular form of adiponectin is suggested to be formed after proteolytic cleavage by a leucocyte elastase secreted by activated macrophages and/or neutrophils (Waki et al., 2005).

Adiponectin receptor 1 and 2 (AdipoR1 and AdipoR2) are encoded by genes located at chromosome 1p36.13-q41 and 12p13.31, respectively (Yamauchi et al., 2003). AdipoR1 is expressed ubiquitously but is most abundant in squeletal muscle whereas AdipoR2 is predominantly expressed in the liver. These two receptors are integral membrane proteins with seven transmembrane domains with an internal N-terminal collagenous domain and an external C-terminal globular structure. AdipoR1 has high affinity for gAPN whereas AdipoR2 mainly recognizes fAPN (Kadowaki & Yamauchi, 2005).

T-cadherin has also been proposed as an adiponectin receptor for the hexameric and HMW forms of adiponectin but not for the trimeric or globular forms (Hug et al., 2004).

2.2. Adiponectinemia in physiological conditions and in situation of obesity

Adiponectin is highly abundant in the circulation with plasma concentrations ranging from 0.5 to 30 μg/ml and accounting for up to 0.01% of total plasma proteins (Barb, Williams, Neuwirth, & Mantzoros, 2007).

A gender difference of adiponectin levels has been detected in humans. Plasma adiponectin was 35% lower in men than women but no differences were seen between pre- and postmenopausal women (Nishizawa et al., 2002). Lower adiponectin levels were observed in boys compared to girls (6.72 μg/ml vs. 7.30 μg/ml) and these values were inversely correlated with body mass index (BMI), age, height, and insulin-like growth factor 1 (IGF1) concentrations (Bottner et al., 2004). Adiponectin decreased in parallel with the progression through puberty in boys but not in girls. This decrease was negatively correlated with testosterone levels in boys, but no association was noted regarding circulating estradiol in girls. The inverse relationship between adiponectin levels and obesity was more pronounced in adolescent than in prepubertal children (Bottner et al., 2004). Taken together, these results suggest that androgens are able to decrease plasma adiponectin levels through direct effect on adipocytes.

Obese women present low serum adiponectin levels compared to women with normal BMI. A study found that serum adiponectin concentrations were 7.6 μg/ml in women with normal weight, 4.6 μg/ml for overweight women, and 4.7 μg/ml in obese women (Alokail, Al-Daghri, Al-Attas, & Hussain, 2009).

In a recent cross-sectional study, Wu, Yu, Stanczyk, Tseng, and Pike (2011) observed that serum adiponectin concentrations in healthy postmenopausal Asian-American women was 14.7 μg/ml which was inversely correlated to BMI, waist circumference, and waist-to-hip ratio (WHR).

Adiponectin levels are positively associated with sex hormone-binding globulin (SHBG) (Gavrila et al., 2003; Onat et al., 2008; Yasui et al., 2007). A recent study showed a positive association between serum levels of adiponectin and SHBG both before and after adjustment for BMI and WHR. Androgens (total and free testosterone, androstenedione, dehydroepiandrosterone), estrogens (estradiol, estrone), and IGF1 were not correlated with adiponectin levels. Interestingly, free estradiol and IGFBP-3 concentrations were inversely associated with adiponectin levels after adjustment for BMI and WHR (Wu et al., 2011).

2.3. Nutritional modulation of adiponectin production

Several studies have investigated the potential role of dietary factors in modulating adiponectin serum levels. Interestingly, it appeared that a closer adherence to a Mediterranean style dietary pattern was associated with higher adiponectin concentrations, independently of variation in age, adiposity, energy intake, and physical activity (Mantzoros, Williams, Manson, Meigs, & Hu, 2006). Similarly, another study showed that a Mediterranean diet accompanied by increased physical activity significantly increased adiponectin concentration in obese postmenopausal women (Esposito et al., 2003). Moreover, higher intakes of fiber and magnesium have been associated with increased plasma adiponectin concentrations in men (Qi, Rimm, Liu, Rifai, & Hu, 2005).

The concentrations of serum adiponectin were neither modified by a consumption of coffee (caffeinated or decaffeinated) nor black tea. In contrast, an increased level of adiponectin was observed among green tea drinkers with 19% higher levels among those who drank one to three times green tea per week and 31% higher among those who drank four or more per week (Wu et al., 2011).

Soy intake was unrelated to adiponectin level. These observations were maintained after adjustment for BMI and WHR (Wu et al., 2011). But some authors have found that an early dietary exposure to soy protein isolate induced mammary tissue adiponectin production in mouse without corresponding effects on systemic adiponectin levels. These experiments revealed a dietary regulation of "local" mammary adiponectin expression, suggesting that mammary adipose-derived adiponectin may be a mediator of the protective effect of soy protein isolate. Moreover, this dietary induction of adiponectin protein levels in the mammary stromal adipocytes induced ERβ expression in mammary epithelial cells (Rahal & Simmen, 2011).

Osmotin, a pathogenesis-related (PR)-5 family of plant defense proteins that induces apoptosis in yeast, was found to activate AMP kinase in C2C12 myocytes through the adiponectin receptors. Osmotin, an ubiquitous protein found in various fruits and vegetables, is extremely stable and remain active after contact with the human digestive or respiratory systems (Kadowaki & Yamauchi, 2005). But further research has to be conducted.

3. ADIPONECTIN AND BREAST CANCER RISK

3.1. Epidemiological data

Miyoshi et al. (2003) were the first team in 2003 to drive a case–control study investigating the link between adiponectin levels and breast cancer risk. They noted low serum adiponectin levels in breast cancer patients (7.57 ± 0.31 μg/ml) compared to controls (8.83 ± 0.38 μg/ml). After adjustment for the classical risk factors (age, family history, age at menarche, parity, BMI, and age at menopause), women in the lower tertile of serum adiponectin levels were at higher risk of breast cancer (odds ratios (OR) = 3.63, 95% confidence interval (CI) = 1.61–8.19). This effect was more pronounced for menopausal women with an OR equal to 3.90 (95% CI = 1.23–12.44). In addition, low serum adiponectin levels were associated with large tumors and high tumor grade. No relation was found regarding the lymph node status and the estrogen-receptor (ER) status (Miyoshi et al., 2003). Other, case–control studies demonstrated an inverse relationship between serum adiponectin and risk of breast cancer in postmenopausal women but not in premenopausal women (Mantzoros et al., 2004).

Prospective studies, such as the prospective case–control study nested within the Nurses' Health Study (NHS) and NHS II cohorts, also showed an inverse relationship between serum adiponectin levels and risk of breast cancer among postmenopausal women that was independent of estrogen levels but no clear association among premenopausal women was found. The relationship between adiponectin and breast cancer risk was higher for ductal versus lobular carcinomas. A linear inverse association was observed among never hormonal treatment users (relative risk (RR) top vs. bottom quartile = 0.57, 95% CI = 0.35–0.93) but no trend among ever users (RR = 0.90, 95% CI = 0.65–1.25) (Tworoger et al., 2007).

The presence of genetic variants in the adiponectin pathways may account in breast cancer risk. The rs1501299 TG and GG polymorphisms, responsible for a decrease in circulating levels of adiponectin, were associated with a 59% (OR = 1.59; 95% CI = 1.03–2.48) and 80% (OR = 1.80; 95% CI = 1.14–2.85) increased risk for breast cancer, respectively.

The rs2241766 TG polymorphism, which increased adiponectin levels, was associated with a 39% (OR = 0.61; 95% CI = 0.46–0.80) decreased risk for breast cancer (Kaklamani et al., 2008). However, other studies did not find such associations (Nyante et al., 2011; Teras et al., 2009).

In conclusion, epidemiological studies have shown that low serum adiponectin levels were related to an increased risk of breast cancer.

3.2. Adiponectin mRNA and protein expression in *in vivo* and *in vitro* approaches

The quantification of adiponectin mRNA in breast cancer tissue and in adjacent adipose tissue revealed a very low expression of adiponectin in breast tumors, and a 3.3-fold increased expression in adipose tissue compared to control tissue from healthy women (Korner et al., 2007).

Immunohistochemical staining of breast cancer tissues showed that adiponectin was expressed in 15% of invasive ductal carcinoma samples. In contrast, 75% of normal tissues adjacent to breast cancer expressed this adipokine, and the expression was mainly detected in myoepithelial cells known as "natural tumor suppressors" (Jarde et al., 2008).

Some authors did not detect adiponectin in different breast cancer cell lines (MCF-7, MDA-MB-231, T47D, SK-BR-3), normal cell line (HMEC), and in breast tumor samples, but this adipokine was observed in adipose tissue close to the breast tumor (Takahata et al., 2007).

The effect of adiponectin is mediated through its two receptors AdipoR1 and AdipoR2, which are expressed in different breast cell lines such as MCF-7, MDA-MB-231, MDA-MB-245, MDA-MB-436, T47D, and MCF10a. A higher expression of AdipoR1 compared to AdipoR2 was observed (2.7- to 4.2-fold), whereas none of the cell lines exhibited an intrinsic expression of adiponectin (Korner et al., 2007; Nakayama, Miyoshi, Ishihara, & Noguchi, 2008; Pfeiler et al., 2008).

The use of laser microdissection allowed the detection of the two adiponectin receptors mRNA in adipose tissues, all cancer samples, normal epithelial cell samples, and stromal cell samples (except for AdipoR2). The expression of AdipoR1 appeared higher in breast tumor and in adipose adjacent tissue compared to normal tissue. In contrast, AdipoR2 did not seem to be differentially expressed in tumor breast tissue compared to healthy tissue (Korner et al., 2007).

By immunohistochemistry, AdipoR1 was more expressed in breast tumor tissues than in healthy mammary samples with a strong signal in ductal cells. No difference was observed for AdipoR2 (Korner et al., 2007).

Takahata et al. found a membranous and cytoplasmic staining for AdipoR1 and AdipoR2 both on normal and cancer cells. AdipoR1, but not AdipoR2, was observed in stromal cells (Takahata et al., 2007).

3.3. Biological activities of adiponectin in experimental models

The presence of AdipoR1 and AdipoR2 and the very low expression of adiponectin by breast cells may suggest that adiponectin acts on cancer cells in an endocrine manner rather than an autocrine manner. The use of siRNA against AdipoR1 completely abrogated the growth inhibitory of adiponectin treatment in T47D breast cancer cell line and limited to about 50% the proliferation of MDA-MB-231 mammary cancer cells (Nakayama et al., 2008). On the contrary, two studies reported that siRNA against AdipoR1 or AdipoR2 or both had no or little effect on adiponectin-mediated inhibition of the proliferation of MDA-MB-231 (Dos Santos et al., 2008; Wang et al., 2006). So siRNA experiments suggested that the effect of adiponectin may be mediated by other receptors or other alternative biological pathways.

Yamauchi et al. (2003), investigating the cloning of adiponectin receptors, found that they mediated increased AMPK, peroxisome proliferator-activated receptor (PPAR) α ligand activity, the fatty acid oxidation, and glucose uptake by adiponectin. They also observed by targeted experiments that AdipoR1 transmitted the signals mainly through AMPK, while AdipoR2 act through PPARα-related pathways (Yamauchi et al., 2007).

Numerous studies evaluated the activities of adiponectin on cell proliferation, cell cycle, and apoptosis in different breast cancer cell lines such as MCF-7, MDA-MB-231, and T47D (Table 15.1; Fig. 15.1).

3.3.1 Proliferation

A 24-h adiponectin treatment (10 μg/ml) reduced the ERα-positive MCF-7 cell proliferation and this inhibition was observed up to 96 h (Jarde, Perrier, Vasson, & Caldefie-Chezet, 2011), whereas this effect was not observed in a serum free medium (Treeck et al., 2008). Interestingly, adiponectin inhibited the proliferative response induced by a leptin treatment, an adipokine suggested to increase breast cancer risk (Jarde et al., 2011) and to reduce the MCF-7 cell growth in the presence or absence of IGF1 (Li et al., 2011). The treatment of ERα-positive breast cancer cells (T47D) by adiponectin (20 μg/ml for 48 h) decreased the proliferation by 86% in the presence or absence of serum in the medium. This inhibition was not associated with an induction of the apoptotic pathway. However,

Table 15.1 Effects of adiponectin on breast cancer cells *in vitro*

Cell line	Incubation time	Concentration (ng/ml)	Antiproliferative effect	References
MCF-7	4–6 days	10,000	No	Treeck et al. (2008)
	48 h	50–5000	Yes	Grossmann, Nkhata, Mizuno, Ray, and Cleary (2008)
	24/48/72/96 h	10,000	Yes	Jarde et al. (2009)
	24 h	2.5–250	Yes	Dieudonne et al. (2006)
	ND	20,000; 40,000	Yes	Li et al. (2011)
T47D	48 h	20,000	Yes	Korner et al. (2007)
	48 h	200–5000	Yes	Grossmann et al. (2008)
	24/48/72 h	10,000	Yes	Nakayama et al. (2008)
SKBR3	48 h	500–5000	Yes	Grossmann et al. (2008)
MDA-MB-231	48 h	50–5000	No	Grossmann et al. (2008)
	24/48/72 h	10,000	Yes	Nakayama et al. (2008)
	24 h	25	Yes	Dos Santos et al. (2008)
MDA-MB-361	48 h	50–50,000	No	Grossmann et al. (2008)
MCF10a	4–6 days	10,000	Yes	Treeck et al. (2008)
	2–6 days	8000	Yes	Rahal and Simmen (2011)

Cell line	Incubation time	Concentration (ng/ml)	Cell cycle inhibition	References
MDA-MB-231	48 h	15,000	Yes	Wang et al. (2006)
	48 h	10,000	Yes	Nakayama et al. (2008)
T47D	48 h	10,000	Yes	Nakayama et al. (2008)

Cell line	Incubation time	Concentration (ng/ml)	Apoptotic effect	References
MCF-7	96 h	250	Yes	Dieudonne et al. (2006)
	48 h	ND	No	Arditi, Venihaki, Karalis, and Chrousos (2007)
	30 min	5000	Yes	Nkhata, Ray, Schuster, Grossmann, and Cleary (2009)
	ND	20,000; 40,000	Yes	Li et al. (2011)
T47D	48 h	15,000	No	Wang et al. (2006)
	48 h	20,000	No	Korner et al. (2007)
	48 h	10,000	No	Nakayama et al. (2008)
MDA-MB-231	96 h	ND	Yes	Kang et al. (2005)
	48 h	15,000	Yes	Wang et al. (2006)
	24 h	25–250	No	Dos Santos et al. (2008)
	48 h	10,000	No	Nakayama et al. (2008)
MCF10a	12 h	8000	Yes	Rahal and Simmen (2011)

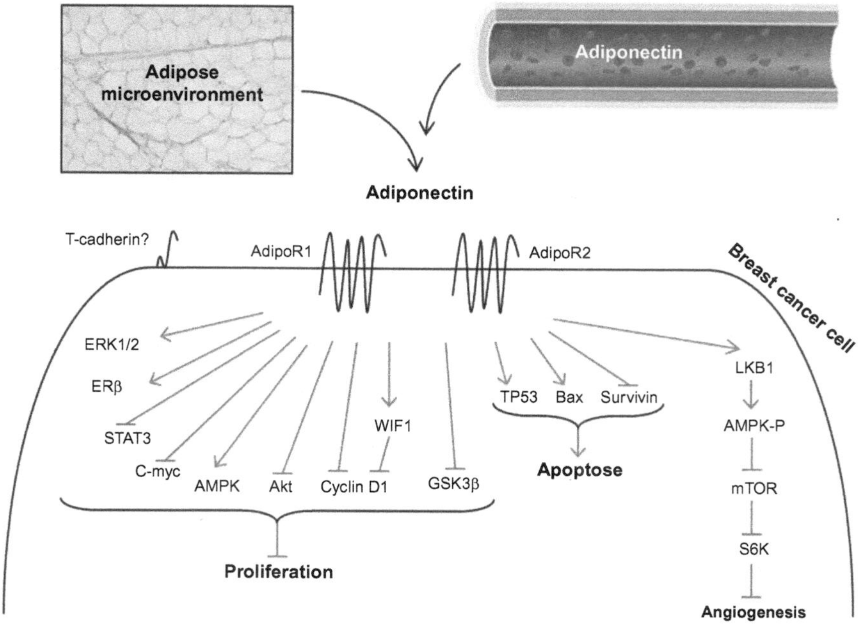

Figure 15.1 Signaling pathways of adiponectin in breast cancer cells. Adiponectin, secreted by adipose cells in the close tumor microenvironment and provided by the circulation, can act on breast cancer cells through its two receptors AdipoR1 and AdipoR2. ERK1/2, extracellular signal-regulated kinases 1/2; ERβ, estrogen-receptor β; STAT3, signal transducer and activator of transcription; AMPK, AMP-activated protein kinase; WIF1, Wnt inhibitory factor-1; GSK3 β, glycogen-synthase kinase-3β; mTOR, mammalian target of rapamycin; S6K, p70 S6 kinase-1. (See Color Insert.)

an activation of ERK1/2 but not AMPK or p38MAPK was noted (Korner et al., 2007). A dose-dependent growth inhibition of ERα-negative MDA-MB-231 proliferation was also reported after adiponectin treatment (10 μg/ml for 24, 48, and 72 h) (Nakayama et al., 2008). Adiponectin attenuated the tumor progression of MDA-MB-231 breast cancer xenograft in athymic nude mice (Wang et al., 2006).

In addition, a recent study showed that the ERα-negative mammary cell line MDA-MB-231 did not respond to adiponectin treatment contrary to the MDA-ERα7 cells produced by the insertion of an ERα transgene into MDA-MB-231 cells (Grossmann et al., 2008). These results suggest a strong relationship between adiponectin signaling and estrogen pathway. Similarly, a transcriptomic analysis demonstrated a downregulation of ERα and aromatase mRNA in adiponectin-treated MCF-7 cells (Jarde et al., 2009).

Treeck et al. (2008) showed that adiponectin inhibited the proliferation of MCF10a mammary epithelial cells. Interestingly, ERβ2 and ERβ5 mRNAs were, respectively, fivefold and a twofold increased in MCF10a cells following adiponectin treatment. High circulating estrogens are known breast cancer risk factors, and epidemiological studies have shown that prolonged exposure to estrogens (early age at menarche, late age at menopause, nulliparity) was associated with a high risk of breast cancer (Collaborative Group on Hormonal Factors in Breast Cancer, 1996). Biological effects of estrogen are mediated through two major receptors, both expressed in 70% of breast tumors. ERα is associated with cell proliferation, and in contrast, ERβ has antiproliferative properties. ERβ2 and ERβ5 are known to form heterodimers with ERα, which results in the inhibition of ERα transcriptional activity (Rahal & Simmen, 2011). The protective role of adiponectin against breast cancer may involve the ERβ pathway.

Because adiponectin decreases the growth of ER-negative MCF10a mammary epithelial cells, this adipokine may inhibit the proliferation in an estrogen-independent manner and notably by decreasing levels of phosphorylated STAT3 (Rahal & Simmen, 2011).

Adiponectin-repressed *c-myc* and *cyclin D1* mRNA expressions suggesting an inhibitory effect of adiponectin on the G1/S cell cycle transition (Dieudonne et al., 2006). Recently, the antiproliferative effect of adiponectin on MCF-7 cells was found to be the result of the significant increase of cells in G(1)/G(0) phase, a concomitant decrease in S phase and with a reduction of cyclin D1 and cyclin E2 expression (Li et al., 2011). However, the expression of cyclin D1 was increased in MCF-7 and T47D after a 24-h adiponectin treatment, suggesting a biphasic effect of adiponectin on cyclin D1 (Grossmann et al., 2008).

Adiponectin-repressed proliferation in breast cancer cells is mediated through an inactivation of p44/42 MAPK protein 1 and 3 expression (Dieudonne et al., 2006; Nkhata et al., 2009; Taliaferro-Smith et al., 2009), a stimulation of AMPK activity involved in the negative control of the cell cycle progression (Dieudonne et al., 2006; Igata et al., 2005; Luo et al., 2005; Rattan, Giri, Singh, & Singh, 2005), and a negative regulation of the Wnt/β-catenin pathway.

In MDA-MB-231 cells, a prolonged exposure to adiponectin blocked serum-induced phosphorylation of Akt and glycogen-synthase kinase-3β, suppressed intracellular accumulation of β-catenin and its nuclear activity, and consequently reduced the expression of cylin D1 (Wang et al., 2006).

Adiponectin may negatively regulate the Wnt/β-catenin pathway in MDA-MB-231 breast cancer cell line. *In vitro* and *in vivo* experiments showed that this effect was mediated through the increase of mRNA and protein expression levels of WIF1 (Wnt inhibitory factor-1), a Wnt antagonist. In MDA-MB-231, overexpression of WIF1 suppressed intracellular accumulation of β-catenin and its nuclear activities, decreased cyclin D1 expression levels and inhibited cell proliferation (Liu et al., 2008).

Similar results were noted for T47D and BT474 breast cancer cell lines.

In MDA-MB-231 tumors implanted in nude mice, the supplementation of both adiponectin and an adenovirus-mediated overexpression of this adipokine substantially enhanced the expression of WIF1 through epigenetic regulations involving the transcription factor specificity protein 1 (Liu et al., 2008).

3.3.2 Apoptosis

A 12-h adiponectin treatment of MCF10a mammary cells increased the number of cells in the sub-G0 (apoptotic) phase, correlated with a decrease in transcript levels of survivin, the antiapoptotic protein normally upregulated in human breast cancer (Rahal & Simmen, 2011). The levels of cleaved caspase-3, caspase-6, and caspase-9 were not altered in MCF-7 cells but adiponectin-activated caspase-8, raising the possibility for this protein to mediate apoptosis through the extrinsic pathway (Grossmann et al., 2008).

Adiponectin induced apoptosis by activating the transcription of proapoptotic genes such as *Bax, TP53*, and *MX1*, and by inhibiting the transcription of antiapoptotic genes such as *Bcl2* and *BAG1*, (Dieudonne et al., 2006; Jarde et al., 2009). However, depending on the cell lines, adiponectin significantly induced apoptosis (Dieudonne et al., 2006; Kang et al., 2005; Nkhata et al., 2009; Wang et al., 2006) or not (Arditi et al., 2007; Dos Santos et al., 2008; Korner et al., 2007; Nakayama et al., 2008).

3.3.3 Angiogenesis

To date, only few studies assessed the antiangiogenic properties of adiponectin in breast carcinogenesis. Adiponectin treatment reduced the invasion and dramatically inhibited the migration of MDA-MB-231 cells. This process was accompanied by a regulation of metastasis-associated genes including the downregulation of the metastasis-promoting gene uPA, and an increase of the expression of TIMP-2, a metastasis-suppressing gene (Wang et al., 2006). Taliaferro-Smith et al. (2009) confirmed the antiangiogenic properties of adiponectin by showing the adiponectin-dependent inhibition of adhesion, invasion, and migration of MCF-7 and T47D breast cancer cells.

Adiponectin inhibited the metastatic properties of breast cancer cells through modulation of the LKB1-AMPK-SK6 axis. Adiponectin stimulated AMPK phosphorylation and activity, resulting in the reduction of mTOR activity, as evidenced by decreased phosphorylation of SK6. Then adiponectin treatment led to an increased expression of tumor suppressor LKB1 (Taliaferro-Smith et al., 2009).

The generation of MMTV-PyVT transgenic mice with a four- to five-fold reduction of adiponectin levels (PyTV(+/−-)/Adiponectin(−/−) vs. PyTV(+/−)/Adiponectin(+/−)) was associated with an earlier tumor onset and an accelerated tumor growth. Moreover, tumor derived from adiponectin haplodeficient mice had enhanced β-catenin nuclear activity and protein stability, partly attributed to the hyperactivation of PI3K/Akt signaling, and the inactivation of PTEN. The coregulation of these signaling pathways is associated with increased breast cancer risk, poor prognosis, and resistance to hormone therapy (Lam et al., 2009).

Surprisingly, a proangiogenic activity of adiponectin toward mammary tumor growth was also described *in vivo*. In adiponectin knockout MMTV-PyTV mice, the rate of tumor growth was decreased at early stages and was followed by a reduction of the angiogenic profile resulting in nutrient deprivation of the tumors and tumor-associated cell death. But in more advanced stages of the tumor, the authors observed that tumors were more aggressive with an increase in circulating endothelial progenitor cells and local VEGF-A production. The loss of adiponectin production in these mice was associated with reduced tumor glucose uptake and increased tumor-associated death. The authors suggested that tumors, after a long-term nutrient deprived environment, may adapt to anaerobic conditions and emerge as very aggressive and metabolically highly active cells at later stages (Landskroner-Eiger et al., 2009).

3.3.4 Autoregulation of adiponectin and impact on other hormones

We recently observed that adiponectin upregulated its own expression and downregulated the expression of leptin and its receptor ObR in MCF-7 breast cancer cells (Jarde et al., 2009). The expression of adiponectin receptors was decreased following a treatment with adiponectin and/or leptin in MDA-MB-231 breast cancer cells (Dos Santos et al., 2008).

In addition, adiponectin may interfere with known pro-proliferative signaling pathways by interacting directly with the ligand (such as FGF and EGF) and preventing the fixation to their receptors (Wang et al., 2005).

A close relationship was also found between adiponectin and lipid metabolism with a downregulation of the acyl-coenzyme A dehydrogenase and the degenerative spermatocyte homologue 1 (Jarde et al., 2009).

4. ADIPONECTIN AS THERAPEUTIC TARGET

Regarding the antiproliferative activity of adiponectin on breast cancer cells, an upregulation of plasma adiponectin, and/or adiponectin receptors, or the development of adiponectin agonists may be of interest in the prevention, and/or treatment of breast cancer, and obesity-associated breast cancer.

As we describe previously, diet is able to modulate adiponectin production and could be of importance in the prevention of such pathology.

PPARγ are highly expressed in adipose tissue but also in benign and malignant breast epithelium. This receptor presents a high affinity for thiazolidinediones which induces insulin-sensitizing effects in animals and in insulin-resistant human subjects (Mueller et al., 1998; Yee et al., 2003). Some studies have shown that the activation of PPARγ signaling pathway by thiazolidinediones inhibited cell proliferation and enhanced tumor cell differentiation (Mueller et al., 1998). Moreover, inhibitory effects of PPARγ treatment were observed in *in vivo* models of mammary tumorigenesis (Elstner et al., 1998; Suh et al., 1999). Interestingly, the expression of adiponectin is mediated through this receptor (Combs et al., 2002). Activating the expression of PPARγ may be an interesting approach to increase adiponectin levels and alter tumor development. To assess this hypothesis, a short-term treatment pilot trial with the thiazolidinedione family member rosiglitazone was conducted in 38 women with early stage breast cancer between the time of diagnostic biopsy and surgery. The treatment increased circulating adiponectin levels but did not alter the tumor cell proliferation assessed by Ki67 immunostaining (Yee et al., 2007). A phase II study using troglitazone in refractory metastatic breast cancer failed to show beneficial effects on disease progression (Burstein et al., 2003). These studies pointed out that a therapy with PPARγ agonists should be conducted on a long-term treatment basis or combined with other nuclear receptor agonists.

The use of adiponectin as a direct therapeutic agent is not yet available because of its expensive production and the difficulty in converting the full-size adiponectin protein into a viable drug. Recently, Otvos et al. proposed another approach consisting in the generation of an adiponectin-based short peptide that mimics adiponectin action which is suitable for clinical development as a therapeutic agent. The identification of a 149–166 amino acid region as the adiponectin active site, and the functional screening of additional 330 peptide analogs covering this region lead to the development of a peptidomimetic called ADP 355. They found that ADP 355 reduced the proliferation of mammary cancer cell lines (MCF-7 and MDA-MB-231) in a dose-dependent manner (at 100 nM–10 μM concentrations). The use of siRNA against AdipoR1 or AdiopR2 confirmed that the effect of ADP 355 was mediated through these two receptors with a greater contribution of AdipoR1. In animal experiments, the intraperitoneal injection of ADP 355 for 28 days suppressed the orthotopic human breast cancer xenograft by 31% (Otvos et al., 2011).

This approach confirms the therapeutic interest of adiponectin agonist, and further studies have to be investigated in order to develop molecules able to induce adiponectin production.

5. CONCLUSION

Obesity is a recognized breast cancer risk factor, and the adipose tissue may play a major role in breast cancer development due to its ability to produce and secrete a lot of molecules called adipokines. Among the most studied adipokines, adiponectin is abundantly produced and secreted by adipocytes. The decreased levels of adiponectin observed both in obese and breast cancer patients leads to the hypothesis that adiponectin may have a protective role against breast cancer development.

The antiproliferative activity of adiponectin on human breast cancer cells and on tumor development has been confirmed by several *in vitro* and *in vivo* studies. Adiponectin effects are mediated through the regulation of multiple pathways, including MAPK, AMPK, Wnt/β-catenin, and ER signaling. Adiponectin can induce apoptosis through the activation of proapoptotic genes such as *Bax*, *TP53*, and *MX1*, and the inhibition of antiapoptotic genes such as *Bcl2* and *BAG1*. Adiponectin is able to reduce the invasion of breast cancer cells *in vitro*. In *in vivo* experiments, contradictory results have been noted with an earlier tumor development and accelerated tumor growth in transgenic mice and on the contrary, very aggressive and metabolically highly active cells at later stages.

Regarding the antiproliferative activities of adiponectin on breast cancer cells, it could be of major interest to develop new approaches in order to increase the concentration of adiponectin and their receptors and to develop adiponectin agonists, permitting to improve the prevention and/or the treatment of breast cancer particularly in obese patients.

REFERENCES

Alokail, M. S., Al-Daghri, N. M., Al-Attas, O. S., & Hussain, T. (2009). Combined effects of obesity and type 2 diabetes contribute to increased breast cancer risk in premenopausal women. *Cardiovascular Diabetology*, *8*, 33.

Arditi, J. D., Venihaki, M., Karalis, K. P., & Chrousos, G. P. (2007). Antiproliferative effect of adiponectin on MCF7 breast cancer cells: A potential hormonal link between obesity and cancer. *Hormone and Metabolic Research*, *39*, 9–13.

Arita, Y., Kihara, S., Ouchi, N., Maeda, K., Kuriyama, H., Okamoto, Y., et al. (2002). Adipocyte-derived plasma protein adiponectin acts as a platelet-derived growth factor-BB-binding protein and regulates growth factor-induced common postreceptor signal in vascular smooth muscle cell. *Circulation*, *105*, 2893–2898.

Barb, D., Williams, C. J., Neuwirth, A. K., & Mantzoros, C. S. (2007). Adiponectin in relation to malignancies: A review of existing basic research and clinical evidence. *The American Journal of Clinical Nutrition*, *86*, s858–s866.

Bottner, A., Kratzsch, J., Muller, G., Kapellen, T. M., Bluher, S., Keller, E., et al. (2004). Gender differences of adiponectin levels develop during the progression of puberty and are related to serum androgen levels. *The Journal of Clinical Endocrinology and Metabolism*, *89*, 4053–4061.

Burstein, H. J., Demetri, G. D., Mueller, E., Sarraf, P., Spiegelman, B. M., & Winer, E. P. (2003). Use of the peroxisome proliferator-activated receptor (PPAR) gamma ligand troglitazone as treatment for refractory breast cancer: A phase II study. *Breast Cancer Research and Treatment*, *79*, 391–397.

Calle, E. E., & Thun, M. J. (2004). Obesity and cancer. *Oncogene*, *23*, 6365–6378.

Collaborative Group on Hormonal Factors in Breast Cancer, (1996). Breast cancer and hormonal contraceptives: Collaborative reanalysis of individual data on 53,297 women with breast cancer and 100,239 women without breast cancer from 54 epidemiological studies. Collaborative Group on Hormonal Factors in Breast Cancer. *The Lancet*, *347*, 1713–1727.

Combs, T. P., Wagner, J. A., Berger, J., Doebber, T., Wang, W. J., Zhang, B. B., et al. (2002). Induction of adipocyte complement-related protein of 30 kilodaltons by PPARgamma agonists: A potential mechanism of insulin sensitization. *Endocrinology*, *143*, 998–1007.

Dieudonne, M. N., Bussiere, M., Dos Santos, E., Leneveu, M. C., Giudicelli, Y., & Pecquery, R. (2006). Adiponectin mediates antiproliferative and apoptotic responses in human MCF7 breast cancer cells. *Biochemical and Biophysical Research Communications*, *345*, 271–279.

Dos Santos, E., Benaitreau, D., Dieudonne, M. N., Leneveu, M. C., Serazin, V., Giudicelli, Y., et al. (2008). Adiponectin mediates an antiproliferative response in human MDA-MB 231 breast cancer cells. *Oncology Reports*, *20*, 971–977.

Elstner, E., Muller, C., Koshizuka, K., Williamson, E. A., Park, D., Asou, H., et al. (1998). Ligands for peroxisome proliferator-activated receptorgamma and retinoic acid receptor inhibit growth and induce apoptosis of human breast cancer cells in vitro and in BNX mice. *Proceedings of the National Academy of Sciences of the United States of America, 95*, 8806–8811.

Esposito, K., Pontillo, A., Di Palo, C., Giugliano, G., Masella, M., Marfella, R., et al. (2003). Effect of weight loss and lifestyle changes on vascular inflammatory markers in obese women: A randomized trial. *JAMA: The Journal of the American Medical Association, 289*, 1799–1804.

Ferlay, J., Shin, H. R., Bray, F., Forman, D., Mathers, C., & Parkin, D. M. (2010). Estimates of worldwide burden of cancer in 2008: GLOBOCAN 2008. *International Journal of Cancer, 127*, 2893–2917.

Gavrila, A., Chan, J. L., Yiannakouris, N., Kontogianni, M., Miller, L. C., Orlova, C., et al. (2003). Serum adiponectin levels are inversely associated with overall and central fat distribution but are not directly regulated by acute fasting or leptin administration in humans: Cross-sectional and interventional studies. *The Journal of Clinical Endocrinology and Metabolism, 88*, 4823–4831.

Grossmann, M. E., Nkhata, K. J., Mizuno, N. K., Ray, A., & Cleary, M. P. (2008). Effects of adiponectin on breast cancer cell growth and signaling. *British Journal of Cancer, 98*, 370–379.

Huang, Z., Hankinson, S. E., Colditz, G. A., Stampfer, M. J., Hunter, D. J., Manson, J. E., et al. (1997). Dual effects of weight and weight gain on breast cancer risk. *JAMA: The Journal of the American Medical Association, 278*, 1407–1411.

Hug, C., Wang, J., Ahmad, N. S., Bogan, J. S., Tsao, T. S., & Lodish, H. F. (2004). T-cadherin is a receptor for hexameric and high-molecular-weight forms of Acrp30/adiponectin. *Proceedings of the National Academy of Sciences of the United States of America, 101*, 10308–10313.

Igata, M., Motoshima, H., Tsuruzoe, K., Kojima, K., Matsumura, T., Kondo, T., et al. (2005). Adenosine monophosphate-activated protein kinase suppresses vascular smooth muscle cell proliferation through the inhibition of cell cycle progression. *Circulation Research, 97*, 837–844.

Jarde, T., Caldefie-Chezet, F., Damez, M., Mishellany, F., Perrone, D., Penault-Llorca, F., et al. (2008). Adiponectin and leptin expression in primary ductal breast cancer and in adjacent healthy epithelial and myoepithelial tissue. *Histopathology, 53*, 484–487.

Jarde, T., Caldefie-Chezet, F., Goncalves-Mendes, N., Mishellany, F., Buechler, C., Penault-Llorca, F., et al. (2009). Involvement of adiponectin and leptin in breast cancer: Clinical and in vitro studies. *Endocrine-Related Cancer, 16*, 1197–1210.

Jarde, T., Perrier, S., Vasson, M. P., & Caldefie-Chezet, F. (2011). Molecular mechanisms of leptin and adiponectin in breast cancer. *European Journal of Cancer, 47*, 33–43.

Jemal, A., Bray, F., Center, M. M., Ferlay, J., Ward, E., & Forman, D. (2011). Global cancer statistics. *CA: A Cancer Journal for Clinicians, 61*, 69–90.

Kadowaki, T., & Yamauchi, T. (2005). Adiponectin and adiponectin receptors. *Endocrine Reviews, 26*, 439–451.

Kaklamani, V. G., Sadim, M., Hsi, A., Offit, K., Oddoux, C., Ostrer, H., et al. (2008). Variants of the adiponectin and adiponectin receptor 1 genes and breast cancer risk. *Cancer Research, 68*, 3178–3184.

Kang, J. H., Lee, Y. Y., Yu, B. Y., Yang, B. S., Cho, K. H., Yoon, D. K., et al. (2005). Adiponectin induces growth arrest and apoptosis of MDA-MB-231 breast cancer cell. *Archives of Pharmacal Research, 28*, 1263–1269.

Korner, A., Pazaitou-Panayiotou, K., Kelesidis, T., Kelesidis, I., Williams, C. J., Kaprara, A., et al. (2007). Total and high-molecular-weight adiponectin in breast cancer: in vitro and in vivo studies. *The Journal of Clinical Endocrinology and Metabolism, 92*, 1041–1048.

Lam, J. B., Chow, K. H., Xu, A., Lam, K. S., Liu, J., Wong, N. S., et al. (2009). Adiponectin haploinsufficiency promotes mammary tumor development in MMTV-PyVT mice by modulation of phosphatase and tensin homolog activities. *PLoS One, 4*, e4968.

Landskroner-Eiger, S., Qian, B., Muise, E. S., Nawrocki, A. R., Berger, J. P., Fine, E. J., et al. (2009). Proangiogenic contribution of adiponectin toward mammary tumor growth in vivo. *Clinical Cancer Research*, *15*, 3265–3276.

Li, G., Cong, L., Gasser, J., Zhao, J., Chen, K., & Li, F. (2011). Mechanisms underlying the anti-proliferative actions of adiponectin in human breast cancer cells, MCF7-dependency on the cAMP/protein kinase-A pathway. *Nutrition and Cancer*, *63*, 80–88.

Liu, J., Lam, J. B., Chow, K. H., Xu, A., Lam, K. S., Moon, R. T., et al. (2008). Adiponectin stimulates Wnt inhibitory factor-1 expression through epigenetic regulations involving the transcription factor specificity protein 1. *Carcinogenesis*, *29*, 2195–2202.

Luo, X. H., Guo, L. J., Yuan, L. Q., Xie, H., Zhou, H. D., Wu, X. P., et al. (2005). Adiponectin stimulates human osteoblasts proliferation and differentiation via the MAPK signaling pathway. *Experimental Cell Research*, *309*, 99–109.

Maccio, A., Madeddu, C., & Mantovani, G. (2009). Adipose tissue as target organ in the treatment of hormone-dependent breast cancer: New therapeutic perspectives. *Obesity Reviews*, *10*, 660–670.

Mantzoros, C., Petridou, E., Dessypris, N., Chavelas, C., Dalamaga, M., Alexe, D. M., et al. (2004). Adiponectin and breast cancer risk. *The Journal of Clinical Endocrinology and Metabolism*, *89*, 1102–1107.

Mantzoros, C. S., Williams, C. J., Manson, J. E., Meigs, J. B., & Hu, F. B. (2006). Adherence to the Mediterranean dietary pattern is positively associated with plasma adiponectin concentrations in diabetic women. *The American Journal of Clinical Nutrition*, *84*, 328–335.

Miyoshi, Y., Funahashi, T., Kihara, S., Taguchi, T., Tamaki, Y., Matsuzawa, Y., et al. (2003). Association of serum adiponectin levels with breast cancer risk. *Clinical Cancer Research*, *9*, 5699–5704.

Mueller, E., Sarraf, P., Tontonoz, P., Evans, R. M., Martin, K. J., Zhang, M., et al. (1998). Terminal differentiation of human breast cancer through PPAR gamma. *Molecular Cell*, *1*, 465–470.

Nakayama, S., Miyoshi, Y., Ishihara, H., & Noguchi, S. (2008). Growth-inhibitory effect of adiponectin via adiponectin receptor 1 on human breast cancer cells through inhibition of S-phase entry without inducing apoptosis. *Breast Cancer Research and Treatment*, *112*, 405–410.

Nishizawa, H., Shimomura, I., Kishida, K., Maeda, N., Kuriyama, H., Nagaretani, H., et al. (2002). Androgens decrease plasma adiponectin, an insulin-sensitizing adipocyte-derived protein. *Diabetes*, *51*, 2734–2741.

Nkhata, K. J., Ray, A., Schuster, T. F., Grossmann, M. E., & Cleary, M. P. (2009). Effects of adiponectin and leptin co-treatment on human breast cancer cell growth. *Oncology Reports*, *21*, 1611–1619.

Nyante, S. J., Gammon, M. D., Kaufman, J. S., Bensen, J. T., Lin, D. Y., Barnholtz-Sloan, J. S., et al. (2011). Common genetic variation in adiponectin, leptin, and leptin receptor and association with breast cancer subtypes. *Breast Cancer Research and Treatment*, *129*, 593–606.

Onat, A., Hergenc, G., Dursunoglu, D., Kucukdurmaz, Z., Bulur, S., & Can, G. (2008). Relatively high levels of serum adiponectin in obese women, a potential indicator of anti-inflammatory dysfunction: Relation to sex hormone-binding globulin. *International Journal of Biological Sciences*, *4*, 208–214.

Otvos, L., Jr., Haspinger, E., La Russa, F., Maspero, F., Graziano, P., Kovalszky, I., et al. (2011). Design and development of a peptide-based adiponectin receptor agonist for cancer treatment. *BMC Biotechnology*, *11*, 90.

Pfeiler, G. H., Buechler, C., Neumeier, M., Schaffler, A., Schmitz, G., Ortmann, O., et al. (2008). Adiponectin effects on human breast cancer cells are dependent on 17-beta estradiol. *Oncology Reports*, *19*, 787–793.

Qi, L., Rimm, E., Liu, S., Rifai, N., & Hu, F. B. (2005). Dietary glycemic index, glycemic load, cereal fiber, and plasma adiponectin concentration in diabetic men. *Diabetes Care*, *28*, 1022–1028.

Rahal, O. M., & Simmen, R. C. (2011). Paracrine-acting adiponectin promotes mammary epithelial differentiation and synergizes with genistein to enhance transcriptional response to estrogen receptor beta signaling. *Endocrinology, 152*, 3409–3421.

Rattan, R., Giri, S., Singh, A. K., & Singh, I. (2005). 5-Aminoimidazole-4-carboxamide-1-beta-D-ribofuranoside inhibits cancer cell proliferation in vitro and in vivo via AMP-activated protein kinase. *The Journal of Biological Chemistry, 280*, 39582–39593.

Renehan, A. G., Soerjomataram, I., & Leitzmann, M. F. (2010). Interpreting the epidemiological evidence linking obesity and cancer: A framework for population-attributable risk estimations in Europe. *European Journal of Cancer, 46*, 2581–2592.

Ryan, A. S., Berman, D. M., Nicklas, B. J., Sinha, M., Gingerich, R. L., Meneilly, G. S., et al. (2003). Plasma adiponectin and leptin levels, body composition, and glucose utilization in adult women with wide ranges of age and obesity. *Diabetes Care, 26*, 2383–2388.

Saito, K., Tobe, T., Minoshima, S., Asakawa, S., Sumiya, J., Yoda, M., et al. (1999). Organization of the gene for gelatin-binding protein (GBP28). *Gene, 229*, 67–73.

Scherer, P. E., Williams, S., Fogliano, M., Baldini, G., & Lodish, H. F. (1995). A novel serum protein similar to C1q, produced exclusively in adipocytes. *The Journal of Biological Chemistry, 270*, 26746–26749.

Suh, N., Wang, Y., Williams, C. R., Risingsong, R., Gilmer, T., Willson, T. M., et al. (1999). A new ligand for the peroxisome proliferator-activated receptor-gamma (PPAR-gamma), GW7845, inhibits rat mammary carcinogenesis. *Cancer Research, 59*, 5671–5673.

Takahata, C., Miyoshi, Y., Irahara, N., Taguchi, T., Tamaki, Y., & Noguchi, S. (2007). Demonstration of adiponectin receptors 1 and 2 mRNA expression in human breast cancer cells. *Cancer Letters, 250*, 229–236.

Taliaferro-Smith, L., Nagalingam, A., Zhong, D., Zhou, W., Saxena, N. K., & Sharma, D. (2009). LKB1 is required for adiponectin-mediated modulation of AMPK-S6K axis and inhibition of migration and invasion of breast cancer cells. *Oncogene, 28*, 2621–2633.

Teras, L. R., Goodman, M., Patel, A. V., Bouzyk, M., Tang, W., Diver, W. R., et al. (2009). No association between polymorphisms in LEP, LEPR, ADIPOQ, ADIPOR1, or ADIPOR2 and postmenopausal breast cancer risk. *Cancer Epidemiology, Biomarkers & Prevention, 18*, 2553–2557.

Trayhurn, P., & Beattie, J. H. (2001). Physiological role of adipose tissue: White adipose tissue as an endocrine and secretory organ. *The Proceedings of the Nutrition Society, 60*, 329–339.

Treeck, O., Lattrich, C., Juhasz-Boess, I., Buchholz, S., Pfeiler, G., & Ortmann, O. (2008). Adiponectin differentially affects gene expression in human mammary epithelial and breast cancer cells. *British Journal of Cancer, 99*, 1246–1250.

Trentham-Dietz, A., Newcomb, P. A., Storer, B. E., Longnecker, M. P., Baron, J., Greenberg, E. R., et al. (1997). Body size and risk of breast cancer. *American Journal of Epidemiology, 145*, 1011–1019.

Tworoger, S. S., Eliassen, A. H., Kelesidis, T., Colditz, G. A., Willett, W. C., Mantzoros, C. S., et al. (2007). Plasma adiponectin concentrations and risk of incident breast cancer. *The Journal of Clinical Endocrinology and Metabolism, 92*, 1510–1516.

Waki, H., Yamauchi, T., Kamon, J., Kita, S., Ito, Y., Hada, Y., et al. (2005). Generation of globular fragment of adiponectin by leukocyte elastase secreted by monocytic cell line THP-1. *Endocrinology, 146*, 790–796.

Wang, Y., Lam, J. B., Lam, K. S., Liu, J., Lam, M. C., Hoo, R. L., et al. (2006). Adiponectin modulates the glycogen synthase kinase-3beta/beta-catenin signaling pathway and attenuates mammary tumorigenesis of MDA-MB-231 cells in nude mice. *Cancer Research, 66*, 11462–11470.

Wang, Y., Lam, K. S., Xu, J. Y., Lu, G., Xu, L. Y., Cooper, G. J., et al. (2005). Adiponectin inhibits cell proliferation by interacting with several growth factors in an oligomerization-dependent manner. *The Journal of Biological Chemistry*, *280*, 18341–18347.

Wu, A. H., Yu, M. C., Stanczyk, F. Z., Tseng, C. C., & Pike, M. C. (2011). Anthropometric, dietary, and hormonal correlates of serum adiponectin in Asian American women. *Nutrition and Cancer*, *63*, 549–557.

Yamauchi, T., Kamon, J., Ito, Y., Tsuchida, A., Yokomizo, T., Kita, S., et al. (2003). Cloning of adiponectin receptors that mediate antidiabetic metabolic effects. *Nature*, *423*, 762–769.

Yamauchi, T., Nio, Y., Maki, T., Kobayashi, M., Takazawa, T., Iwabu, M., et al. (2007). Targeted disruption of AdipoR1 and AdipoR2 causes abrogation of adiponectin binding and metabolic actions. *Nature Medicine*, *13*, 332–339.

Yasui, T., Tomita, J., Miyatani, Y., Yamada, M., Uemura, H., Irahara, M., et al. (2007). Associations of adiponectin with sex hormone-binding globulin levels in aging male and female populations. *Clinica Chimica Acta*, *386*, 69–75.

Yee, L. D., Guo, Y., Bradbury, J., Suster, S., Clinton, S. K., & Seewaldt, V. L. (2003). The antiproliferative effects of PPARgamma ligands in normal human mammary epithelial cells. *Breast Cancer Research and Treatment*, *78*, 179–192.

Yee, L. D., Williams, N., Wen, P., Young, D. C., Lester, J., Johnson, M. V., et al. (2007). Pilot study of rosiglitazone therapy in women with breast cancer: Effects of short-term therapy on tumor tissue and serum markers. *Clinical Cancer Research*, *13*, 246–252.

Yokota, T., Oritani, K., Takahashi, I., Ishikawa, J., Matsuyama, A., Ouchi, N., et al. (2000). Adiponectin, a new member of the family of soluble defense collagens, negatively regulates the growth of myelomonocytic progenitors and the functions of macrophages. *Blood*, *96*, 1723–1732.

CHAPTER SIXTEEN

Adiponectin: A Novel Link Between Adipocytes and COPD

Yoshito Takeda[1], Kaori Nakanishi, Isao Tachibana, Atsushi Kumanogoh

Department of Respiratory Medicine, Allergy and Rheumatic Diseases, Osaka University Graduate School of Medicine, Suita, Osaka, Japan

[1]Corresponding author: e-mail address: yoshito@imed3.med.osaka-u.ac.jp

Contents

Abstract

Adiponectin (APN) is a unique adipokine with multiple salutary effects such as anti-apoptotic, anti-inflammatory, and anti-oxidative activities in numerous organs and cells. Chronic obstructive pulmonary disease (COPD), a growing cause of mortality and morbidity worldwide, often results from the smoking habit and is considered a lifestyle-related disease. COPD is frequently complicated with comorbidities, such as cardiovascular disease, diabetes mellitus, and osteoporosis; however, the molecular mechanisms linking COPD and the associated comorbidities are poorly understood. Recent data have revealed a role for APN in the lung; mice lacking APN spontaneously develop a COPD-like phenotype with extrapulmonary effects, including systemic inflammation, body weight loss, and osteoporosis. This finding highlights the key role of APN in lung pathology and the novel cross talk between lung and adipose tissues. This review summarizes recent advances in understanding the physiological and pathological role of APN in the lung.

Vitamins and Hormones, Volume 90
ISSN 0083-6729
http://dx.doi.org/10.1016/B978-0-12-398313-8.00016-6

1. INTRODUCTION

Adipose tissue communicates with multiple organs via a number of adipokines and influences physiological and pathological processes (Kadowaki et al., 2006; Tilg & Moschen, 2006). Adiponectin (APN) is an adipocyte-specific plasma protein with a structure similar to that of complement factor (C1q) and surfactant proteins (SP-A and SP-D), which play important roles in inflammation and immunity (Yokota et al., 2000). While APN exists in trimers, hexamers, a high-molecular weight form, and a small proteolytic cleavage product, the biological activities of these different forms remain poorly understood (Kadowaki & Yamauchi, 2005). The production of APN is decreased by smoking, genetic factors, and obesity-linked disorders, such as cardiovascular disease (CVD), hyperlipidemia (HL), and diabetes mellitus (DM), indicating that hypoadiponectinemia might underlie the etiology of numerous diseases (Arita et al., 1999; Iwashima et al., 2005; Kadowaki et al., 2006). APN has attracted much attention because of its pleiotropic effects and insulin-sensitizing, anti-inflammatory, and antiapoptotic properties (Kadowaki & Yamauchi, 2005; Kobayashi et al., 2004). Of note, APN-knockout (APN-KO) mice were found to show insulin resistance under a high-fat diet, whereas double deletion of the APN receptor 1 (AdipoR1) and AdipoR2 in mice abrogated APN's metabolic actions (Maeda et al., 2002; Yamauchi et al., 2007). Therefore, APN has been considered a key adipokine in the development of metabolic syndrome.

Although APN has been extensively studied for its anti-diabetic and anti-atherogenic effects in various organs (Tilg & Moschen, 2006), little is known about the role of APN in the lung. Given its anti-protease, anti-inflammatory, and anti-oxidant functions, which have been regarded as pivotal mechanisms for numerous lung diseases, one might predict that APN could also participate in the pathogenesis of lung diseases.

COPD is characterized by poorly reversible limited airflow that is usually progressive and associated with an abnormal inflammatory response of the lungs to noxious particles or gasses, particularly cigarette smoke. COPD is a growing cause of mortality and morbidity worldwide and is predicted to be the third leading cause of death and the fifth most common cause of disability by 2020 (Lopez & Murray, 1998). Recently, increased attention has been paid to the comorbidities associated with COPD, such as CVD, DM, body weight loss, and osteoporosis (Agustí et al., 2003; Fabbri, Luppi, Beghé, & Rabe, 2008), because the majority of COPD patients die of non-respiratory diseases, including CVD and cancer, and other causes (McGarvey et al., 2007). Moreover, epidemiologic studies have confirmed that comorbidities and airflow limitations in COPD patients markedly affect health outcomes (Chatila, Thomashow, Minai, Criner, & Make, 2008; Sin & Man, 2003). Hence, it is of great importance and urgency to identify the mechanism linking COPD and these comorbidities, since doing so may reduce COPD mortality.

Of great interest, APN-KO mice spontaneously showed alveolar enlargement and increased lung compliance, diagnosed as COPD, through age-dependent endothelial dysfunction (Nakanishi et al., 2011). Furthermore, APN-KO mice exhibited not only systemic inflammation but also extrapulmonary phenotypes, such as body weight loss and osteoporosis. Finally, adenovirus-mediated APN supplementation ameliorated the emphysematous phenotype. Taking into consideration the fact that such major phenotypes have never been reported without some stimulus in organs other than the lung, the unexpected phenotype in APN-KO mice highlights an important role for APN in the lungs. Herein, we review the recent findings on APN in lung disease, with a special focus on COPD.

2. STRUCTURE OF APN: SIMILARITY TO SURFACTANT PROTEINS

APN is composed of 244 amino acid residues containing a signal sequence, a collagen-like domain at the N-terminus, and a C1q-like globular domain at the C-terminus (Nishida, Funahashi, & Shimomura, 2007). On the basis of structural analyses of its murine homolog Acrp30, APN is considered to form a homotrimeric subunit with a collagen-like triple-helical structure. These structural features suggest that APN belongs to a family of proteins identified as soluble defense collagens that includes complement C1q and the collectins mannose-binding lectin, lung surfactant protein A (SP-A), and lung surfactant protein D (SP-D). These molecules are thought to function as components of the innate immune system, acting to protect the host by recognizing pathogen-associated molecular patterns on microorganisms (Kuroki, Takahashi, & Nishitanim, 2007). By binding to microorganisms through their globular heads, the collectin family can enhance or initiate phagocytosis and inflammatory effects (Haczku, 2008). Of interest, Takemura et al. (2007) found that like SP-D, APN was capable of opsonizing apoptotic cells (Gardai et al., 2003) and that APN protects the organism from systemic inflammation by promoting the clearance of early apoptotic cells by macrophages through a receptor-dependent pathway involving calreticulin. In addition, SP-D-KO mice developed emphysema with increased inflammation in the lung (Wert et al., 2000) as did APN-KO mice. Because COPD patients showed increased concentrations of SP-D (Sin, Leung, Gan, & Man, 2007), SP-D has recently been regarded as one of the biomarkers for COPD (Sin, Pahlavan, & Man, 2008). Given that SP-D plays a pivotal role in processes, such as infection, lung injury, and immune response in the lung, and that SP-D is produced not only in the lung but also in the other tissues (Snyder et al., 2008), these similarities to APN in function and structure led us to consider whether some interplay might exist between them. Intriguingly, epithelial cells transfected with T-cadherin did show less production of SP-D, indicating that SP-D production was regulated through T-cadherin (Takeuchi, Misaki, Fujita, Sonobe, & Ohtsuki, 2001). Further investigation of the interplay between APN and SP-D might provide some hints to help us understand the unrecognized functions of APN.

3. EXPRESSION AND REGULATION OF APN IN THE LUNG

While it has long been assumed that APN was produced and secreted exclusively by adipose tissue (Maeda et al., 1996), recent reports demonstrated that the hormone is produced by several tissues and cells, such as cardiomyocytes, skeletal muscle, osteoblasts, placenta, endothelium, and pituitary cells (Nishida et al., 2007). It is of note that in skeletal muscle, APN production, which probably is regulated through an oxidation–reduction mechanism, is further enhanced by an inflammatory process; it functions in the repair of muscle damage (Fiaschi et al., 2009). Because APN from various tissues may have different autocrine and paracrine effects, it should be of interest to clarify the role of APN produced by cells other than adipocytes in the pathogenesis of COPD. APN and all of its known receptors (APNR1, AdipoR2, APNR2, and T-cadherin) are expressed on numerous cell types in the lung (Miller, Cho, Pham, Ramsdell, & Broide, 2009; Nakanishi et al., 2011; Summer et al., 2008, 2009). Miller et al. (2009) reported that airway epithelial cells expressed significant levels of APN and APNR1 and that APNR1 on lung epithelial cells released IL-8 after stimulation with APN. On the other hand, in the elastase-induced emphysema model, expression of APNR1 and APNR2 was barely detected in whole lungs before stimulation, whereas they later became upregulated on macrophages and on epithelial cells (Nakanishi et al., 2011).

It should be emphasized that this localization does not necessarily imply that APN is produced in non-adipose tissues and is consistent with adherence of APN to the tissue. In this connection, previous reports suggest that APN specifically binds to collagen types I, III, and V, which are present in the vascular intima, and that APN was localized in the injured vessels but not in the intact ones (Okamoto et al., 2000).

While smoking was known to correlate with hypoadiponectinemia, the dynamics of the partitioning of APN between lung and blood during the progression of COPD has remained poorly understood. In the elastase-induced emphysema model, the concentration of APN in the bronchoalveolar lavage fluid (BALF) and in the plasma has been monitored. The plasma concentration of APN was significantly decreased after stimulation, whereas the concentration of APN in the BALF was first reciprocally increased and later returned to basal levels (Nakanishi et al., 2011). The marked decrease in plasma APN concentration may be associated with the reciprocal increase in APN in the BALF, which could in turn result from enhanced permeability after instillation of elastase. In contrast, Summer et al. reported that while smokers without COPD showed decreased APN in BALF, COPD patients showed increased APN in BALF (Summer et al., 2008). This discrepancy may be due to variations in the species, the timing of the assay, or different types of stimulation, such as tobacco and elastase. Further study will be required to elucidate the dynamics of the distribution of APN between the lung and the blood stream.

While the overall anti-inflammatory properties of APN have been established, the regulation of APN production is complex and poorly understood. For example, hypoadiponectinemia can induce a variety of inflammatory conditions, such as CVD, DM, HL, and renal dysfunction (Kadowaki & Yamauchi, 2005), whereas patients with systemic lupus erythematosus, chronic kidney diseases, and chronic heart failure paradoxically show hyperadiponectinemia (Beige, Heipmann, Stumvoll, Körner, & Kratzsch, 2009; Kistorp et al., 2005; Sada et al., 2006). Similarly, smokers without COPD show hypoadiponectinemia (Abbasi et al., 2006; Iwashima et al., 2005; Kawamoto et al., 2010), whereas hyperadiponectinemia, not hypoadiponectinemia, is observed in patients with COPD (Chan et al., 2010; Krommidas et al., 2010; Tomoda et al., 2007). Indeed, Iwashima et al. (2005) revealed that H_2O_2 and nicotine, both of which are considered to be major oxidation–reduction agents, reduced the mRNA expression and secretion of adiponectin from 3T3-L1 adipocytes. According to the accumulating evidence, stimulation by cigarette smoking is likely to diminish APN production.

In contrast to the fact that hypoadiponectinemia causes a COPD-like phenotype in mice, the dynamics of APN in humans are puzzling. Indeed, APN levels are positively correlated with not only the severity of COPD, but also its exacerbation (Chan et al., 2010; Krommidas et al., 2010). Furthermore, Chan et al. (2010) revealed that plasma APN was inversely correlated with 1-s forced expiratory volume after adjustment for age, body mass index (BMI), and smoking history. These phenomena have been considered to reflect "reverse epidemiology" which refers to alterations in the relations between risk factors and clinical outcomes (Kopple, 2005). So far, there is no widely accepted hypothesis to account for this observation. One possible explanation for this paradox is that while APN secretion can be increased by weight loss, end-stage disease causing cachexia could cause APN upregulation. However, Tomoda et al. (2007) found that the APN level was significantly higher than that in controls even after adjusting for body weight, although the APN level was inversely correlated with BMI. Hence, weight loss could be partially attributable to APN upregulation. Alternatively, alterations in the APN protein that affect multimerization or receptor affinity as well as dysregulation of the APN receptors may attenuate APN activity (so-called APN resistance). Otherwise, hyperadiponectinemia may be the attempt to counterregulate a systemic inflammatory stress in systemic proinflammatory conditions. Further investigation is warranted to elucidate this phenomenon to better understand the significance of APN.

4. COPD-LIKE PHENOTYPE IN APN-KO MICE

Although APN has been extensively studied for its antidiabetic and antiatherogenic effects (Tilg & Moschen, 2006), less is known regarding the role of APN in the lung. We provided some important evidence regarding the involvement of APN in the pathogenesis of COPD. Surprisingly, APN-KO mice showed an enlarged airspace and developed progressive emphysema from week 3 to week 80 (Fig. 16.1) (Nakanishi et al., 2011). In parallel, the chord length, reflecting alveolar enlargement, increased during the same period (Fig. 16.1). Consistently, functional analysis as well as computed tomography measurements revealed a significant increase in lung compliance (Fig. 16.1). In one previous study, however, APN-KO mice showed abnormal alveolarization and increases in both proteolytic and tumor necrosis factor-alpha activity, which resulted in an emphysematous phenotype (Summer et al., 2008). Given that the vascular endothelial

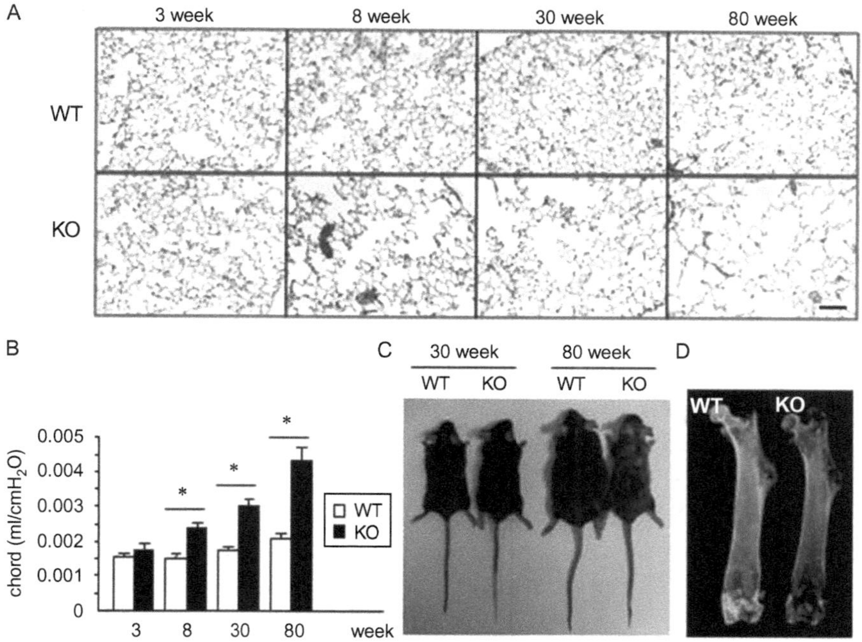

Figure 16.1 APN-KO mice develop a progressive COPD-like phenotype with systemic manifestations. (A) Lung sections from APN-KO mice and WT littermates were stained with hematoxylin and eosin. Scale bar, 100 μm. (B) Lung compliance (c chord) of mice at 3, 8, 30, and 80 weeks of age was measured with a whole body plethysmograph. (C) Shown are whole body images of WT and APN-KO male mice at 30 weeks (left) and 80 weeks (right) of age. (D) Representative radiographs of femurs from WT and APN-KO mice at 80 weeks of age. Bars represent the mean $\pm$ SE; *, $P < 0.05$ versus WT. (See Color Insert.)

growth factor (VEGF) signaling pathway is critical for lung development (Shi, Bellusci, & Warburton, 2007) and that the expression of VEGFR2, a major receptor of VEGF, is downregulated in APN-KO lungs, abnormal development could partially contribute to the pathology of APN-KO lungs. Indeed, a recent paper suggests that a gene necessary for lung development is a candidate for development of COPD (Hersh et al., 2011). The differences in the reported lung phenotypes, however, may originate from differences in the strains, environments, and the KO mice themselves.

COPD can no longer be regarded as a disease of the lungs only, but its association with a wide variety of comorbidities must be considered. Indeed, it has been proposed that COPD be viewed as a chronic systemic inflammatory syndrome (Fabbri & Rabe, 2007). Of importance, APN-KO mice exhibited not only systemic inflammation as assessed by elevated C-reactive protein but also extrapulmonary phenotypes, such as body weight loss and osteoporosis. Although previous papers provide credible but contradictory results about the involvement of APN in the regulation of body weight (Kim et al., 2007; Qi et al., 2004) and bone density (Jürimäe and Jürimäe, 2007; Oshima et al., 2005), it is critical to determine whether deletion of APN itself affects these phenotypes, or whether the COPD-like phenotypes produce these systemic manifestations through spill-over effects from the lung.

As one vascular phenotype reported in the lung, APN-KO mice showed pulmonary hypertension (PH) as determined by both transthoracic echocardiography and hemodynamic measurements (Summer et al., 2009). It is likely that PH in the APN-KO lung would be very mild relative to that which can be detected by physiologic methods alone. Because APN-KO mice spontaneously cause a COPD-like phenotype at a relatively younger age, much care should be taken in interpreting whether deletion of APN causes PH itself or whether the COPD-like phenotype induces the PH. Medoff et al. (2009) also observed that APN-KO mice developed increased allergic inflammation as well as severe pulmonary arterial hypertension in a murine model of chronic asthma. These reports suggest that hypoadiponectinemia is closely associated with PH.

Although several animal models have been reported to show COPD-like phenotypes (Brusselle et al., 2006; Wert et al., 2000; Zhang, Shan, Jiang, Cohn, & Lee, 2006), only a few mice have shown such systemic manifestations as body weight loss and osteoporosis (Hoshino et al., 2007; Kuro-o et al., 1997; Takeda et al., 2008). Because increasing attention is being paid to COPD and its comorbidities, the APN-KO mouse is a valuable model because of its close resemblance to the systemic nature of human COPD.

5. MECHANISMS OF COPD-LIKE PHENOTYPE IN APN-KO MICE

Although the mechanisms of COPD have been extensively studied, much remains to be elucidated (Barnes, Shapiro, & Pauwels, 2003; Yoshida & Tuder, 2007). While the antiprotease and antioxidant functions of APN have been traditionally viewed as key mechanisms in COPD, apoptosis of septal epithelial and endothelial cells has recently been proposed to be part of the pathogenesis of emphysema (Aoshiba, Yokohori, & Nagai, 2003; Kasahara et al., 2000; Tuder, Yoshida, Fijalkowka, Biswal, & Petrache, 2006). Notably, an underlying mechanism to explain the COPD-like phenotype in APN-KO lungs was increased endothelial apoptosis. Consistently, expression of VEGFR-2 and platelet endothelial cell adhesion molecule-1 (PECAM-1) on endothelial cells was decreased in the APN-KO mice. PECAM-1/CD31 is a 130-kDa transmembrane glycoprotein member of the immunoglobulin superfamily expressed on endothelial cells, leukocytes, and platelets and has been recognized to be an efficient signaling molecule with diverse roles in vascular biology including angiogenesis, platelet function, and apoptosis (Newman & Newman, 2003). Of special interest is the absence of PECAM-1-impaired murine alveolarization that occurs as a result of inhibition of endothelial function (DeLisser et al., 2006). Thus, deletion of APN causes endothelial apoptosis, probably through diminished expression of VEGFR-2 and PECAM-1 on lung endothelial cells, which could lead emphysema in APN-KO lungs. So far, VEGF is believed to be an obligatory lung maintenance factor; however, APN may become a new player in lung maintenance.

The importance of endothelial dysfunction during hypoadiponectinemia has been further validated in mice and humans (Antonopoulos, Lee, Margaritis, & Antoniades, 2011; Shimabukuro et al.; 2003 Wong et al., 2011). Given that endothelial dysfunction has been viewed as one of the key mechanisms linking COPD and its comorbidities (Barr, 2011; Maclay et al., 2009), it is pivotal to elucidate how APN ameliorates endothelial dysfunction (Fig. 16.2). Considering the multifunctional nature of APN, however, various alternative mechanisms for the COPD-like phenotype could be involved. First, circulating endothelial progenitor cells are reduced in patients with COPD (Palange et al., 2006). Because APN can upregulate expression of the endothelial cell marker proteins VEGFR-2 and PECAM-1 in cultured progenitor cell populations, thereby promoting the differentiation and function of the endothelial progenitor cells (Shibata et al., 2008), their dysfunction may partially contribute to the pathogenesis of COPD progression in the APN-KO mice. Second, apoptosis and the removal of apoptotic cells, known as efferocytosis, are closely related to the regulation of normal lung structure, while impaired efferocytosis is associated with the pathogenesis of COPD (Vandivier, Henson, & Douglas, 2006). Because APN can facilitate the removal of early apoptotic cells by macrophages and modulate the processes of inflammation and autoimmunity (Takemura et al., 2007), impaired efferocytosis may contribute to the pathogenesis of the COPD-like phenotype in APN-KO mice.

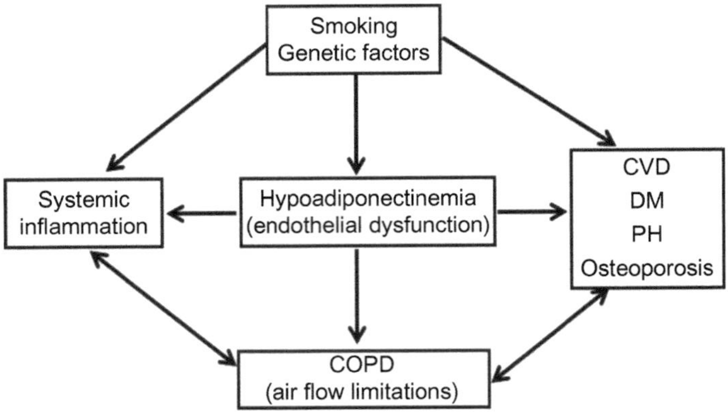

Figure 16.2 Hypothetical schema of the association between hypoadiponectinemia, COPD, and comorbidities. Hypoadiponectinemia is closely associated with smoking, genetic factors, and lifestyle-related disease, such as CVD, DM, and PH. In addition, hypoadiponectinemia causes not only systemic inflammation, but also COPD through endothelial dysfunction. Hypoadiponectinemia could therefore be one of the underlying key mechanisms in COPD and the key link between COPD and various comorbidities.

Third, increasing attention has been paid to involvement of aging in the progression of COPD (MacNee & Tuder, 2009; Tsuji, Aoshiba, & Nagai, 2006), and COPD has recently been regarded as an accelerating lung diseases (Ito & Barnes, 2009). Given that APN transgenic mice have been reported to show longer survival (Otabe et al., 2007) and that centenarians show increased concentrations of APN compared with younger persons (Bik et al., 2006), the phenotype seen in aged APN-KO mice might reflect the signs of premature aging (Nakanishi et al., 2011). Taking into consideration, the possible involvement of APN in the progression of aging, APN could be a novel target for antiaging.

6. THERAPEUTIC POTENTIAL OF APN

Since recently, the primary aim of COPD treatment is to relieve respiratory symptoms, a more comprehensive approach to management of COPD and its comorbidities might provide an opportunity to affect the natural history of COPD (Martinez, Donohue, & Rennard, 2011). Because APN plays an important role in the regulation of metabolic, inflammatory, and apoptotic pathways, it improves insulin sensitivity, promotes vascular health, and increases cell survival. These effects indicate that APN may have a potential therapeutic role not only in treatment of DM, CVD, and cancer, but also in lung disease. Indeed, adenovirus-mediated APN supplementation reverses the progression of the emphysematous phenotype by inhibiting endothelial apoptosis through maintaining the expression of VEGFR-2 and PECAM-1 (Nakanishi et al., 2011). In line with this finding, Schweitzer et al. (2011) reported that treatment with adipose stem cells would ameliorate not only emphysema associated with endothelial cell apoptosis but also extrapulmonary effects, such as body weight loss in the cigarette smoke-induced emphysema model. Notably, they revealed that application of cell-free conditioned medium alone could reproduce the vascular protective effects of adipose stem cells. These beneficial effects may result from the function of APN.

In view of the multifunctional nature of APN, however, it is possible that other entities or phenomena, such as endothelial progenitor cells and oxidant/antioxidant imbalance, may account for APN's beneficial effect. Petrache et al. (2005) reported that increased ceramide, a lipid second messenger, caused emphysema and that inhibition of *de novo* ceramide synthesis blocked the development of emphysema in mice. Recently, APN has been reported to potently stimulate ceramidase activity via activation of its two known receptors, APNR1 and APNR2, thus connecting sphingolipid metabolism with APN functions and providing a potential mechanism explaining the different actions of APN (Holland et al., 2011). Of interest, these authors also identified a pathway involving APN and insulin receptor substrate 2, which is independent of the APNR1/APNR2 pathway. Therefore, these findings suggest that APN and APNRs could regulate a balance between ceramide and S1-P levels, thereby maintaining cellular survival and homeostasis even in the lungs. It should also be noted that APN has been reported to act as a paracrine/autocrine factor during tissue regeneration, as it could be a stem cell factor in hematopoiesis and vasculogenesis, as well as in bone and muscle regeneration (Chiarugi & Fiaschi, 2010).

Taking into consideration, the high abundance, complex tertiary structure, and rapid turnover of APN, however, chronic administration of this protein might be challenging. Although recombinant APN has been used extensively in preclinical models, large-scale production of native forms of APN with all posttranslational modifications and proper assembly into the different complexes is quite difficult. Therefore, an alternative approach would be to identify agents that can enhance the secretion or expression of APN. Among the drugs currently in clinical use, the PPARr agonists and angiotensin receptor blockers (ARBs) offer an inexpensive approach to upregulate the concentration of APN. Intriguingly, statins and ARBs have been shown to possess potent anti-inflammatory properties that positively affect COPD and CVD (Mancini et al., 2006). Because these drugs have been reported to affect APN concentration (Maeda et al., 2001), their beneficial effects may partly result from the influence of APN. Indeed, one of the ARBs has been reported to ameliorate emphysema and improve exercise capacity in the elastase-induced emphysema model (Raupach et al., 2011). Further study will be required, however, to evaluate the potential of APN as a novel therapeutic option.

Finally, it would be intriguing and important to investigate whether APN supplementation rescues the concomitant extrapulmonary effects of COPD, such as body weight loss and systemic inflammation. As stated above, adipose stem cells have been reported to improve body weight loss after smoking exposure; APN might also possess the potential to increase the body weight. Of interest, APN has been reported to increase appetite and reduce energy expenditure by stimulating AMP-activated protein kinase in the hypothalamus (Kubota et al., 2007). Thus, APN would also be able to rescue loss of body weight not only through endothelial maintenance but also through appetite improvement.

Given the pleiotropic properties of APN that include insulin-sensitization and anti-inflammatory and antiapoptotic effects, overcoming hypoadiponectinemia as a treatment could be beneficial not only for COPD but also for its comorbidities (Fig. 16.2).

7. CONCLUSION

The phenotype of APN-KO mice is barely apparent unless metabolically challenged. However, recent findings of the phenotype in APN-KO mice have revealed the unexpected pivotal role of APN in the lung. Given the anti-protease, anti-inflammatory, and anti-oxidant functions of APN, which have been regarded as pivotal mechanisms for numerous lung diseases, further research will be warranted to clarify the role of APN in the lung. While the negative correlation of APN level with multiple aspects of lifestyle-related diseases remains poorly understood, reversing hypoadiponectinemia or overcoming APN resistance via existing drugs or novel therapeutic strategies seems to be promising against various lung diseases including COPD.

REFERENCES

Abbasi, F., Farin, H. M., Lamendola, C., McLaughlin, T., Schwartz, E. A., Reaven, G. M., et al. (2006). The relationship between plasma adiponectin concentration and insulin resistance is altered in smokers. *The Journal of Clinical Endocrinology and Metabolism, 91*(12), 5002–5007.

Agustí, A. G., Noguera, A., Sauleda, J., Sala, E., Pons, J., & Busquets, X. (2003). Systemic effects of chronic obstructive pulmonary disease. *The European Respiratory Journal, 21*(2), 347–360 (Review).

Antonopoulos, A. S., Lee, R., Margaritis, M., & Antoniades, C. (2011). Adiponectin as a regulator of vascular redox state: Therapeutic implications. *Recent Patents on Cardiovascular Drug Discovery, 6*(2), 78–88 (Review).

Aoshiba, K., Yokohori, N., & Nagai, A. (2003). Alveolar wall apoptosis causes lung destruction and emphysematous changes. *American Journal of Respiratory Cell and Molecular Biology, 28*(5), 555–562.

Arita, Y., Kihara, S., Ouchi, N., Takahashi, M., Maeda, K., Miyagawa, J., et al. (1999). Paradoxical decrease of an adipose-specific protein, adiponectin, in obesity. *Biochemical and Biophysical Research Communications, 257*(1), 79–83.

Barnes, P. J., Shapiro, S. D., & Pauwels, R. A. (2003). Chronic obstructive pulmonary disease: Molecular and cellular mechanisms. *The European Respiratory Journal, 22*(4), 672–688 (Review).

Barr, R. G. (2011). The epidemiology of vascular dysfunction relating to chronic obstructive pulmonary disease and emphysema. *Proceedings of the American Thoracic Society, 8*(6), 522–527.

Beige, J., Heipmann, K., Stumvoll, M., Körner, A., & Kratzsch, J. (2009). Paradoxical role for adiponectin in chronic renal diseases? An example of reverse epidemiology. *Expert Opinion on Therapeutic Targets, 13*(2), 163–173 (Review).

Bik, W., Baranowska-Bik, A., Wolinska-Witort, E., Martynska, L., Chmielowska, M., Szybinska, A., et al. (2006). The relationship between adiponectin levels and metabolic status in centenarian, early elderly, young and obese women. *Neuro Endocrinology Letters, 27*(4), 493–500.

Brusselle, G. G., Bracke, K. R., Maes, T., D'hulst, A. I., Moerloose, K. B., Joos, G. F., et al. (2006). Murine models of COPD. *Pulmonary Pharmacology & Therapeutics, 19*(3), 155–165 (Review).

Chan, K. H., Yeung, S. C., Yao, T. J., Ip, M. S., Cheung, A. H., Chan-Yeung, M. M., et al. (2010). Elevated plasma adiponectin levels in patients with chronic obstructive pulmonary disease. *The International Journal of Tuberculosis and Lung Disease, 14*(9), 1193–1200.

Chatila, W. M., Thomashow, B. M., Minai, O. A., Criner, G. J., & Make, B. J. (2008). Comorbidities in chronic obstructive pulmonary disease. *Proceedings of the American Thoracic Society, 5*(4), 549–555 (Review).

Chiarugi, P., & Fiaschi, T. (2010). Adiponectin in health and diseases: From metabolic syndrome to tissue regeneration. *Expert Opinion on Therapeutic Targets, 14*(2), 193–206 (Review).

DeLisser, H. M., Helmke, B. P., Cao, G., Egan, P. M., Taichman, D., Fehrenbach, M., et al. (2006). Loss of PECAM-1 function impairs alveolarization. *The Journal of Biological Chemistry, 281*(13), 8724–8731.

Fabbri, L. M., Luppi, F., Beghé, B., & Rabe, K. F. (2008). Complex chronic comorbidities of COPD. *The European Respiratory Journal, 31*(1), 204–212.

Fabbri, L. M., & Rabe, K. F. (2007). From COPD to chronic systemic inflammatory syndrome? *The Lancet, 370*(9589), 797–799.

Fiaschi, T., Cirelli, D., Comito, G., Gelmini, S., Ramponi, G., Serio, M., et al. (2009). Globular adiponectin induces differentiation and fusion of skeletal muscle cells. *Cell Research, 19*(5), 584–597.

Gardai, S. J., Xiao, Y. Q., Dickinson, M., Nick, J. A., Voelker, D. R., Greene, K. E., et al. (2003). By binding SIRPalpha or calreticulin/CD91, lung collectins act as dual function surveillance molecules to suppress or enhance inflammation. *Cell, 115*(1), 13–23.
Haczku, A. (2008). Protective role of the lung collectins surfactant protein A and surfactant protein D in airway inflammation. *The Journal of Allergy and Clinical Immunology, 122*(5), 861–879.
Hersh, C. P., Silverman, E. K., Gascon, J., Bhattacharya, S., Klanderman, B. J., Litonjua, A. A., et al. (2011). SOX5 is a candidate gene for chronic obstructive pulmonary disease susceptibility and is necessary for lung development. *American Journal of Respiratory and Critical Care Medicine, 183*(11), 1482–1489.
Holland, W. L., Miller, R. A., Wang, Z. V., Sun, K., Barth, B. M., Bui, H. H., et al. (2011). Receptor-mediated activation of ceramidase activity initiates the pleiotropic actions of adiponectin. *Nature Medicine, 17*(1), 55–63.
Hoshino, T., Kato, S., Oka, N., Imaoka, H., Kinoshita, T., Takei, S., et al. (2007). Pulmonary inflammation and emphysema: Role of the cytokines IL-18 and IL-13. *American Journal of Respiratory and Critical Care Medicine, 176*(1), 49–62.
Ito, K., & Barnes, P. J. (2009). COPD as a disease of accelerated lung aging. *Chest, 135*(1), 173–180 (Review).
Iwashima, Y., Katsuya, T., Ishikawa, K., Kida, I., Ohishi, M., Horio, T., et al. (2005). Association of hypoadiponectinemia with smoking habit in men. *Hypertension, 45*(6), 1094–1100.
Jürimäe, J., & Jürimäe, T. (2007). Plasma adiponectin concentration in healthy pre- and post-menopausal women: Relationship with body composition, bone mineral, and metabolic variables. *American Journal of Physiology. Endocrinology and Metabolism, 293*(1), E42–E47.
Kadowaki, T., & Yamauchi, T. (2005). Adiponectin and adiponectin receptors. *Endocrine Reviews, 26*(3), 439–451 (Review).
Kadowaki, T., Yamauchi, T., Kubota, N., Hara, K., Ueki, K., & Tobe, K. J. (2006). Adiponectin and adiponectin receptors in insulin resistance, diabetes, and the metabolic syndrome. *The Journal of Clinical Investigation, 116*(7), 1784–1792 (Review).
Kasahara, Y., Tuder, R. M., Taraseviciene-Stewart, L., Le Cras, T. D., Abman, S., Hirth, P. K., et al. (2000). Inhibition of VEGF receptors causes lung cell apoptosis and emphysema. *The Journal of Clinical Investigation, 106*(11), 1311–1319.
Kawamoto, R., Tabara, Y., Kohara, K., Miki, T., Ohtsuka, N., Kusunoki, T., et al. (2010). Smoking status is associated with serum high molecular adiponectin levels in community-dwelling Japanese men. *Journal of Atherosclerosis and Thrombosis, 17*(4), 423–430.
Kim, J. Y., van de Wall, E., Laplante, M., Azzara, A., Trujillo, M. E., Hofmann, S. M., et al. (2007). Obesity-associated improvements in metabolic profile through expansion of adipose tissue. *The Journal of Clinical Investigation, 117*(9), 2621–2637.
Kistorp, C., Faber, J., Galatius, S., Gustafsson, F., Frystyk, J., Flyvbjerg, A., et al. (2005). Plasma adiponectin, body mass index, and mortality in patients with chronic heart failure. *Circulation, 112*(12), 1756–1762.
Kobayashi, H., Ouchi, N., Kihara, S., Walsh, K., Kumada, M., Abe, Y., et al. (2004). Selective suppression of endothelial cell apoptosis by the high molecular weight form of adiponectin. *Circulation Research, 94*(4), e27–e31.
Kopple, J. D. (2005). The phenomenon of altered risk factor patterns or reverse epidemiology in persons with advanced chronic kidney failure. *The American Journal of Clinical Nutrition, 81*(6), 1257–1266.
Krommidas, G., Kostikas, K., Papatheodorou, G., Koutsokera, A., Gourgoulianis, K. I., Roussos, C., et al. (2010). Plasma leptin and adiponectin in COPD exacerbations: Associations with inflammatory biomarkers. *Respiratory Medicine, 104*(1), 40–46.
Kubota, N., Yano, W., Kubota, T., Yamauchi, T., Itoh, S., Kumagai, H., et al. (2007). Adiponectin stimulates AMP-activated protein kinase in the hypothalamus and increases food intake. *Cell Metabolism, 6*(1), 55–68.

Kuroki, Y., Takahashi, M., & Nishitanim, C. (2007). Pulmonary collectins in innate immunity of the lung. *Cellular Microbiology*, *9*(8), 1871–1879.

Kuro-o, M., Matsumura, Y., Aizawa, H., Kawaguchi, H., Suga, T., Utsugi, T., et al. (1997). Mutation of the mouse klotho gene leads to a syndrome resembling ageing. *Nature*, *390* (6655), 45–51.

Lopez, A. D., & Murray, C. C. (1998). The global burden of disease, 1990-2020. *Nature Medicine*, *4*(11), 1241–1243.

Maclay, J. D., McAllister, D. A., Mills, N. L., Paterson, F. P., Ludlam, C. A., Drost, E. M., et al. (2009). Vascular dysfunction in chronic obstructive pulmonary disease. *American Journal of Respiratory and Critical Care Medicine*, *180*(6), 513–520.

MacNee, W., & Tuder, R. M. (2009). New paradigms in the pathogenesis of chronic obstructive pulmonary disease I. *Proceedings of the American Thoracic Society*, *6*(6), 527–531 (Review).

Maeda, K., Okubo, K., Shimomura, I., Funahashi, T., Matsuzawa, Y., & Matsubara, K. (1996). cDNA cloning and expression of a novel adipose specific collagen-like factor, apM1 (AdiPose Most abundant Gene transcript 1). *Biochemical and Biophysical Research Communications*, *221*(2), 286–289.

Maeda, N., Shimomura, I., Kishida, K., Nishizawa, H., Matsuda, M., Nagaretani, H., et al. (2002). Diet-induced insulin resistance in mice lacking adiponectin/ACRP30. *Nature Medicine*, *8*(7), 731–737.

Maeda, N., Takahashi, M., Funahashi, T., Kihara, S., Nishizawa, H., Kishida, K., et al. (2001). PPARgamma ligands increase expression and plasma concentrations of adiponectin, an adipose-derived protein. *Diabetes*, *50*(9), 2094–2099.

Mancini, G. B., Etminan, M., Zhang, B., Levesque, L. E., FitzGerald, J. M., & Brophy, J. M. (2006). Reduction of morbidity and mortality by statins, angiotensin-converting enzyme inhibitors, and angiotensin receptor blockers in patients with chronic obstructive pulmonary disease. *Journal of the American College of Cardiology*, *47*(12), 2554–2560.

Martinez, F. J., Donohue, J. F., & Rennard, S. I. (2011). The future of chronic obstructive pulmonary disease treatment—Difficulties of and barriers to drug development. *The Lancet*, *378*(9795), 1027–1037 (Review).

McGarvey, L. P., John, M., Anderson, J. A., Zvarich, M., Wise, R. A. & TORCH Clinical Endpoint Committee (2007). Ascertainment of cause-specific mortality in COPD: Operations of the TORCH Clinical Endpoint Committee. *Thorax*, *62*(5), 411–415.

Medoff, B. D., Okamoto, Y., Leyton, P., Weng, M., Sandall, B. P., Raher, M. J., et al. (2009). Adiponectin deficiency increases allergic airway inflammation and pulmonary vascular remodeling. *American Journal of Respiratory Cell and Molecular Biology*, *41*(4), 397–406.

Miller, M., Cho, J. Y., Pham, A., Ramsdell, J., & Broide, D. H. (2009). Adiponectin and functional adiponectin receptor 1 are expressed by airway epithelial cells in chronic obstructive pulmonary disease. *Journal of Immunology*, *182*(1), 684–691.

Nakanishi, K., Takeda, Y., Tetsumoto, S., Iwasaki, T., Tsujino, K., Kuhara, H., et al. (2011). Involvement of endothelial apoptosis underlying chronic obstructive pulmonary disease-like phenotype in adiponectin-null mice: Implications for therapy. *American Journal of Respiratory and Critical Care Medicine*, *183*(9), 1164–1175.

Newman, P. J., & Newman, D. K. (2003). Signal transduction pathways mediated by PECAM-1: New roles for an old molecule in platelet and vascular cell biology. *Arteriosclerosis, Thrombosis, and Vascular Biology*, *23*(6), 953–964 (Review).

Nishida, M., Funahashi, T., & Shimomura, I. (2007). Pathophysiological significance of adiponectin. *Medical Molecular Morphology*, *40*(2), 55–67 (Review).

Okamoto, Y., Arita, Y., Nishida, M., Muraguchi, M., Ouchi, N., Takahashi, M., et al. (2000). An adipocyte-derived plasma protein, adiponectin, adheres to injured vascular walls. *Hormone and Metabolic Research*, *32*(2), 47–50.

Oshima, K., Nampei, A., Matsuda, M., Iwaki, M., Fukuhara, A., Hashimoto, J., et al. (2005). Adiponectin increases bone mass by suppressing osteoclast and activating osteoblast. *Biochemical and Biophysical Research Communications*, *331*(2), 520–526.

Otabe, S., Yuan, X., Fukutani, T., Wada, N., Hashinaga, T., Nakayama, H., et al. (2007). Overexpression of human adiponectin in transgenic mice results in suppression of fat accumulation and prevention of premature death by high-calorie diet. *American Journal of Physiology. Endocrinology and Metabolism*, *293*(1), E210–E218.

Palange, P., Testa, U., Huertas, A., Calabrò, L., Antonucci, R., Petrucci, E., et al. (2006). Circulating haemopoietic and endothelial progenitor cells are decreased in COPD. *The European Respiratory Journal*, *27*(3), 529–541.

Petrache, I., Natarajan, V., Zhen, L., Medler, T. R., Richter, A. T., Cho, C., et al. (2005). Ceramide upregulation causes pulmonary cell apoptosis and emphysema-like disease in mice. *Nature Medicine*, *11*(5), 491–498.

Qi, Y., Takahashi, N., Hileman, S. M., Patel, H. R., Berg, A. H., Pajvani, U. B., et al. (2004). Adiponectin acts in the brain to decrease body weight. *Nature Medicine*, *10*(5), 524–529.

Raupach, T., Lüthje, L., Kögler, H., Duve, C., Schweda, F., Hasenfuss, G., et al. (2011). Local and systemic effects of angiotensin receptor blockade in an emphysema mouse model. *Pulmonary Pharmacology & Therapeutics*, *24*(2), 215–220.

Sada, K. E., Yamasaki, Y., Maruyama, M., Sugiyama, H., Yamamura, M., Maeshima, Y., et al. (2006). Altered levels of adipocytokines in association with insulin resistance in patients with systemic lupus erythematosus. *The Journal of Rheumatology*, *33*(8), 1545–1552.

Schweitzer, K. S., Johnstone, B. H., Garrison, J., Rush, N. I., Cooper, S., Traktuev, D. O., et al. (2011). Adipose stem cell treatment in mice attenuates lung and systemic injury induced by cigarette smoking. *American Journal of Respiratory and Critical Care Medicine*, *183*(2), 215–225.

Shi, W., Bellusci, S., & Warburton, D. (2007). Lung development and adult lung diseases. *Chest*, *132*(2), 651–656 (Review).

Shibata, R., Skurk, C., Ouchi, N., Galasso, G., Kondo, K., Ohashi, T., et al. (2008). Adiponectin promotes endothelial progenitor cell number and function. *FEBS Letters*, *582*(11), 1607–1612.

Shimabukuro, M., Higa, N., Asahi, T., Oshiro, Y., Takasu, N., Tagawa, T., et al. (2003). Hypoadiponectinemia is closely linked to endothelial dysfunction in man. *The Journal of Clinical Endocrinology and Metabolism*, *88*(7), 3236–3240.

Sin, D. D., Leung, R., Gan, W. Q., & Man, S. P. (2007). Circulating surfactant protein D as a potential lung-specific biomarker of health outcomes in COPD: A pilot study. *BMC Pulmonary Medicine*, 7, 13.

Sin, D. D., & Man, S. F. (2003). Why are patients with chronic obstructive pulmonary disease at increased risk of cardiovascular diseases? The potential role of systemic inflammation in chronic obstructive pulmonary disease. *Circulation*, *107*(11), 1514–1519.

Sin, D. D., Pahlavan, P. S., & Man, S. F. (2008). Surfactant protein D: A lung specific biomarker in COPD? *Therapeutic Advances in Respiratory Disease*, *2*(2), 65–74.

Snyder, G. D., Oberley-Deegan, R. E., Goss, K. L., Romig-Martin, S. A., Stoll, L. L., Snyder, J. M., et al. (2008). Surfactant protein D is expressed and modulates inflammatory responses in human coronary artery smooth muscle cells. *American Journal of Physiology. Heart and Circulatory Physiology*, *294*(5), H2053–H2059.

Summer, R., Fiack, C. A., Ikeda, Y., Sato, K., Dwyer, D., Ouchi, N., et al. (2009). Adiponectin deficiency: A model of pulmonary hypertension associated with pulmonary vascular disease. *American Journal of Physiology. Lung Cellular and Molecular Physiology*, *297*(3), L432–L438.

Summer, R., Little, F. F., Ouchi, N., Takemura, Y., Aprahamian, T., Dwyer, D., et al. (2008). Alveolar macrophage activation and an emphysema-like phenotype in adiponectin-deficient mice. *American Journal of Physiology. Lung Cellular and Molecular Physiology*, *294*(6), L1035–L1042.

Takeda, Y., He, P., Tachibana, I., Zhou, B., Miyado, K., Kaneko, H., et al. (2008). Double deficiency of tetraspanins CD9 and CD81 alters cell motility and protease production of macrophages and causes chronic obstructive pulmonary disease-like phenotype in mice. *The Journal of Biological Chemistry*, *283*(38), 26089–26097.

Takemura, Y., Ouchi, N., Shibata, R., Aprahamian, T., Kirber, M. T., Summer, R. S., et al. (2007). Adiponectin modulates inflammatory reactions via calreticulin receptor-dependent clearance of early apoptotic bodies. *The Journal of Clinical Investigation*, *117* (2), 375–386.

Takeuchi, T., Misaki, A., Fujita, J., Sonobe, H., & Ohtsuki, Y. (2001). T-cadherin (CDH13, H-cadherin) expression downregulated surfactant protein D in bronchioloalveolar cells. *Virchows Archiv*, *438*(4), 370–375.

Tilg, H., & Moschen, A. R. (2006). Adipocytokines: Mediators linking adipose tissue, inflammation and immunity. *Nature Reviews. Immunology*, *6*(10), 772–783 (Review).

Tomoda, K., Yoshikawa, M., Itoh, T., Tamaki, S., Fukuoka, A., Komeda, K., et al. (2007). Elevated circulating plasma adiponectin in underweight patients with COPD. *Chest*, *132* (1), 135–140.

Tsuji, T., Aoshiba, K., & Nagai, A. (2006). Alveolar cell senescence in patients with pulmonary emphysema. *American Journal of Respiratory and Critical Care Medicine*, *174*(8), 886–893.

Tuder, R. M., Yoshida, T., Fijalkowka, I., Biswal, S., & Petrache, I. (2006). Role of lung maintenance program in the heterogeneity of lung destruction in emphysema. *Proceedings of the American Thoracic Society*, *3*(8), 673–679 (Review).

Vandivier, R. W., Henson, P. M., & Douglas, I. S. (2006). Burying the dead: The impact of failed apoptotic cell removal (efferocytosis) on chronic inflammatory lung disease. *Chest*, *129*(6), 1673–1682 (Review).

Wert, S. E., Yoshida, M., LeVine, A. M., Ikegami, M., Jones, T., Ross, G. F., et al. (2000). Increased metalloproteinase activity, oxidant production, and emphysema in surfactant protein D gene-inactivated mice. *Proceedings of the National Academy of Sciences of the United States of America*, *97*(11), 5972–5977.

Wong, W. T., Tian, X. Y., Xu, A., Yu, J., Lau, C. W., Hoo, R. L., et al. (2011). Adiponectin is required for PPARγ-mediated improvement of endothelial function in diabetic mice. *Cell Metabolism*, *14*(1), 104–115.

Yamauchi, T., Nio, Y., Maki, T., Kobayashi, M., Takazawa, T., Iwabu, M., et al. (2007). Targeted disruption of AdipoR1 and AdipoR2 causes abrogation of adiponectin binding and metabolic actions. *Nature Medicine*, *13*(3), 332–339.

Yokota, T., Oritani, K., Takahashi, I., Ishikawa, J., Matsuyama, A., Ouchi, N., et al. (2000). Adiponectin, a new member of the family of soluble defense collagens, negatively regulates the growth of myelomonocytic progenitors and the functions of macrophages. *Blood*, *96*(5), 1723–1732.

Yoshida, T., & Tuder, R. M. (2007). Pathobiology of cigarette smoke-induced chronic obstructive pulmonary disease. *Physiological Reviews*, *87*(3), 1047–1082.

Zhang, X., Shan, P., Jiang, G., Cohn, L., & Lee, P. J. (2006). Toll-like receptor 4 deficiency causes pulmonary emphysema. *The Journal of Clinical Investigation*, *116*(11), 3050–3059.

INDEX

Note: Page numbers followed by "*f*" indicate figures, and "*t*" indicate tables.

B

C

D

E

H

J

L

M

N

V

W

X

Y

Z

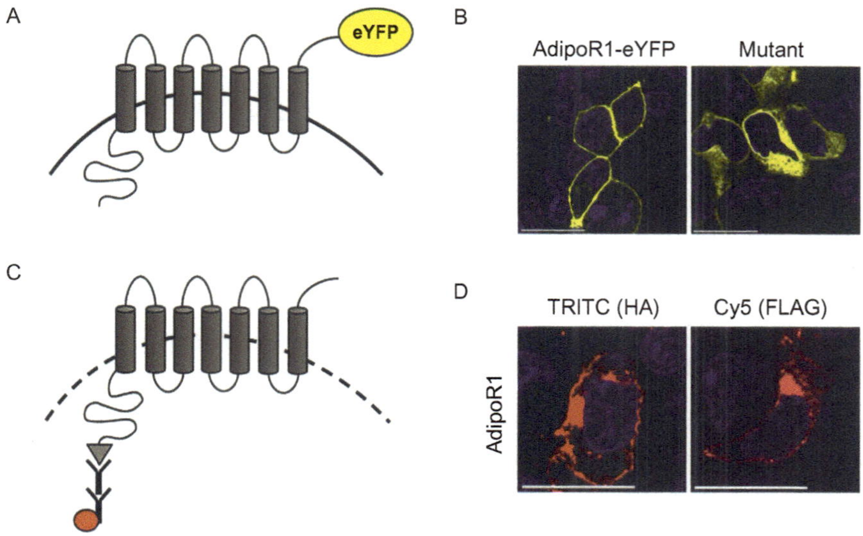

Cathleen Juhl and Annette G. Beck-Sickinger, Figure 2.2 Visualization of AdipoR. (A) AdipoR fusion to eYFP enables direct monitoring in living cells. (B) Live-cell imaging of the AdipoR1 and a receptor mutant which is located intracellularly. (C) Immunostaining of the AdipoR using fluorophore-conjugated antibodies. (D) Immunostaining of HA-AdipoR1 and Flag-AdipoR1 using TRITC- and Cy5-conjugated antibodies, respectively.

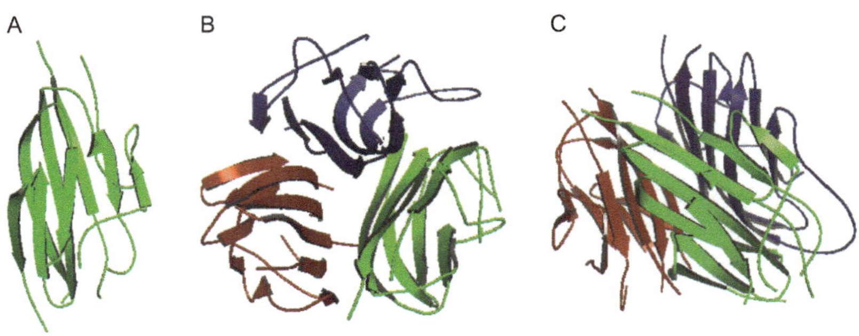

Cathleen Juhl and Annette G. Beck-Sickinger, Figure 2.3 Structure of the murine gAd based on PDB entry 1C28. (A) Ribbon model of the gAd monomer. (B) Top view of the bell-shaped gAd trimer. The chains a, b, and c are colored in green, blue, and red, respectively. (C) Side view of the gAd trimer. All structures were created with the program PyMOL.

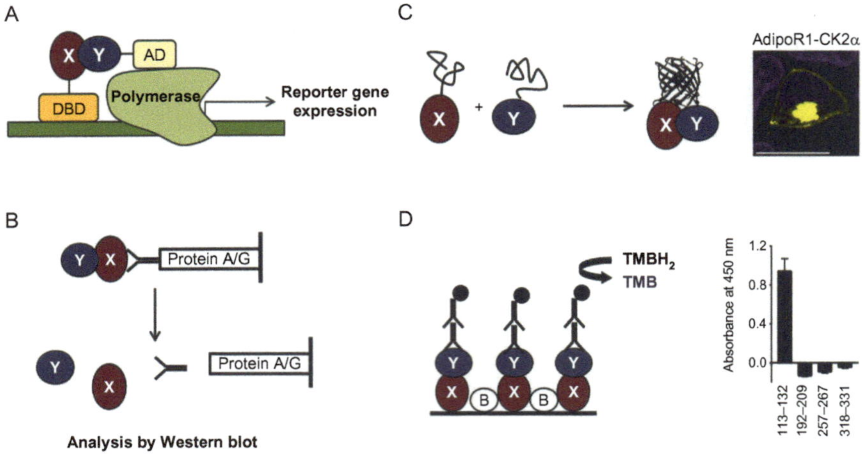

Cathleen Juhl and Annette G. Beck-Sickinger, Figure 2.4 Most common approaches to detect AdipoR interactions (X AdipoR, Y interacting protein). (A) Y2H systems allow the identification of novel interaction partners (DBD DNA-binding domain, AD activation domain). (B) Coimmunoprecipitation of AdipoR complexes. (C) BiFC experiment and visualization of the AdipoR1 interaction with CK2α (Juhl, Mörl, & Beck-Sickinger, 2011). (D) ELISA and identification of the interaction site between the AdipoR1 and CK2β (Heiker et al., 2009).

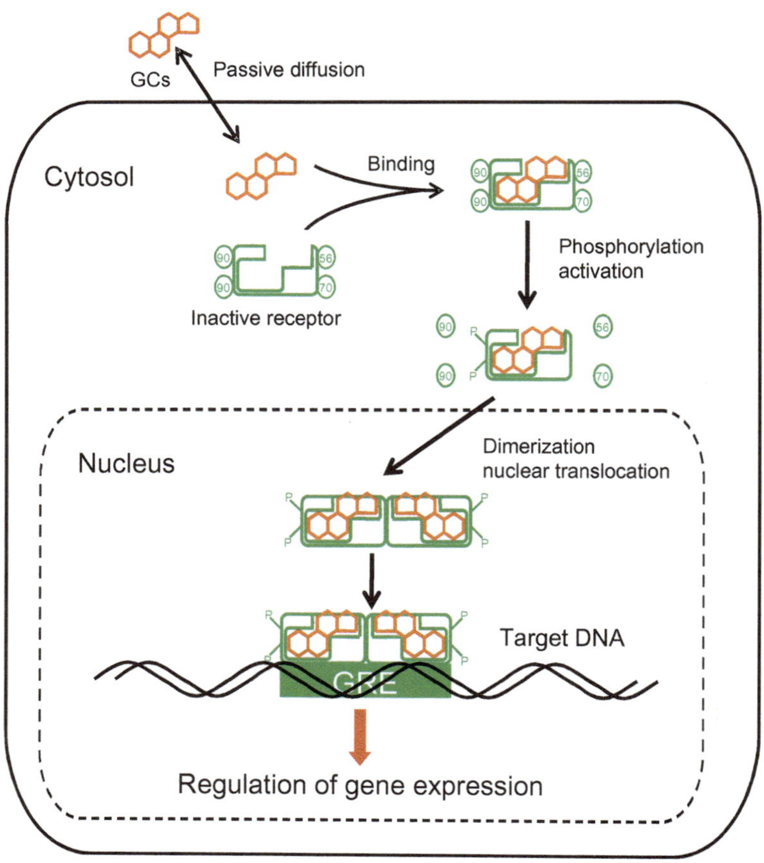

Siddharth Sukumaran *et al.*, Figure 7.2 Molecular mechanism for the regulation of gene expression by glucocorticoids.

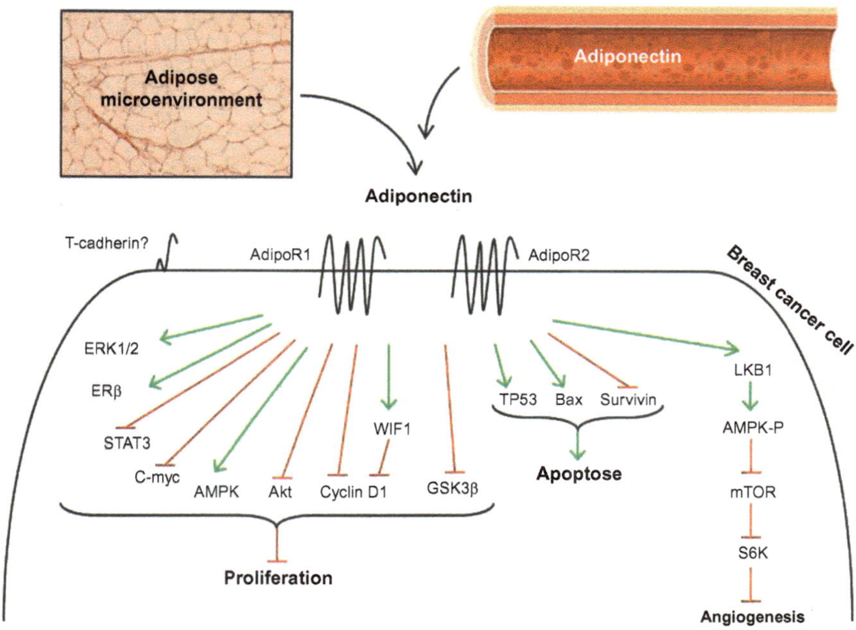

Laetitia Delort *et al.*, Figure 15.1 Signaling pathways of adiponectin in breast cancer cells. Adiponectin, secreted by adipose cells in the close tumor microenvironment and provided by the circulation, can act on breast cancer cells through its two receptors AdipoR1 and AdipoR2. ERK1/2, extracellular signal-regulated kinases 1/2; ERβ, estrogen-receptor β; STAT3, signal transducer and activator of transcription; AMPK, AMP-activated protein kinase; WIF1, Wnt inhibitory factor-1; GSK3 β, glycogen-synthase kinase-3β; mTOR, mammalian target of rapamycin; S6K, p70 S6 kinase-1.

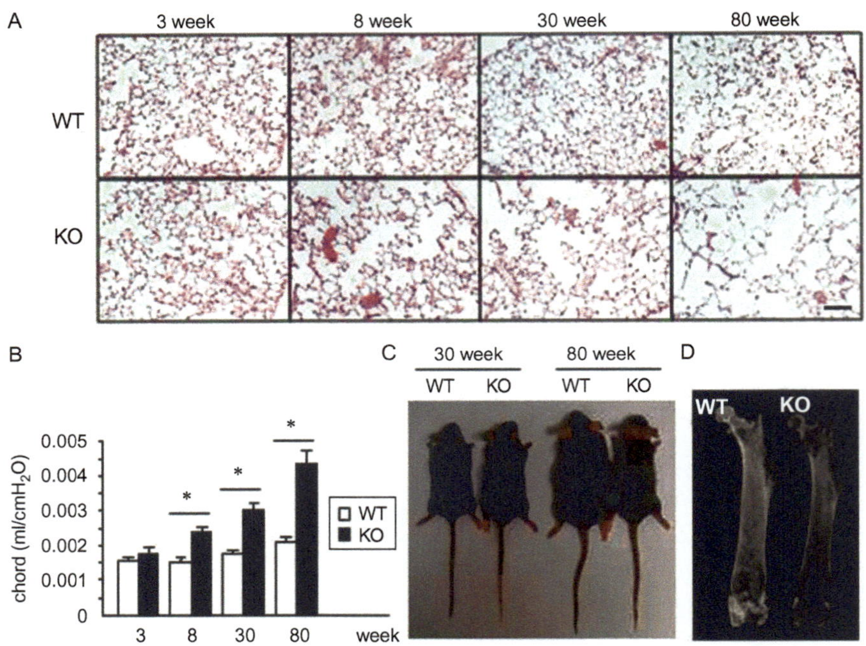

Yoshito Takeda *et al.*, Figure 16.1 APN-KO mice develop a progressive COPD-like phenotype with systemic manifestations. (A) Lung sections from APN-KO mice and WT littermates were stained with hematoxylin and eosin. Scale bar, 100 μm. (B) Lung compliance (c chord) of mice at 3, 8, 30, and 80 weeks of age was measured with a whole body plethysmograph. (C) Shown are whole body images of WT and APN-KO male mice at 30 weeks (left) and 80 weeks (right) of age. (D) Representative radiographs of femurs from WT and APN-KO mice at 80 weeks of age. Bars represent the mean ± SE; *, $P < 0.05$ versus WT.

Kolibri 360 GmbH
Ferdinand-Jühlke-Straße 7
99095 Erfurt, Deutschland
produktsicherheit@kolibri360.de